Özalp Babaoğlu Keith Marzullo (Eds.)

Distributed Algorithms

10th International Workshop, WDAG '96
Bologna, Italy, October 9-11, 1996
Proceedings

Series Editors

Gerhard Goos, Karlsruhe University, Germany

Juris Hartmanis, Cornell University, NY, USA

Jan van Leeuwen, Utrecht University, The Netherlands

Volume Editors

Özalp Babaoğlu
Department of Computer Science, University of Bologna
Piazza Porta S. Donato, 5, I-40127 Bologna, Italy
E-mail: ozalp@cs.unibo.it

Keith Marzullo
Department of Computer Science and Engineering
University of California, San Diego
9500 Gilman Drive, La Jolla, CA 92093-0114, USA
E-mail: marzullo@cs.ucsd.edu

Cataloging-in-Publication data applied for

Die Deutsche Bibliothek - CIP-Einheitsaufnahme

Distributed algorithms : 10th international workshop ; proceedings / WDAG '96, Bologna, Italy, October 9 - 11, 1996. Özalp Babaoğlu ; Keith Marzullo (ed.). - Berlin ; Heidelberg ; New York ; Barcelona ; Budapest ; Hong Kong ; London ; Milan ; Paris ; Santa Clara ; Singapore ; Tokyo : Springer, 1996
(Lecture notes in computer science ; Vol. 1151)
ISBN 3-540-61769-8
NE: Babaoğlu, Özalp [Hrsg.]; WDAG <10, 1996, Bologna>; GT

CR Subject Classification (1991): F.1. D1.3, C.2.2, C.2.4, D.4.4-5, F.2.2

ISSN 0302-9743
ISBN 3-540-61769-8 Springer-Verlag Berlin Heidelberg New York

Printed in Germany

Typesetting: Camera-ready by author
SPIN 10513788 06/3142 – 5 4 3 2 1 0 Printed on acid-free paper

Lecture Notes in Computer Science 1151

Edited by G. Goos, J. Hartmanis and J. van Leeuwen

Springer
Berlin
Heidelberg
New York
Barcelona
Budapest
Hong Kong
London
Milan
Paris
Santa Clara
Singapore
Tokyo

Preface

The *Workshop on Distributed Algorithms* is intended to be a forum covering all aspects of distributed algorithms for computation and communication. It is devoted both to the presentation of new research results and to the identification of future research directions.

The year 1996 marks the tenth WDAG conference. It also marks a time at which the field of distributed computing is undergoing a major transition from being the domain of a small research community to that of industry and the public at large. Distributed computing is rapidly becoming the principal computing paradigm for doing business in widely diverse areas, and their requirements present new opportunities and challenges for research in distributed algorithms.

For the tenth WDAG conference, papers were solicited describing original results in all areas of distributed algorithms and their applications, including algorithms for control and communication, distributed searching including resource discovery and retrieval, network protocols, applications and services, fault tolerance and high availability, real-time distributed systems, algorithms for dynamic topology management, mobile computing, distributed intelligent agents, issues in synchrony, asynchrony, scalability and real-time, replicated data management, security in distributed systems, self stabilization, wait-free algorithms, and techniques and paradigms for the design and analysis of distributed systems.

In response to the call for papers, 75 submissions were received from 22 different countries, and 21 papers were accepted for presentation at the workshop. The selection was based on perceived originality and quality. There was considerable interest in papers that examined practical distributed algorithms and their applications. The program committee consisted of:

D. Agrawal (UC Santa Barbara)
J. Anderson (UNC Chapel Hill)
H. Attiya (Technion, Haifa)
Ö. Babaoğlu (U. of Bologna, co-chair)
E. N. Elnohazy (CMU, Pittsburgh)
A. Fekete (U. of Sydney)
P. Jayanti (Dartmouth College, Hanover)
S. Kutten (IBM, Yorktown Heights)
D. Malkhi (AT&T Bell Labs, Murray Hill)
K. Marzullo (UC San Diego, co-chair)
A. Panconesi (Freie U., Berlin)
M. Raynal (IRISA, Rennes)
A. Ricciardi (UT Austin)
R. Segala (U. of Bologna)
G. Tel (Utrecht U.)

We wish to express our gratitude to the Program Committee members and to the external referees who assisted them during the reviewing process. CaberNet, the ESPRIT Network of Excellence in Distributed Systems, kindly sponsored WDAG 96, and CIRFID at the University of Bologna graciously hosted the meeting. We would also like to thank the two invited speakers, Domenico Ferrari (Università Cattolica, Piacenza) and Butler Lampson (MIT, Cambridge and Microsoft, Redmond) for joining us in Bologna.

Bologna, October 1996

Özalp Babaoğlu
Keith Marzullo

List of Referees

Yehuda Afek
Mustaque Ahamad
Lorenzo Alvisi
Alessandro Amoroso
Anish Arora
James Aspnes
Alberto Bartoli
Rozoy Brigitte
Harry Buhrman
Nadia Busi
Tushar Chandra
Craig Chase
Manhoi Choy
Renzo Davoli
Luca de Alfaro
Roberto de Prisco
Shlomi Dolev
Paulo Ferreira
Riccardo Focardi
Roy Friedman
Juan Garay
Vijay K. Garg
Brian Grayson
Rachid Guerraoui
Sibsankar Haldar
Jean-Michel Helary
Maurice Herlihy
Ted Herman
Jaap-Henk Hoepman
Michel Hurfin
Valerie Issarny
Alon Itai
Jeremiah W. James
Kevin Jeffay
Theodore Johnson
Rajeev Joshi
Idit Keidar
Sanjay Khanna
Sandeep Kulkarni
Akhil Kumar
Wei-Cheng Lai
Raimondas Lencevicius
Jon C. Little
Victor Luchangco
Bernard Mans
James Roger Mitchell
Alberto Montresor
Yoram Moses
Achour Mostefaoui
Venkat Murty
Nancy Norris
Graziano Obertelli
Michael Ogg
Gerard Padiou
Marina Papatriantafilou
David Peleg
Evelyn Pierce
Ian Piumarta
David Plainfosse
Noel Plouzeau
Philippe Queinnec
Tal Rabin
Sergio Rajsbaum
Haritha Ramavarjula
Sampath Rangarajan
Lihu Rappoport
K. V. Ravikanth
Michael Reiter
Arend Rensink
Eric J. Rothfus
Andre Schiper
Marco Schneider
Marc Shapiro
Nir Shavit
Alex Shvartsman
Gurdip Singh
C. Skawratananond
Shaune R. Stark
John Hamilton Slye
Scott Stoller
David Stotts
Jeremy Sussman
Yu-Pang Tang
Gadi Taubenfeld
Oliver E. Theel
Dan Touitou
John Tromp
Philippas Tsigas
Hong Va Leong
Robbert VanRenesse
George Varghese
Avishai Wool
Daniel Wu
Shmuel Zaks
Asaph Zemach

Table of Contents

How to Build a Highly Available System Using Consensus

Butler W. Lampson[1]

Microsoft
180 Lake View Av., Cambridge, MA 02138

Abstract. Lamport showed that a replicated deterministic state machine is a general way to implement a highly available system, given a consensus algorithm that the replicas can use to agree on each input. His Paxos algorithm is the most fault-tolerant way to get consensus without real-time guarantees. Because general consensus is expensive, practical systems reserve it for emergencies and use leases (locks that time out) for most of the computing. This paper explains the general scheme for efficient highly available computing, gives a general method for understanding concurrent and fault-tolerant programs, and derives the Paxos algorithm as an example of the method.

1 Introduction

A system is available if it provides service promptly on demand. The only way to make a highly available system out of less available components is to use redundancy, so the system can work even when some of its parts are broken. The simplest kind of redundancy is replication: make several copies or 'replicas' of each part.

This paper explains how to build efficient highly available systems out of replicas, and it gives a careful specification and an informal correctness proof for the key algorithm. Nearly all of the ideas are due to Leslie Lamport: replicated state machines [5], the Paxos consensus algorithm [7], and the methods of specifying and analyzing concurrent systems [6]. I wrote the paper because after I had read Lamport's papers, it still took me a long time to understand these methods and how to use them effectively. Surprisingly few people seem to know about them in spite of their elegance and power.

In the next section we explain how to build a replicated state machine that is both efficient and highly available, given a fault-tolerant algorithm for consensus. Section 3 gives some background on the consensus problem and its applications. Section 4 reviews the method we use for writing specifications and uses it to give a precise specification for consensus in several forms. In section 5 we introduce the basic idea behind the Paxos algorithm for consensus and derive the algorithm from this idea and the specification. Finally we explain some important optimizations and summarize our conclusions.

---

[1] Email: blampson@microsoft.com. This paper is at http://www.research.microsoftcom.

2 Replicated State Machines

Redundancy is not enough; to be useful it must be coordinated. The simplest way to do this is to make each non-faulty replica do the same thing. Then any non-faulty replica can provide the outputs; if the replicas are not fail-stop, requiring the same output from f replicas will tolerate $f - 1$ faults. More complicated kinds of redundancy (such as error-correcting codes) are cheaper, but they depend on special properties of the service being provided.

In this section we explain how to coordinate the replicas in a fully general and highly fault tolerant way. Then we explore an optimization called 'leases' that makes the coordination very efficient in nearly all cases.

2.1 Coordinating the Replicas

How can we arrange for each replica to do the same thing? Adopting a scheme first proposed by Lamport [5], we build each replica as a deterministic state machine; this means that the transition relation is a function from (state, input) to (new state, output). It is customary to call one of these replicas a 'process'. Several processes that start in the same state and see the same sequence of inputs will do the same thing, that is, end up in the same state and produce the same outputs. So all we need for high availability is to ensure that all the non-faulty processes see the same inputs. The technical term for this is 'consensus' (sometimes called 'agreement' or 'reliable broadcast'). Informally, we say that several processes achieve consensus if they all agree on some value; we give a formal definition later on.

So if several processes are implementing the same deterministic state machine and achieve consensus on the values and order of the inputs, they will do the same thing. In this way it's possible to replicate an *arbitrary* computation and thus make it highly available. Of course we can make the order a part of the input value by defining some total order on the set of inputs, for instance by numbering them 1, 2, 3, ...

In many applications the inputs are requests from clients to the replicated service. For example, a replicated storage service might have *Read*(a) and *Write*(a, d) inputs, and an airplane flight control system might have *ReadInstrument*(i) and *RaiseFlaps*(d) inputs. Different clients usually generate their requests independently, so it's necessary to agree not only on what the requests are, but also on the order in which to serve them. The simplest way to do this is to number them with consecutive integers, starting at 1. This is done in 'primary copy' replication, since it's easy for one process (the primary) to assign consecutive numbers. So the storage service will agree on Input 1 = *Write*(x, *3*) and Input 2 = *Read*(x).

There are many other schemes for achieving consensus on the order of requests when their total order is not derived from consecutive integers; see Schneider's survey [11]. These schemes label each input with some value from a totally ordered set (for instance, (client UID, timestamp) pairs) and then devise a way to be certain that you have seen all the inputs that can ever exist with labels smaller than a given value. This is complicated, and practical systems usually use a primary to sequence the inputs instead.

2.2 Leases: Efficient Highly Available Computing

Fault-tolerant consensus is expensive. Exclusive access by a single process (also known as locking) is cheap, but it is not fault-tolerant—if a process fails while it is holding a lock, no one else can access the resource. Adding a timeout to a lock makes a fault-tolerant lock or 'lease'. Thus a process holds a lease on a state component or 'resource' until an expiration time; we say that the process is the 'master' for the resource while it holds the lease. No other process will touch the resource until the lease expires. For this to work, of course, the processes must have synchronized clocks. More precisely, if the maximum skew between the clocks of two processes is ε and process P's lease expires at time t, then P knows that no other process will touch the resource before time $t - \varepsilon$ on P's clock.

While it holds the lease the master can read and write the resource freely. Writes must take bounded time, so that they can be guaranteed either to fail or to precede any operation that starts after the lease expires; this can be a serious problem for a resource such as a SCSI disk, which has weak ordering guarantees and a long upper bound on the time a write can take.

Locks in transaction processing systems are usually leases; if they expire the transaction is aborted, which means that its writes are undone and the transaction is equivalent to **skip**. A process that uses leases outside the scope of a transaction must take care to provide whatever atomicity is necessary, for example, by ensuring that the resource is in a good state after every atomic write (this is called 'careful writes'), or by using standard redo or undo methods based on logs [4]. The latter is almost certainly necessary if the resource being leased is itself replicated.

A process can keep control of a resource by renewing its lease before it expires. It can also release its lease, perhaps on demand. If you can't talk to the process that holds the lease, however (perhaps because it has failed), you have to wait for the lease to expire before touching its resource. So there is a tradeoff between the cost of renewing a lease and the time you have to wait for the lease to expire after a (possible) failure. A short lease means a short wait during recovery but a higher cost to renew the lease. A long lease means a long wait during recovery but a lower cost to renew.

A lease is most often used to give a process the right to cache some part of the state, for instance the contents of a cache line or of a file, knowing that it can't change. Since a lease is a kind of lock, it can have a 'mode' which determines what operations its holder can do. If the lease is exclusive, then its process can change the leased state freely. This is like 'owner' access to a cache line or ownership of a multi-ported disk.

2.3 Hierarchical Leases

In a fault-tolerant system leases must be granted and renewed by running consensus. If this much use of consensus is still too expensive, the solution is hierarchical leases. Run consensus once to elect a czar C and give C a lease on a large part of the state. Now C gives out sub-leases on x and y to masters. Each master controls its own resources. The masters renew their sub-leases with the czar. This is cheap since it doesn't require any coordination. The czar renews its lease by consensus. This costs more, but there's only one czar lease. Also, the czar can be simple and less likely to fail, so a longer lease may be acceptable.

Hierarchical leases are commonly used in replicated file systems and in clusters.

By combining the ideas of consensus, leases, and hierarchy, it's possible to build highly available systems that are also highly efficient.

3 Consensus

Several processes achieve consensus if they all agree on some allowed value called the 'outcome' (if they could agree on any value the solution would be trivial: always agree on 0). Thus the interface to consensus has two actions: allow a value, and read the outcome. A consensus algorithm terminates when all non-faulty processes know the outcome.

There are a number of specialized applications for consensus in addition to general replicated state machines. Three popular ones are:

Distributed transactions, where all the processes need to agree on whether a transaction commits or aborts. Each transaction needs a separate consensus on its outcome.

Membership, where a group of processes cooperating to provide a highly available service need to agree on which processes are currently functioning as members of the group. Every time a process fails or starts working again there must be a new consensus.

Electing a leader of a group of processes without knowing exactly what the members are.

Consensus is easy if there are no faults. Here is a simple implementation. Have a fixed *leader* process. It gets all the *Allow* actions, chooses the outcome, and tells everyone. If it were to fail, you would be out of luck. Standard two-phase commit works this way: the allowed value is *commit* if all participants are prepared, and *abort* if at least one has failed. If the leader fails the outcome may be unknown.

Another simple implementation is to have a set of processes, each choosing a value. If a majority choose the same value, that is the outcome (the possible majorities are subsets with the property that any two majorities have a non-empty intersection). If the processes choose so that there is no majority for a value, there is no outcome. If some members of the majority were to fail the outcome would be unknown.

Consensus is tricky when there are faults. In an asynchronous system (in which a non-faulty process can take an arbitrary amount of time to make a transition) with perfect links and even one faulty process, there is no algorithm for consensus that is guaranteed to terminate [3]. In a synchronous system consensus is possible even with processes that have arbitrary or malicious faults (Byzantine agreement), but it is expensive in messages sent and in time [9].

4 Specifications

We are studying systems with a state which is an element of some (not necessarily finite) state space, and a set of actions (not necessarily deterministic) that take the system from one state to another. Data abstractions, concurrent programs, distributed systems, and fault-tolerant systems can all be modeled in this way. Usually we describe the state spaces as the Cartesian product of smaller spaces called 'variables'.

4.1 How to Specify a System with State

In specifying such a system, we designate some of the actions or variables as 'external' and the rest as 'internal'. What we care about is the sequence of external actions (or equivalently, the sequence of values of external variables), because we assume that you can't observe internal actions or variables from outside the system. We call such a sequence a 'trace' of the system. A specification is a set of traces, or equivalently a predicate on traces. Such a set is called a 'property'.

We can define two special kinds of property. Informally, a 'safety' property asserts that nothing bad ever happens; it is the generalization of partial correctness for sequential programs. A 'liveness' property asserts that something good eventually happens; it is the generalization of termination. You can always tell that a trace does not have a safety property by looking at some finite prefix of it, but you can never do this for a liveness property. Any property (that is, any set of sequences of actions) is the intersection of a safety property and a liveness property [2].

In this paper we deal only with safety properties. This seems appropriate, since we know that there is no terminating algorithm for asynchronous consensus. It is also fortunate, because liveness properties are much harder to handle.

It's convenient to define a safety property by a state machine whose actions are also divided into external and internal ones. All the sequences of external actions of the machine define a safety property. Do not confuse these specification state machines with the replicated state machines that we implement using consensus.

We define what it means for one system Y to implement another system X as follows:

- Every trace of Y is a trace of X; that is, X's safety property implies Y's safety property.
- Y's liveness property implies X's liveness property.

The first requirement ensures that you can't tell by observing Y that it isn't X; Y never does anything bad that X wouldn't do. The second ensures that Y does all the good things that X is supposed to do. We won't say anything more about liveness.

Following this method, to specify a system with state we must first define the state space and then describe the actions. We choose the state space to make the spec clear, not to reflect the state of the implementation. For each action we say what it does to the state and whether it is external or internal. We model an action with parameters and results such as *Read*(x) by a family of actions, one of which is *Read*(x) **returning** 3; this action happens when the client reads x and the result is 3.

Here are some helpful hints for writing these specs.

- Notation is important, because it helps you to think about what's going on. Invent a suitable vocabulary.
- Less is more. Fewer actions are better.
- More non-determinism is better, because it allows more implementations.

I'm sorry I wrote you such a long letter; I didn't have time to write a short one.
Pascal

4.2 Specifying Consensus

We are now ready to give specifications for consensus. There is an *outcome* variable initialized to nil, and an action *Allow*(*v*) that can be invoked any number of times. There is also an action *Outcome* to read the *outcome* variable; it must return either nil or a *v* which was the argument of some *Allow* action, and it must always return the same *v*.

More precisely, we have two requirements:

Agreement: Every non-nil result of *Outcome* is the same.

Validity: A non-nil *outcome* equals some allowed value.

Validity means that the outcome can't be any arbitrary value, but must be a value that was allowed. Consensus is reached by choosing some allowed value and assigning it to *outcome*. This spec makes the choice on the fly as the allowed values arrive.

Here is a precise version of the spec, which we call C. It gives the state and the actions of the state machine. The state is:

outcome : *Value* ∪ {nil} **initially** nil

The actions are:

Name	Guard	Effect
**Allow*(*v*)		**choose if** *outcome* = nil **then** *outcome* := *v* **or** **skip**
**Outcome*		**choose return** *outcome* **or** **return** nil

Here the external actions are marked with a *. The guard is a precondition which must be true in the current state for the action to happen; it is true (denoted by blank) for both of these actions. The **choose** ... **or** ... denotes non-deterministic choice, as in Dijkstra's guarded commands.

Note that *Outcome* is allowed to return nil even after the choice has been made. This reflects the fact that in an implementation with several replicas, *Outcome* is often implemented by talking to just one of the replicas, and that replica may not yet have learned about the choice.

Next we give a spec T for consensus with termination. Once the internal *Terminate* action has happened, the outcome is guaranteed not to be nil. You can find out whether the algorithm has terminated by calling *Done*. We mark the changes from C by boxing them.

State: *outcome* : *Value* ∪ {nil} **initially** nil

done	: Bool	**initially** false

Name	Guard	Effect
**Allow(v)*		**choose if** *outcome* = nil **then** *outcome* := *v* **or** **skip**
**Outcome*		**choose return** *outcome* **or** **if not** *done* **then** **return** nil
**Done*		**return** *done*
Terminate	outcome ≠ nil	*done* := true

Note that the spec T says nothing about whether termination will actually occur. An implementation in which *Outcome* always returns nil satisfies T. This may seem unsatisfactory, but it's the best we can do with an asynchronous implementation. In other words, a stronger spec would rule out an asynchronous implementation.

Finally, here is a more complicated spec D for 'deferred consensus'. It accumulates the allowed values and then chooses one of them in the internal action *Agree*.

State:		
outcome	: *Value* ∪ {nil}	**initially** nil
done	: Bool	**initially** false
allowed	: **set** *Value*	**initially** {}

Name	Guard	Effect
**Allow(v)*		*allowed* := *allowed* ∪ {*v*}
**Outcome*		**choose return** *outcome* **or** **if not** *done* **then return** nil
**Done*		**return** *done*
Agree(v)	*outcome* = nil **and** *v* **in** *allowed*	*outcome* := *v*
Terminate	outcome ≠ nil	*done* := true

It should be fairly clear that D implements T. To prove this using the abstraction function method described in the next section, however, requires a prophecy variable or backward simulation [1, 10], because C and T choose the outcome as soon as the see the allowed value, while an implementation may make the choice much later. We have two reasons for giving the D spec. One is that some people find it easier to understand than T, even though it has more state and more actions. The other is to move the need for a prophecy variable into the proof that D implements T, thus simplifying the much more subtle proof that Paxos implements D.

5 Implementation

In this section we first explain the abstraction function method for showing that an implementation meets a specification. This method is both general and practical. Then we discuss some hints for designing and understanding implementations, and illustrate the method and hints with abstraction functions for the simple implementations given earlier. The next section shows how to use the method and hints to derive Lamport's Paxos algorithm for consensus.

5.1 Proving that Y implements X

The definition of 'implements' tells us what we have to do (ignoring liveness): show that every trace of Y is a trace of X. Doing this from scratch is painful, since in general each trace is of infinite length and Y has an infinite number of traces. The proof will therefore require an induction, and we would like a proof method that does this induction once and for all. Fortunately, there is a general method for proving that Y implements X without reasoning explicitly about traces in each case. This method was originally invented by Hoare to prove the correctness of data abstractions. It was generalized by Lamport [6] and others to arbitrary concurrent systems.

The method works like this. First, define an *abstraction function* f from the state of Y to the state of X. Then show that Y *simulates* X:

1) f maps an initial state of Y to an initial state of X.
2) For each Y-action and each reachable state y there is a sequence of X-actions (perhaps empty) that is the same externally, such that this diagram commutes.

X-actions
$f(y)$ → $f(y')$
f ↑ f ↑
Y-action
y → y'

A sequence of X-actions is the same externally as a Y-action if they are the same after all internal actions are discarded. So if the Y-action is internal, all the X-actions must be internal (perhaps none at all). If the Y-action is external, all the X-actions must be internal except one, which must be the same as the Y-action.

A straightforward induction shows that Y implements X: For any Y-behavior we can construct an X-behavior that is the same externally, by using (2) to map each Y-action into a sequence of X-actions that is the same externally. Then the sequence of X-actions will be the same externally as the original sequence of Y-actions.

If Y implements X, is it always to possible to prove it using this method. The answer is "Almost". It may be necessary to modify Y by adding auxiliary 'history' and 'prophecy' variables according to certain rules which ensure that the modified Y has exactly the same traces as Y itself. With the right history and prophecy variables it's always possible to find an abstraction function [1]. An equivalent alternative is to use an abstraction relation rather than an abstraction function, and to do 'backward' simulation as well as the 'forward' simulation we have just described [10]. We mention these complications for completeness, but avoid them in this paper.

In order to prove that Y simulates X we usually need to know what the reachable states of Y are, because it won't be true that every action of Y from an arbitrary state of Y simulates a sequence of X-actions; in fact, the abstraction function might not even be defined on an arbitrary state of Y. The most convenient way to characterize the reachable states of Y is by an *invariant*, a predicate that is true of every reachable state. Often it's helpful to write the invariant as a conjunction; then we call each conjunct an invariant. It's common to need a stronger invariant than the simulation requires; the extra strength is a stronger induction hypothesis that makes it possible to establish what the simulation does require.

So the structure of a proof goes like this:

- Define an abstraction function.
- Establish invariants to characterize the reachable states, by showing that each action maintains the invariants.
- Establish the simulation, by showing that each Y-action simulates a sequence of X-actions that is the same externally.

This method works only with actions and does not require any reasoning about traces. Furthermore, it deals with each action independently. Only the invariants connect the actions. So if we change (or add) an action of Y, we only need to verify that the new action maintains the invariants and simulates a sequence of X-actions that is the same externally.

In the light of this method, here are some hints for deriving, understanding, and proving the correctness of an implementation.

- Write the specification first.
- Dream up the idea of the implementation. This is the crucial creative step. Usually you can embody the key idea in the abstraction function.
- Check that each implementation action simulates some spec actions. Add invariants to make this easier. Each action must maintain them. Change the implementation (or the spec) until this works.
- Make the implementation correct first, then efficient. More efficiency means more complicated invariants. You might need to change the spec to get an efficient implementation.

An efficient program is an exercise in logical brinksmanship
Dijkstra

In what follows we give abstraction functions for each implementation we consider, and invariants for the Paxos algorithm. The actual proofs that the invariants hold and that each Y-action simulates a suitable sequence of X-actions are routine, and we omit them.

5.2 Abstraction Functions for the Simple Implementations

Recall our two simple, non-fault-tolerant implementations. In the first a single *leader* process, with the same state as the specification, tells everyone else the outcome (this is how two-phase commit works). The abstraction function to C is:

outcome = the *outcome* of the coordinator.

done = everyone has gotten the outcome.

This is not fault-tolerant—it fails if the leader fails.

In the second there is set of processes, each choosing a value. If a majority choose the same value, that is the outcome. The abstraction function to C is:

outcome = the choice of a majority, or nil if there's no majority.

done = everyone has gotten the outcome.

This is not fault-tolerant—it fails if a majority doesn't agree, or if a member of the majority fails.

6 The Paxos Algorithm

In this section we describe Lamport's Paxos algorithm for implementing consensus [7]. This algorithm was independently invented by Liskov and Oki as part of a replicated data storage system [8]. Its heart is the best known asynchronous consensus algorithm. Here are its essential properties:

It is run by a set of *leader* processes that guide a set of *agent* processes to achieve consensus.

It is correct no matter how many simultaneous leaders there are and no matter how often leader or agent processes fail and recover, how slow they are, or how many messages are lost, delayed, or duplicated.

It terminates if there is a single leader for a long enough time during which the leader can talk to a majority of the agent processes twice.

It may not terminate if there are always too many leaders (fortunate, since we know that guaranteed termination is impossible).

To get a complete consensus algorithm we combine this with a sloppy timeout-based algorithm for choosing a single leader. If the sloppy algorithm leaves us with no leader or more than one leader for a time, the consensus algorithm may not terminate during that time. But if the sloppy algorithm ever produces a single leader for long enough the algorithm will terminate, no matter how messy things were earlier.

We first explain the simplest version of Paxos, without worrying about the amount of data stored or sent in messages, and then describe the optimizations that make it reasonably efficient. To get a really efficient system it's usually necessary to use leases as well.

6.1 The Idea

First we review the framework described above. There is a set of agent processes, indexed by a set I. The behavior of an agent is deterministic; an agent does what it's told. An agent has 'persistent' storage that survives crashes. The set of agents is fixed for a single run of the algorithm (though it can be changed using the Paxos algorithm itself). There are also some leader processes, indexed by a totally ordered set L, that tell the agents what to do. Leaders can come and go freely, they are not deterministic, and they have no persistent storage.

The key idea of Paxos comes from the non-fault-tolerant majority consensus algorithm described earlier. That algorithm gets into trouble if the agents can't agree on a majority, or if some members of the majority fail so that the rest are unsure whether consensus was reached. To fix this problem, Paxos has a sequence of *rounds* indexed by a set N. Round n has a single leader who tries to get a majority for a single value v_n. If one round gets into trouble, another one can make a fresh start. If round n achieves a majority for v_n then v_n is the outcome.

Clearly for this to work, any two rounds that achieve a majority must have the same value. The tricky part of the Paxos algorithm is to ensure this property.

In each round the leader

- *queries* the agents to learn their status for past rounds,
- chooses a value and *commands* the agents, trying to get a majority to accept it, and
- if successful, distributes the value as the outcome to everyone.

It takes a total of 2½ round trips for a successful round. If the leader fails repeatedly, or several leaders fight it out, it may take many rounds to reach consensus.

6.2 State and Abstraction Function

The state of an agent is a persistent 'status' variable for each round; persistent means that it is not affected by a failure of the agent. A status is either a *Value* or one of the special symbols *no* and *neutral*. The agent actions are defined so that a status can only change if it is *neutral*. So the agent state is defined by an array s:

State: $s_{i,n}$: *Value* $\cup$ {*no*, *neutral*} **initially** *neutral*

A round is *dead* if a majority has status *no*, and *successful* if a majority has status which is a *Value*.

The state of a leader is the round it is currently working on (or nil if it isn't working on a round), the value for that round (or nil if it hasn't been chosen yet), and the leader's idea of the *allowed* set.

State:			
	n_l	: $N \cup$ {nil}	**initially** nil
	u_l	: *Value* $\cup$ {nil}	**initially** nil
	$allowed_l$	: **set** *Value*	**initially** {}

The abstraction function for *allowed* is just the union of the leaders' sets:

AF: $allowed = \bigcup_{l \in L} allowed_l$

We define the value of round n as follows:

$v_n \equiv$ **if** some agent i has a *Value* in $s_{i,n}$ **then** $s_{i,n}$
else nil

For this to be well defined, we must have

Invariant 1: A round has at most one value.

That is, in a given round all the agents with a value have the same value. Now we can give the abstraction function for *outcome*:

AF: $outcome = v_n$ for some successful round n
or nil if there is no successful round.

For this to be well defined, we must have

Invariant 2: Any two successful rounds have the same value.

We maintain invariant 1 (a round has at most one value) by ensuring that a leader works on only one round at a time and never reuses a round number, and that a round has at most one leader. To guarantee the latter condition, we make the leader's identity part of the round number by using (sequence number, leader identity) pairs as round numbers. Thus $N = (J, L)$, where J is some totally ordered set, usually the integers. Leader l chooses (j, l) for n_l, where j is a J that l has not used before. For instance, j might be the current value of local clock. We shall see later how to avoid using any stable storage at the leader for choosing j.

6.3 Invariants

We introduce the notion of a *stable* predicate on the state, a predicate which once true, remains true henceforth. This is important because it's safe to act on the truth of a stable predicate. Anything else might change because of concurrency or crashes.

Since a non-*neutral* value of $s_{i,n}$ can't change, the following predicates are stable:

$s_{i,n} = no$
$s_{i,n} = v$
$v_n \quad = v$

n is dead
n is successful

Here is a more complex stable predicate:

n is anchored $\equiv$ **for all** $m \leq n$, m is dead **or** $v_n = v_m$

In other words, n is anchored iff when you look back at rounds before n, skipping dead rounds, you see the same value as n. If all preceding rounds are dead, n is anchored no matter what its value is. For this to be well-defined, we need a total ordering on N's, and we use the lexicographic ordering.

Now all we have to do is to maintain invariant 2 while making progress towards a successful round. To see how to maintain the invariant, we strengthen it until we get a form that we can easily maintain with a distributed algorithm, that is, a set of actions each of which uses only the local state of a process.

Invariant 2: Any two successful rounds have the same value.

follows from

Invariant 3: **for all** n and $m \leq n$, **if** m is successful **then** $v_n = \text{nil}$ **or** $v_n = v_m$

which follows from

Invariant 4: **for all** n and $m \leq n$, **if** m is not dead **then** $v_n = \text{nil}$ **or** $v_n = v_m$

and we rearrange this by predicate calculus

$\equiv$ **for all** n and $m \leq n$, m is dead **or** $v_n = \text{nil}$ **or** $v_n = v_m$

$\equiv$ **for all** n, $v_n = \text{nil}$ **or** (**for all** $m \leq n$, m is dead **or** $v_n = v_m$)

$\equiv$ **for all** n, $v_n = \text{nil}$ **or** n is anchored

So all we have to do is choose each non-nil v_n so that
there is only one, and
n is anchored.
Now the rest of the algorithm is obvious.

6.4 The Algorithm

A leader has to choose the value of a round so that the round is anchored. To accomplish this, the leader l chooses a new n_l and *queries* all the agents to learn their status in *all* rounds with numbers less than n_l (and also the values of those rounds). Before an agent responds, it changes any *neutral* status for a round earlier than n_l to *no*, so that the leader will have enough information to anchor the round. Responses to the query from a majority of agents give the leader enough information to make round n_l anchored, as follows:

The leader looks back from n_l, skipping over rounds for which no *Value* status was reported, since these must be dead (remember that l has heard from a majority, and the reported status is a *Value* or *no*). When l comes to a round n with a *Value* status, it chooses that value v_n as u_l. Since n is anchored by invariant 4, and all the rounds between n and n_l are dead, n_l is also anchored if this u_l becomes its value.

If all previous rounds are dead, the leader chooses any allowed value for u_l. In this case n_l is certainly anchored.

Because 'anchored' and 'dead' are stable properties, no state change can invalidate this choice.

Now in a second round trip the leader *commands* everyone to accept u_l for round n_l. Each agent that is still neutral in round n_l (because it hasn't answered the query of a later round) *accepts* by changing its status to u_l in round n_l; in any case it reports its status to the leader. If the leader collects u_l reports from a majority of agents, then it knows that round n_l has succeeded, takes u_l as the agreed outcome of the algorithm, and sends this fact to all the processes in a final half round. Thus the entire process takes five messages or 2½ round trips.

Note that the round succeeds (the *Agree* action of the spec happens and the abstract *outcome* changes) at the instant that some agent forms a majority by accepting its value, even though no agent or leader knows at the time that this has happened. In fact, it's possible for the round to succeed without the leader knowing this fact, if some agents fail after accepting but before getting their reports to the leader, or if the leader fails.

An example may help your intuition about why this works. The table below shows three rounds in two different runs of the algorithm with three agents *a*, *b*, and *c* and *allowed* = {7, 8, 9}. In the left run all three rounds are dead, so if the leader hears from all three agents it knows this and is free to choose any allowed value. If the leader hears only from *a* and *b* or *a* and *c*, it knows that round 3 is dead but does not know that round 2 is dead, and hence must choose 8. If it hears only from *b* and *c*, it does not know that round 3 is dead and hence must choose 9.

In the right run, no matter which agents the leader hears from, it does not know that round 2 is dead. In fact, it was successful, but unless the leader hears from *a* and *c* it doesn't know that. Nonetheless, it must choose 9, since it sees that value in the latest non-dead round. Thus a successful round such as 2 acts as a barrier which prevents any later round from choosing a different value.

					Status			
	v_n	$s_{a,n}$	$s_{b,n}$	$s_{c,n}$	v_n	$s_{a,n}$	$s_{b,n}$	$s_{c,n}$
Round 1	7	7	no	no	8	8	no	no
Round 2	8	8	no	no	9	9	no	9
Round 3	9	no	no	9	9	no	no	9
Leader's choices for round 4	7, 8, or 9 if *a, b, c* report 8 if *a, b* or *b, c* report 9 if *b, c* report				9 no matter what majority reports			

Presumably the reason there were three rounds in both runs is that at least two different leaders were involved, or the leader failed before completing each round. Otherwise round 1 would have succeeded. If leaders keep overtaking each other and forcing the agents to set their earlier status to *no* before an earlier rounds reach the command phase, the algorithm can continue indefinitely.

Here are the details of the algorithm. Its actions are the ones described in the table together with boring actions to send and receive messages.

Leader l	**Message**	**Agent** i
Choose a new n_l		
Query a majority of agents for their status	query(n_l) $\rightarrow$	for all $m < n_l$, **if** $s_{i,m} = \mathit{neutral}$ **then** $s_{i,m} := \mathit{no}$
	$\leftarrow$ report(i, s_i)	
Choose u_l to keep n_l anchored. If all $m < n_l$ are dead, choose any v in *allowed*$_l$		
Command a majority of agents to accept u_l	command(n_l, u_l) $\rightarrow$	**if** $s_{i,n_l} = \mathit{neutral}$ **then** $s_{i,n_l} := u_l$
	$\leftarrow$ report(i, n_l, s_{i,n_l})	
If a majority accepts, publish the outcome u_l	outcome(u_l) $\rightarrow$	

The algorithm makes minimal demands on the properties of the network: lost, duplicated, or reordered messages are OK. Because nodes can fail and recover, a better network doesn't make things much simpler. We model the network as a broadcast medium from leader to agents; in practice this is usually implemented by individual messages to each agent. Both leaders and agents can retransmit as often as they like; in practice agents retransmit only in response to the leader's retransmission.

A complete proof requires modeling the communication channels between the processes as a set of messages which can be lost or duplicated, and proving boring invariants about the channels of the form "if a message report(i, s) is in the channel, then s_i agrees with s except perhaps at some *neutral* components of s."

6.5 Termination: Choosing a Leader

When does the algorithm terminate? If no leader starts another round until after an existing one is successful, then the algorithm definitely terminates as soon as the leader succeeds in both querying and commanding a majority. It doesn't have to be the same majority for both, and the agents don't all have to be up at the same time. Therefore we want a single leader, who runs one round at a time. If there are several leaders, the one running the biggest round will eventually succeed, but if new leaders keep starting bigger rounds, none may ever succeed. We saw a little of this behavior in the example above. This is fortunate, since we know from the Fischer-Lynch-Paterson result [3] that there is no algorithm that is guaranteed to terminate.

It's easy to keep from having two leaders at once if there are no failures for a while, the processes have clocks, and the *usual* maximum time to send, receive, and process a message is known:

Every potential leader that is up broadcasts its name.

You become the leader one round-trip time after doing a broadcast, unless you have received the broadcast of a bigger name.

Of course this algorithm can fail if messages are delayed or processes are late in responding to messages. When it fails, there may be two leaders for a while. The one running the largest round will succeed unless further problems cause yet another leader to arise.

7 Optimizations

It's not necessary to store or transmit the complete agent state. Instead, everything can be encoded in a small fixed number of bits, as follows. The relevant part of s_i is just the most recent *Value* and the later *no* states:

$s_{i,\, last_i} = v$

$s_{i,\, m} = no$ for all m between $last_i$ and $next_i$

$s_{i,\, m} = neutral$ for all $m \geq next_i$.

We can encode this as $(v,\, last_i,\, next_i)$. This is all that an agent needs to store or report.

Similarly, all a leader needs to remember from the reports is the largest round for which a value was reported and which agents have reported. It is not enough to simply count the reports, since report messages can be duplicated because of retransmissions.

The leaders need not be the same processes as the agents, although they can be and usually are. A leader doesn't really need any stable state, though in the algorithm as given it has something that allows it to choose an n that hasn't been used before. Instead, it can poll for the $next_i$ from a majority after a failure and choose an n_l with a bigger j. This will yield an n_l that's larger than any from this leader that has appeared in a command message so far (because n_l can't get into a command message without having once been the value of $next_i$ in a majority of agents), and this is all we need.

If Paxos is used to achieve consensus in a non-blocking commit algorithm, the first round-trip (query/report) can be combined with the prepare message and its response.

The most important optimization is for a sequence of consensus problem, usually the successive steps of a replicated state machine. We try to stay with the same leader, usually called the 'primary', and run a sequence of instances of Paxos numbered by another index p. The state of an agent is now

State: $s_{p,\, i,\, n}$: $Value \cup \{no,\, neutral\}$ **initially** *neutral*

Make the fixed size state $(v,\, last,\, next,\, p)$ for agent i encode

$s_{q,\, i,\, m} = no$ for all $q \leq p$ and $m < next$.

Then a query only needs to be done once each time the leader changes. We can also piggyback the outcome message on the command message for the next instance of Paxos. The result is 2 messages (1 round trip) for each consensus.

8 Conclusion

We showed how to build a highly available system using consensus. The idea is to un a replicated deterministic state machine, and get consensus on each input. To make it efficient, use leases to replace most of the consensus steps with actions by one process.

We derived the most fault-tolerant algorithm for consensus without real-time guarantees. This is Lamport's Paxos algorithm, based on repeating rounds until you get a majority, and ensuring that every round after a majority has the same value. We saw how to implement it with small messages, and with one round-trip for each consensus in a sequence with the same leader.

Finally, we explained how to design and understand a concurrent, fault-tolerant system. The recipe is to write a simple spec as a state machine, find the abstraction function from the implementation to the spec, establish suitable invariants, and show that the implementation simulates the spec. This method works for lots of hard problems.

9 References

1. M. Abadi and L. Lamport. The existence of refinement mappings. *Theoretical Computer Science* **82**, 2, May 1991.
2. B. Alpern and F. Schneider. Defining liveness. *Information Processing Letters* **21**, 4, 1985.
3. M. Fischer, N. Lynch, and M. Paterson. Impossibility of distributed consensus with one faulty process. *J. ACM* **32**, 2, April 1985.
4. J. Gray and A. Reuter. *Transaction Processing: Concepts and Techniques*. Morgan Kaufmann, 1993.
5. L. Lamport. The implementation of reliable distributed multiprocess systems. *Computer Networks* **2**, 1978.
6. L. Lamport. A simple approach to specifying concurrent systems. *Comm. ACM*, **32, 1,** Jan. 1989.
7. L. Lamport. The part-time parliament. Technical Report 49, Systems Research Center, Digital Equipment Corp., Palo Alto, Sep. 1989.
8. B. Liskov and B. Oki. Viewstamped replication, *Proc. 7th PODC*, Aug. 1988.
9. N Lynch. *Distributed Algorithms*. Morgan Kaufmann, 1996.
10. N. Lynch and F. Vaandrager. Forward and backward simulations for timing-based systems. *Lecture Notes in Computer Science* 600, Springer, 1992.
11. F. Schneider. Implementing fault-tolerant services using the state-machine approach: A tutorial. *Computing Surveys* **22** (Dec 1990).

Distributed Admission Control Algorithms for Real-Time Communication

INVITED PAPER

Domenico Ferrari

CRATOS
Universita` Cattolica del Sacro Cuore
Via Emilia Parmense, 84
I-29100 Piacenza, Italy
E-mail: dferrari@pc.unicatt.it

Abstract. The distributed algorithms that constitute the basis of several popular approaches to admission control for real-time (i.e., guaranteed-performance) connections in packet-switching networks are described. Separate descriptions are given for the cases of unicast connections and multicast connections, which require different algorithms for some of the parts of the procedure. Real-time connections are expected to be needed for good-quality transmission of continuous-media (audio and video) streams, and will be subject to controlled admission so as to avoid congestion.

1. Introduction

The emergence of integrated-services networks is one of the most important current phenomena in the fields (which are rapidly becoming a single field) of computing and communications. These networks, which for economic reasons must exploit the packet-switching (or cell-switching) approach, are characterized by their ability to carry all existing types of trafffic while providing each type with the network services it requires. At the levels within a network's architecture which we focus on in this paper, i.e., the internetwork and transport layers, such services have to do primarily with performance and reliability aspects. Different types of traffic have different requirements in terms of performance and reliability. For example, while data traffic requires completeness and correctness of the delivered packets, it does not usually have stringent performance (throughput, delay, delay variance) requirements. On the other hand, video traffic needs bounds on performance indices to be guaranteed for quality of reception to be acceptable at all times, but can normally tolerate a few errors in the packets delivered and even a few losses (or deadline violations, which are equivalent to losses) of packets in the network [1].

The main challenge for integrated-services network designers is to build networks that will satisfy to the largest possible extent all of these conflicting requirements as the various types of traffic share the network's resources and contend with one another for them.

The protocols currently in use on packet-switching networks have not been designed to meet this challenge, and are therefore inadequate for integrated-services networks. They may still be used in such networks to transfer data traffic, but must be supplemented with *real-time protocols* (also called, though somewhat improperly, *multimedia protocols*), which will deal with the needs of real-time (e.g., audio, video) traffic [2]. An alternative approach is, of course, to replace them with *integrated-services protocols* designed to provide each type of traffic with the performance and reliability it requires.

In either case, a packet-switching network cannot offer any guaranteed performance bounds if admission to it is not controlled. No matter how sophisticated its congestion control algorithms are, overloads will inevitably cause congestion, and this will in turn result in lower throughputs, longer delays, and larger delay variations than any pre-established bound if congestion lasts long enough (note that the maximum duration of an overload beyond which a reasonable bound is violated is usually quite short). Thus, admission control is an essential component of any multimedia or integrated-services protocol.

How should we control admission to a packet-switching network? If we wish to retain for data traffic the connectionless approach that has been so successful in the Internet, we cannot apply admission control to such traffic since the shipment of a packet (or *datagram*) is done without preambles. In principle, admission control could be done when the first packet of a new communication is given to the network to transmit. The source would then be informed by the network whether the rest of the communication will be accepted, and would notify the network when it ships the last packet. However, this would be philosophically wrong and practically inconvenient because (a) in a connectionless context, packets should not be considered by the network as related to one another, (b) they may follow different routes, which will make it very hard or outright impossible to check whether a new communication can be admitted, and (c) admission control would cause much useless work to be done when the communication consists of one or a few short packets. Even packet-by-packet admission control would be philosophically wrong and practically hard because of similar reasons.

We should therefore limit admission control to the real-time portion of the traffic. Can we do this while preventing congestion from occurring? The answer is positive if the scope of the question is limited to real-time traffic, which is the one that must really be preserved from congestion. As long as we assign a higher priority to real-time traffic in the network, congestion caused by non-real-time traffic overloads should not affect real-time traffic, which can be shielded from real-time traffic overloads by controlling its admission into the network. It should be noted that, besides getting a higher priority, real-time traffic, to be completely protected from congestion, must also be allocated in all routers sufficient buffer space that cannot be taken away by other traffic. For instance, this objective can be reached if buffer space is assigned to real-time communications statically, or if non-real-time packets can be expelled from the buffers they occupy by real-time packets.

Thus, admission of real-time traffic is to be controlled. This can be easily done if a connection-oriented approach is chosen for this portion of the total traffic. Since real-time traffic usually consists of streams (e.g., video streams, audio streams), and since the requirements are not per-packet but per-stream, establishing a connection for a stream allows us to specify, in the request for the connection, the needs of the stream in terms of performance (as well as of reliability), and allows the network to verify, during the establishment of the connection, that the new stream can be admitted without introducing congestion, or introducing only a tolerable amount of congestion in the worst case. A connection-oriented solution is reasonable also because, as will be seen in more detail below, guaranteeing performance (especially delays) of a multi-packet transmission is much much easier when the route of the packets is fixed, a property that is usually enforced by the connection-oriented approach.

We conclude that both the connectionless and the connection-oriented approach are to be used side-by-side in an integrated-services network [3]. Their coexistence without undesirable interferences has been demonstrated to be possible in several research experiments. The new model of communication this conclusion leads to is the *real-time connection* (which has been given such names as *channel, flow, stream,* and a few others), i.e., a connection with pre-specified performance bounds. When the setup of such a connection is requested by a real-time client of the network, the network must decide whether to accept or reject the request. It does so by employing an *admission control algorithm.* This paper discusses the characteristics these algorithms have to exhibit when they are implemented in a fully distributed way.

2. Distributed Admission Control: The Unicast Case

The verification of the acceptability of a request for a new real-time connection could be done by a centralized decision-maker. This server must have complete and up-to-date knowledge of the state of the network, that is, the relevant information about the real-time connections that exist (i.e., that have been set up and not yet torn down) in each router. The relevant information is the worst-case amounts each connection may contribute to the loads of the routers it traverses. Such a centralized approach to admission control could be used in a small network or even in a large one (e.g., an internetwork) partitioned into smaller subnetworks, provided that (a) the route of the real-time connection to be set up is chosen by a *routing server* at the outset, (b) the residual capacities of the routers on the piece of the route that falls into a given subnetwork are verified by that subnetwork's admission controller, and (c) the replies of all the controllers involved are assembled by a *master controller* that gives the final answer to the client.

Even with this technique, which makes the centralized approach to admission control usable in a large network by eliminating potential performance and reliability bottlenecks, this approach is vulnerable to failures in the controllers, which may make substantial subnetworks inaccessible to real-time traffic for relatively long periods of time. A distributed solution seems to be more attractive, as it avoids the problems of route partitioning, collecting replies, and sensitivity to failures. It is also more in tune with the need to establish a connection, an operation that is usually done by a message visiting in sequence all the routers along the path of the new connection. This does not require complete knowledge of the route at the outset: the route may be constructed hop by hop, as the connection is being set up. The only drawback of not knowing the whole route at the outset is that no parallel admission testing along various route pieces, hence no speedup, is possible.

The speed with which a real-time connection is set up, or with which a negative reply is given to the requesting client, is crucial. Real-time clients cannot usually wait a long time, unless they are reserving connections in advance of their use [4]. The connection-oriented approach will be much more easily accepted even by its enemies if the network's response to a connection request will be perceived as being instantaneous. For a very long connection (say, between 10,000 and 20,000 kilometers), the propagation latency for one terrestrial round trip will be close to the threshold of humanly noticeable delays. For this reason, we must (a) fully establish a new real-time connection in no more than one round trip; (b) keep the admission tests simple and fast; and (c) reduce or eliminate the probability that the stablishment request messages will be delayed by congestion in the non-real-time portion of the traffic (their packets should, if possible, get a priority higher than that of non-real-time packets).

In this section, we consider the establishment of *unicast* (i.e., one-source, one-destination) connections. Before giving a description of how such a connection can be set up, we must specify the connection model we will refer to in our discussion. The path of a connection consists of a series of *nodes* connected by *links*. Each node contains a finite queue (corresponding to the amount of buffer space available to the connection's packets in the node; note that we assume, for simplicity, static per-real-time-connection allocation of buffers) and a queueing server (representing the node's processor, which will either process packet headers or transmit packets to the next node). Node processors are shared among all the real-time connections existing in the node, and are characterized by their processing rate (which is either CPU speed or the bandwidth of the link downstream) as well as by their service discipline. Links are characterized by the propagation delay they introduce. Clearly, with this model, a router or gateway may be represented by more than one node.

Since each node traversed by a setup message must test the request for possible acceptance, a router may have to run more than one set of admission tests.

As a virtual circuit connecting the source to the destination is set up, the establishment message pauses in each node along the route a bit longer to allow admission tests to be performed there. These tests must determine whether it might be impossible for the new connection to obtain what it wants, and whether its addition might cause violations of the performance and reliability bounds that have been guaranteed in each node to the existing real-time connections traversing it.

For the primary resources involved (router CPU time, buffer space, link bandwidth), testing is relatively straightforward, at least conceptually. Since these resources are additive, we add the amounts needed by the new connection to those already earmarked for the existing connections, so that we can keep track of how much of each resource would be needed in the worst case (i.e., if all connections were flooded with their worst-case traffic). This bookkeeping operation is often called, somewhat improperly, *resource reservation*, a term suggesting an exclusive dedication of resources to connections that is not possible in a packet-switching context. In any event, when the total amount of a resource that has been "reserved" in a node equals the amount of that resource available for real-time communication in that node, no additional real-time connections can be admitted until one or more of the existing connections is torn down. Non-real-time packets, however, can use any of the "reserved" resources that are not being used at that very moment.

What is not always straightforward is calculating the amount of a resource that is needed to satisfy probabilistic guarantees [5]. Also, there is a resource that, unlike those we just mentioned, is not additive: the *delay resource* or *schedulability resource*, which we had to bring into existence when we found that some real-time connections could not guarantee given delay bounds even when all the other resources were plentifully available. The reason for such a situation is intuitively easy to understand: there are cases in which the scheduler in a node may be unable to construct a schedule that will satisfy all packet deadlines in that node (note that the relationship between the end-to-end delay bound and the per-node delay bounds is discussed below). What is needed for this resource is a schedulability test, which is generally more complicated than the others, and which involves every time each of the connections traversing the node and not just a single quantity that summarizes the *real-time load* on a resource in the node. This test, of course, varies with the scheduling algorithm the node uses to select packets, whereas the other tests do not have such direct dependencies. Since all the tests to be performed in a node are local to that node, i.e., they do not entail contacting other nodes and/or getting data from them, such an admission control algorithm works also with nodes implementing different scheduling algorithms, as long as each node performs the tests that are the correct ones for it.

It is not obvious that an admission control algorithm can be distributed by breaking it into a sequence of *local* admisssion tests. Indeed, to be done without endangering the guarantees being offered, this subdivision requires some assumptions and a careful organization of the operations. Some QoS (i.e., *quality of service*) parameters, such as delay and delay jitter, are additive. The end-to-end bounds of these parameters can be broken into sums of local (per-node) bounds, which are to be used in local admission tests (e.g., the schedulability ones). Some other QoS parameters, the typical example of which is throughput, have the same value at all points of a connection's path, and this is the value to be used in local tests involving those parameters. Finally, there are QoS parameters, i.e., the probabilities, that are neither constant nor additive, but, if statistical independence can be assumed to exist among the local probabilistic bounds, then their values are multiplicative. One such parameter is the probability that a packet will not be delayed beyond its deadline in a node: a very conservative, worst-case approach computes the probability that a packet will not be late at its destination as the product of the probabilities that the packet will not miss its deadline (its local delay bound) in the first node *and* will not miss its deadline in the second node *and* so on. In reality, even when a few deadlines are missed, it is still possible that the packet will not be late at the destination; thus, the approach we suggest is quite pessimistic, but greatly facilitates testing. The same approach is not pessimistic but quite realistic when applied to the probability that a packet is not lost due to buffer overflow, as long as statistical independence can be assumed. The *localization of bounds* is important not only because it allows us to distribute admission testing to all nodes on a connection's path, but also because it summarizes in each node traversed by a connection all the characteristics of the connection that are relevant to admission testing of requests for further connections having in common with it as much as its entire path or as little as that node. It isolates nodes from one another, and increases substantially their autonomy with respect to admission control decisions.

The single round-trip requirement for the establishment procedure has the crucial consequence that the final decision about acceptance or rejection of a request must be made by the destination. (We assume, for simplicity of discussion, that the setup message is issued by the source of the real-time connection to be created. The message, however, can be issued by the destination and travel from the destination to the source in those cases where receiver-initiated establishment is preferred, without any problems for the admission control procedure described in this paper.) The admission decision cannot be made before having run admission tests in all nodes along the path. Thus, the destination is the earliest point such a decision can be made (in the case of "reverse setup", this point is the source). It is also the latest, as any later changes to the decision would require an additional message to be sent to inform nodes already visited by the return message, and this might take as long as one and a half additional round trips. A total number of round trips larger than 1 would generally improve the result (i.e., better distribute the "load" due to the connection, reduce the resources needed by some connections), but these benefits are almost always negligible with respect to the decrease in setup speed the additional round trips would entail. The only final decision that might be made during the forward trip is one of rejection if the request fails a test in a node. If it passes all tests in a node, the request moves to the next node, but it does not yet know whether it will be accepted or not. Nor does it know the states of the downstream nodes. To maximize the probability of success (i.e., that the request will be accepted), it makes worst-case assumptions about these states. For instance, it assumes that the minimum local delay bounds the downstream nodes will be able to offer to the new connection will be large because of schedulability restrictions (the minimum local delay bound in a node can be computed as a function of the local delay bounds and maximum packet service time of the already established connections as well as of the new one); therefore, each node will offer, to contribute as little as possible to the end-to-end delay bound and leave as much "room" as possible for the bounds of the other nodes, its minimum local delay bound. The destination receives the sum of all these minimum delay bounds, which is the minimum bound that can be guaranteed to the new connection by the chosen path in its current state. An alternative that has merit (see the next paragraph) is to have each node contribute a delay bound larger than the minimum by a moderate amount.

In the same vein, ignorance about the downstream states causes most other resources to be (tentatively) overallocated to a new connection during the forward pass of the setup message. The reverse message has full knowledge of the final decision and of all the states, and either releases all the resources "reserved" during the forward pass (if the request has been rejected) or downsizes their amounts (if it has been accepted). In both cases, during the interval between the forward and the reverse visits to a node, the resources available in the node for other new connections are scarce or nil, and some requests arriving at the node in that interval may be rejected, while they would be accepted if they were to arrive before the beginning or after the end of that interval. One solution, which unfortunately may increase the setup time of some connections appreciably, is to lock nodes during the interval between visits, so that requests will have to wait at locked nodes until the time they are unlocked by the end of the reverse visit. Another serious problem of this approach is that it may introduce deadlocks, which will require the addition of mechanisms for detecting and breaking them. All in all, it seems more reasonable to adopt a moderate overallocation policy and immediately reject those requests that find less resources available than they need.

The reduction to their final values of the amounts of resources allocated to a new real-time connection being set up during the forward pass is called *relaxation*. There are two ways in which relaxation can be performed: one approach consists of calculating at the destination all the amounts on the basis of the end-to-end bounds specified by the client and of the states of all the nodes traversed by the connection's path (note that these states have to be effectively summarized by the values of a few variables picked up by the establishment message during the forward pass); the other approach amounts to letting each of the nodes traversed by the reverse message take responsibility for as much of each bound as it can, and adjust local allocations accordingly. With the first approach, the reverse message has to carry the values of the local bounds and possibly of some of the allocations (e.g., buffer sizes) for all the nodes on the path, though some of the results may be computed locally by each node and do not have to be transmitted. The second approach is a distributed one: it requires much less information to be transmitted, and is more flexible; for example, it can adapt even to cases in which the state of a node on the path has changed (due to a connection having been torn down or modified in some of its parameter values [6]).

3. Distributed Admission Control: The Multicast Case

When the real-time connection to be created is a multicast (i.e., single-source, multiple-destination) one, admission control becomes more difficult. It is still possible, and, for the reasons mentioned at the beginning of Section 2, advantageous to resort to a distributed algorithm. It is also still possible to set up a real-time connection in a single round trip, but only if QoS parameter values differing from the best obtainable ones can be tolerated. We can establish a real-time connection by a receiver-initiated process as well as by a source-initiated one. However, here we describe only the latter, since the former can be obtained as a combination of unicast establishment procedures, which may follow the approach discussed in Section 2 [7].

A distributed source-initiated process consists of sending out of the source an establishment message, which visits in sequence the nodes on the path. For simplicity of description, we assume that the connection's route (topologically, a multicast tree) has been pre-determined, e.g., by a routing server. Whenever the message reaches a branch node, it splits into as many copies as there are branches coming out of that node. The copies are not identical, however: the sets of the destinations they are for are disjoint, and the union of these sets is the set of all the leaves of the subtree rooted in the branch node.

Admission tests are performed in each node when it is reached by a setup message. The tests are not fundamentally different from those used in the unicast case (which is a special case of multicast establishment anyway). However, since each destination should be allowed to have its own QoS parameter values, otherwise we would have to restrict the validity of our approach to the unrealistic case of homogeneous networks and similar distances from the source, a rejection message can be immediately returned when a bound is violated in a node only if the bounds of *all* destinations are violated. If the bounds of only a fraction of them are violated, to save time and resources we may at most inhibit any further testing and the propagation of message instances up to those destinations.

These observations suggest that multicast connection establishment is not an operation with a binary result (success or failure) like the establishment of unicast connections. In general, the operation will be both a partial success and a partial failure: some destinations will be reachable by a real-time multicast tree with the specified QoS, others will not. The requesting client will not get a yes-or-no answer, but a list of the destinations that will be able to receive from the source real-time streams satisfying the client-provided worst-case description and with guarantees on the desired QoS. Again, the last point on the path from the source to a destination where the acceptance or rejection decision can be made is the destination. Each destination that has been reached by an instance of the setup message returns an *accept* or *reject* message to the source. (Note that, if the final decision concerning a destination is made before the final point, the message instance for it may be stopped at the last branch node before the destination, and the destination will therefore not be reached by it.) The reverse messages from the leaves of a subtree have to wait for all the reverse messages from the other leaves of the same subtree in the branch node, so that they can be combined into a single message, which summarizes the replies pertaining to that subtree. In other words, there must be *barrier synchronization* in each branch node. The source receives a single reverse message containing a summary of the replies from all the leaves of the multicast tree. Since the reverse pass would block if any of the reverse messages were to get lost, it is necessary to transmit both the request and the reply messages using a reliable protocol.

One of the consequences of the presence of multiple destinations is that centralized relaxation (i.e., relaxation fully computed at the destination) is no longer a viable option. In the multicast case, relaxation must be distributed. The same policies adopted in the unicast case can be used along the parts of the tree where there are no branch nodes. The reverse message instances coming into a branch node from its outgoing links will generally produce different values for the local bounds (e.g., different local delay bounds). Since there is only one node, it cannot have more than one local bound. In the case of delay bounds, the smallest value is to be chosen. If a larger one were chosen, the destinations corresponding to values smaller than the chosen one would not be able to satisfy their client-specified end-to-end delay requirement. On the other hand, if a value for the local delay bound lower than expected by a destination is assigned in a node, the end-to-end delay bound for that destination will be smaller than that requested by the client. One consequence of this result is that more resources than strictly necessary are reserved (in the delay bound case, more of the node's "delay resource"), and less capacity is left for additional real-time connections through the same node.

Re-adjusting the situation is possible, but can only be done by changing the local bounds and the resource allocations in some nodes of the appropriate subtrees downstream, since the destinations that would be getting more stringent bounds than needed can only be differentiated from the others along the part of the tree exclusively serving them, which is downstream, close to the destination. For delay bounds, for example, it is sufficient to increase the local bounds in the nodes on the path that serves only the destination being considered. This re-adjustment operation requires a message to travel from the branch node where the conflict among values has arisen to those nodes. Unfortunately, the extra messages, besides making the procedure no longer single-round-trip, may encounter serious difficulties. For example, an increase in the local delay bound in a node requires the allocation of a larger buffer to the corresponding connection. If that additional amount of space is not available, the request for a connection to that destination may have to be rejected in spite of the determination made earlier by the destination itself that such a connection was feasible. This case clearly shows that resource requirements in various nodes (even on different branches of a multicast tree) are interdependent.

The additional messages are to be followed by replies, which the sending node must wait for before forwarding the reverse message towards the source with the final decision concerning the leaves of the subtree rooted in the node itself. Once the appropriate changes have been made to the local bounds and to the allocations in that subtree, these are "frozen", and any re-adjustments needed to resolve a conflict among local bound values in an upstream branch node are made immediately upstream of the root of that subtree. Thus, it is as though the subtree's root had become a leaf of another subtree closer to the source. For instance, the extra messages to be sent from an upstream branch node will not propagate downstream beyond this "new leaf" into the frozen subtree.

An alternative to this complicated and possibly time-consuming relaxation scheme is *dynamic resource reallocation* or *migration* [8], which leaves the task of readjusting allocations and local bounds to the post-establishment, data delivery period. The relaxation process in this alternative approach is primitive, even to the point of being non-existent. Allocations and bounds are refined later, using mechanisms that can also be exploited to modify the values of the QoS parameters of real-time connections "on the fly". Not surprisingly, even these mechanisms contain some interesting distributed algorithms.

4. Conclusion

This paper has described in qualitative and generic terms a distributed procedure that may be used to decide whether a request for a real-time connection in a packet-switching network can be accepted. The description has treated the two cases of unicast connections and multicast connections separately since they differ from each other to a substantial degree. The main goal of the description is to stimulate interest in the distributed algorithms that constitute such an admission control procedure, so that, if they deserve the theoreticians' attention, they may be studied and possibly improved. Their future practical importance certainly makes this effort worthwhile.

Acknowledgment

The author is grateful to the Fondazione della Cassa di Risparmio di Piacenza e Vigevano for its generous support.

References

[1] D. Ferrari. "Client Requirements for Real-Time Communication Services," *IEEE Communications Magazine,* vol. 28. n. 11. pp. 65-72, November 1990.

[2] D. Ferrari, "The Tenet Experience and the Design of Protocols for Integrated-Services Internetworks," preprint published on the Web at URL http: //www.tenet.berkeley.edu, July 1995.

[3] D. Ferrari, "Should an Integrated-Services Internetwork be Connectionless or Connection-Oriented? A Position Paper," *Proc. 6th Int. Workshop on Network and Operating System Support for Digital Audio and Video,* Zushi, Japan, pp. 3-4, April 1996.

[4] D. Ferrari, A.Gupta, and G. Ventre, "Distributed Advance Reservation of Real-Time Connections," *Proc. 5th Int. Workshop on Network and Operating System Support for Digital Audio and Video,* Durham, NH, pp. 15-26, April 1995.

[5] D. Ferrari and D. C. Verma, "A Scheme for Real-Time Channel Establishment in Wide-Area Networks," *IEEE J. Selected Areas in Communications,* vol. 8, n. 3, pp. 368-379, April 1990.

[6] C. Parris, H. Zhang, and D. Ferrari, "Dynamic Management of Guaranteed Performance Multimedia Connections," *Multimedia Systems,* vol. 1, pp. 267-283, July 1994.

[7] R. Bettati, D. Ferrari, A. Gupta, W. Heffner, W. Howe, M. Moran, Q. Nguyen, and R. Yavatkar, "Connection Establishment for Multi-Party Real-Time Communication," *Proc. 5th Int. Workshop on Network and Operating System Support for Digital Audio and Video,* Durham, NH, pp. 255-266, April 1995.

[8] R. Bettati and A. Gupta, "Dynamic Resource Migration for Multiparty Real-Time Communication," *Proc. 16th Int. Conf. on Distributed Computing Systems,* Hong Kong, pp. 646-655, May 1996.

Randomization and Failure Detection: A Hybrid Approach to Solve Consensus*

Marcos Kawazoe Aguilera and Sam Toueg

Cornell University, Computer Science Department, Ithaca NY 14853-7501, USA

Abstract. We present a Consensus algorithm that combines randomization and unreliable failure detection, two well-known techniques for solving Consensus in asynchronous systems with crash failures. This hybrid algorithm combines advantages from both approaches: it guarantees deterministic termination if the failure detector is accurate, and probabilistic termination otherwise. In executions with no failures or failure detector mistakes, the most likely ones in practice, Consensus is reached in only two asynchronous rounds.

1 Background

It is well-known that Consensus cannot be solved in asynchronous systems with failures, even if communication is reliable, at most one process may fail, and it can only fail by crashing. This "impossibility of Consensus", shown in a seminal paper by Fischer, Lynch and Paterson [FLP85], has been the subject of intense research seeking to "circumvent" this negative result, e.g., [Ben83, BT83, Rab83, DDS87, DLS88, CT96, CHT96].

We focus on two of the major techniques to circumvent the impossibility of Consensus in asynchronous systems: randomization and unreliable failure detection. The first one assumes that each process has an oracle (denoted *R-oracle*) that provides *random bits* [Ben83]. The second technique assumes that each process has an oracle (denoted *FD-oracle*) that provides *a list of processes suspected to have crashed* [CT96]. Each approach has some advantages over the other, and we seek to combine advantages from both.

With a randomized Consensus algorithm, every process can query its R-oracle, and use the oracle's random bit to determine its next step. With such an algorithm, termination is achieved with probability 1, within a finite expected number of steps (for a survey of randomized Consensus algorithms see [CD89]).

With a failure-detector based Consensus algorithm, every process can query its local FD-oracle (which provides a list of processes that are suspected to have crashed) to determine the process's next step. Consensus can be solved with FD-oracles that make an infinite number of mistakes. In particular, Consensus can be solved with any FD-oracle that satisfies two properties, *strong completeness* and *eventual weak accuracy*. Roughly speaking, the first property states that every process that crashes is eventually suspected by every correct process, and the second one states that some correct process

* Research partially supported by NSF grant CCR-9402896, by DARPA/NASA Ames grant NAG-2-593, and by an Olin Fellowship.

is eventually not suspected. These properties define the weakest class of failure detectors that can be used to solve Consensus [CHT96].

In this paper we describe a hybrid Consensus algorithm with the following properties. Every process has access to both an R-oracle and an FD-oracle. If the FD-oracle satisfies the above two properties, the algorithm solves Consensus (no matter how the R-oracle behaves). If the FD-oracle loses its accuracy property, but the R-oracle works, the algorithm still solves Consensus, albeit "only" with probability 1. In executions with no failures or failure detector mistakes, the most likely ones in practice, the algorithm reaches Consensus in two asynchronous rounds. A discussion of the relative merits of randomization, failure detection, and this hybrid approach is postponed to Sect. 6.

The idea of combining randomization and failure detection to solve Consensus in asynchronous systems first appeared in [DM94]. A related idea, namely, combining randomization and deterministic algorithms to solve Consensus in synchronous systems was explored in [GP90, Zam96]. A brief comparison with our results is given in Sect. 7.

2 Informal Model

Our model of asynchronous computation is patterned after the one in [FLP85], and its extension in [CHT96]. We only sketch its main features here. We consider *asynchronous* distributed systems in which there is no bound on message delay, clock drift, or the time necessary to execute a step. To simplify the presentation of our model, we assume the existence of a discrete global clock. This is merely a fictional device: the processes do not have access to it. We take the range $\mathcal{T}$ of the clock's ticks to be the set of natural numbers $\mathbb{N}$.

The system consists of a set of n *processes*, $\Pi = \{p_0, p_1, \dots, p_{n-1}\}$. Every pair of processes is connected by a reliable communication channel. Up to f processes can fail by *crashing*. A failure pattern indicates which processes crash, and when, during an execution. Formally, a *failure pattern* F is a function from $\mathbb{N}$ to 2^{Π}, where $F(t)$ denotes the set of processes that have crashed through time t. Once a process crashes, it does not "recover", i.e., $\forall t : F(t) \subseteq F(t+1)$. We define crashed$(F) = \bigcup_{t \in \mathbb{N}} F(t)$ and correct$(F) = \Pi -$ crashed(F). If $p \in$ crashed(F) we say *p crashes (in F)* and if $p \in$ correct(F) we say *p is correct (in F)*.

Each process has access to two oracles: a failure detector, henceforth denoted the *FD-oracle*, and a random number generator, henceforth denoted the *R-oracle*. When a process queries its FD-oracle, it obtains a list of processes. When it queries its R-oracle it obtains a bit. The properties of these oracles are described in the two next sections.

A distributed algorithm $\mathcal{A}$ is a collection of n deterministic automata (one for each process in the system) that communicate by sending messages through reliable channels. The execution of $\mathcal{A}$ occurs in *steps* as follows. For every time $t \in \mathcal{T}$, at most one process takes a step. Each step consists of receiving a message; querying the FD-oracle; querying the R-oracle; changing state; and optionally sending a message to one process. We assume that messages are never lost. That is, if a process does not crash, it eventually receives every message sent to it.

A schedule is a sequence $\{s_j\}_{j \in \mathbb{N}}$ of processes and a sequence $\{t_j\}_{j \in \mathbb{N}}$ of strictly increasing times. A schedule indicates which processes take a step and when: for each j, process s_j takes a step at time t_j. A schedule is *consistent (with respect to a failure pattern F)* if a process does not take a step after it has crashed (in F). A schedule is *fair (with respect to a failure pattern F)* if each process that is correct (in F) takes an infinite number of steps. We consider only schedules that are consistent and fair.

2.1 FD-oracles

Every process p has access to a local FD-oracle module that outputs a list of processes that are suspected to have crashed. If some process q belongs to such list, we say that p *suspects* q.[2] FD-oracles can make mistakes: it is possible for a process p to be suspected by another even though p did not crash, or for a process to crash and never be suspected. FD-oracles can be classified according to properties that limit the extent of such mistakes. We focus on one of the eight classes of FD-oracles defined in [CT96], namely, the class of *Eventually Strong* failure detectors, denoted $\diamond S$. An FD-oracle belongs to $\diamond S$ if and only if it satisfies two properties:

Strong completeness: Eventually every process that crashes is permanently suspected by *every* correct process (formally, $\exists t \in \mathcal{T}, \forall p \in \text{crashed}(F), \forall\, q \in \text{correct}(F)$, $\forall t' \geq t : p \in \text{FD}_q^{t'}$, where $\text{FD}_q^{t'}$ denotes the output of q's FD-oracle module at time t').

Eventual weak accuracy: There is a time after which some correct process is never suspected by any correct process (formally, $\exists t \in \mathcal{T}, \exists p \in \text{correct}(F), \forall\, t' \geq t$, $\forall q \in \text{correct}(F) : p \notin \text{FD}_q^{t'}$).

It is known that $\diamond S$ is the weakest class of FD-oracles that can be used to solve Consensus.

2.2 R-oracles

Each process has access to a local R-oracle module that outputs one bit each time it is queried. We say that the R-oracle is *random* if it outputs an independent random bit for each query. For simplicity, we assume a uniform distribution, i.e., a random R-oracle outputs 0 and 1, each with probability $1/2$.

2.3 Adversary Power

When designing fault-tolerant algorithms, we often assume that an intelligent adversary has some control on the behavior of the system, e.g., the adversary may be able to control the occurrence and the timing of process failures, the message delays, and the scheduling of processes. Adversaries may have limitations on their computing power and on the information that they can obtain from the system. Different algorithms are designed to defeat different types of adversaries [CD89].

We now describe the adversary that our hybrid algorithm defeats. The adversary has unbounded computational power, and full knowledge of all process steps that already

[2] In general, processes do not have to agree on the list of suspects at any one time or ever.

occurred. In particular, it knows the contents of all past messages, the internal state of all processes in the system,[3] and all the previous outputs of both the R-oracle and FD-oracle. With this information, at any time in the execution, the adversary can dynamically select which process takes the next step, which message this process receives (if any), and which processes (if any) crash. The adversary, however, operates under the following restrictions: the final schedule must be consistent and fair, every message sent to a correct process must be eventually received, and at most f processes may crash over the entire execution.

In addition to the above power, we allow the adversary to initially select *one* of the two oracles to control, and possibly corrupt.[4] If the adversary selects to control the R-oracle, it can predict and even determine the bits output by that oracle. For example, the adversary can force some local R-oracle module to always output 0, or it can dynamically adjust the R-oracle's output according to what the processes have done so far.

If the adversary selects to control the FD-oracle, it can ensure that the FD-oracle does not satisfy eventual weak accuracy. In other words, at *any* time the adversary can include *any* process (whether correct or not) in the output of the local FD-oracle module of any process. The adversary, however, does not have the power to disrupt the strong completeness property of the FD-oracle. This is not a limitation in practice: most failure detectors are based on time-outs and eventually detect all process crashes.

If the adversary does not control the R-oracle then the R-oracle is random. If the adversary does not control the FD-oracle then the FD-oracle is in $\diamond S$. We stress that the algorithm does *not* know which one of the two oracles (FD-oracle or R-oracle) is controlled by the adversary.

3 The Consensus Problem

Uniform Binary Consensus is defined in terms of two primitives, propose(v) and decide(v), where $v \in \{0, 1\}$. When a process executes propose(v), we say that it *proposes* v; similarly, when a process executes decide(v), we say that it *decides* v. The Uniform Binary Consensus problem is specified as follows:

Uniform agreement: If processes p and p' decide v and v', respectively, then $v = v'$;

Uniform validity: If a process decides v, then v was proposed by some process;

Termination: Every correct process eventually decides some value.

For probabilistic Consensus algorithms, Termination is weakened to

Termination with probability 1: With probability 1, every correct process eventually decides some value.

[3] This is in contrast to the assumptions made by several algorithms, e.g., those that use cryptographic techniques.

[4] From the definitions of these oracles, it is clear that we can allow the adversary to control the behavior of both oracles for an arbitrary but finite amount of time. The only restriction is that it must eventually stop controlling one of the oracles.

4 Hybrid Consensus Algorithm

The hybrid Consensus algorithm shown in Fig. 1 combines Ben-Or's algorithm [Ben83] with failure-detection and the rotating coordinator paradigm used in [CT96]. With this paradigm, we assume that all processes have a priori knowledge that, during phase k, one selected process, namely $p_{k \bmod n}$, is the coordinator. The algorithm works under the assumption that a majority of processes are correct (i.e., $n > 2f$). It is easy to see that this requirement is necessary for any algorithm that solves Consensus in asynchronous systems with crash failures, even if all processes have access to a random R-oracle and an FD-oracle that belongs to $\Diamond S$.

In the hybrid algorithm, every message contains a tag (R, P, S or E), a phase number, and a value which is either 0 or 1 (for messages tagged P or S, it could also be "?"). Messages tagged R are called *reports*; those tagged with P are called *proposals*; those with tag S are called *suggestions [to the coordinator]*; those with tag E are called *estimates [from the coordinator]*. When p sends (R, k, v), (P, k, v) or (S, k, v) we say that p *reports*, *proposes* or *suggests* v in phase k, respectively. When the coordinator sends (E, k, v) we say that the coordinator sends estimate v in phase k.

Each execution of the **while** loop is called a *phase*, and each phase consists of four asynchronous rounds. In the first round, processes report to each other their current estimate (0 or 1) for a decision value.

In the second round, if a process receives a majority of reports for the *same* value then it proposes that value to all processes, otherwise it proposes "?". Note that it is impossible for one process to propose 0 and another process to propose 1. At the end of the second round, if a process receives $f + 1$ proposals for the same value different than ?, then it decides that value. If it receives at least one value different than ?, then it adopts that value as its new estimate, otherwise it adopts ? for estimate.

In the third round, processes suggest their estimate to the current coordinator. If the coordinator receives a value different than ? then it sends that value as its estimate. Otherwise, the coordinator queries the R-oracle, and sends the random value that it obtains as its estimate.

In the fourth round, processes wait until they receive the coordinator's estimate or until their FD-oracle suspects the coordinator. If a process receives the coordinator's estimate, it adopts it. Otherwise, if its current estimate is ?, it adopts a random value obtained from its R-oracle.

To simplify the presentation, the algorithm in Fig. 1 does not include a halt statement. Moreover, once a correct process decides a value, it will keep deciding the same value in all subsequent phases. However, it is easy to modify the algorithm so that every process decides at most once, and halts at most one round after deciding.

This algorithm always satisfies the safety properties of Consensus. This holds no matter how the FD-oracle or the R-oracle behave, that is, even if these oracles are totally under the control of the adversary. On the other hand, the algorithm satisfies liveness properties only if the FD-oracle satisfies strong completeness. Strong completeness is easy to achieve in practice: most failure-detectors use time-out mechanisms, and every process that crashes eventually causes a time-out, and therefore a permanent suspicion.

Every process p executes the following:

```
0  procedure propose(v_p)                      {v_p is the value proposed by process p}
1    x ← v_p                                   {x is p's current estimate of the decision value}
2    k ← 0

3    while true do
4      k ← k + 1                               {k is the current phase number}
5      c ← p_{k mod n}                         {c is the current coordinator}
6      send (R, k, x) to all processes

7      wait for messages of the form (R, k, *) from n − f processes    {'*' can be 0 or 1}
8      if received more than n/2 (R, k, v) with the same v
9      then send (P, k, v) to all processes
10     else send (P, k, ?) to all processes

11     wait for messages of the form (P, k, *) from n − f processes    {'*' can be 0, 1 or ?}
12     if received at least f + 1 (P, k, v) with the same v ≠ ? then decide(v)
13     if at least one (P, k, v) with v ≠ ? then x ← v else x ← ?
14     send (S, k, x) to c

15     if p = c then
16       wait for messages of the form (S, k, *) from n − f processes
17       if received at least one (S, k, v) with v ≠ ?
18       then send (E, k, v) to all processes
19       else
20         random_bit ← R-oracle                 {query R-oracle}
21         send (E, k, random_bit) to all processes

22     wait until receive (E, k, v_coord) from c or c ∈ FD-oracle      {query FD-oracle}
23     if received (E, k, v_coord)
24     then x ← v_coord
25     else if x = ? then x ← R-oracle           {query R-oracle}
```

Fig. 1. Hybrid Consensus algorithm

Theorem 1. *Assume $n > 2f$, i.e., a majority of processes is correct.*

(Safety) The hybrid algorithm always satisfies validity and uniform agreement.

(Liveness) Suppose that the FD-oracle *satisfies strong completeness.*

- *If the* FD-oracle *satisfies eventual weak accuracy, i.e., it is in $\Diamond S$, then the algorithm satisfies termination.*
- *If the* R-oracle *is random then the algorithm satisfies termination with probability 1.*

Proof. See [AT96]. □

If the *R-oracle* is random, the expected number of rounds for termination is $O(2^{2n})$. However, it can be shown that, as in [Ben83], termination is reached in constant expected number of rounds if $f = O(\sqrt{n})$. In Sect. 6, we outline a similar hybrid algorithm that terminates in constant expected number of rounds even for $f = O(n)$.

5 An Optimization

The algorithm in Fig. 1 was designed to be simple rather than efficient, because our main goal here is to demonstrate the viability of a "robust" hybrid approach (one in which termination can occur in more than one way: by "good" failure detection or by "good" random draws). The following optimization suggests that such hybrid algorithms can also be efficient in practice.

In many systems, failures are rare, and failure detectors can be tuned to seldom make mistakes (i.e., erroneous suspicions). The algorithm in Fig. 1 can be optimized to perform particularly well in such systems. The optimized version ensures that all correct processes decide by the end of two asynchronous rounds when the first coordinator does not crash and no process erroneously suspects it.[5]

$c \leftarrow p_0$ {p_0 is the first coordinator}
if $p = c$ **then send** $(E, 0, v_p)$ to all processes {if p is the first coordinator}

wait until receive $(E, 0, v\_coord)$ **from** c **or** $c \in$*FD-oracle* {query FD-oracle}
if received $(E, 0, v\_coord)$
then send $(P, 0, v\_coord)$ **to** all processes
else send $(P, 0, ?)$ **to** all processes

wait for messages of the form $(P, 0, *)$ **from** $n - f$ processes {"*" can be 0, 1 or ?}
if received at least $f + 1$ $(P, 0, v)$ with the same $v \neq ?$ **then** *decide*(v)
if received at least one $(P, 0, v)$ with $v \neq ?$ **then** $x \leftarrow v$

Fig. 2. Optimization for the hybrid algorithm

This optimization is obtained by inserting some extra code between lines 2 and 3 of the hybrid algorithm. This code, given in Fig. 2, consists of a phase (phase 0) with two asynchronous rounds. In the first round, p_0 sends a message to all processes; in the second round, every process sends a message to all processes. We claim that: (1) the optimization code preserves the correctness of the original algorithm; and (2) processes decide quickly in the absence of failures and erroneous suspicions. To see (1) note that:

- No correct process blocks during the execution of the optimization code (phase 0), i.e., all correct processes start phase 1;

[5] Processes decide in two rounds even if up to $n - 2f - 1$ processes erroneously suspect it.

- Any process p that starts phase 1 does so with x_p set to the initial value of some process;
- If some process decides v in phase 0 then all processes that start round 1 do so with their variable x set to v.

To see (2), note that if p_0 is correct and no process suspects p_0, then all processes wait for its estimate v and propose v in phase 0; so every process receives $n - f$ proposals for v and thus decides v in phase 0. Thus we have:

Theorem 2. *Theorem 1 holds for the optimized hybrid algorithm. Moreover, in executions with no crashes or false suspicions, all processes decide in two rounds.*

6 Discussion

In practice, many systems are well-behaved most of the time: few failures actually occur, and most messages are received within some predictable time. Failure-detector based algorithms (whether "pure" ones like in [CT96] or hybrid ones like in this paper) are particularly well-suited to take advantage of this: (time-out based) failure detectors can be tuned so that the algorithms perform optimally when the system behaves as predicted, and performance degrades gracefully as the system deviates from its "normal" behavior (i.e., if failures occur or messages take longer than expected). For example, the optimized version of our hybrid algorithm solves Consensus in only two asynchronous rounds in the executions that are most likely to occur in practice, namely, runs with no failures or erroneous suspicions.

The above discussion suggests that using this hybrid approach is better than using the randomized approach alone. In fact, randomized Consensus algorithms for asynchronous systems tend to be inefficient in practical settings.[6] Typically, their performance depends more on "luck" (e.g., many processes happen to start with the same initial value or happen to draw the same random bit) than on how "well-behaved" the underlying system is (e.g., on the number of failures that actually occur during execution). The fact that randomized algorithms are extremely "robust", i.e., they do not depend on how the system behaves, may also be an inherent source of inefficiency.

Note that our hybrid algorithm terminates with probability 1 even if the FD-oracle is completely inaccurate (in fact even if every process suspects every other process all the time). So it is more robust than algorithms that are simply failure-detector based.

An important remark is now in order about the expected termination time of our hybrid algorithm. We developed this algorithm by combining Ben-Or's randomized algorithm [Ben83] with the failure detection ideas in [CT96]. We selected Ben-Or's algorithm because it is the simplest, and thus the most appropriate to illustrate this approach, even though its expected number of rounds is exponential in n for $f = O(n)$. By starting from an efficient randomized algorithm, due to Chor et al. [CMS89], we can obtain a hybrid algorithm that terminates in constant expected number of rounds, as we now briefly explain.

---

[6] Algorithms that assume that processes a priori agree on a long sequence of random bits [Rab83, Tou84] are more efficient than others. But this assumption may be too strong for some systems.

Roughly speaking, the randomized asynchronous Consensus algorithm in [CMS89] is obtained from Ben-Or's algorithm by replacing each coin toss with the toss of a "weakly global coin" computed by a *coin_toss* procedure. We can do exactly the same: replace the coin tosses of the algorithm in Fig. 1 with those obtained by using the *coin_toss* procedure. More precisely, in each phase, every process: (a) invokes this procedure between the second and third rounds (i.e., between lines 13 and 14) to obtain a random bit, and (b) uses this random bit rather than querying the R-oracle (in lines 20 and 25).[7]

As in [CMS89], this modified hybrid algorithm terminates[8] in constant expected number of rounds for $f \leq n(3-\sqrt{5})/2 \approx 0.38n$. But also as in [CMS89], and in contrast to the algorithm in Sect. 4, it assumes that the adversary cannot see the internal state of processes or the content of messages. With the optimization of Fig. 2, this modified hybrid algorithm also terminates in two rounds in failure-free and suspicion-free runs.

7 Related Work

The idea of combining randomization with a deterministic Consensus algorithm appeared in [GP90], and was further developed in [Zam96]. These works, however, are for *synchronous* systems only and do not involve failure detection.

Dolev and Malki were the first to combine randomization and unreliable failure detection to solve Consensus in asynchronous systems with process crashes only [DM94]. That work differs from ours in many aspects:

- The hybrid algorithms given in [DM94] assume that *both* the R-oracle and the FD-oracle always work correctly. If the failure detector loses it accuracy property, processes may decide differently; if the random source of bits is corrupted, processes may never decide.
- Two goals of [DM94] are to use failure detection to increase the resiliency and ensure the deterministic termination of randomized Consensus algorithms. The hybrid Consensus algorithms given in [DM94] achieve the first goal, by increasing the resiliency from $f < n/2$ to $f < n$, but not the second one. It is stated, however, that a future version of the paper will give an algorithm that achieves both goals.
- The two hybrid algorithms in [DM94] use failure detectors that are stronger than $\diamond S$. The first one — which supposes that the same sequence of random bits is shared by all the processes, as in [Rab83] — assumes that some correct process is *never* suspected by any process. The second algorithm — which drops the assumption of a common sequence of bits — assumes that $\Omega(n)$ correct processes are never suspected by any process.

[7] As in [CMS89], another simple modification is necessary: the addition of a "synchronization round" just before the *coin_toss* procedure. In this round, processes broadcast "wait" messages, then wait until $n - f$ such messages are received.

[8] Provided, of course, that the FD-oracle satisfies strong completeness.

Acknowledgement

We are grateful to Vassos Hadzilacos and the anonymous referees for their valuable comments.

References

[AT96] Marcos Kawazoe Aguilera and Sam Toueg. Randomization and failure detection: A hybrid approach to solve consensus. Technical Report 96-1592, Department of Computer Science, Cornell University, June 1996. Available by anonymous ftp from ftp.cs.cornell.edu in pub/sam/hybrid.consensus.algorithm.ps.gz.

[Ben83] Michael Ben-Or. Another advantage of free choice: Completely asynchronous agreement protocols. In *Proceedings of the Second ACM Symposium on Principles of Distributed Computing*, pages 27–30, August 1983.

[BT83] Gabriel Bracha and Sam Toueg. Resilient consensus protocols. In *Proceedings of the Second ACM Symposium on Principles of Distributed Computing*, pages 12–26, August 1983. An extended and revised version appeared as "Asynchronous consensus and broadcast protocols" in the *Journal of the ACM*, 32(4):824-840, October 1985.

[CD89] Benny Chor and Cynthia Dwork. Randomization in Byzantine Agreement. *Advances in Computer Research (JAI Press Inc.)*, 4:443–497, 1989.

[CHT92] Tushar Deepak Chandra, Vassos Hadzilacos, and Sam Toueg. The weakest failure detector for solving consensus. In *Proceedings of the Tenth ACM Symposium on Principles of Distributed Computing*, pages 147–158, August 1992.

[CHT96] Tushar Deepak Chandra, Vassos Hadzilacos, and Sam Toueg. The weakest failure detector for solving consensus. *Journal of the ACM*, 43(4), July 1996. An earlier version appeared in [CHT92].

[CMS89] Benny Chor, Michael Merritt, and David B. Shmoys. Simple constant-time consensus protocols in realistic failure models. *Journal of the ACM*, 36(3):591–614, 1989.

[CT91] Tushar Deepak Chandra and Sam Toueg. Unreliable failure detectors for asynchronous systems. In *Proceedings of the Tenth ACM Symposium on Principles of Distributed Computing*, pages 325–340. ACM Press, August 1991.

[CT96] Tushar Deepak Chandra and Sam Toueg. Unreliable failure detectors for reliable distributed systems. *Journal of the ACM*, 43(2):225–267, March 1996. An earlier version appeared in [CT91].

[DDS87] Danny Dolev, Cynthia Dwork, and Larry Stockmeyer. On the minimal synchronism needed for distributed consensus. *Journal of the ACM*, 34(1):77–97, January 1987.

[DLS88] Cynthia Dwork, Nancy A. Lynch, and Larry Stockmeyer. Consensus in the presence of partial synchrony. *Journal of the ACM*, 35(2):288–323, April 1988.

[DM94] Danny Dolev and Dalia Malki. Consensus made practical. Technical Report CS94-7, The Hebrew University of Jerusalem, March 1994.

[FLP85] Michael J. Fischer, Nancy A. Lynch, and Michael S. Paterson. Impossibility of distributed consensus with one faulty process. *Journal of the ACM*, 32(2):374–382, April 1985.

[GP90] Oded Goldreich and Erez Petrank. The best of both worlds: guaranteeing termination in fast randomized Byzantine Agreement protocols. *Information Processing Letters*, 36(1):45–49, October 1990.

[Rab83] Michael Rabin. Randomized Byzantine Generals. In *Proceedings of the Twenty-Fourth Symposium on Foundations of Computer Science*, pages 403–409. IEEE Computer Society Press, November 1983.

[Tou84] Sam Toueg. Randomized Byzantine Agreements. In *Proceedings of the Third ACM Symposium on Principles of Distributed Computing*, pages 163–178, August 1984.

[Zam96] Arkady Zamsky. A randomized Byzantine Agreement protocol with constant expected time and guaranteed termination in optimal (deterministic) time. In *Proceedings of the Fifteenth ACM Symposium on Principles of Distributed Computing*, pages 201–208, May 1996.

Levels of Authentication in Distributed Agreement

Malte Borcherding

Institute of Computer Design and Fault Tolerance
University of Karlsruhe
76128 Karlsruhe, Germany
malte.borcherding@informatik.uni-karlsruhe.de

Abstract. Reaching agreement in the presence of Byzantine (arbitrary) faults is a fundamental problem in distributed systems. It has been shown that message authentication is a useful tool in designing protocols with high fault tolerance, but it imposes the additional problem of key distribution.
In the past, agreement protocols using message authentication required complete agreement on all public keys. Because this pre-agreement has to rely on techniques outside the system (e.g., trusted servers which never fail), it is useful to consider lower levels of key distribution which need as few assumptions as possible.
In this paper, we identify several levels of key distribution and describe their properties with regard to the achievable fault tolerance in two agreement problems.

Keywords: Byzantine agreement, crusader agreement, authentication, distributed systems, fault tolerance

1 Introduction

The problem of distributed agreement arises when a set of nodes in a distributed system need to have a consistent view of a message sent by one of them, despite the presence of arbitrarily faulty nodes. Several kinds of agreement have been defined in the past. The most stringent kind of agreement is *Byzantine agreement* (BA) as defined in [LSP82]. It requires that the following three conditions be met:

(B1) All correct nodes agree on the same value.
(B2) If the sender is correct, all correct nodes agree on the value of the sender.
(B3) Each correct node eventually decides on a value.

One variant of this agreement is *crusader agreement* (CA), introduced in [Dol82]. Here, it is not necessary that all nodes agree on the same value if the sender is faulty. It is required, though, that those nodes which do not agree *know* that the sender is faulty:

(C1) All correct nodes that do not explicitly know that the sender is faulty agree on the same value.
(C2) If the sender is correct, all correct nodes agree on the value of the sender.
(C3) Each correct node eventually decides on a value or knows that the sender is faulty.

Protocols for distributed agreement are generally divided into two classes: authenticated protocols and non-authenticated protocols. In authenticated protocols, all messages are signed digitally in a way that the signatures cannot be forged and a signed message can be unambiguously assigned to its signer. This mechanism allows a node to prove to others that it has received a certain message from a certain node. Authenticated protocols can tolerate an arbitrary number of faulty nodes. In non-authenticated protocols, no messages are signed. For BA and CA, these protocols require more than two thirds of the participating nodes to be correct ([LSP82, Dol82]).

While authenticated protocols offer the best fault tolerance, it is not at all trivial to distribute the public keys of all participants of an agreement protocol consistently amongst each other. Typically, key distribution requires a single trusted entity, or a group of entities which is completely reliable as a whole ([Gon93]). Since these assumptions restrict the usual assumptions about the participant's behaviour, they should be kept as weak as possible. Hence, it is useful to have a look at possible scenarios with different kinds of key distribution and different common knowledge about these distributions.

2 Model of Computation

In this section we describe the model of computation used throughout this paper. Our world consists of a fully interconnected network with n nodes (processors), t of which may be faulty. In order to avoid special cases, we assume that $n \geq 3$ and $n > t+1$. For $n < 3$ there are trivial solutions, and for $n \leq t+1$ agreement always holds by definition.

The nodes operate at a known minimal speed, and messages are transmitted reliably in bounded time. Furthermore, a receiver of a message can identify its immediate sender. Communication takes place in successive *rounds*. In each round a node may send messages to other nodes and receives all messages sent to it in the current round. The actions a node takes in the next round depend solely on the messages it has received so far. We make no assumptions about the *type of failures* that occur. If a node is faulty, it may behave in an arbitrary manner. This type of behaviour is usually referred to as *Byzantine fault*.

In addition, we assume the existence of an unforgeable signature scheme. Examples for signature schemes which are unforgeable with a sufficiently high probability (given today's state of the art) are DSA and RSA [Nat92, RSA78]. In these schemes, a prospective signer has a pair of keys, namely a private key and a public key. The private key is used for signing, while the public key can be used to verify a signature made with the private key.

The assumption of a signature scheme alone is not very strong (see section 3.2). It is often necessary to make assumptions about the distribution of the public keys. The strongest assumption is that of *complete authentication.* It comprises the following four properties:

(A1) If a correct node assigns a signed message to a correct node P, then P has signed the message.
(A2) A message signed by a correct node P is assigned to P by all correct nodes.
(A3) If a message is assigned to a (possibly faulty) node P by a correct node, then all correct nodes assign it to P.
(A4) All correct nodes can sign messages.

In terms of private/public keys, properties (A1) and (A2) state that there is agreement on the public keys of the correct nodes, and the correct nodes keep their secret keys secret. Property (A3) extends the agreement to the public keys of the faulty nodes. Owing to the Byzantine failures, it cannot be assumed that the faulty nodes keep their secret keys secret. Fault models which assume that faulty nodes do not give their secrets away (or sign messages on behalf of others) are used in [GLR95, EM96].

3 Levels of Authentication

Solutions for agreement problems usually assume either no authentication or complete authentication. These assumptions are only two extreme points in a spectrum of possible scenarios. In this section, we will identify several different levels of authentication, give situations in which they arise, and present their properties with regard to the achievable fault tolerance.

3.1 No Authentication

In this situation, no means of key distribution and no signature scheme is available or wanted, e.g., due to lack of processor speed, lack of private local storage or insufficient trust into existing methods. Here applies the long-known requirement $n > 3t$ ([PSL80, LSP82]) which makes agreement between three nodes impossible if one may fail.

3.2 Local ("Byzantine") Authentication

This type of agreement can be reached when no means of agreement on the keys is provided and each node distributes its public key by itself, using a signature scheme. With a challenge-response key distribution protocol, properties (A1) and (A2) can be enforced if (A4) holds and a signature scheme exists ([Bor95]). That is, a faulty node can distribute different public keys to different nodes, but it can not claim a public key of a correct node for itself. This makes local authentication strictly stronger than no authentication: A faulty node can not forge messages

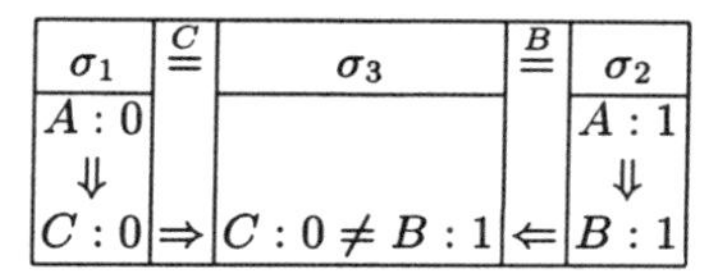

Fig. 1. Proof of Theorem 1

sent by correct nodes without invalidating the signature. Furthermore, a message signed by a correct node cannot be misattributed to a faulty node.

Although no complete agreement on the public keys of the *faulty* nodes can be guaranteed, this level of authentication has been shown to be useful for Failure Discovery, a sub-problem of Byzantine agreement ([HH93, Bor95]). Using local authentication, Byzantine agreement with few messages in the failure-free runs is possible.

Unfortunately, in this setting the impossibility of Byzantine agreement for $n \leq 3t$ cannot be overcome, as stated in the following theorem.

Theorem 1. *It is not possible to reach Byzantine agreement if one third of the nodes is faulty and only local authentication is assumed.*

Proof. This proof is a variation of the proof in [LSP82]: Let $n \leq 3t$. Then it is possible to partition the nodes into three nonempty sets A, B, and C, such that each set contains at most t elements. The members of these sets will be denoted a_i, b_i, and c_i, respectively. The sender will always be a_1.

We consider the most general protocol with an arbitrary number of rounds: If a correct node receives a message in round r, it signs this message and sends it to all nodes in round $r+1$. Thus, the maximum possible information flow is achieved. We do not specify how the information is used to reach the final decision. We will only observe whether or not a node receives different messages in different runs of a protocol. If the messages are the same, the node has to draw the same conclusions in the respective runs.

We will consider three possible scenarios σ_1, σ_2, and σ_3. In σ_1, the nodes in B are faulty, and a_1 sends 0. In σ_2, the nodes in C are faulty, and a_1 sends 1. In σ_3, the nodes in A are faulty, and a_1 sends send 0 to C and 1 to B.

Hence, in σ_1, the nodes in C have to decide for 0, and in σ_2, the nodes in B have to decide for 1. We will show that in σ_3, the nodes in B receive the same messages as in σ_2, while the nodes in C receive the same messages as in σ_1. So, in σ_3, the nodes in B and C have to decide for different values, which contradicts (B1). An example with tree nodes will be given below.

We will use the following notation for the messages: 0_{a_1BCB} means that the value 0 was first signed by a_1, then by a node in B, followed by a node in C, and then again by a node in B. Small letters denote actual signatures, while capitals represent the signature of some member of the respective set. For nodes in A there will be two sets of signatures. They will be denoted A_B (a_{iB}) and A_C (a_{iC}), respectively.

The three scenarios will be constructed in a way that the signatures on all messages seen by correct nodes in B and C are of the following form:

- The first signature is a_{1B} if the value is 1, and a_{1C} if the value is 0.
- A_B is followed by B or A_B.
- A_C is followed by C or A_C.
- B and C are followed by A_B, A_C, B, or C.

Signatures A_B and A_C cannot be recognized by correct nodes in C and B, respectively. The behaviour of the nodes in the three scenarios is as follows:

Scenario σ_1 (B faulty): All nodes distribute consistent keys in the key distribution protocol. The nodes in A sign all messages with A_C. In the first round of the agreement protocol, a_1 signs the value 0 and sends it to all nodes. In the following rounds, the nodes in A and C sign all messages correctly and send them to all nodes.
The nodes in B treat all messages from B and C correctly. When a node b_i receives a message from A, it replaces those signatures A_C, which are consecutively at the end, with A_B. The signatures A_B are chosen arbitrarily by the nodes in B and cannot be recognized by nodes in A or C. If all signatures on the message are from nodes in A, the value is set to 1 before substituting the signatures. Finally, b_i signs the message correctly and sends it to all nodes. If, for example, b_i receives $0_{a_{1C}c_2a_{4C}a_{2C}}$, it will echo the message as $0_{a_{1C}c_2a_{4B}a_{2B}b_i}$, while a message $0_{a_{1C}}$ becomes $1_{a_{1B}b_i}$.

Scenario σ_2 (C faulty): All nodes distribute consistent keys in the key distribution protocol. The nodes in A sign all messages with A_B. In the first round of the agreement protocol, a_1 signs the value 1 and sends it to all nodes. In the following rounds, the nodes in A and B sign all messages correctly and send them to all nodes.
The nodes in C treat all messages from B and C correctly. When a node c_i receives a message from A, it replaces those signatures A_B, which are consecutively at the end, with A_C. The signatures A_C are chosen arbitrarily by the nodes in C and cannot be recognized by nodes in A or B. If all signatures on the message are from nodes in A, the value is set to 0 before substituting the signatures. Finally, c_i signs the message correctly and sends it to all nodes. If, for example, c_i receives $1_{a_{1B}b_2a_{4B}a_{2B}}$, it will echo the message as $1_{a_{1B}b_2a_{4C}a_{2C}c_i}$, while a message $1_{a_{1B}}$ becomes $0_{a_{1C}c_i}$.

Scenario σ_3 (A faulty): In the key distribution protocol, the nodes in B and C distribute consistent keys. The nodes in A send different keys A_B and A_C to the nodes in B and C.
In the first round of the agreement protocol, a_1 sends $1_{a_{1B}}$ to B and $0_{a_{1C}}$ to C. In the following rounds, the nodes in B and C sign all messages correctly and send them to all nodes. The nodes in A also sign all messages and send them to all nodes. Before sending a message to B, though, they replace those signatures A_C which are consecutively at the end, with A_B. If there are only signatures by nodes in A, the value is set to 1. Likewise, in

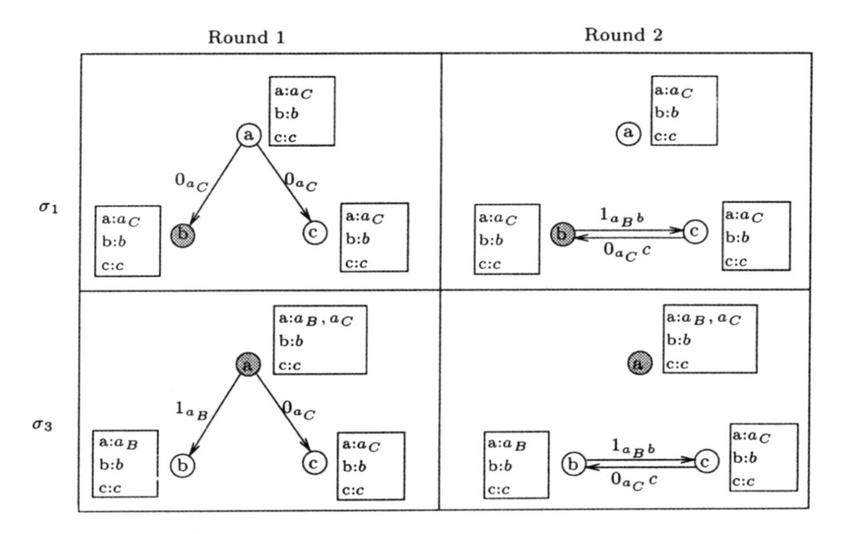

Fig. 2. Impossibility of Byzantine agreement with local authentication

messages to C, signatures A_B are replaced by A_C, and the value is set to 0. This is possible since the nodes in A may cooperate. If, for example, a_i receives $1_{a_{1B}b_2a_{4B}a_{2B}}$, it will forward this message to C as $1_{a_{1B}b_2a_{4C}a_{2C}a_{iC}}$.

It can easily be seen that for each message received by a node in B in scenario σ_2, there is an indistinguishable message in scenario σ_3, and vice versa. The same holds for the nodes of C in scenarios σ_1 and σ_3. Since a node b_i has to decide for 1 in σ_2, and a node c_i has to decide for 0 in σ_1, they decide for different values in σ_3. This violates (B1). This reasoning is depicted in Fig. 1. □

Example 1. The idea of the proof will be demonstrated at the most simple example: There are three nodes a, b, and c. One of them may be faulty, and the sender is a. Scenarios σ_1 and σ_3 are depicted in Fig. 2. Messages to and from a in the second round are omitted, as well as self-addressed messages. The boxes at each node show the respective views of the public keys, and faulty nodes are shadowed.

If c is correct, it will receive 0_{a_C} in the first round of σ_1 and σ_3. In the second round, it will receive $0_{a_C a_C}$, $0_{a_C c}$, and $1_{a_B b}$, where a_B is not recognizable to c. From these messages, c can deduce that one of the following cases must hold:

1. a is correct and has sent 0. The unrecognizable signature has been produced by b. In this case, c has to decide for 0.
2. b is correct, but a has sent different values to b and c. Furthermore, a has sent different public keys to b and c. Then, c has to decide for the same value as b.

In order to fulfill (B2), c has to decide for 0. If a is faulty, b sees messages

Protocol for Crusader Agreement

1. The sender signs its value and sends it to all others.
2. If a node recognizes the signature, it transmits the received value and signature to all others. Otherwise, it decides for a faulty sender.
3. If a node has not decided in the second step, it looks at the values signed by the sender it has seen so far. If there is exactly one distinct value, it decides for that value. Otherwise, it decides for a faulty sender.

Fig. 3. Crusader agreement for crusader authentication

consistent with σ_2 and decides for 1, which violates (B1). If b and c always decide for some default value, one of them violates (B1) in σ_1 or σ_2.

The proof for the achievable fault tolerance for crusader agreement is similar to the proof given above. Hence, to reach crusader agreement under local authentication, more than two thirds of correct nodes are necessary.

3.3 Sound, but Incomplete Authentication

For this kind of authentication we assume that no two correct nodes have different public keys of a third node. However, a node may not know the public keys of all other nodes. In this setting, two cases can be distinguished. In the first case, there is no agreement on who knows which keys, while in the second case, this agreement is assumed.

Crusader Authentication Here, we assume that only the public keys of the faulty nodes are distributed incompletely, but it is not known who actually knows whose public key.

This requires a more benign behaviour of the faulty nodes during the key distribution process. A faulty node may still choose not to send its public key to some other nodes, but it does not distribute different keys. Apart from that, it may behave arbitrarily. Instead of (A3), (A3)′ holds in this context:

(A3)′ All correct nodes which assign a certain message to a node assign it to the same node.

We will refer to this level of authentication as *crusader authentication*, since the public keys of the nodes are agreed upon like values sent by crusader agreement: If a correct node A has a key of B, it is the correct one. Otherwise, A knows that B is faulty.

In such a setting, *crusader agreement* is possible for any number of faulty nodes. It can be reached in only two rounds (see Fig. 3).

Theorem 2. *The protocol in Fig. 3 reaches crusader agreement for any $n \geq t+2$ if crusader authentication holds.*

Proof. (C1) is trivially fulfilled for those nodes which decide that the sender is faulty. Now assume that two correct nodes decide for different values. Then both must have received their values with the sender's signature in the first round. Furthermore, they must not have received a different signed value in the second round. But that is impossible, since both must have sent their values to all others in the second round.

(C2) is fulfilled, because a correct sender signs its value in the first round and sends it to all nodes. Since the sender is correct, all nodes assign the message correctly and send it to all others in the second round. There exists exactly one value signed by the sender, so all correct nodes will decide for that value. (C3) is fulfilled by the limited number of rounds. □

Byzantine agreement can be reached with $n \geq 2t + 1$. The protocol in Fig. 4 has this fault tolerance. It is a variation of the Exponential Information Gathering *(EIG)* protocol which was introduced by Bar-Noy *et al.* [BDDS87], based on the protocol in [LSP82]. In this protocol, the sender starts by sending its value to all other nodes. In the following t rounds, each node signs and forwards messages received in the previous round to the other nodes.

During protocol execution, each node maintains an *EIG* tree which contains the received information in a structured manner. Such a tree has $t + 1$ levels, one level per communication round. The root has $n - 1$ children, and in each of the following levels, the vertices have one child less than those of the previous level. Hence, on level r, each vertex has $n - r$ children (we consider the root level as level 1). A part of such a tree for $n = 7$ is shown in Fig. 5.

The vertices have labels which are assigned in the following manner: The root is labeled with the sender's name. In the following levels, the children of a vertex are labeled with the names of the nodes not yet on the path from the root. We identify a vertex in the tree by the the sequence of labels from the root to the respective vertex. Note that in no such sequence a node's name appears twice. A vertex labeled with the name of a correct (faulty) node will be called a correct (faulty) vertex. From the construction, a vertex on level r has at most t faulty children and at least $n - r - t$ correct children.

In the first round of the protocol, each node stores the value received from the sender in the root of its *EIG* tree. In the following rounds, each correct node broadcasts the contents of the level of its tree most recently filled in, and fills the next level with messages it receives. If a node X receives a message from Y claiming that it has stored v in vertex $ABCDE$, X stores v in vertex $ABCDEY$ of its *EIG* tree. Hence, a value v in vertex $ABCDEY$ is interpreted as "Y said E said ... B said A said v". If a node failed to send a value, a default value is stored.

Due to the structure of the tree, not all received messages are stored. Those messages in which a node reports about a message which was once sent by itself are ignored. In Fig. 5, labels and stored values are separated by a colon. The faulty vertices are shadowed.

When a node has completed its tree (after round $t + 1$), it uses the collected data to decide for the outcome of the protocol. This is done by ***resolving*** each

Protocol for Byzantine Agreement

1. The nodes fill their EIG trees for $t+1$ rounds. They only consider messages which carry the recognizable signature of the immediate sender.
2. A leaf is resolved to its value with the last signature removed.
3. For a non-leaf on level $r < t+1$ with label X, two cases are distinguished:
   - The signature of X is known: Consider the set of resolved children which are signed correctly. If it has at least $t-r+1$ members, take the (relative) majority value. The vertex is then resolved to that value without the last signature. If the set has fewer members, the vertex is resolved to a default value.
   - Signature of the X is unknown: Take a set of maximum size of resolved children which carry the same signature. If it contains at least $t-r+1$ elements, take the relative majority. The vertex is then resolved to that value, with the last signature removed. If no such set exists, the vertex is resolved to a default value.
4. The result of the protocol is the resolved value of the root.

Fig. 4. Byzantine agreement for crusader authentication

vertex to a certain value, depending on the resolved values of its children. The exact rules are given in steps 2 and 3 of the protocol. A vertex which is resolved to the same value by all correct nodes will be called *common.* The following example will demonstrate the rules for resolving.

Example 2. Let $n = 7$ and $t = 3$ (see Fig. 5). Let us further assume that the five children of a vertex at level $r = 2$ are resolved to the following values: default, default, $1_{ab}, 2_{ab'}, 2_{ab'}$. A reducing node who knows the signature of b notices that only one of these values has been signed correctly. Since $1 < t-r+1 = 2$, it will choose the default value.

A node which does not know the signature of b uses values $2_{ab'}$ and $2_{ab'}$ for its decision, since they constitute the largest set of values with the same signature, and the set has two elements[1]. The majority value is $2_{ab'}$, and the vertex will be resolved to 2_a. Since two correct nodes can reduce this vertex to different values, it is not common.

The following two lemmas will be used to prove the correctness of the protocol:

Lemma 3. *Assuming crusader authentication and $n \geq 2t+1$, the following holds: A correct vertex is resolved to its stored value, with the last signature removed.*

[1] We assume implicitly that it is possible to decide whether two signatures were made with the same private key. This can be achieved when each signature is required to come with the respective public key.

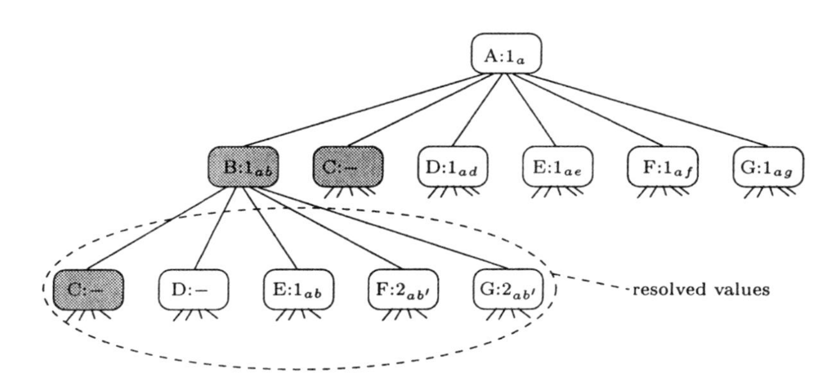

Fig. 5. Part of an EIG tree

Proof. The proof is by induction on the levels of the tree, from the leaves to the root. For a leaf, the lemma is trivially correct. Now suppose that for all vertices above level r the lemma is true. Then it is also true for a correct vertex on level r, because it has at least $t - r + 1$ correct children (due to $n \geq 2t + 1$).

From the induction hypothesis, all of them will be resolved to their stored values. Since the vertex on level r is correct, its stored value is the same as the resolved values of the children. Furthermore, the signatures on the values are correct. Hence, the vertex will be resolved as stated in the lemma. □

Lemma 4. *Assuming crusader authentication and $n \geq 2t + 1$, the following holds: A faulty vertex which has only faulty ascendants*[2] *is common.*

Proof. The proof is again by induction on the levels from the leaves to the root. On the last level $(t+1)$, the lemma is true, because there is no such vertex. Now suppose that for all vertices above level r the lemma is true. We will now show that the lemma is true for the vertices on level r.

Consider a faulty vertex on level r which has only faulty ascendants. With the induction hypothesis and Lemma 3, all children of this vertex are common. Then all correct nodes resolve that vertex to the same value, because they will base their decision on the same set of values. This can be shown as follows: A set of $t - r + 1$ elements can not contain only values from faulty vertices, since a faulty vertex with faulty ascendants in level r has at most $t - r$ faulty children. But if there is one value from a correct vertex in the set, then all values carry the correct signature, since a correct node will only forward values with correct signatures. □

Theorem 5. *Assuming crusader authentication and $n \geq 2t + 1$, the following holds: The algorithm in Fig. 4 reaches Byzantine agreement.*

Proof. If the sender is correct, the root of the tree will be resolved to the value received in the first round by all correct nodes, due to Lemma 3. Hence, all correct nodes agree on the correct value.

[2] An ascendant is a vertex on the direct path to the root.

If the sender is faulty, the root is common due to Lemma 4. Hence, all correct nodes agree on the same value. □

The next theorem shows that the fault tolerance of $n \geq 2t + 1$ cannot be improved.

Theorem 6. *Byzantine agreement cannot be reached for $n \leq 2t$, if only crusader authentication holds.*

Proof. Suppose that $n \leq 2t$ holds. Then the nodes can be partitioned in four nonempty sets A, B, C, and D, such that no set contains more than $t/2$ members. Now consider four scenarios, where in each scenario the members of two sets behave faulty. The sender is always a_1.

In all four scenarios, the nodes in C will play the role of those who do not know the public keys of nodes in A. If A is correct, they will only pretend not to know the keys. If A if faulty, they will actually not know the keys.

Scenario σ_1 (C and D faulty): All public keys are known to all nodes. In the first round, a_1 sends 0_{a_1} to all, and the nodes in A and B forward all messages correctly.
The nodes in C behave correctly, except that they pretend not to know the public keys of the nodes in A. The nodes in D behave towards the nodes in C as follows: They pretend to receive from nodes in A messages of the form $m_{A'...A'}$, where A' is recognizable to D. Furthermore, they forward messages $1_{A...A}$ to C as $0_{A'...A'D}$. Apart from this, they behave correctly.

Scenario σ_2 (B and C faulty): All public keys are known to all nodes. In the first round, a_1 sends 1_{a_1} to all, and the nodes in A and D forward all messages correctly.
The nodes in C behave correctly, except that they pretend not to know the public keys of the nodes in A. The nodes in B behave towards the nodes in C as follows: They pretend to receive from nodes in A messages of the form $m_{A'...A'}$, where A' is recognizable to B. Furthermore, they forward messages $0_{A...A}$ to C as $1_{A'...A'B}$. Apart from this, they behave correctly.

Scenario σ_3 (A and B faulty): All public keys except those from nodes in A are known to all nodes. The keys of A are known to all except the nodes in C.
During the agreement protocol, the nodes behave exactly as in σ_2.

Scenario σ_4 (A and D faulty): All public keys except those from nodes in A are known to all nodes. The keys of A are known to all except the nodes in C.
During the agreement protocol, the nodes behave exactly as in σ_1.

The nodes in the respective sets have the following views (see Fig. 6):

- The nodes in B cannot distinguish σ_1 from σ_4. Hence, they decide in both scenarios for 0.
- The nodes in C cannot distinguish σ_3 and σ_4. Since the nodes in B decide for 0 in σ_4, the nodes in C have to decide for 0 in σ_3 and σ_4.

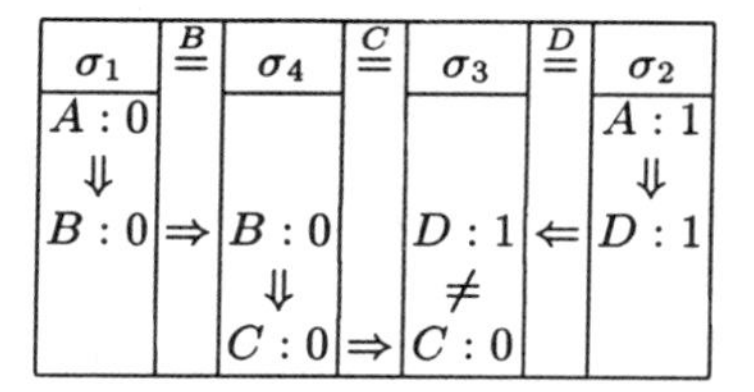

Fig. 6. Proof of Theorem 6

- The nodes in D cannot distinguish σ_2 and σ_3. Hence, they decide in both scenarios for 1.

With this reasoning, the nodes in C and D decide for different values in σ_3, which contradicts (B1). □

Partial Authentication In this setting it is common knowledge whose public keys are distributed completely. For simplicity, we will only regard the case in which a node's public key is either known to all nodes or to none. This situation arises when the keys are distributed by a globally trusted server which does not know all keys, or when some of the nodes are not able to sign messages. There is no relationship between the state of a node (faulty/correct) and the distribution of its public key. Here, (A3) holds, but (A4) becomes

(A4)″ Only a known set of s nodes can sign messages.

We will call this level of authentication *partial authentication.* A node whose public key is known is called *signer.* We can distinguish whether or not the sender is a signer.

Sender is no signer If $n > s \geq 2t + 1$, the sender sends its value to the signers in the first round. Then the signers distribute this value via an authenticated protocol with all other non-signing nodes as bystanders. The final result is the majority of the results of these protocols.

Such a protocol, where some node do not send any messages (and hence do not sign), is described in [DS83] (Theorem 6). In this protocol, the non-signing nodes have to draw their conclusions from the messages they receive from the signers. The protocol runs for $t+1$ rounds. For our purposes, the protocols could also be run for only t rounds.

A sketch of the proof is as follows: If the sender is faulty, there are at most $t - 1$ faulty signers left, and all correct nodes will agree on the values of the s sub-protocols. Hence, all correct nodes will eventually agree on the same value.

If the sender is correct, all correct signers will send the same values in their respective sub-protocols. These protocols have the property that the nodes reach agreement on the values of the correct nodes, even if the number of rounds is not greater than the number of faulty nodes. Hence, a majority of the results of the sub-protocols will be the correct value.

For $s \leq 2t$, no better fault tolerance than in the unsigned case can be achieved. This is shown in the following theorem:

Theorem 7. *If the sender of a protocol does not sign its messages and $s \leq 2t$ holds, then Byzantine agreement cannot be reached with $n \leq 3t$, assuming partial authentication.*

Proof. Suppose $n \leq 3t$ and $s = 2t$. Then it is possible to partition the nodes into three nonempty sets A, B, and C, such that A contains exactly the non-signers and no set contains more than t elements. The sender is a_1. Now consider three scenarios σ_1, σ_2, and σ_3:

Scenario σ_1 (B faulty): a_1 sends 0 in the first round. When a node b_i receives a message from A without any signatures (i.e., the message passed only nodes in A), it forwards 1_{b_i} to all nodes. All other messages are correctly signed and forwarded.

Scenario σ_2 (C faulty): a_1 sends 1 in the first round. When a node c_i receives a message from A without any signatures (i.e., the message passed only nodes in A), it forwards 0_{c_i} to all nodes. All other messages are correctly signed and forwarded.

Scenario σ_3 (A faulty): In the first round, a_1 sends 0 to C and 1 to B. In the remainder of the protocol, a node a_i forwards a signed message unchanged to all nodes. Unsigned messages are passed to B as 1 and to C as 0.

In σ_1, the nodes in C have to decide for 0, since the sender is correct. For the same reason, the nodes in B have to decide for 1 in σ_2. Scenario σ_3 has been constructed in a way that it is indistinguishable from σ_1 by C and from σ_2 by B. Hence, the nodes in B and C decide for different values in σ_3, contradicting (B1). □

With a similar argument, it can be shown that the same restrictions apply for crusader agreement.

Sender is signer For $n \geq s \geq 2t + 1$, the protocol from [DS83] mentioned above could be used. With the (less efficient) EIG-protocol in Fig. 7, one signer can be omitted, such that Byzantine agreement is possible for $n > s \geq 2t$.

Theorem 8. *Assuming partial authentication, the following holds: If the sender signs and $n > s \geq 2t$ holds, the protocol in Fig. 7 reaches Byzantine agreement.*

Proof. We distinguish whether the sender is correct or faulty:

Sender correct: There is exactly one value signed by the sender. Since there are at least $t + 1$ correct nodes, at least t of which are signers, at least one leaf is reduced to the sender's value signed by t correct nodes. This value will be considered recursively in the process of resolving. Hence, there is exactly one value for the result of the protocol, which is the value signed by the sender.

Protocol for Byzantine Agreement

1. The nodes fill their EIG trees for $t+1$ rounds.
2. A leaf is resolved to its value. If the last sender was a signer, the signature is removed.
3. For a vertex on level $r < t+1$ with label X, two cases are distinguished:
   - X is signer: Take the majority of the vertice's resolved children which carry X's signature.
   - X is no signer: Take the majority of the vertice's resolved children.
4. The result of the protocol is the resolved value of the root.

Fig. 7. Byzantine agreement with partial authentication and a signing sender

Sender faulty: Here, we show that all vertices in the second level are common. The vertices which correspond to correct signers are common with the same argumentation as in the case of a correct sender.

The vertices which correspond to correct non-signers are common, because they have at least t correct and signed children (as opposed to at most $t-1$ faulty children). Since these vertices are common (same argumentation as above), there is always a majority of correctly resolved children.

Finally, there are the faulty vertices (signed or not). Their correct children are common, as can be shown as above. For their faulty children to be common, it is (by recursion) necessary that all faulty vertices at level $t+1$ with only faulty ascendants be common. This is the case, since there are no such vertices.

Hence, all correct nodes take the same set of values as a basis for the final decision. This leads to a common value. □

For $n = s = 2t$, any protocol for complete authentication can be used. Hence, Byzantine agreement is possible for any $n \geq s \geq 2t$. For $0 < s \leq 2t-1$, $n \geq 3t-1$ (or $n = s$) is necessary:

Theorem 9. *Under the assumption of partial authentication the following holds: If the sender signs its messages and* $0 < s \leq 2t - 1$ *holds, then Byzantine agreement is only possible if* $n \geq 3t - 1$ *or* $n = s$.

Proof (Sketch). Byzantine agreement is always possible with $n = s$ ([LSP82]). Now suppose $n > s = 2t - 1$ and $n \leq 3t - 2$ holds. From Theorem 7 follows that it is generally impossible to reach Byzantine agreement on the messages sent by the non-signers. Hence, the non-signers may as well send no messages at all. Now let the faulty signers, except the sender, be in A, the correct signers in B, and the non-signers in C.

If the sender is faulty, A can behave towards C as if the sender sends “1”, while B receives (and sends) messages consistent with the sender saying “2”. The nodes in C cannot distinguish whether A or C is faulty, although they have to agree with the correct nodes. □

Crusader agreement is possible for any number of faulty nodes. It can easily be verified that the protocol of Fig. 3 reaches crusader agreement under the assumption of partial authentication and a signing sender.

3.4 Complete Authentication

In this setting it is assumed that all nodes agree on the public keys of all nodes. If $n \leq 3t$, this agreement may be reached using a trusted entity at the time of set-up (e.g., a trusted person traveling from site to site) or dynamically by using a trusted server. It is worth noting that if a trusted server would exist during execution of the protocols, it could also help solving Byzantine agreement very easily. With complete authentication, $n > t$ is possible. This case is dealt with to a great extent in the original papers ([PSL80, LSP82]), and in [DS83].

If $n \geq 3t+1$, one should install complete authentication by agreement on the public keys at the very beginning. Once this level of authentication is reached, an arbitrary number of nodes may become faulty without disturbing agreement. In addition, with complete authentication, very simple and message-efficient protocols become possible.

4 Summary

In this paper, we have focused on protocols for Byzantine agreement and crusader agreement in the presence of incomplete and incorrect authentication. We have identified several possible scenarios which yield different degrees of maximum fault tolerance.

	BA	CA
No auth.	$3t+1$ [LSP82]	$3t+1$ [Dol82]
Byz. auth.	$3t+1$	$3t+1$
Crus. auth.	$2t+1$	t
Partial auth., Sender doesn't sign	$2t+2$ if $s \geq 2t+1$ $3t+1$ else	$2t+2$ if $s \geq 2t+1$ $3t+1$ else
Partial auth., Sender signs	$2t$ if $s \geq 2t$ $3t-1$ if $s < 2t \wedge s < n$	t
Complete auth.	t [LSP82]	t

Table 1. Minimal n for different levels of authentication and types of agreement

Table 1 gives an overview of the results. It shows minimum values for n (number of nodes) with regard to t (maximum number of faulty nodes). In the results for partial authentication, s denotes the number of signing nodes.

In the past, only the two extreme levels "no authentication" and "complete authentication" have been investigated. As can be seen, there is a trade-off between the level of authentication and the possible fault tolerance. The stronger the authentication, the higher the possible fault tolerance. Furthermore, we have shown that there are environments where the two agreement problems under consideration do not have the same fault tolerance.

References

[BDDS87] Amotz Bar-Noy, Danny Dolev, Cynthia Dwork, and H. Raymond Strong. Shifting gears: Changing algorithms on the fly to expedite Byzantine agreement. In *Proceedings of the 6th ACM Symposium on Principles of Distributed Computing (PODC)*, pages 42–51, Vancouver, Canada, 1987.

[Bor95] Malte Borcherding. Efficient failure discovery with limited authentication. In *Proceedings of the 15th International Conference on Distributed Computing Systems (ICDCS)*, pages 78–82, Vancouver, Canada, 1995. IEEE Computer Society Press.

[Dol82] Danny Dolev. The Byzantine generals strike again. *Journal of Algorithms*, 3(1):14–30, 1982.

[DS83] Danny Dolev and Raymond Strong. Authenticated algorithms for Byzantine agreement. *SIAM Journal of Computing*, 12(5):656–666, November 1983.

[EM96] Klaus Echtle and Asif Masum. A mutiple bus braodcast protocol resilient to non-cooperative Byzantine faults. In *Proceedings of the 26th International Symposium on Fault-Tolerant Computing (FTCS)*. IEEE Computer Society Press, 1996.

[GLR95] Li Gong, Patrick Lincoln, and John Rushby. Byzantine agreement with authentication: Observations and applications in tolerating hybrid and link faults. In *Proceedings of the Fifth Dependable Computing for Critical Applications (DCCA-5)*, 1995.

[Gon93] Li Gong. Increasing availability and security of an authentication service. *IEEE Journal on Selected Areas in Comunications*, 11(5):657–662, June 1993.

[HH93] Vassos Hadzilacos and Joseph Y. Halpern. The Failure Discovery problem. *Math. Systems Theory*, 26:103–129, 1993.

[LSP82] Leslie Lamport, Robert Shostak, and Marshall Pease. The Byzantine Generals problem. *ACM Transactions on Programming Languages and Systems*, 4(3):382–401, 1982.

[Nat92] National Institute of Standards and Technology. The Digital Signature Standard. *Communications of the ACM*, 35(7):36–40, July 1992.

[PSL80] M. Pease, R. Shostak, and L. Lamport. Reaching agreement in the presence of faults. *Journal of the ACM*, 27(2):228–234, April 1980.

[RSA78] R. Rivest, A. Shamir, and L. Adleman. A method for obtaining digital signatures and public-key cryptosystems. *Communications of the ACM*, 21(2):120–126, February 1978.

Efficient and Robust Sharing of Memory in Message-Passing Systems*

(Extended Abstract)

Hagit Attiya**

Department of Computer Science, The Technion
Haifa 32000, Israel

Abstract. An emulation of a wait-free, atomic, single-writer multi-reader register in an asynchronous message passing system is presented. The emulation can withstand the failure of up to half of the processors, and requires $O(n)$ messages (for each read or write operation), assuming there are $n+1$ processors in the system. It improves on the previous emulation, which required $O(n^2)$ messages (for each read or write operation). The message complexity of the new emulation is within a constant factor of the optimum.
The new emulation implies improved algorithms to solve the following problems in the message-passing model in the presence of processor failures: multi-writer multi-reader registers, concurrent time-stamp systems, ℓ-exclusion, atomic snapshots, randomized consensus, implementation of data structures, as well as improved fault-tolerant algorithms for any solvable decision task.

1 Introduction

Two major interprocessor communication models in distributed systems have attracted much attention and study: the *shared-memory* model and the *message-passing* model. In the shared-memory model, processors communicate by writing and reading to shared registers. In the message-passing model, processors are located at the nodes of a network and communicate by sending messages over communication links. In both models, we consider asynchronous unreliable systems in which failures may occur; a processor fails by stopping and a slow processor cannot be distinguished from a failed processor.

Originally, the two models were investigated separately; algorithms and impossibility results were designed and proved for each of the models individually. However, Attiya, Bar-Noy and Dolev have shown that the message-passing variation of the asynchronous model can be reduced to the shared-memory variation [2]. This was proved by presenting an emulation of read/write registers in

---

* This research was partially supported by grant No. 92-0233 from the United States-Israel Binational Science Foundation (BSF), Jerusalem, Israel, Technion V.P.R. funds, and the fund for the promotion of the research at the Technion.

** Email: hagit@cs.technion.ac.il.

the message-passing model. The emulation implies that any wait-free algorithm in the shared-memory model that is based on *atomic, single-writer multi-reader* registers can be executed in the message-passing model.

This emulation immediately implies that many algorithms designed for the shared-memory model can be automatically employed in the message-passing model; see the many references in [2]. Given this emulation, the recent flourishing research on wait-free solvability in asynchronous systems (e.g., [6, 12, 13, 14, 18]) solely addresses the shared-memory model. These papers rely on the emulation of [2] to translate the results into the message-passing model. Unfortunately, the emulation of [2] has a relatively high message complexity: each read or write operation requires $O(n^2)$ messages, for a system with $n+1$ processors.

This paper presents an improved algorithm for the emulation of an atomic single-writer multi-reader register in message-passing systems in the presence of processor failures. Each emulated operation (read or write) requires $O(n)$ messages, each with $O(n^3)$ bits, constant time, and $O(n^5)$ bits of local memory. This improves on the emulation of [2] in which each emulated operation (read or write) requires $O(n^2)$ messages, each with $O(n^5)$ bits, constant time, and $O(n^6)$ bits of local memory.[3]

Wait-free protocols in shared-memory systems allow a processor to complete its operation regardless of the speed of other processors. In message-passing systems, it can be shown, following the proof in [1], that for many problems requiring global coordination, there is no solution that can prevail over a "strong" adversary—an adversary that can stop a majority of the processors. Such an adversary can cause two groups of fewer than a majority of the processors to operate separately by suspending all the messages from one group to the other. For many global coordination problems this leads to contradicting and inconsistent operations by the two groups. As mentioned in [1], similar arguments show that processors cannot halt after deciding. Thus, in our emulation a processor that is disconnected (permanently) from a majority of the processors is considered *faulty* and is blocked.[4] A wait-free algorithm will run correctly under our emulation if at least a majority of the processors are non-faulty.

Intuitively, $\Omega(n)$ messages are necessary in order to implement a shared register. If a write operation sends less than $n/2$ messages then at most $n/2$ processors can store the latest value written. Since these processors may later fail, by stopping, later read operations will not be able to return the latest value written. A similar argument can be used to prove a lower bound on the read operation. (We do not elaborate on the details of this proof in this extended abstract.)

In the emulation of [2], atomicity of the reads is guaranteed by communicating timestamped values between the readers. In contrast, our emulation relies on a handshake mechanism between the writer and the readers to guarantee

[3] The complexity analysis in [2] states that message size is $O(n)$ bits, and local memory requires $O(n^3)$ bits. This analysis is incorrect, as it does not account correctly for the size of bounded timestamps.

[4] Such a processor will not be able to terminate its operation but will never produce erroneous results.

atomicity of the reads, following the ideas of Haldar and Vidyasankar [11]. However, since communication between the writer and the readers is not by directly writing and reading to registers, but rather by sending messages to a majority of processors, we still need timestamps. As in [2], bounded timestamps ([16, 8]) are used to bound the size of messages and local memory. Since timestamps are used in a different manner, the number of messages needed for maintaining them correctly is reduced. Furthermore, it allows a smaller set of timestamps to be outstanding and therefore, reduces the size of the messages and the local memory requirements.

Two other papers investigated the relationships between shared-memory and message-passing systems. Bar-Noy and Dolev ([5]) provide translations between protocols in the shared-memory and the message-passing models. These translations apply only to protocols that use restricted form of communication. Chor and Moscovici ([7]) present a hierarchy of resilience for problems in shared-memory systems and complete networks. They show that for some problems, the wait-free shared-memory model is not equivalent to the complete network model, where up to half of the processors may fail. This result, however, assumes that processors *halt* after deciding.

The rest of this paper is organized as follows. In Section 2, we define the expected behavior of a single-writer multi-reader atomic register (Section 2.1), and the message-passing model (Section 2.2). Section 3 contains an overview of the emulation. Section 4 presents the basic mechanism for maintaining information in our emulation, while Section 5 includes the main algorithm for emulating an atomic single-writer multi-reader register, together with its correctness proof and complexity analysis. We conclude in Section 6, with a discussion of the results.

2 Definitions

In this section, we discuss the models addressed in this paper. We always consider a set of $n+1$ independent and asynchronous processors, $p_0, \ldots, p_n$. Each processor is a (possibly infinite) state machine, with a unique initial state, and a transition function.

2.1 The Problem: Single-Writer Multi-Reader Atomic Registers

An atomic register, which allows a single processor to write and a number of processors to read, was defined using axioms by Lamport [17]. The definition presented here is an equivalent one (see [17, Proposition 3]) that is simpler to use.

An *atomic, single-writer multi-reader register* is an abstract data structure. Each register is accessed by two procedures, $\mathsf{write}_w(v)$ that is executed only by a specific processor p_0 the *writer*, which we also denote p_w, and $\mathsf{read}_i(v)$ that can be executed by any *reader* processor p_i, $1 \leq i \leq n$.

An emulation supplies code for these procedures, written in the low-level computation model, which in our case is the message passing model as defined below. Each sequence of invocations of the read and write procedures generates an execution, which is required to be *linearizable* [15].

We concentrate on the emulation of a single register. Since linearizability is local (cf. [15]), multiple copies of the emulation can be composed to implement any number of registers.

We associate computation steps with the invocation and the response of read and write procedures in a natural manner. An operation op_1 *precedes* an operation op_2, if the response of op_1 is generated before op_2 is invoked. Two operations are *overlapping* if neither of them precedes the other.

Linearizability requires that each interleaved execution of the operations is equivalent to an execution in which the operations are executed sequentially, and the order of non-overlapping operations is preserved. In the specific case of read and write procedures, this condition translates into the following two properties:

1. A read operation r returns either the value written by the most recent preceding write operation (the initial value if there is no such write) or a value written by a write operation that overlaps r. (This is the *regularity* condition.)
2. If a read operation r_1 reads a value from a write operation w_1, and a read operation r_2 reads a value from a write operation w_2 and r_1 precedes r_2, then w_2 does not precede w_1. (This is the *atomicity* condition.)

2.2 The Computation Model: Message Passing Systems

In a message-passing system, processors communicate by sending *messages* (taken from some alphabet $\mathcal{M}$) to each other. Processors are located at the nodes of a complete network, and any processor can send to any other processor.

We model computations of the system as sequences of steps. Each step is either a *message delivery step*, representing the delivery of a message to a processor, or a *computation step* of a single processor.

In each message delivery step, a single message is placed in the incoming buffer of the processor. Formally, a message delivery step is a pair (i, m), where $m \in \mathcal{M}$. In each indivisible computation step, a processor receives all messages delivered to it since its last computation step, performs some local computation and sends some messages, and possibly changes its local state. Formally, a computation step of processor p_i is a tuple (i, s, s', M), where s and s' are the old and new states of p_i (respectively), and $M \subseteq \mathcal{M} \times \{0, \ldots n\}$ is the set of messages sent by processor p_i. The set M specifies a set of *send events* that occur in a computation step. A message m is *delivered* to processor p_i when the step (i, m) happens; the message is *received* by p_i when p_i takes a computation step following the delivery of that message.

An *execution* is an infinite sequence of steps satisfying the following conditions:

1. The old state in the first computation step of processor p_i is p_i's initial state,
2. the old state of each subsequent computation step of processor p_i is the new state of the previous step,

3. the new state of any computation step of processor p_i is obtained by applying the transition function of p_i to the old state and the messages delivered since the last computation step, and
4. there is a one-to-one mapping from delivery steps to corresponding send events (with the same message).

The network is not explicitly modeled; however, the last condition guarantees that messages are not duplicated or corrupted. The network is allowed to not deliver some messages or to deliver them out of order.

A faulty processor simply stops operating. More formally, processor p_i is *nonfaulty* in an execution if the execution contains an infinite number of computation steps by p_i; otherwise, it is *faulty*. It is assumed that all messages are eventually delivered. Note that messages may be delivered to a faulty processor, even after it stops taking steps.

The quality of an emulation is evaluated by the worst case, over all possible executions of the emulation, of the following quantities:

Message complexity: The number of messages sent in an execution of a write or read operation.

Message size: The size of the messages, in bits.

Time complexity; The time to execute a write or a read operation, under the assumption that any message is either received within one time unit, or never at all (cf. [3]).

Local memory size: The amount of the local memory used by a processor, in bits.

3 Overview of the Emulation

The emulation of [2] is based on the construction of Israeli and Li [16], in which the writer writes a copy of the new value for each reader. Timestamps are added to the values to distinguish the most recent value; these timestamps are bounded using ideas of [16, 8]. To guarantee atomicity of the reads, the readers also forward to each other the most recent value they have read; the bounded timestamp system requires the writer to keep track of the timestamps currently in use. Since copies of the writer's values are forwarded around, keeping track of them causes the quadratic message complexity (see the analysis in [2]).

In our emulation, we side-step the cost of keeping track of the timestamps forwarded around, by avoiding the forwarding altogether. Our emulation follows the ideas of Haldar and Vidyasankar [11]. In their construction, atomicity of the reads is guaranteed since reads do not return values of writes that overlap them. Instead, when a reader detects that a write operation is in progress, it copies the value of a previous write operation. Since values are obtained directly from the processor which wrote them, and not from other processors which read them, it is simpler for the writing processor to know which values are in use.

Since the high-level algorithm we use is almost the same as the algorithm of [11], we do not elaborate further about it here. Instead, we explain how processors exchange information in our model, where message passing is required to replace the regular shared registers used in the construction of [11].

In a message passing system, unlike the shared memory model, there is no way for one processor to make sure that another processor will see a message sent to it. Furthermore, the sender cannot wait to receive an acknowledgment from the receiver, since the receiver may fail. Therefore, instead of sending messages directly to each other, processors have to communicate via intermediates. The idea, used in several previous papers ([1, 2, 5]), is to "write" a value by making sure that at least a majority of the processors store it, and to "read" a value by requesting it from a majority of processors. To allow a processor to know which of the values received from a majority of the processors is the most recent value "written", the values are accompanied by a timestamp. It can be seen (as proved formally later) that such "read" operation returns the latest value "written".

The timestamps are taken from a *bounded sequential time-stamp system* ([16]), which is a finite domain, $\mathcal{L}$, of timestamps together with a total order relation, $\succ$. Whenever the writer needs a new timestamp it produces a new one, larger (with respect to the $\succ$ order) than all the timestamps that exist in the system. This is done by invoking a special procedure called LABEL, whose output is a new timestamp that is greater than all the timestamps in the set given as input to the procedure. In addition, it is assumed that we can compare two timestamps according to the ordering $\succ$. Bounded sequential time-stamp systems with these properties were presented in [16, 8].

The key for the correctness of a bounded sequential time-stamp system is to guarantee that the set of timestamps that exist in the system is contained in the set of timestamps given as input to the LABEL procedure. As discussed above, we have to be very careful about the way these timestamps are passed around, and to make sure they are not forwarded. In our emulation, a timestamped value is sent only from the writing processor to the reading processors, through a majority of the processors. The reading processors handle the timestamp in an "un-interpreted" manner, and do not forward it, thus avoiding the need to keep track of timestamps that are forwarded around.

In our emulation, each processor plays two roles: first, it acts as a reader or as a writer in the main algorithm; in addition, it acts as an intermediate, storing and echoing values for the readers and the writers. By running both tasks in parallel, a processor can play each role independently (without passing information between the tasks). The exposition in the rest of the paper is as if there are two separate sets of processors, one containing the *participants* of the high-level emulation, as readers, denoted $p_1, \ldots, p_n$, and the writer, denoted p_w (or p_0), and the other containing the *intermediates*, denoted $q_0, \ldots, q_n$.

4 The Low Level Communication Mechanism

This section we explain the low level communication mechanisms used to exchange information: the procedure maj-write used to store a value at a majority of the intermediates, and the procedure maj-read used to get a value from a majority of the intermediates. Our description in this section concentrates on single shared variable, *var*, which is sometimes left implicit. However, the communication mechanisms for several variables can be combined together to save messages. (See below in Section 5.3.)

Each variable can be written by a single processor, and read by all processors. However, there are no atomicity requirements on concurrent reads of several processors; essentially, the variable is similar in its properties to a single-writer multi-reader regular register (cf. [17]).[5] Note that in this section we refer to the *writing processor*, which is not necessarily the writer of the register implemented by the high-level algorithm, and to the *reading processors*, which are not necessarily the readers of the register implemented by the high-level algorithm.

For simplicity, we assume that there is a "virtual" edge, $\langle i, j \rangle$, between each participant p_i and each intermediate q_j. A *ping-pong* rule is employed on every edge $\langle i, j \rangle$: p_i sends the first message on $\langle i, j \rangle$; afterwards, p_i and q_j alternate turns in sending further messages on $\langle i, j \rangle$. Processor p_i manages the ping-pong on its edges to the intermediates by maintaining a vector *turn* of length $n + 1$, with an entry for each intermediate q_j, that can get the values *here* or *there*. If $turn(j) = here$ then it is p_i's turn on $\langle i, j \rangle$, and only then p_i may send a message to q_j. If $turn(j) = there$ then either p_i's message is in transit, or q_j received p_i's message and has not replied yet (it might be that q_j failed), or q_j's message is in transit. Initially, $turn(j) = here$. Hereafter, we assume that the vector *turn* is updated automatically by the **send** and **receive** operations.

In addition to storing the most recent value for the variable, the intermediate also keeps track which timestamps it has most recently sent to the reading participants. The participant writing the variable also keeps track of the latest two timestamps it has sent to each intermediate. These sets allow the processor writing the variable to know which of the timestamps it has generated so far are still outstanding (as we prove below).

Each intermediate holds, in the variable *last*, the last value received from the writing processor for this variable. In addition, each intermediate holds an array *pending*[0..*n*]; *pending*[*i*] contains the latest value sent to p_i.

The processor writing the variable holds a local array *pending*[0..*n*]; *pending*[*j*] contains the timestamps suspected to be in use by the intermediate q_j:
pending[*j*].*last* is the latest timestamp sent to q_j,
pending[*j*].*prev* is the previous timestamp sent to q_j, and
pending[*j*].*forward* is the latest *pending* array received from q_j.

As described above, each invocation of maj-read or maj-write procedures involves sending one message to all intermediates and waiting for a response from

[5] The variable can be easily modified to be atomic for read operations of a single processor.

a majority of them. To manage the communication, each of the maj-write and maj-read procedures maintains a local array *status*[0..n]; *status*[j] may contain one of the following values:

notsent: the message was not yet sent to q_j (since $turn(j) = there$);
notack: the message was sent but not yet acknowledged by q_j;
ack: the message was acknowledged by q_j.

In addition, an integer counter, *#acks*, counts the number of responses (from intermediates) received so far.

The participant's code for maj-write and maj-read appears in Figure 1, while the intermediate's code appears in Figure 2.

The key feature of the communication is that the *pending* variable at the writing processor keeps track of values forwarded by intermediates to the reading processors.

Lemma 1. *Assume the last timestamp forwarded by an intermediate* q_j *to a reading processor* p_i *is* x, *then* $x \in pending_k[j]$, *where* p_k *is the writing processor.*

Proof. Clearly, by the code of the algorithm, $pending_j[i] = x$.
Denote x'=$pending_k[j].prev$ and x''=$pending_k[j].last$. If $x' = x$ or $x'' = x$, then the lemma clearly holds. Otherwise, it follows that p_k has sent at least two timestamps to q_j, since q_j has forwarded x to p_i. Note that $pending_j[i] = x$ when q_j received x', and therefore, in q_j's response to x', $pending_j[i] = x$. Since p_k has already sent another timestamp, x'', it has received q_j's response to x' and has set $pending_k[j]$.forward[i] $= x$. □

We now prove the two important properties of the low-level communication mechanism: *(1)* The *pending* variable (at the writing processor) contains all the values that may be viable (in a sense defined precisely below). *(2)* If a maj-write operation mw completes before a maj-read operation, mr, starts, then mr returns a value which was written by mw or by a later maj-write operation. The proof of the second property relies on the first one.

To make the first property more precise, consider a finite execution prefix, α. A timestamp x is *viable* after α if:

1. for some intermediate q_i, the value of *var* is x after α, or
2. for some participant p_i, the value of *var* is x after α, or
3. a message containing x is sent from the writing processor to some intermediate in α, and is received in some execution extending α, or
4. a message containing x is sent from some intermediate q_i to some participant p_j in α, and is received in some execution extending α.

The following lemma states the first property of the low-level communication mechanism; it is an immediate consequence of the code and of Lemma 1.

Lemma 2. *For any finite execution prefix* α, *if* x *is viable after* α *then* $x \in pending_w$.

```
Procedure maj-write(var,val) ;
    add LABEL(pending) to val
    #acks := 0
    for all j, 0 ≤ j ≤ n do status(j) := notsent
    for all j, 0 ≤ j ≤ n do
        if turn(j) = here then
            send ⟨val⟩ to q_j
            pending[j].prev := pending[j].last
            pending[j].last := val
            status(j) := notack
    while #acks < ⌈n/2⌉ + 1 do for any message M received from q_j:
            pending[j].forward := M
            if status(j) = notsent then      /* M is a response to an old message */
                send ⟨val⟩ to q_j
                pending[j].prev := pending[j].last
                pending[j].last := val
                status(j) := notack
            else if status(j) = notack then
                status(j) := ack
                #acks := #acks+1
end procedure maj-write

Procedure maj-read(var) returns val;
    #acks := 0
    for all j, 0 ≤ j ≤ n do status(j) := notsent
    for all j, 0 ≤ j ≤ n do
        if turn(j) = here then
            send ⟨Request(var)⟩ to q_j
            status(j) := notack
    while #acks < ⌈n/2⌉ + 1 do for any message ⟨val'⟩ received from q_j:
            if status(j) = notsent then   /* received a response to an old request */
                send ⟨Request(var)⟩ to q_j
                status(j):= notack
            else if status(j) = notack then
                if val' ≻ val then val := val'
                status(j) := ack
                #acks := #acks+1
    return val
end procedure maj-read
```

Fig. 1. *Procedures* maj-write *and* maj-read.

```
upon receiving a message ⟨val⟩ from p_i:                      /* sent by maj-write */
    last := val
    send pending to p_i
upon receiving a message ⟨Request(var)⟩ from p_i:    /* request from maj-read */
    send ⟨last⟩ to p_i
    pending[i] := val
```

Fig. 2. *Code for an intermediate processor q_j.*

The input of the LABEL procedure is the value of the variable *pending* at the writing processor. By Lemma 2, it contains all the viable timestamps. By the properties of a bounded sequential time-stamp system ([16, 8]), we have:

Lemma 3. *A timestamp generated in the call to the* LABEL *procedure is greater than any viable timestamp in the system.*

Note that the ping-pong mechanism guarantees that the responses received were indeed sent in reply to the message sent in the maj-read or maj-write procedures. It follows that at least $\lceil \frac{n}{2} \rceil + 1$ processors either store the recent value for the variable (in maj-write) or respond with their current value for the variable (in maj-read).

The next lemma deals with the ordering of the value returned by a maj-read operation and the value written by a maj-write operation which completely precedes it. This lemma proves the second key property of the low-level communication mechanism.

Lemma 4. *If a* maj-write *operation $mw(v, x)$ completes before a* maj-read *operation $mr(v)$ which returns y, then $y \succeq x$.*

Proof. By the code of the algorithm, when $mw(v, x)$ completes, at least a majority of the intermediates store a value with timestamp greater than or equal to x. Similarly, at least a majority of the processors send their current value in $mr(v)$. Thus, there must be at least one intermediate, say q_j, that stored a value $x_j' \succeq x$ and replied in $mr(v)$. Since y is maximal among the values obtained in $mr(v)$, it follows that $y \succeq x_j' \succeq x$. □

We now analyze the complexity of the low-level communication mechanism. The ping-pong mechanism used in maj-read and maj-write implies:

Lemma 5. *At most $2n$ messages are sent as the result of each invocation of a* maj-read *or* maj-write *procedure.*

If each message takes at most one time unit, then, since maj-read and maj-write involve sending one message to a majority of the processors and receiving their response, the existence of a nonfaulty majority implies:

Lemma 6. *Each invocation of a* maj-read *or* maj-write *takes at most* 2 *time units.*

It is slightly more complicated to analyze the size of the messages and the local memory, in bits. Obviously, this depends on the number of timestamps maintained in each processor, in the array *pending*, and the size of the bounded timestamps. By the structure of the bounded timestamp system, the size of the bounded timestamps depends on the number of timestamps that are given as input to procedure LABEL, that is, on the number of timestamps in $pending_k$, at the writing processor, p_k. The *pending* variable at the writing processor contains at most n^2 timestamps. The constructions of bounded sequential time-stamp systems ([16, 8]) imply that a timestamp can be represented using $O(n^2)$ bits. Therefore, we have:

Lemma 7. *For a single variable var, the messages require $O(n^3)$ bits, the local memory at the writing processor requires $O(n^4)$ bits and the local memory at each intermediate requires $O(n^3)$ bits.*

5 The Main Algorithm and its Correctness Proof

The algorithm presented here has exactly the same structure as the algorithm of Haldar and Vidyasankar [11]; however, invocations to maj-read and maj-write replace reading and writing regular variables. The translation is not immediate, since the reading of several variables is combined in order to save on the communication costs.

5.1 Details of the Algorithm

The algorithm uses several variables, each of which is accessed using the low level maj-write and maj-read procedures. All variables are written by a single processor, and read by all other processors.[6]

There are two copies of the value, *main*[1], and *main*[2], as well as a ternary variable, *pointer*, telling what is the current version that should be read. The writing processor of these values is the writer. Two arrays are used for the hand-shake between the writer and the readers: *writing*[1..*n*], written by the writer, and *reading*[1..*n*], in which the *i*th entry is written by reader p_i. There is an additional array, *overlap*[1..*n*], in which the *i*th entry is written by reader p_i, and all readers can read.

In addition, several local variables are used to store copies of the shared variables; their usage should be clear from the code.

The pseudo-code for the algorithm appears in Figure 3.

[6] We assume the writing processor can read the variable without accessing the shared memory, by keeping a local copy; for simplicity, we omit the details of how this is done.

```
Procedure write_w(v) ;                      /* executed by the writer p_w to write v */
1:   maj-write(pointer,1)
2:   maj-write(main[1],v)
3:   maj-write(pointer,2)
4:   maj-write(pointer,0)
5:   maj-write(main[2],v)
6:   lr := maj-read(reading)                /* read a complete array of handshake bits */
7:   for j := 1 to n do lw[j] := ¬lr[j]                        /* negate handshake bits */
8:   maj-write(writing,lw)                              /* set handshake bits unequal */
end procedure write_w

Procedure read_i(v) ;                         /* executed by a reader p_i and returns v */
1:   val := maj-read(main[1])
2:   lw := maj-read(writing[i])                                   /* read handshake bit */
3:   if (lw≠reading[i]) then
4:        maj-write(overlap[i],0)
5:        maj-write(reading[i],lw)                           /* set handshake bits equal */
6:   lp := maj-read(pointer)
7:   if lp = 0 then return val
8:   else if lp =2 then
9:        maj-write(overlap[i],1)
10:       return val
11:  else                                                          /* in this case, lp =0 */
12:       lr := maj-read(reading)                         /* read readers' handshake bits */
13:       lo := maj-read(overlap)                                     /* read overlap bits */
14:       lw := maj-read(writing)                          /* read writer's handshake bits */
15:       if for some j, (lr[j] = lw[j]) and (lo =1) then
16:            maj-write(overlap[i],1)
17:            return val
18:       else val := maj-read(main[2])
19:       return val
end procedure read_i ;
```

Fig. 3. *The emulation.*

In the write procedure, the writer signals it is about to write the first copy, by setting *pointer* to be 1 (Line 1), and then writes (Line 2). Then it sets the pointer to the intermediate value (Line 3), and afterwards signals it is about to write the second copy, by setting *pointer* to be 0 (Line 4), and then writes (Line 5). Finally, the writer negates all handshake variables with the readers (Lines 6-8).

The read procedure is significantly more complicated. The reader reads the first copy (Line 1), and then checks if the writer is overlapping it by reading the handshake variable (Line 2). If the writer is not overlapping, then the reader unsets the *overlap* flag (Line 4) and equates the handshake variable (Line 5). Then the reader checks the current value of the pointer (Line 6). If the pointer is 0, then the write of the first copy has completed, and the reader can safely return the value read from it (Line 7). If the pointer is 2, then the writer is in between the first copy and the second copy; in this case, the reader sets its *overlap* flag, and returns the value it read from the first copy. If the pointer is 1, then it is not clear if the writer has written to the first copy, so the reader checks if other readers have overlapped the write. This is done by reading the *reading*, *overlap* and *writing* arrays (Lines 12, 13 and 14, respectively). If for some other reader p_j, the handshake bits are equal and the *overlap* flag is set, then the reader sets its *overlap* flag (Line 16) and returns the value it read from the first copy (Line 17); otherwise, it returns the value of the second copy (Lines 18-19).

5.2 Partial Correctness

The partial correctness proof, i.e., showing that any execution of the algorithm is linearizable, follows the corresponding proof of [11], and we do not detail it here.

Informally, the regularity property, i.e., that every read operation returns either the value written by the most recent preceding write operation (the initial value if there is no such write) or a value written by a write operation that overlaps this read operation, follows since reads cannot return values not yet written, or values that were overwritten (by Lemma 4).

The atomicity property, i.e., if a read operation r_1 reads a value from a write operation w_1, and a read operation r_2 reads a value from a write operation w_2 and r_1 precedes r_2, then w_2 does not precede w_1, is proved by case analysis, depending on the index of the main copy whose value is returned by r_1 and r_2. That is, we check the four possible combinations of r_1 returning the value from *main*[1] or *main*[2], and r_2 returning the value from *main*[1] or *main*[2].

5.3 Complexity Analysis

The complexities of the emulation are dominated by the complexities of maj-read and maj-write, as shown in the proof of the following theorem, stating the main result of this paper.

Theorem 8. *There exists a bounded emulation of an atomic, single-writer multi-reader register in a complete network, in the presence of* $\lfloor \frac{n}{2} \rfloor$ *processor failures. Each invocation of a* read *operation or a* write *operation requires* $O(n)$ *messages each of size* $O(n^3)$ *bits, and* $O(1)$ *time; the emulation requires a local memory with* $O(n^5)$ *bits.*

Proof. Let us evaluate the complexities of the algorithm.

Since each invocation of a read or a write operation contains a constant number of invocations of maj-read and maj-write, Lemma 5 implies that each invocation of a read or a write operation requires $O(n)$ messages.

Similarly, Lemma 6 implies that each execution of a read or a write operation terminates within $O(1)$ time units.

Finally, let us calculate the bit complexity required by the algorithm. The longest messages sent by the algorithm are vectors of n values. Regarding each entry as a separate variable (since, in two of the cases, *overlap* and *reading*, they are written by different readers), a simple application of Lemma 7 implies that the longest message requires $O(n^4)$. However, we can note that, in response to the writing processor, the intermediates send only the *pending* vector of its entry ($O(n^3)$ bits). Similarly, the intermediates send only the set of n timestamps (again, $O(n^3)$ bits) to the reading processor. These facts imply the tighter bound.

A simple application of Lemma 7 implies that the local memory size is $O(n^5)$, at the writer, since it is the writing processor for $O(n)$ variables, and $O(n^4)$, at each reader, since it is the writing processor for $O(1)$ variables. In addition, each processor holds $O(n^4)$ local memory bits as an intermediate for $O(n)$ variables. □

6 Conclusion

We have presented an emulation of atomic, single-writer multi-reader registers in message-passing systems (networks), in the presence of processor failures. Each operation (read or write) to the register requires $O(n)$ messages and constant time. The size of the messages and the local memory (in bits) is polynomial in n. We have also shown that at least $\frac{n}{2}$ messages per operation, has to be sent by any emulation of an atomic register.

Clearly, we cannot improve on the time complexity of our emulation. However, there is clearly place for improvement is in the size (in bits) of the messages and the local memory. One possibility to improve the efficiency using the current scheme is to reduce the size of the bounded timestamps used by the emulation, either by reducing the number of outstanding timestamps, or by employing other timestamp systems, e.g., [10].

Acknowledgments: The author thanks K. Vidyasankar for helpful discussions on implementing atomic registers, and the WDAG referees for useful comments on the previous version of the paper.

References

1. H. Attiya, A. Bar-Noy, D. Dolev, D. Peleg, and R. Reischuk, Renaming in an Asynchronous Environment, *Journal of the ACM,* Vol. 37, No. 3 (July 1990), pp. 524–548.
2. H. Attiya, A. Bar-Noy and D. Dolev, Sharing Memory Robustly in Message-Passing Systems, *Journal of the ACM,* Vol. 42, No. 1 (1995), pp. 124–142.
3. B. Awerbuch, Optimal Distributed Algorithms for Minimum Weight Spanning Tree, Counting, Leader Election and Related Problems, in *Proceeding of the 19th ACM Symposium on Theory of Computing,* pp. 230–240, 1987.
4. B. Awerbuch, Y. Mansour, and N. Shavit, Polynomial End-To-End Communication, in *Proceedings of the 30th IEEE Symposium on Foundations of Computer Science,* pp. 358–363, 1989.

5. A. Bar-Noy, and D. Dolev, Shared-Memory vs. Message-Passing in an Asynchronous Distributed Environment, *Mathematical Systems Theory*, Vol. 26 (1993), pp. 21–39.
6. E. Borowsky and E. Gafni, Generalized FLP Impossibility Result for t-Resilient Asynchronous Computations, in *Proceedings of the 1993 ACM Symposium on Theory of Computing*, May 1993.
7. B. Chor, and L. Moscovici, Solvability in Asynchronous Environments, in *Proceedings of the 30th IEEE Symposium on Foundations of Computer Science*, pp. 422–427, 1989.
8. D. Dolev, and N. Shavit, *unpublished manuscript*, July 1987. Appears in [4].
9. D. Dolev, and N. Shavit, Bounded Concurrent Time-Stamp Systems are Constructible, *SIAM Journal on Computing,* to appear. Also: in *Proceedings of the 21st ACM Symposium on Theory of Computing*, pp. 454–466, 1989.
10. C. Dwork and O. Waarts, Simple and Efficient Bounded Concurrent Timestamping or: Bounded Concurrent Timestamp Systems are Comprehensible!, in *Proceedings of the 24th ACM Symposium on Theory of Computing*, pp. 655–666, 1992.
11. S. Haldar and K. Vidyasankar, Constructing 1-Writer Multireader Multivalued Atomic Variables from Regular Variables, *Journal of the ACM*, Vol. 42, No. 1 (1995), pp. 186–203.
12. M.P. Herlihy and S. Rajsbaum, Algebraic Spans, in *Proceedings of the 14th Annual ACM Symposium on Principles of Distributed Computing*, pages 90-99, August 1995.
13. M.P. Herlihy and N. Shavit, The Asynchronous Computability Theorem for t-Resilient Tasks, in *Proceedings of the 1993 ACM Symposium on Theory of Computing*, May 1993.
14. M.P. Herlihy and N. Shavit, A Simple Constructive Computability Theorem for Wait-Free Computation, in *Proceedings of the 1994 ACM Symposium on Theory of Computing*, May 1994.
15. M.P. Herlihy and J.M. Wing, Linearizability: A Correctness Condition for Concurrent Objects, *ACM Trans. on Programming Languages and Systems*, Vol. 12, No. 3 (1990), pp. 463–492.
16. A. Israeli, and M. Li, Bounded Time-stamps, *Distributed Computing,* Vol. 6, No. 4 (1993), pp. 205–209.
17. L. Lamport, On Interprocess Communication, Parts I and II, *Distributed Computing*, Vol. 1, No. 1 (1986), pp. 77–101.
18. N. Lynch and S. Rajsbaum, On the Borowsky-Gafni Simulation, in *Proceedings of the 4th Israeli Symposium on Theory of Computing and Systems*, (June 1996), pp. 4–15.

Plausible Clocks: Constant Size Logical Clocks for Distributed Systems

Francisco J. Torres-Rojas and Mustaque Ahamad
{torres, mustaq}@cc.gatech.edu
College of Computing
Georgia Institute of Technology, USA

Abstract. In a Distributed System with N sites, the detection of causal relationships between events can only be done with vector clocks of size N. This gives rise to scalability and efficiency problems for accurate logical clocks. In this paper we propose a class of logical clocks called plausible clocks that can be implemented with a number of components not affected by the size of the system and yet they provide good ordering accuracy. We develop rules to combine plausible clocks to produce more accurate clocks. Several examples of plausible clocks and their combination are presented. Using a simulation model, we evaluate the performance of these clocks.

1 Introduction

In large scale distributed systems, efficient access to shared information requires the use of caching and replication. In such an environment, it is necessary to order read and update operations on an object to determine its most recent value. Logical clocks have been explored for ordering events in distributed systems. These clocks do not require synchronized physical clocks and can be implemented by including additional information with messages exchanged in the system. Although vector clocks can precisely order events of a distributed system and detect concurrent events, they are expensive to maintain and manipulate since vectors of integers must be included in messages and it is necessary to compare them to determine the order between operations. Besides, since vector clocks have a component for each node in the system, they are not scalable.

Scalar clocks can be implemented efficiently (e.g. Lamport Clocks), but when events are timestamped with these clocks, two events may appear to be ordered according to their timestamps even when they are concurrent. In a distributed object system, this could lead to unnecessary consistency operations. For example, in causal consistency, if a new object value written by operation *o* is fetched at a node, existing object copies at the node are removed from its cache if the operations that produced them causally precede *o*. This is done because the cached objects may potentially be overwritten by more recent operations that precede *o*. With scalar clocks, an object produced by an operation which is concurrent with *o* may unnecessarily be removed from the node cache. Such unnecessary removals and extra communications on accessing such objects in the future can be avoided if the timestamps associated with object copies are derived from more precise clocks.

In this paper, we explore logical clocks that can be implemented with a size which is independent of the number of nodes in the distributed system and yet they provide ordering accuracy close to vector clocks. Such clocks are useful in systems where ordering of concurrent events only impacts performance and not correctness. This is true for many consistency maintenance schemes, therefore performance is not affected significantly if these clocks order only a small number of concurrent events. We call these **plausible** clocks. We describe several such clocks and present rules to combine them to produce more accurate clocks. We study the performance of these clocks using a simulation model for two different distributed systems. For example, the experiments show that, with a storage overhead of 6 components and a message overhead of 4 components in a system with 76 nodes, a clock system that we develop obtains the same results as vector clocks in more than 92% of the event pairs (vector clocks would have required 76 components).

Section **2** presents the background material on logical clocks. We introduce the concept of plausible clocks in Section **3**. Section **4** presents how such clocks can be combined and demonstrates that the combination rules preserve correctness and provide improved accuracy. Some examples of constant size clocks are presented in Section **5**. Section **6** describes the simulation model used to evaluate the proposed clocks together with the obtained results. In Section **7**, we present the conclusions of this paper.

2 Logical Clocks

2.1 Lamport Clocks

The *local history* of site *i* is a sequence of events $H_i = e_{i1}e_{i2}...$ that are executed on site *i*. There is a total order on the local events of each site. The *global history* **H** of the Distributed System is the set of all the events occurring at all the sites of the system.

Definition 1. [2] defines the *causality relation* "$\rightarrow$" over events such that:

If e_{ij} and $e_{ik} \in H_i$ and $j < k$, then $e_{ij} \rightarrow e_{ik}$.

If e_{im} is **send** (M) and e_{jn} is **receive** (M) and M is the same message in both cases, with arbitrary i, j, m, and n, then $e_{im} \rightarrow e_{jn}$.

If $\mathbf{x} \rightarrow \mathbf{y}$ and $\mathbf{y} \rightarrow \mathbf{z}$, then $\mathbf{x} \rightarrow \mathbf{z}$.

If none of the above conditions hold between two different events **x** and **y**, we will say that **x** and **y** are "concurrent". We will denote this situation as $\mathbf{x} \leftrightarrow \mathbf{y}$. The relation $\rightarrow$ is not reflexive and the relation $\leftrightarrow$ is not necessarily transitive. ❑

In 1978, Leslie Lamport proposed the concept of *logical clocks* to define the order between events in Distributed Systems [2]. It consists of a mapping τ from events to the set of integers that in principle captures the causal order between events. Lamport Clocks exhibit the *weak clock condition*, that states that $\forall \mathbf{x}, \mathbf{y} \in \mathbf{H}$:

$$x \rightarrow y \Rightarrow \tau(x) < \tau(y)$$

Lamport Clocks capture the order between causally related events but they do not detect concurrency between events and just by inspecting two timestamps, we are not able to decide if the associated events are causally related (hence the name weak clock condition).

2.2 Vector Clocks

The *vector clocks* technique consists of a mapping $\mathbf{T}$ from events to integer vectors. Each site i keeps an integer vector v_i of N entries (number of sites), where the entry p represents the current knowledge that the site has of the activity at site p. Site i keeps its own logical clock in entry i of its vector clock. Vector clocks are updated when events occur at the local site or when messages are received (all messages are tagged with the current vector time of the sender). If **x** is a event of H_i, then $\mathbf{T}(\mathbf{x})$ is the value of v_i when **x** is executed. Site i updates v_i according to the rules:

V1) When an event is generated:

$v_i[i] = v_i[i] + 1$

V2) When a message with timestamp w is received:

$1 \leq k \leq N$: $v_i[k] = \max(v_i[k], w[k])$

$v_i[i] = v_i[i] + 1$

Definition 2. Given two vector times T_1 and T_2, we define the following tests:

$T_1 = T_2 \Leftrightarrow \forall k\; T_1[k] = T_2[k]$

$T_1 \leq T_2 \Leftrightarrow \forall k\; T_1[k] \leq T_2[k]$

$T_1 < T_2 \Leftrightarrow T_1 \leq T_2$ and $\exists k$ such that $T_1[k] < T_2[k]$

$T_1 \leftrightarrow T_2 \Leftrightarrow \exists k$ such that $T_1[k] < T_2[k]$ and $\exists j$ such that $T_1[j] > T_2[j]$. ❑

It is proved in [3] that there is an isomorphism between vector clocks and the causality relation between events in **H**, since $\forall$ **x,y** $\in$ **H**:

$$\mathbf{x} = \mathbf{y} \Leftrightarrow \mathbf{T}(\mathbf{x}) = \mathbf{T}(\mathbf{y})$$
$$\mathbf{x} \rightarrow \mathbf{y} \Leftrightarrow \mathbf{T}(\mathbf{x}) < \mathbf{T}(\mathbf{y})$$
$$\mathbf{x} \leftrightarrow \mathbf{y} \Leftrightarrow \mathbf{T}(\mathbf{x}) \leftrightarrow \mathbf{T}(\mathbf{y}) \quad (1)$$

Condition (1) is called the *strong clock condition.*

2.3 Disadvantages of Vector Clocks

Vector clocks are useful in understanding the behavior of Distributed Systems. However, they have the major disadvantage of not being constant in size: the implementation of vector clocks requires an entry for each one of the N sites in the system. If N is large, several problems will arise. There are growing storage costs because each site must reserve space to keep the local version of the vector clock and, depending on the particular system, the vector times associated with certain events must be stored as well. All messages of a distributed computation are tagged with timestamps read from the vector clock, which adds considerable overhead to the communication in the system. Thus, vector clocks have poor scalability.

Charron-Bost [1] proved that given a Distributed System with N sites, there is always a possible combination of events occurring in the system whose causality can only be captured by vector clocks with N-entries. It may be possible to design a different mechanism to determine causality between events, however the previous result indicates that if such mechanism captures completely the causality relation it would have a size $\Theta(N)$. This discourages any attempt to define some kind of clock that, while constant in size, captures completely the causality relation.

2.4 Reducing the Size of Timestamps

A technique to reduce the size of the timestamps appended to messages is proposed in [6]. It is based on the observation that a given site tends to interact frequently with only a small set of other sites and that the timestamps assigned to two consecutive events differ in just a few entries. This technique reduces the communications overhead but, as [4] and [5] mention, there are cases when the causality relationship between different messages sent to the same site is lost and even though the size of the timestamps is reduced, the size of several data structures depends on the number of sites in the Distributed System. A different technique is described in [8], where each site maintains a vector v with N entries. Site i tags the messages that it sends with just $v[i]$, eventually this value will update entry i of the vector of the receiver. This technique fails to represent transitive dependencies and is more useful for applications where the causal dependencies are performed off-line [9].

3 Plausible Clocks

In this section we propose a class of clocks that do not characterize causality completely, but are scalable because they can be implemented using constant size structures.

A *timestamp* is a structure that represents an instant in time as observed by some site. The particular details of this structure are left open. A *time-tag* is a structure that is appended to each message sent in a Distributed System; the format of a time-tag could be identical to the format of a timestamp or, preferably, it could be simpler.

Definition 3. For a Distributed System with global history **H**, a *Time Stamping System* (TSS) X is defined as a six-tuple (**S**, **G**, X.**stamp**, X.**comp**, X.**tag**, s_0) where:

S is a set of timestamps with a particular structure.
G is a set of time-tags with a particular structure.
X.**stamp** is the *timestamping function* mapping **H** to **S.**
X.**comp** is the *comparison function* mapping **S**×**S** to the set **R**={'→','←','=',

'↔'}.
X.**tag** is the *tagging function* mapping **S** to **G.**
$s_0 \in$ **S** is the initial timestamp of the TSS.

X.**stamp** assigns timestamps to each event of the global history **H**. When a message is sent, a tag is created applying *X*.**tag** to the current logical time. *X*.**comp** allows us to compare two timestamps $\mathbf{t}_1$, $\mathbf{t}_2 \in$ **S**.We define the auxiliary function *X*.**rel** such that:

X.**rel** (**x**,**y**) = *X*.**comp** (*X*.**stamp(x)**, *X*.**stamp(y)**)

The obvious meaning of the results of *X*.**rel** with **x**, **y** ∈ **H** are:

X.**rel** (**x**,**y**) = '=' ⇔ TSS *X* believes that **x** and **y** are the same event.
X.**rel** (**x**,**y**) = '→' ⇔ TSS *X* believes that **x** causally precedes **y**.
X.**rel** (**x**,**y**) = '←' ⇔ TSS *X* believes that **y** causally precedes **x**.
X.**rel** (**x**,**y**) = '↔' ⇔ TSS *X* believes that **x** and **y** are concurrent.

Notice that *X*.**comp** compares timestamps, while *X*.**rel** reports the causal relation between two events from the point of view of *X*. ❑

As an example, we can define a TSS *V* = (**S**, **G**, *V*.**stamp**, *V*.**comp**, *V*.**tag**, s_0) based on vector clocks such that:

S is a set of N-dimensional vectors of integers.
G ⊆ **S** is a set of N-dimensional vectors of integers.
V.**stamp** is defined according to rules **V1** and **V2** of section **2**.

$$V.\mathbf{comp}(\mathbf{t}_1,\mathbf{t}_2) = \begin{cases} '\rightarrow' & \text{if } (\mathbf{t}_1 < \mathbf{t}_2) \\ '\leftarrow' & \text{if } (\mathbf{t}_1 > \mathbf{t}_2) \\ '=' & \text{if } (\mathbf{t}_1 = \mathbf{t}_2) \\ '\leftrightarrow' & \text{if } (\mathbf{t}_1 \leftrightarrow \mathbf{t}_2) \end{cases} \quad \text{with } \mathbf{t}_1, \mathbf{t}_2 \in \mathbf{S}. \text{ (See Definition 2)}$$

V.**tag**(t) = t. (i.e. the tag is identical to the current time when the message is sent).
s_0 = <0,0,0,...,0>.

Definition 4. A TSS *X characterizes causality* [5] if ∀x, **y** ∈ **H**:

x = **y** ⇔ *X*.**rel** (**x**,**y**) = '='
x → **y** ⇔ *X*.**rel** (**x**,**y**) = '→'
x ← **y** ⇔ *X*.**rel** (**x**,**y**) = '←'
x ↔ **y** ⇔ *X*.**rel** (**x**,**y**) = '↔' ❑

Notice that this is equivalent to the strong clock condition.

Theorem 1. *V characterizes causality.*

Proof. This result follows from property **(1)** and is proved in [5]. ❑

Definition 5. A TSS P is *plausible* if $\forall \mathbf{x}, \mathbf{y} \in \mathbf{H}$:

$$(\mathbf{x} = \mathbf{y}) \Leftrightarrow P.\mathbf{rel}\ (\mathbf{x},\mathbf{y}) = \text{'='}$$
$$(\mathbf{x} \rightarrow \mathbf{y}) \Rightarrow P.\mathbf{rel}\ (\mathbf{x},\mathbf{y}) = \text{'}\rightarrow\text{'}$$
$$(\mathbf{x} \leftarrow \mathbf{y}) \Rightarrow P.\mathbf{rel}\ (\mathbf{x},\mathbf{y}) = \text{'}\leftarrow\text{'}$$

❑

A plausible TSS satisfies the weak clock condition and also assigns unique timestamps to events.

Theorem 2. *If a TSS P is plausible then:*

$$P.\mathbf{rel}\ (\mathbf{x},\mathbf{y}) = \text{'}\rightarrow\text{'} \Rightarrow (\mathbf{x} \rightarrow \mathbf{y}) \vee (\mathbf{x} \leftrightarrow \mathbf{y})$$
$$P.\mathbf{rel}\ (\mathbf{x},\mathbf{y}) = \text{'}\leftarrow\text{'} \Rightarrow (\mathbf{x} \leftarrow \mathbf{y}) \vee (\mathbf{x} \leftrightarrow \mathbf{y})$$
$$P.\mathbf{rel}\ (\mathbf{x},\mathbf{y}) = \text{'}\leftrightarrow\text{'} \Rightarrow (\mathbf{x} \leftrightarrow \mathbf{y})$$

Proof. If P reports $\mathbf{x} \rightarrow \mathbf{y}$, by definition we know that it is impossible that $\mathbf{x} \leftarrow \mathbf{y}$ or $\mathbf{x} = \mathbf{y}$, so the only possibility left is $(\mathbf{x} \rightarrow \mathbf{y}) \vee (\mathbf{x} \leftrightarrow \mathbf{y})$. The case $\mathbf{x} \leftarrow \mathbf{y}$ is equivalent. If P reports $\mathbf{x} \leftrightarrow \mathbf{y}$, this must be true, because if the actual causal relation were '=', '←' or '→', it would have been reported as such. ❑

A plausible TSS P never confuses the direction of the causality between any two ordered events. If in fact **x** causally precedes **y**, P will always report $\mathbf{x} \rightarrow \mathbf{y}$, or if **y** causally precedes **x**, P will always report $\mathbf{x} \leftarrow \mathbf{y}$. At the same time if P states that $\mathbf{x} \leftrightarrow \mathbf{y}$ this necessarily is correct. In a plausible TSS the timestamps are unique. It is proved in [7] that it is possible to decide if a *cut* is *not consistent* [3] using plausible clocks. Vector clocks are plausible clocks, but not every plausible TSS X characterizes causality since it is possible that $\mathbf{x} \leftrightarrow \mathbf{y}$, but that instead X reports $\mathbf{x} \rightarrow \mathbf{y}$ or $\mathbf{x} \leftarrow \mathbf{y}$.

4 Combination of TSSs

Given two plausible TSSs A and B, they can be easily combined to design a new plausible TSS X. The objective of this combination is to produce TSSs where the ordering between more pairs of events is correctly established.

Theorem 3. (Rule of Contradiction in TSSs) *Let A and B be two plausible TSSs.* $\forall \mathbf{x}, \mathbf{y} \in \mathbf{H}$ *it holds that:*

$$(A.\mathbf{rel}\ (\mathbf{x},\mathbf{y}) \neq B.\mathbf{rel}\ (\mathbf{x},\mathbf{y})) \Rightarrow \mathbf{x} \leftrightarrow \mathbf{y}$$

Proof. By definition, A and B determine with precision if $\mathbf{x} = \mathbf{y}$, then obviously their results about equality of events must coincide in every case. Whenever that $\mathbf{x} \rightarrow \mathbf{y}$, both A and B produce '→', i.e. their results in these cases are always identical. The same is true when $\mathbf{x} \leftarrow \mathbf{y}$. From Theorem 2 we know that if a plausible TSS reports that $\mathbf{x} \leftrightarrow \mathbf{y}$, this must be true. Therefore, if either A or B determines that $\mathbf{x} \leftrightarrow \mathbf{y}$, this is true even if the other disagrees.

Let's assume that (A.**rel** (x,y) = '$\leftarrow$') $\wedge$ (B.**rel** (x,y) = '$\rightarrow$'). Since A and B are plausible TSSs we can use Theorem 2 to affirm that:

$$A.\mathbf{rel}\ (\mathbf{x,y}) = `\leftarrow' \Rightarrow (\mathbf{x} \leftarrow \mathbf{y}) \vee (\mathbf{x} \leftrightarrow \mathbf{y}) \text{ and}$$
$$B.\mathbf{rel}\ (\mathbf{x,y}) = `\rightarrow' \Rightarrow (\mathbf{x} \rightarrow \mathbf{y}) \vee (\mathbf{x} \leftrightarrow \mathbf{y})$$

Definition **1** rules out the case ($\mathbf{x} \rightarrow \mathbf{y}$) $\wedge$ ($\mathbf{x} \leftarrow \mathbf{y}$), therefore $\mathbf{x} \leftrightarrow \mathbf{y}$. Thus, if two plausible TSSs disagree in the causal relation between two events **x** and **y**, then necessarily, these events are concurrent. ❑

Definition 6. We say that the TSS X is a **combination** of A and B if we use them both simultaneously to timestamp the events of **H** and to include tags in the messages sent in the Distributed System. The comparison function of X is defined as shown in Figure 1 (the timestamps of X are ordered pairs that include the timestamps of the original TSSs).

```
X.comp (<A_timestamp1,B_timestamp1>,<A_timestamp2, B_timestamp2>)
{
  char result, A_result, B_result;

  A_result = A.comp  (A_timestamp1,A_timestamp2);
  B_result = B.comp  (B_timestamp1,B_timestamp2);

  if (A_result == B_result)
     result = A_result; /* Uncertain */
  else
     result = '↔';        /* Certain */

  return (result);
}
```

Figure 1. Comparison Algorithm for a Combination of TSSs.

Definition 7. Let X be an arbitrary TSS and V be a TSS that characterizes causality. For a finite history **H**, we define the parameter $\rho(X)$ in this way:

$$\rho(X) = |\{[\mathbf{x,y}] \in \mathbf{H} \times \mathbf{H} \mid X.\mathbf{rel}\ (\mathbf{x,y}) \neq V.\mathbf{rel}\ (\mathbf{x,y})\}| \,/\, |\,\mathbf{H} \times \mathbf{H}\,|$$

This quantity, called the *rate of errors of* X, is the proportion of all the pairs of events in the global history **H**, whose causality relation is wrongly established by the TSS X. When it is clear from the context which TSS is X, we jut use ρ. The better the TSS the lower is ρ (for vector clocks, ρ is 0.0). ❑

Theorem 4. (Plausible + Plausible = Plausible) *Let A and B be two plausible TSSs. If X is the combination of A and B then X is a plausible TSS and it holds that* $(\rho(A) \geq \rho(X)) \wedge (\rho(B) \geq \rho(X))$.

Proof. From Theorem 3 and Definition **6** it is evident that X inherits the properties of A and B concerning the detection of the cases $\mathbf{x} = \mathbf{y}$, $\mathbf{x} \rightarrow \mathbf{y}$ and $\mathbf{x} \leftarrow \mathbf{y}$ $\forall \mathbf{x}, \mathbf{y} \in \mathbf{H}$ Therefore X is plausible. Because of the same reason, the parameter $\rho(X)$ can't be greater than $\rho(A)$ or $\rho(B)$. It is possible (and desirable) that X detects more pairs of concurrent events than A or B. ❑

5 Examples of Constant Size Clocks

We explore three groups of plausible clocks (*R-entries vector*, *K-Lamport clocks* and *Combined TSS*). The first one is a variant of the standard vector clocks where the vectors have a fixed number of entries. The second group is an extension of Lamport clocks, where each site keeps its logical clock and a collection of the maximum message tags received by itself and by sites that directly or indirectly have communicated with this site. The third group combines TSSs from the previous two groups. Obviously, these are not the only possible plausible clocks, but they are simple and efficient to implement.

5.1 R-Entries Vector TSS

R-Entries Vector TSS (*REV*) is a variant of *vector clocks*, where vectors have a fixed size $\mathbf{R} \leq \mathbf{N}$, independent of the number of sites in the Distributed System. Site *i* will update the entry *i* modulo **R**, which implies that multiple sites share the same entry in the vector[1]. The mechanisms for timestamp comparison and messages tagging are almost identical to the ones defined for vector clocks. Let's define *REV* = (**S**, **G**, *REV*.**stamp**, *REV*.**comp**, *REV*.**tag**, s_0), where:

> **S** is a set of pairs of the form <*O*, *v*> where *O* is an integer, and *v* is a R-dimensional vector of integers.
>
> **G** is a set of R-dimensional vectors of integers.
>
> *REV*.**stamp** is defined with the rules:
>
> **RV1)** When an event is generated:
>
> ```
> v [Site % R]= v [Site % R] + 1;
> ```
>
> **RV2)** When a message with timestamp *w* is received:
>
> $1 \leq i \leq R$: $v[i] = \max(v[i], w[i])$
>
> ```
> v [Site % R]= v [Site % R] + 1;
> ```
>
> *REV*.**comp** is defined with the code of Figure 2. The function *Compare_vectors* is assumed to be defined as the function *V*.**comp** in the TSS *V*.
>
> *REV*.**tag** (<*O*, *v*>) = *v*. Only the vector is appended to each message.
>
> s_0 = <*Site identifier*, <0, 0,..,0>>..

Theorem 5. *REV is plausible.*

Proof. *REV* must satisfy Definition **5**. Let $\mathbf{x}, \mathbf{y} \in \mathbf{H}$ be two arbitrary events such that *REV*.**stamp(x)** = $<O_x, v>$ and *REV*.**stamp(y)** =$<O_y, w>$. If **x** and **y** were generated at the same site, *Compare_vectors* directly defines the order of these events. Now, let's consider the case where they are generated at different sites.

---

1. Other mappings between sites and entries of the vector are possible, however we just consider the modulo **R** mapping in this paper.

```
REV.comp(<Ox,v>,<Oy,w>)
{
  char vector_result,result;
  int Entry_x, Entry_y;

  Entry_x = Ox % R;
  Entry_y = Oy % R;

  vector_result = Compare_Vectors (v,w);

  if (Ox != Oy)
    { /* Events occur at different sites */
     if ((w[Entry_y] > v[Entry_y]) && (vector_result == '→'))
        result = '→';  /* Uncertain */
     else
        if ((v[Entry_x] >  w[Entry_x]) && (vector_result == '←'))
           result = '←'; /* Uncertain */
        else
           result = '↔'; /* Certain */
    }
  else /* Same site */
     result = vector_result; /* Certain */

  return (result);
}
```

Figure 2. Comparison Algorithm for *REV.*

If **x** $\rightarrow$ **y** then necessarily $v < w$ (i.e. `vector_result` is '$\rightarrow$'). This is result of the definition of *REV*.**tag** and rule **RV2**. The analysis is equivalent for the case **x** $\leftarrow$ **y**. In any of the situations described above *REV*.**comp** will report '$\rightarrow$' or '$\leftarrow$', respectively. Therefore:

$$\mathbf{x} \rightarrow \mathbf{y} \Rightarrow REV.\mathbf{rel}\,(\mathbf{x},\mathbf{y}) = \text{`}\rightarrow\text{'}$$
$$\mathbf{x} \leftarrow \mathbf{y} \Rightarrow REV.\mathbf{rel}\,(\mathbf{x},\mathbf{y}) = \text{`}\leftarrow\text{'}$$

Notice that when **x** $\rightarrow$ **y**, $w[O_y \ \% \ \mathbf{R}]$ must be strictly greater than $v[O_y \ \% \ \mathbf{R}]$ because $w[O_y \ \% \ \mathbf{R}]$ is increased when **y** is executed. Thus, if $v < w$ but $v[O_y \ \% \ \mathbf{R}] = w[O_y \ \% \ \mathbf{R}]$ then **x** and **y** are concurrent events. Similarly, if $v = w$ or $v \leftrightarrow w$ then necessarily **x** $\leftrightarrow$ **y**.

In conclusion, *REV* is plausible. ❑

5.2 K-Lamport Time Stamping System

This family of TSSs uses the same data structures as *REV*: a site identification and a vector of integers. However, the rules to update this vector are different. Each site keeps a Lamport clock together with information about the maximum time-tag received by itself and by the previous sites that directly or indirectly have had communications with this site. In order to better understand the dynamics of the *K-Lamport TSS*, we present and analyze the properties of the basic case *2-Lamport TSS.*

2-Lamport Time Stamping System

2-Lamport TSS (*2LA*) is an extension of Lamport clocks where a site remembers its local clock and the maximum tag received as well. There is a difference between knowing, for instance, that events **x** and **y** have Lamport clocks 6 and 10 and knowing that the maximum tag received at **y**'s site is 5. With just the Lamport Clocks we'd conclude that **x** → **y**, but with the extra information we know that **x** could not causally precede **y**, and that therefore **x** and **y** are concurrent events. In *2LA* each node keeps a local Lamport Clock in entry 0 of its vector and saves the maximum tag appended to any received message in entry 1. When a message is sent, it is tagged with just entry 0.

Let's define *2LA* = (**S**, **G**, *2LA*.**stamp**, *2LA*.**comp**, *2LA*.**tag**, s_0), where:

S is a set of pairs of the form <*O*, *v*> where *O* is an integer, and *v* is a 2-dimensional vector of integers.

G is a set of integers.

2LA.**stamp** is defined with the rules:

2L1) When an event is generated:

$v[0] = v[0] + 1;$

2L2) When a message with timestamp *Y* is received:

$v[0] = \max(v[0], Y); v[0] = v[0] + 1$
$v[1] = \max(v[1], Y);$

2LA.**comp** is defined with the code of Figure 3.

2LA.**tag** (<*O*, *v*>) = *v* [0]. Only one integer is appended to each message.

s_0 = <*Site identifier*, 0, 0>.

```
2LA.comp(<Ox,v>,<Oy,w>)
{
  char result;

  if (Ox != Oy)
     {/* Events didn't occur at the same site */
      if (v[0] <= w[1])
         result = '→'; /* Uncertain */
      else
         if (w[0] <= v[1])
            result = '←';   /* Uncertain */
         else
            result = '↔'; /* Certain */
     }
  else /* Same site */
     if        (v[0]< w[0]) result = '→'; /* Certain */
     else if (v[0]> w[0]) result = '←'; /* Certain */
     else                  result = '='; /* Certain */

  return (result);
}
```

Figure 3. Comparison Algorithm for *2LA*.

Notice that *v* [0] > *v* [1] for any timestamp assigned to an event.

Theorem 6. *2LA is plausible.*

Proof. Let **x**, **y** ∈ **H** be two arbitrary events such that *2LA*.**stamp(x)** = <O_x, *v*> and *2LA*.**stamp(y)** =<O_y, *w*>. If **x** and **y** occur at the same site the causal relation is correctly established just by comparing $v[0]$ and $w[0]$. Now, consider the case when **x** and **y** have been generated at different sites.

If **x** is causally before than **y**, then $v[0]$ would have been communicated to the site where **y** occurs and the entry $w[1]$ would have been updated. Therefore:

$$\mathbf{x} \rightarrow \mathbf{y} \Rightarrow (v\,[0] \leq w[1])$$
$$\mathbf{x} \leftarrow \mathbf{y} \Rightarrow (w[0] \leq v\,[1])$$

If $(v[0] > w[1])$ **x** doesn't causally precede **y** because of rule **2L2** and the definition of *2LA*.**tag**. Conversely, if $(w[0] > v[1])$, **y** doesn't causally precede **x**. Therefore,

$$(v[0] > w[1]) \wedge (w[0] > v[1]) \Rightarrow (\mathbf{x} \leftrightarrow \mathbf{y})$$

Since we know that $(v[1] < v[0]) \wedge (w[1]< w[0])$, we can notice that:

$(v[0]< w[1]) \wedge (w[0]< v[1]) \Rightarrow v[0]< w[1]< w[0]< v[1] \Rightarrow v[0]< v[1] \Rightarrow$ Contradiction.
$(v[0]< w[1]) \wedge (w[0]= v[1]) \Rightarrow v[0]< w[1]< w[0]= v[1] \Rightarrow v[0]< v[1] \Rightarrow$ Contradiction.
$(v[0]= w[1]) \wedge (w[0]< v[1]) \Rightarrow v[0]= w[1]< w[0]< v[1] \Rightarrow v[0]< v[1] \Rightarrow$ Contradiction.
$(v[0]= w[1]) \wedge (w[0]= v[1]) \Rightarrow v[0]= w[1]< w[0]= v[1] \Rightarrow v[0]< v[1] \Rightarrow$ Contradiction.

Table 1 summarizes all the previous relations. Using the information on this table, *2LA* detects correctly all the cases where $\mathbf{x} \rightarrow \mathbf{y}$ and all the case where $\mathbf{x} \leftarrow \mathbf{y}$. Therefore,

$\mathbf{x} \rightarrow \mathbf{y} \Rightarrow$ *2LA*.**rel** (**x**,**y**) = '→'
$\mathbf{x} \leftarrow \mathbf{y} \Rightarrow$ *2LA*.**rel** (**x**,**y**) = '←'

Table 1. Possible relations between v and w (*2LA*).

	$v[0] < w[1]$	$v[0] = w[1]$	$v[0] > w[1]$
$w[0] < v[1]$	Impossible	Impossible	$\mathbf{x} \leftarrow \mathbf{y}$ *uncertain*
$w[0] = v[1]$	Impossible	Impossible	$\mathbf{x} \leftarrow \mathbf{y}$ *uncertain*
$w[0] > v[1]$	$\mathbf{x} \rightarrow \mathbf{y}$ *uncertain*	$\mathbf{x} \rightarrow \mathbf{y}$ *uncertain*	$\mathbf{x} \leftrightarrow \mathbf{y}$ *certain!*

Thus, *2LA* is plausible. ❑

General Case: K-Lamport TSS

Consider briefly what 3-Lamport TSS (*3LA*) would be. Timestamps are of the form <*O*, *v*> where *O* is a site id and *v* is a 3-entries vector, whose entry 0 is a Lamport clock. Entries $v[0]$ and $v[1]$ become the tag of each message sent. When a message with tag y is received, $v[0]$ is updated with $y[0]$ in a standard Lamport clock fashion and $v[1]$ and $v[2]$ are max-ed with $y[0]$ and $y[1]$. The results shown in Table 1 are valid for *3LA* and entries 1 and 2 of any two timestamps generated by this TSS exhibit these same relations. Thus, given two events **x** and **y** with timestamps <*O*, *v*> and <*P*, *w*>:

$\mathbf{x} \rightarrow \mathbf{y} \Rightarrow (v[0] \leq w[1]) \wedge (v[1] \leq w[2])$
$\mathbf{x} \leftarrow \mathbf{y} \Rightarrow (v[1] \geq w[0]) \wedge (v[2] \geq w[1])$

When *3LA* encounters the first situation it returns '→', detecting correctly all the cases where **x** → **y**. In the second case it returns '←' and detects all the cases where **x** ← **y** correctly. If none of these tests is satisfied *3LA* reports **x** ↔ **y**.

K-Lamport TSS (*KLA*) is a generalization of *2LA* and *3LA*, where we extend the evident pattern shown in these TSSs to a vector with K entries. Let's define *KLA* = (**S**, **G**, *KLA*.**stamp**, *KLA*.**comp**, *KLA*.**tag**, s_0), where:

S is a set of pairs of the form <*O*, *v*> where *O* is an integer, and *v* is a K-dimensional vector of integers.

G is a set of (K-1)-dimensional vectors of integers.

KLA.**stamp** is defined with the rules:

KL1) When an event is generated:

$v[0] = v[0] + 1;$

KL2) When a message with timestamp *w* is received:

$v[0] = \max(v[0], w[0]);\ v[0] = v[0] + 1;$
$1 \leq i \leq \text{K-1}: v[i] = \max(v[i], w[i-1]);$

KLA.**comp** is defined with the code of Figure 5.

KLA.**tag** (<*O*, *v*>) = <*v*[0],...,*v*[K-2]>. K-1 integers are appended to each message.

s_0 = <*Site identifier*, 0, 0,...,0>.

```
KLA.comp(<Ox,v>,<Oy,w>)
{
  char result;

  if (Ox != Oy)
     {/* Events didn't occur at the same site */
      if ((v[0] <= w[1])&&(v[1] <= w[2])&& ... &&(v[K-2] <= w[K-1]))
         result = '→'; /* Uncertain */
      else
         if ((v[1] >= w[0])&&(v[2] >= w[1])&& ... &&(v[K-1] >= w[K-2]))
            result = '←'; /* Uncertain */
         else
            result = '↔'; /* Certain */
     }
  else /* Same site */
     if      (v[0]< w[0]) result = '→'; /* Certain */
     else if (v[0]> w[0]) result = '←'; /* Certain */
     else                 result = '='; /* Certain */

  return (result);
}
```

Figure 4. Comparison Algorithm for *KLA*

Notice that for the non-zero entries of *v*, *v* [i]> *v* [i+1]. The rightmost entries of the vector could contain zeroes. However, this does not affect the correctness of the algorithm.

Theorem 7. *KLA is plausible.*

Proof. If **x** is causally before **y**, then v[i] would have been communicated to the site where **y** occurs and the entry w[i+1] would have been updated. Therefore $\forall$ i < K-1:

$$\mathbf{x} \rightarrow \mathbf{y} \Rightarrow (v\,[i] \leq w[i+1])$$
$$\mathbf{x} \leftarrow \mathbf{y} \Rightarrow (w[i] \leq v\,[i+1])$$

If (v[i] > w[i+1]), **x** doesn't causally precede **y** because of rule **KL2** and the definition of *KLA*.**tag**. Conversely, if (w[i] > v[i+1]), **y** doesn't causally precede **x**. Therefore,

$$(v[i] > w[i+1]) \wedge (w[i] > v[i+1]) \Rightarrow (\mathbf{x} \leftrightarrow \mathbf{y})$$

We know that (v [i+1] < v [i]) $\wedge$ (w [i+1] < w [i]), then:

$(v[i]<w[i+1])\wedge(w[i]<v[i+1])\Rightarrow v[i]<w[i+1]<w[i]<v[i+1]\Rightarrow v[i]<v[i+1]\Rightarrow$ Contradiction.
$(v[i]<w[i+1])\wedge(w[i]=v[i+1])\Rightarrow v[i]<w[i+1]<w[i]=v[i+1]\Rightarrow v[i]<v[i+1]\Rightarrow$ Contradiction.
$(v[i]=w[i+1])\wedge(w[i]<v[i+1])\Rightarrow v[i]=w[i+1]<w[i]<v[i+1]\Rightarrow v[i]<v[i+1]\Rightarrow$ Contradiction.
$(v[i]=w[i+1])\wedge(w[i]=v[i+1])\Rightarrow v[i]=w[i+1]<w[i]=v[i+1]\Rightarrow v[i]<v[i+1]\Rightarrow$ Contradiction.

Table 2 summarizes all the previous relations. Using the information on this table, *KLA* detects correctly all the cases where **x** $\rightarrow$ **y** and all the cases where **x** $\leftarrow$ **y** correctly. So,

$$\mathbf{x} \rightarrow \mathbf{y} \;\Rightarrow KLA.\mathbf{rel}(\mathbf{x},\mathbf{y}) = `\rightarrow'$$
$$\mathbf{x} \leftarrow \mathbf{y} \;\Rightarrow KLA.\mathbf{rel}(\mathbf{x},\mathbf{y}) = `\leftarrow'$$

Table 2. Possible relations between v and w (*KLA*).

	v[i] < w[i+1]	v[i] = w[i+1]	v[i] > w[i+1]
w[i] < v[i+1]	Impossible		**x** $\leftarrow$ **y** *uncertain*
w[i] = v[i+1]			**x** $\leftarrow$ **y** *uncertain*
w[i] > v[i+1]	**x** $\rightarrow$ **y** *uncertain*	**x** $\rightarrow$ **y** *uncertain*	**x** $\leftrightarrow$ **y** *certain!*

Since the timestamps are unique, *KLA* satisfies all the affirmations of Definition **5** and therefore it is plausible. ❑

Every case where **x** $\leftrightarrow$ **y** that is recognized by *(K-1)LA*, is also recognized by *KLA*, while the converse is not always true. Figure 5 shows the dynamics of *KLA*. Figure 5(a) presents the update of the clock after a normal event. Figure 5(b) shows the **send** and **receive** operations; the Lamport clock element is updated in a normal way, while the rest of the elements of the receivers' clocks are *max*-ed with the shifted value of the clocks of the senders. Figures 5(c) and 5(d) summarize the tests made in Figure 4.

5.3 Combined Time Stamping System

This TSS is a combination of *REV* and *KLA*. Let's define *Comb* = (**S**, **G**, *Comb*.**stamp**, *Comb*.**comp**, *Comb*.**tag**, s_0), where:

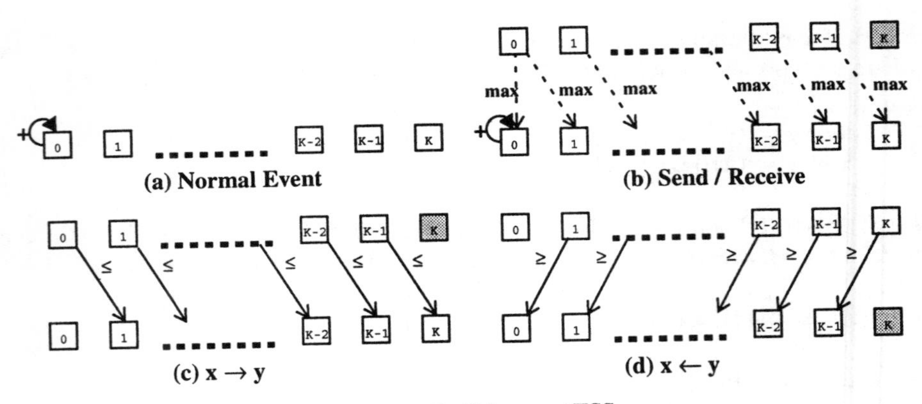

Figure 5. *K-Lamport* TSS

S is a set of elements of the form <*O*, *v*, *l*>, with *O* an integer, *v* a R-Dimensional vector of integers and *l* a K-Dimensional vector of integers

G is a set of elements of the form <*v*, *l*>, with *v* a R-Dimensional vector of integers and *l* a (K-1)-Dimensional vector of integers

Comb.**stamp** is defined with the rules:

Co1) When an event is generated:

- Apply **RV1** on *v*.
- Apply **KL1** on *l*.

Co2) When a message with timestamp < *w*, *m*> is received:

- Apply **RV2** on *v* and *w*.
- Apply **KL2** on *l* and *m*.

Comb.**comp** is defined with the code of Figure 6.

Comb.**tag** (<*O*, *v*, *l*>) = < *v*, *l*[0],*l*[1],...,*l*[K-2]>.

s_0 = <*Site identifier*, 0, 0, .., 0>.

```
Comb.comp(<Ox,v,l>,<Oy,w,m>)
{
  char result, R_result, K_result;

  R_result = REV.comp (<Ox,v>,<Oy,w>);
  K_result = KLA.comp (<Ox,l>,<Oy,m>);

  if (R_result == K_result)
     result = R_result; /* Uncertain */
  else
     result = '↔';      /* Certain */

  return (result);
}
```

Figure 6. **Comparison Algorithm for** ***COMB*****.**

Theorem 8. *Comb is plausible.*

Proof. This result follows from Theorem 4 since each component clock is plausible. ❑

6 Performance Evaluation

We'll ponder the number of cases for which plausible clocks such as the ones defined in this paper, fail to report the correct causal relations. We generate random global histories **H** and use the proposed TSSs to timestamp all the events of these histories. Using the function **comp** of each TSS, we determine the causal relation between each pair of events of **H**×**H** and compare this result with the one produced by vector clocks under the same circumstances.

6.1 Simulation

In the first part of the simulation, we generate a sample history of a Distributed System with N sites. In the second part, this history is executed, collecting timestamps and statistics about each one of the TSSs. A sample is a set of N sequences of events.There are 3 types of events: private event, *send* a message and *receive* a message. The samples were of two types: Random communication pattern and Client/Server communication pattern. In the first type, any pair of sites can communicate with each other. The probability of site i sending a message to site j is the same $\forall\ i, j \leq$ N. In the second type of sample, the sites are divided into *clients* and *servers*. Client sites can communicate only with server sites in a *request/reply* fashion, where the client first sends a message to a server and the next event is a receive from this server. We assume that servers don't send unsolicited messages (e.g. callbacks) to the clients. On the other hand, servers are free to communicate among themselves in a random fashion, but for each message that they receive from a client, the next event must be a send to this client. The simulation "executes" the particular history of each site, sending and receiving messages and keeping the timestamps assigned to each event by each of the TSSs that we are evaluating, together with standard vector clocks. After that, we evaluate the functions **comp** of each TSS on all the ordered pairs from the set **H**×**H** and compute the parameter ρ.

6.2 Results

A total of 348,008,157 pairs of events distributed in 4 groups of 6 samples each (A, B, C and D), were used to evaluate the proposed TSSs. The first group of samples exhibit a random communication pattern. The other 3 groups have a Client/Server communication pattern with 1, 2 and 3 servers respectively.The upper part of Table 3 shows statistics for each group of samples and the lower part presents the rate of errors (ρ) that was obtained for each of the samples when timestamped by each TSS described in section 5. The final column of the table shows weighted averages of this parameter. As it was predicted by Theorem 4, *Comb* consistently produces the minimum values of ρ for exactly the same samples. The evaluations indicate that *Comb* has an excellent performance when the communication pattern is Client/Server. In particular, the best results are obtained for Group B (1 server and up to 75 clients), where this TSS correctly determined the causal relation between a 92.8% of the

95,938,833 pairs of events considered (with just a storage overhead of 6 elements and a message overhead of 4 elements). If we take into account all the 348,008,157 pairs of events we find that *Combined TSS* is correct in 84% of the cases.

Table 3. Values of ρ for the evaluated TSSs.

	A	B	C	D	Aver.
Samples	6	6	6	6	
Sites (min-max)	10 ~ 100	11 ~ 76	12 ~ 77	13 ~ 76	
Servers	N.A.	1	2	3	
Events	13956	17447	17445	17303	
Pairs	62,767,490	95,938,833	95,741,981	93,559,853	
REV (R=2)	0.558	0.154	0.166	0.193	0.241
KLA (K=3)	0.522	0.076	0.133	0.167	0.197
Comb (R=2, K=3)	0.405	0.072	0.111	0.138	0.160

6.3 Effects of the values of R and K

By decreasing the number of components in the vector clocks, we are improving the efficiency of clock operations but decreasing the accuracy with which they detect orderings between events. With a simulation study (7078 events and 50,098,084 pairs of events), we relate the ordering accuracy of *REV* and *KLA* with the size of the vector clock. The sample has a Client/Server communication pattern and simulates a Distributed System with 100 sites, 60 events per site and 35 messages between sites. Figure 7 plots the obtained values of ρ for this sample when R and K are varied from 2 to 99 entries. In general, the rate of errors reduces when the size of the vector is increased. The *REV* curve presents a fast reduction of ρ during the initial increase in R, e.g. it is cut down from 0.167 to 0.097 when R is increased from 2 to 15 entries. However, after that point the pace of reduction of ρ slows down. In order to move ρ down from 0.097 to 0.02, R must be increased from 15 to 77. In certain cases an increment in R produces an increment in the number of errors. Obviously, a value of R=100 makes *REV* equivalent to standard vector clocks and therefore ρ would be 0.0. The *KLA* curve shows an excellent start, with a rate of errors of 0.156 for *2LA* which is reduced to 0.083 for *3LA* and to 0.079 for *5LA*. The minimum value of ρ is 0.078 with 15 entries, from that point on there are no improvements when more entries are added to the clock, in fact even with as many entries as sites the rate of errors never gets to zero. It is interesting to consider the problem of how to distribute a given number of entries between *REV* and *KLA*. This simulation experiment demonstrates that with a modest size vector clocks a large number of orderings can be captured correctly.

6.4 Some applications of Plausible Clocks

In general, plausible clocks are useful for any application where imposing orderings on some pairs of concurrent events has no effect on the correctness of the results. Currently, we are using plausible clocks for the development of mutual consistency protocols for shared objects in Distributed Systems. Notice that given the imperfection

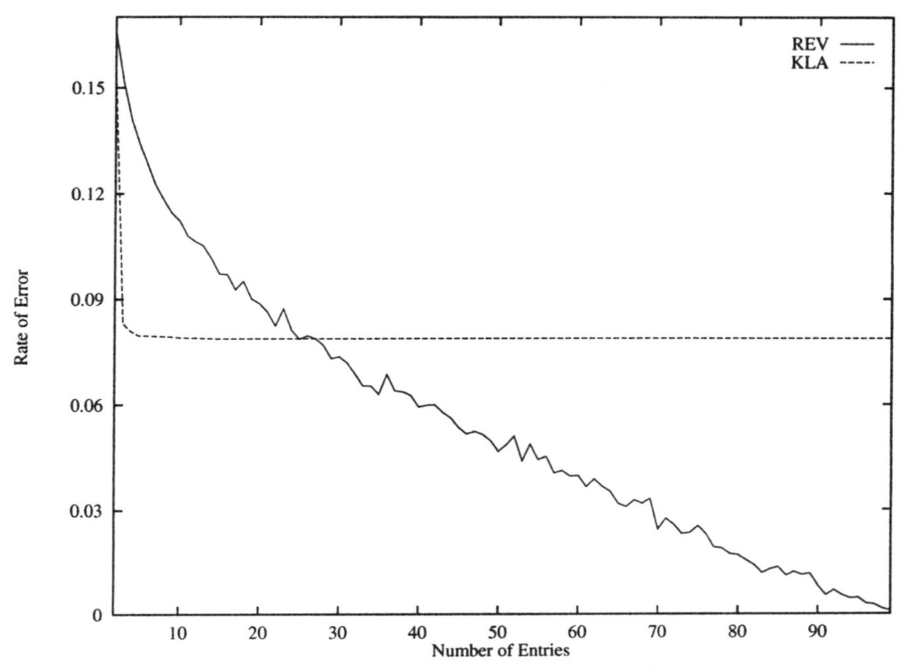

Figure 7. Varying the values of R and K.

of plausible clocks, some applications could incur inefficiencies from time to time (e.g. unnecessarily invalidate an object that was in no conflict with other objects). However, such decisions will not induce wrong results, and if the frequency of these invalidations is relatively low, the cost of this unnecessary work is compensated by the potential for scalability and the savings in communications overhead, storage costs and timestamps processing.

7 Conclusions

Property (1) states that there is an isomorphism between vector clocks and the causality relation of events in a Distributed System. Therefore, vector clocks are useful in understanding the behavior of these systems. However, they have the major disadvantage of not being constant in size: the implementation requires the presence of an entry for each one of the N sites in the distributed system. The results of [1] discourage any attempt to define some kind of clock that, while constant in size, captures completely the causality relation.

In order to reason about clocks without paying attention to their particular implementations, we define the formalism of Time Stamping Systems (TSS). We propose a class of logical clocks called **plausible clocks** that can be implemented with a constant number of components and yet they provide ordering accuracy close to vector clocks. We develop rules to combine known plausible clocks to produce more accurate clocks. We presented several implementations of constant size plausible clocks: *REV*, that needs R+1 entries to store a timestamp, and tags each message with R integers; *KLA*, that needs K+1integers for storage but only K-1 integers per message and *Comb*, that requires R+K+1 integers for storage and R+K-1 integers per message.These implementations were evaluated using a simulation model and we found that even with a random communication pattern, in more than 84% of the cases, the causal relation between events was correctly established (this result goes up to almost 93% of accuracy when we consider a Client/Server model).

We claim that any constant size clock must be plausible in order to be useful, but evidently, there are many other possible implementations of plausible clocks that would be interesting to consider. We have to evaluate the effects of different communication patterns on the performance of plausible clocks. Also, we're interested in analyzing how to distribute a given number of entries of a vector between *REV* and *KLA* clocks.

8 References

1. B. Charron-Bost, "Concerning the size of logical clocks in Distributed Systems", Information Processing Letters 39, pp. 11-16. 1991.
2. L. Lamport, "Time, clocks and the ordering of events in a Distributed System", Communications of the ACM, vol 21, pp. 558-564, July 1978.
3. F. Mattern, "Virtual Time and Global States in Distributed Systems", Conf. (Cosnard et al (eds)) Proc. Workshop on Parallel and Distributed Algorithms, Chateau de Bonas, Elsevier, North Holland, pp. 215-226. October 1988.
4. S. Meldal, S. Sankar and J. Vera, "Exploiting locality in maintaining potential causality". Proc. 10th Annual ACM Symposium on Principles of Distributed Computing, Montreal, Canada, pp. 231-239, 1991.
5. R. Schwarz and F. Mattern, "Detecting causal relationships in distributed computations: in search of the holy grial", Distributed Computing, Vol. 7, 1994.
6. M. Singhal, A. Kshemkalyani, "An efficient implementation of vector clocks", Inf. Process Lett. Vol 43, pp. 47-52, 1992.
7. F. Torres-Rojas, "Efficient Time Representation in Distributed Systems", MSc. Thesis, Georgia Institute of Technology, 1995.
8. J. Fowler and W. Zwaenepoel, "Causal distributed breakpoints", Proc. of 10th Int'l. Conf. on Distributed Computing Systems, pp. 134-141, 1990.
9. M. Raynal and M. Singhal, "Logical Time: Capturing Causality in Distributed Systems", IEEE Computer, Vol. 29, No. 2, Feb., 1996.

Abstracting Communication to Reason about Distributed Algorithms

Michel Charpentier, Mamoun Filali, Philippe Mauran,
Gérard Padiou, Philippe Quéinnec

LIMA - IRIT, ENSEEIHT
2, rue Charles Camichel
31071 Toulouse cedex, FRANCE
e-mail: {charpov,mauran,padiou,queinnec}@enseeiht.fr, filali@irit.fr

Abstract. In distributed systems, message passing is a low level representation of communication resulting in intricate designs and proofs. This paper presents a new abstraction to express communication: the observation. This notion provides a more concise expression of programs and properties, and consequently is an effective help in understanding and reasoning about distributed algorithms. Observations are formalized in the UNITY framework.
We define the observation relation and state its main properties. Then, we present the description and the proof of a generic problem. The abstract level of description separates concerns between the algorithm and its communication pattern. Thus, the topology of observations can be changed while preserving the algorithm properties.

1 Introduction

Designing and reasoning about distributed systems is more often a matter of ingenuity than a matter of engineering. This difficulty seems to stem from the very nature of distributed systems:

- distributed systems are *parallel* systems, which results in a combinatorial explosion of the states of the global system (when compared to the states of its component sites);
- distributed systems are *asynchronous* systems, which introduces one more level into the combinatorics: the current state of a site depends on the history of remote sites; this difficulty comes from the impossibility for a site to maintain an accurate knowledge about the current (global) state of the system.

In this context, it seems crucial to support the design of distributed algorithms by providing conceptual tools that:

- ensure the correctness of algorithms, to achieve a safe design, if not a systematic one;
- make it easier to reason about distributed systems by proposing an efficient (i.e. abstract and concise) representation of such systems.

In this purpose, we use the UNITY formalism [2] which provides a sound formal framework [7, 8] and appears to be well suited to describing distributed systems. Our contribution provides an abstraction of interactions between sites. The construct we propose is intended to be in the spirit of UNITY minimalism: it solely aims at expressing what is – according to us – the essential feature of distributed systems: the (a priori) impossibility for a site to maintain an up-to-date global knowledge, due to the lack of a common memory (or clock).

More precisely, we define an ***observation*** as a relation between two variables, the *source* and the *image*. It states that the values of the image are some of the former values of the source, in a chronological order. We present a set of theorems to relate the properties of the images to the properties of the sources on the one hand, and to link the observation relation to UNITY logical operators on the other hand.

Thus, a first interest of this construct is to provide a set of results that can be used for reasoning about a wide range of distributed algorithms. A second contribution of this construct is to offer a declarative and progressive approach to distributed algorithms development: the observation relations specify the "interactions topology", independently from the specification of the algorithm itself. The observation relation bridges a gap between the UNITY model, where interaction is described with global shared variables, and the message passing model, close to the physical world.

The declarative aspect results in a more concise expression of programs and properties, and consequently is an effective help in understanding and reasoning about distributed algorithms. We illustrate the contribution of observations from this point of view as well as the interest of a sound formal framework through the expression of a solution (based on [15]) to a well-known problem: distributed mutual exclusion.

In the next section, we present the UNITY formalism. The third section introduces the observation relation and states its main properties. Then, we illustrate the use of observations to express and to prove a distributed mutual exclusion algorithm. The last section concludes and describes our current work.

2 UNITY

The purpose of the UNITY formalism [2] is to provide a framework to reason about the *correctness* of programs. It is a general model that allows the specification of parallel programs from an operational point of view (writing program statements) as well as from a declarative point of view (stating program properties). Besides, this formalism is designed to make it easier to check the consistency of these two points of view. The minimalism of the UNITY formalism is one of its strong points: it ensures its generality, while guaranteeing its simplicity of use.

The UNITY formalism consists of two parts: a programming language, based on transition systems, and a specification language, based on a linear temporal logic [12], to express safety and progress properties of the corresponding transition systems.

This section describes parts of UNITY syntax and semantics useful to understand the remainder of the paper. A more comprehensive presentation can be found in [2] or [14].

2.1 The UNITY Language

A UNITY program F describes a state transition system. It consists of a declaration of variables, a specification of their initial values, and a set of multiple-assignement statements:

- the **declare** section defines the set of state variables along with their type;
- the **initially** section consists of a predicate on state variables, called $init.F$, expressing the possible initial states for the program;
- the **assign** section, denoted by $assign.F$, consists of a set of multiple guarded assignments separated by the $[\!]$ mark.
  UNITY also allows quantification over statements to describe a family of assignments, written as $\langle [\!] v : condition :: instructions(v) \rangle$.

Semantically, executing a UNITY program consists in choosing non-deterministically one assignment and then executing it atomically. This choice is repeated infinitely, assuming weak fairness: each statement is executed infinitely often.

2.2 Operational Semantics

To formalize the semantics of observations in accordance with the UNITY formalism, we use the notion of operational semantics of a program. These semantics models define the non-empty set of infinite sequences of states corresponding to possible computations of the program. Given an operational semantics $\mathcal{O}$ and a UNITY program F, $\mathcal{O}.F$ is the set of possible computations σ. For each sequence σ, the state σ_i is a mapping between the state variables and their value.[1] In this paper, variables are handled as state functions and the value of variable v in state σ_i is denoted by $v.\sigma_i$. By extension, $v.\sigma$ represents the successive values of the variable v in the computation σ, and $e.\sigma_i$ denotes the value of the *expression* e in the state σ_i.

For example, a well known operational semantics is $\mathcal{O}_1$, presented in [17]. For a program F, $\mathcal{O}_1.F$ is the set of sequences of states σ for which the following properties hold:

- $init.F.\sigma_0$
- $\langle \forall i :: \langle \exists s : s \in assign.F :: \sigma_i \; s \; \sigma_{i+1} \rangle \rangle$

[1] To denote the i th element of a sequence σ, we use the mathematical notation σ_i instead of $\sigma.i$.

- $\langle \forall s : s \in assign.F :: \langle \exists_i^\infty :: \sigma_i \; s \; \sigma_{i+1} \rangle \rangle$

In the following, all the definitions and theorems are valid given any operational semantics.

2.3 The UNITY Logic

We now define the basic operators of the UNITY logic in an operational semantics $\mathcal{O}$.

- p **unless** q : in any computation, if p becomes true, then it remains true at least until q becomes true. In semantics $\mathcal{O}$, it becomes:

$$p \textbf{ unless } q \textbf{ in } F \equiv \langle \forall \sigma, i : \sigma \in \mathcal{O}.F :: \; (p \wedge \neg q).\sigma_i \Rightarrow (p \vee q).\sigma_{i+1} \rangle$$

- **stable** p : once p becomes true, it remains true[2]:

$$\textbf{stable } p \textbf{ in } F \equiv \langle \forall \sigma, i : \sigma \in \mathcal{O}.F \;\; :: \;\; p.\sigma_i \Rightarrow p.\sigma_{i+1} \rangle$$

- $p \mapsto q$: if p is true, then q will eventually become true:

$$p \mapsto q \textbf{ in } F \equiv \langle \forall \sigma, i : \sigma \in \mathcal{O}.F :: \; p.\sigma_i \Rightarrow \langle \exists j : j \geq i :: \; q.\sigma_j \rangle \rangle$$

In the operational semantics $\mathcal{O}_1$, these operators are equivalent to Sanders' [16].

2.4 Representing distributed algorithms

In distributed computing, a program is structured as a set of processes (or nodes) and communication is usually described in terms of message passing [1, 6, 10].

In the UNITY model, each node is described as a program component and the whole system is implemented as a quantified union:

$$DC = \langle [\!] s : 0 < s \leq N :: Node(s) \rangle$$

A common approach to program message exchanges consists in translating send and receive primitives into append and extract operations on sequences. All components share the variables of type sequence that implement communications. However, readability is poor since message passing is not part of the syntax, and proofs are difficult since message semantics is not part of the logic. We propose another description using observations.

[2] **stable** p is equivalent to p **unless** $false$.

3 Observation Semantics

An observation defines a relation over the variables of a distributed program: $`x \prec\!\cdot\; x$ (read $`x$ observes x) states that the history of the variable $`x$ is a sub-history of the (remote) variable $x$. The variable $x$ (resp. $`x$) is called the *source* (resp. the *image*).

In the message passing paradigm, communication and synchronization cannot be described separately. Observation aims at describing solely communication. If a specific scheme of synchronization is required, it can be explicitly programmed. For instance, a handshake protocol can be implemented with two mutual observations.

Before describing the observation relation, we define a abstract model of time.

3.1 Clock Functions

We assume that an ideal discrete totally ordered time exists. In the following, such a time is represented by the set of natural numbers. A clock function is then defined as an approximation of this ideal time. This approximation satisfies safety and liveness constraints.

More precisely, a clock function is a mapping C from natural numbers to natural numbers, satisfying the following properties:

- **safety 1:** a clock never outgrows the ideal time: $\langle \forall t :: \; C.t \leq t \rangle$
- **safety 2:** a clock is monotonously increasing: $\langle \forall t :: \; C.t \leq C.(t+1) \rangle$
- **liveness:** a clock eventually increases: $\langle \forall t :: \; \langle \exists t' :: \; C.t < C.t' \rangle\rangle$

We write $Clock(C)$ when the mapping C satisfies these properties.

Properties.

- the value of any clock at the initial time is 0;
- the identity function is a representation of the ideal time;
- the composition of two clocks is a clock.

3.2 The Observation Relation $\prec\!\cdot$

Definition. The observation relation between two variables of the same program is defined by:

$$`v \prec\!\cdot\; v \textbf{ in } F \equiv \langle \forall \sigma : \sigma \in \mathcal{O}.F :: \langle \exists\, C \;:\; Clock(C) :: \langle \forall t :: `v.\sigma_t = v.\sigma_{C.t} \rangle\rangle\rangle \quad (1)$$

This definition states that any value of $`v$ is a previous value of $v$ (clock safety 1), that $`v$ is assigned its values in a chronological order (clock safety 2) and that $`v$ is eventually assigned more recent values of v (clock liveness).

The variable v is called the *observed variable*, or, in short, the *source*, and $`v$ is the *image variable*, or, in short, the *image of v*.

If $\langle \forall t : t \neq 0 :: C.t < t \rangle$, the observation is *strict*: the image cannot reflect immediately a change in the value of the source.

Generalization. In the previous definition, an observation is a relation between variables of a program. This definition is generalized to take into account any expression defined on the variables, or on the state of the program. Then, the formula:

$$`e \prec\!\!\cdot\; e \ \mathbf{in}\ F$$

expresses that the value of the expression $`e$ is always a former value of the expression $e$, that $`e$ takes (part of) e values in a chronological order, and that $`e$ is eventually updated with a more recent value of e.

Now, we can define the observation of a tuple of variables. The relation:

$$(`u_1, `u_2, \cdots, `u_n) \prec\!\!\cdot\; (u_1, u_2, \cdots, u_n) \ \mathbf{in}\ F$$

states that the variables u_i are observed simultaneously, that is with the same clock function.

We can also consider the observation relation to be a new operator in the UNITY logic. Then, the observation property deals with state predicates, i.e. boolean expressions.

3.3 Properties

We consider two kinds of properties: properties of the observation relation itself, and properties about the observation relation and the UNITY logic. Proofs of the following properties are given in [3].

Notation. Properties are stated as deduction rules:

$$\frac{p}{q}\ \texttt{prop} \quad \text{means that} \quad \forall F :: \ p \ \mathbf{in}\ F \ \Rightarrow q \ \mathbf{in}\ F$$

where

- F is a UNITY program;
- prop is the symbolic name of the property.

u and v stand for expressions of any type, and p is a predicate.

Some Properties of the Relation $\prec\!\!\cdot$

- The observation relation is reflexive, antisymmetric and transitive:

$$\frac{\mathtt{true}}{u \prec\!\!\cdot\; u}\ \texttt{OBS\_REFL}$$

$$\frac{u \prec\!\!\cdot\; v \ ,\ v \prec\!\!\cdot\; u}{\mathbf{invariant}\ u = v}\ \texttt{OBS\_ANTISYM}$$

$$\frac{u \prec\!\!\cdot\; v \ ,\ v \prec\!\!\cdot\; w}{u \prec\!\!\cdot\; w}\ \texttt{OBS\_TRANS}$$

- Compatibility with function application:

$$\frac{u \prec\!\!\cdot\; v}{f.u \prec\!\!\cdot\; f.v}\ \texttt{OBS\_F} \qquad \frac{f.u \prec\!\!\cdot\; f.v \ ,\ f \text{ injection}}{u \prec\!\!\cdot\; v}\ \texttt{OBS\_F\_INJ}$$

The Observation Relation and the UNITY Logic

$$\frac{(`u, `v) \prec (u, v)}{`u \prec u}\ \texttt{OBS\_L} \qquad \frac{(`u, `v) \prec (u, v)}{`v \prec v}\ \texttt{OBS\_R}$$

$$\frac{`p \prec p\ ,\ \textbf{stable}\ p}{\textbf{stable}\ `p}\ \texttt{OBS\_STABLE} \qquad \frac{`p \prec p\ ,\ \textbf{stable}\ p}{`p\ \textbf{detects}\ p}\ \texttt{OBS\_DETECTS}$$

$$\frac{`p \prec p\ ,\ p\ \textbf{unless}\ `p}{p \mapsto `p}\ \texttt{OBS\_LEADSTO1} \qquad \frac{`p \prec p\ ,\ \textbf{stable}\ p}{p \mapsto `p}\ \texttt{OBS\_LEADSTO2}$$

$$\frac{`u \prec u\ ,\ \forall k :: \textbf{stable}\ u > k}{\textbf{invariant}\ `u \leq u}\ \texttt{OBS\_UPB}$$

The first two theorems state that the projections of an image tuple are images of the projections of the source tuple; the third one states that the image of a stable predicate is stable; OBS_DETECTS states that a stable predicate can be detected by its image[3]; the two LEADSTO theorems ensures the occurrence of $`p$ given some properties of p; OBS_UPB means that the image of a non-decreasing counter is always smaller than the source.

In the framework of reactive programs, which are characterized by an interaction with their environment, the properties between the observation relation and the UNITY logical predicates can be used in two different ways:

- either to deduce properties of the observations based on environment properties, e.g. OBS_STABLE,
- or to deduce properties of the environment based on properties of the observation, e.g. OBS_DETECTS.

3.4 Observations as a Mechanism

In the preceding section, an observation is defined as a relation between variables (or expressions) and we have stated several theorems relating the relation to other UNITY predicates. From this point of view, $`e \prec e$ is a property that, for a given program, is either true or false. We now intend to use observations to avoid an explicit description of the communication between nodes of a distributed system.

[3] p **detects** q means **invariant** $p \Rightarrow q\ \wedge\ q \mapsto p$.

The `observe` Section. An observation is now seen as a mechanism which satisfies the properties of the observation relation. UNITY is extended with a new section called `observe` which describes observation relations assumed by the program. The declaration of these observations replaces an explicit implementation of the mechanism, including the initialization of the images.

When the `observe` section contains $`x \prec\!\!\cdot\, x$, a mechanism is assumed to update $`x$ so that $`x$ is assigned former values of $x$ in a chronological order. Moreover, only this mechanism is allowed to modify $`x$.

The UNITY language extended with the `observe` section must be given an operational semantics. Such a semantics, defined as an extension of $\mathcal{O}_1$, can be found in [4].

An Implementation of an Observation. We present here a simple implementation of an observation based on the non-deterministic assignment of the current value of the source to the image. The liveness property is ensured by the weak fairness of the UNITY execution model.

Program *Obs*
declare
    $`v : T;$ $\{ \; T$ is the type of $v \; \}$
initially
    $`v = v$ $\{$ due to $C.0 = 0 \; \}$
assign
    $`v := v$
end

The proof of $`v \prec\!\!\cdot\, v$ **in** $Obs [\!] G$, where $G$ does not modify $`v$, is given in [3].

4 A Mutual Exclusion Algorithm Using Observations

This section gives an extended example of a mutual exclusion algorithm based on observations.

We consider a system consisting of a fixed number of nodes (or processes) and a critical resource that at most one process is allowed to use at any time. We present here an algorithm that ensures this mutual exclusion property. More precisely, the following properties must be achieved:

- safety property: at most one process has permission to use the resource;
- liveness property: every request for the resource is eventually granted.

With respect to this problem, the behavior of a node is a three state automaton. A variable records the current state of each node. Initially, in the "thinking" state, the node is outside the critical section. After a request, it remains in a "requesting" state until executing its critical section in the "mutex" state. Each node completes its critical section in a finite time and returns to the "thinking" state.

A solution to this problem is based on logical clocks. Such an algorithm has been proposed by Ricart and Agrawala [15]. When a node requires mutual exclusion, its request is assigned a timestamp. A node is allowed to enter mutual exclusion when its request is the oldest one. For the thinking nodes, a default request is assumed with a maximal value (infinite value).

In the following, we give a solution based on the same principle, using a slightly different approach for handling logical clocks and request timestamping.

4.1 An Observation Oriented Solution

To solve this problem by means of observations, we assign a clock vector[4] $C[1..N]$ to each node called $Mutex(s)$. In this vector, the component $Mutex(s).C[s]$ is the local clock of the node s and the other components are clock images of the other nodes. More precisely, $Mutex(s).C[s']$ with $s' \neq s$ is an image of the remote clock $Mutex(s').C[s']$ (Fig. 1). We specify this observation relation as[5]:

$$\forall s, s' \ :: \ Mutex(s).C[s'] \prec\!\cdot\, Mutex(s').C[s']$$

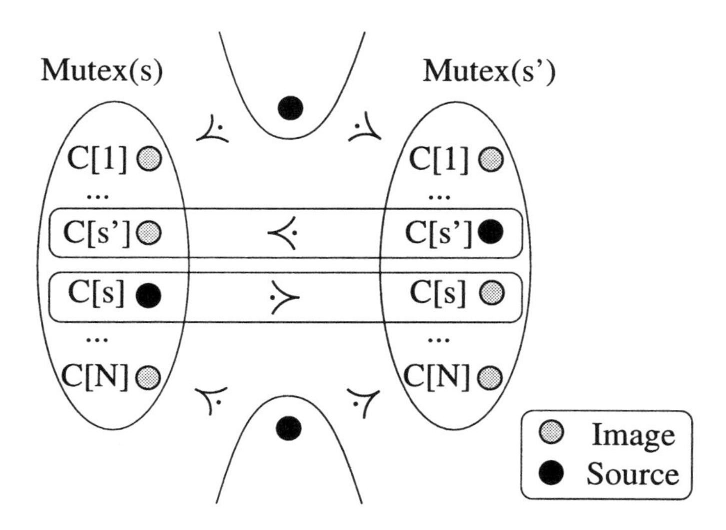

Fig. 1. Node observations

The node behavior is described by the initial conditions and the actions taken at each state.

---

[4] Here, clocks are variables; they must not be confused with the clock functions of the previous section.

[5] This specification includes the relation $Mutex(s).C[s] \prec\!\cdot\, Mutex(s).C[s]$. This relation is redundant ($\prec\!\cdot$ is a reflexive relation) but simplifies the writing of observations.

- Initially, a node is thinking.
- When a node $Mutex(s)$ is thinking, it cannot prevent another node from entering. For this purpose, it infinitely often increases its local clock $C[s]$.
- When a node is requesting, it stops its local clock. It will enter mutual exclusion when its "suspended" clock becomes smaller than all the local images.

Consequently, a local clock $Mutex(s).C[s]$ is updated only by the node s and is observed by the other nodes.

The following UNITY program is a model of this behavior.

Program $Mutex(s\ :\ 1..N)$
$C\ \ :$ **array** $1..N$ **of** Int;
$state$: $\{thinking, requesting, mutex\}$;
initially
$C[s] = s\ \wedge\ state = thinking$;
always
$Minimum = \langle \forall s'\ :\ 0 < s' \leq N\ \wedge\ s' \neq s\ ::\ (C[s] < C[s'])\rangle$
observe
$\langle \forall s'\ :\ 0 < s' \leq N ::\ C[s'] \prec Mutex(s').C[s']\rangle$
assign
{ think and tick }
$C[s] := C[s] + N$ **if** $state = thinking$
$[\!]$
{ request and wait }
$state := requesting$ **if** $state = thinking$
$[\!]$
{ enter and use }
$state := mutex$ **if** $(state = requesting) \wedge Minimum$
$[\!]$
{ leave and tick }
$state, C[s] := thinking, C[s] + N$ **if** $state = mutex$
end $Mutex$

Then, the distributed configuration DC is instantiated as the union of $Mutex$ component programs:

$$DC\ =\ \langle [\!]\ s\ :\ 0 < s \leq N\ ::\ Mutex(s)\rangle$$

The **observe** clause of this union is the conjunction of the corresponding clauses in each component program. For this configuration we obtain the previously specified observations:

$$\forall s, s' ::\ Mutex(s).C[s'] \prec Mutex(s').C[s']$$

Notation. In the following, $s.anything$ is an abbreviation for $Mutex(s).anything$. Moreover, $s.mutex$, $s.requesting$ and $s.thinking$ are used for, respectively, $Mutex(s).state = mutex$, $Mutex(s).state = requesting$ and $Mutex(s).state = thinking$.

Specification. This algorithm must satisfy the following safety and liveness properties:

- safety:

$$\textbf{invariant } \langle \forall s, s' :: s.mutex \wedge s'.mutex \ \Rightarrow \ (s = s') \rangle \textbf{ in } DC \tag{2}$$

- liveness:

$$\forall s \ :: \ s.resquesting \mapsto s.mutex \textbf{ in } DC \tag{3}$$

4.2 Proof of the Algorithm

Clock Properties. In this solution, a local clock is implemented by means of an integer. Its initial value is the number of the corresponding node. A tick consists in incrementing the clock by N. Such an implementation can be seen as a simplified version of Lamport's logical clocks [11] and ensures the following properties:

- the value domain of two clocks are disjoint:

$$\langle \forall s, s' : \ s \neq s' \ :: Dom(s.C[s]) \cap Dom(s'.C[s']) = \emptyset \rangle$$

- a clock is an increasing counter:

$$\forall s, k :: \textbf{ stable } C[s] > k$$

- clock values are totally ordered:

$$\langle \forall s, s' : \ s \neq s' \ :: \ (s.C[s] < s'.C[s']) \vee (s.C[s] > s'.C[s']) \rangle$$

Mutual Exclusion Property.

Lemma 1. *If a local clock is smaller than the local images, it is also smaller than all the remote clocks.*

Proof. Let us consider a node $Mutex(s)$ such that its clock is the minimum of the images:

$$\langle \forall s' : \ s' \neq s \ :: s.C[s] < s.C[s'] \rangle$$

Since local clocks are increasing counters, clock images satisfy (theorem OBS_UPB):

$$\langle \forall s' : \ s' \neq s \ :: \ s.C[s'] \leq s'.C[s'] \rangle$$

From both preceding properties, it follows:

$$\langle \forall s' : \ s' \neq s \ :: s.C[s] < s.C[s'] \leq s'.C[s'] \rangle$$

Consequently, $C[s]$ is also a global minimum:

$$\langle \forall s, s' : \ s' \neq s \ :: s.C[s] < s'.C[s'] \rangle$$

□

When a node is in mutual exclusion, its clock is the minimum of the images and remains constant. From the previous lemma, we deduce that its clock remains the global minimum.

Since the local clocks have disjoint domains and the clock values are totally ordered, this minimum is unique. If the invariant was satisfied for two different nodes s_1 and s_2, this would imply:

$$s_1.Minimum \land s_2.Minimum$$

Then, by the unicity of the minimum, it follows that $s_1 = s_2$ which contradicts the hypothesis.
□

Liveness Property. Let s_0 be the node with the smallest clock:

$$s_0.C[s_0] = \langle \min s :: s.C[s] \rangle$$

We show through the next three lemmata that the local clock of s_0 eventually increases. Next, we show that the global minimum eventually increases. Then we deduce the liveness property.

Lemma 2. *If the node s_0 is in the mutex state, it eventually ticks:*

$$\forall k :: \quad s_0.mutex \land s_0.C[s_0] = k \;\mapsto\; s_0.C[s_0] > k \tag{4}$$

Proof. This is ensured by the weak fairness of the UNITY execution model, which eventually executes the "leave and tick" statement. □

Lemma 3. *If the node s_0 is in the requesting state, it eventually ticks :*

$$\forall k :: \quad s_0.requesting \land s_0.C[s_0] = k \;\mapsto\; s_0.C[s_0] > k \tag{5}$$

Proof. The program DC satisfies:

$$\forall k :: s_0.requesting \land s_0.C[s_0] = k \textbf{ unless } s_0.mutex \tag{6}$$

With OBS_LEADSTO2, we deduce that if a local clock is greater than l, its image eventually becomes greater than l:

$$\forall s :: \frac{s_0.C[s] \prec s.C[s], \; \forall m :: \textbf{stable } s.C[s] > m}{\forall l :: s.C[s] > l \mapsto s_0.C[s] > l}$$

Thanks to the PSP (Progress-Safety-Progress) theorem[6], we deduce:

$$\begin{array}{c} \forall s, k, l : s \neq s_0 :: \\ s_0.requesting \land s_0.C[s_0] = k \land s.C[s] > l \\ \mapsto \\ s_0.mutex \lor (s_0.requesting \land s_0.C[s_0] = k \land s_0.C[s] > l) \end{array}$$

[6] PSP Theorem [2]: $\frac{p \mapsto q \;,\; r \textbf{ unless } b}{p \land r \;\mapsto\; (q \land r) \lor b}$

Let $k = l = s_0.C[s_0]$, then this formula simplifies into:

$$\begin{array}{c}\forall s : s \neq s_0 :: \\ s_0.requesting \mapsto s_0.mutex \vee (s_0.requesting \wedge s_0.C[s_0] < s_0.C[s]) \end{array} \tag{7}$$

Using OBS_STABLE and OBS_F, it follows that $s_0.C[s]$ is non-decreasing. With (6), we prove that:

$$\forall s : s \neq s_0 :: s_0.requesting \wedge s_0.C[s_0] < s_0.C[s] \textbf{ unless } s_0.mutex \tag{8}$$

We can then apply the completion theorem[7] to (7) and (8) and deduce:

$$\begin{array}{c} s_0.requesting \\ \mapsto \\ s_0.mutex \vee (s_0.requesting \wedge \langle \forall s : s \neq s_0 :: s_0.C[s_0] < s_0.C[s]\rangle) \end{array} \tag{9}$$

As the program DC satisfies:

$$s_0.requesting \wedge \langle \forall s : s \neq s_0 :: s_0.C[s_0] < s_0.C[s]\rangle \textbf{ unless } s_0.mutex$$

the instruction "enter and use" is eventually chosen and performed. Consequently, we have:

$$s_0.requesting \wedge \langle \forall s : s \neq s_0 :: s_0.C[s_0] < s_0.C[s]\rangle \;\mapsto\; s_0.mutex$$

We deduce by cancelling[8] the right hand of the disjunction in (9):

$$s_0.requesting \mapsto s_0.mutex \tag{10}$$

And using lemma 2, we conclude:

$$\forall k :: \; s_0.requesting \wedge s_0.C[s_0] = k \;\mapsto\; s_0.C[s_0] > k$$

□

Lemma 4. *If the node s_0 is in the thinking state, it eventually ticks.*

Proof. We have (from the program text):

$$\begin{array}{c} \forall k :: s_0.thinking \wedge s_0.C[s_0] = k \\ \mapsto \\ (s_0.thinking \wedge s_0.C[s_0] > k) \vee (s_0.requesting \wedge s_0.C[s_0] = k) \end{array}$$

Using lemma 3 and cancellation, we deduce:

$$\forall k :: \; s_0.thinking \wedge s_0.C[s_0] = k \;\mapsto\; s_0.C[s_0] > k \tag{11}$$

□

[7] Completion theorem [2]: $\frac{\langle \forall i :: p_i \mapsto q_i \vee b\rangle \;,\; \langle \forall i :: q_i \textbf{ unless } b\rangle}{\langle \forall i :: p_i\rangle \mapsto \langle \forall i :: q_i\rangle \vee b}$, for any finite set of predicates p_i, q_i.

[8] Cancellation theorem: $\frac{p \mapsto q \vee b \;,\; b \mapsto r}{p \mapsto q \vee r}$. Here we have $r = q$.

Lemma 5. *The global minimum* $\langle \min s :: s.C[s] \rangle$ *is unbounded.*

Proof. Applying the leadsto finite disjunction to the three lemmata 2, 3 and 4, we obtain:

$$\forall k :: \ s_0.C[s_0] = k \mapsto s_0.C[s_0] > k$$

Therefore, as s_0 was the global minimum, this global minimum is increased:

$$\forall k :: \ \langle \min s :: s.C[s] \rangle = k \mapsto \ \langle \min s :: s.C[s] \rangle > k$$

By induction, we deduce:

$$\forall k, l :: \ \langle \min s :: s.C[s] \rangle = k \mapsto \ \langle \min s :: s.C[s] \rangle > l$$

Using the leadsto infinite disjunction, we finally get:

$$\forall l :: \ \mathtt{true} \ \mapsto \ \langle \min s :: s.C[s] \rangle > l \tag{12}$$

□

Liveness Property. since the global minimum is unbounded, all clocks are unbounded. Since the clock of a requesting node is stopped unless it enters mutual exclusion, it eventually does. □

4.3 Observation Topology

In the preceding algorithm, each node observes all the other nodes. Therefore, the underlying network topology is a mesh. We can also regard this mesh as an observation topology. We can change the observation topology without modifying the algorithm properties. In other words, some topology transformations preserve the algorithms.

For instance, a ring topology may replace the mesh. Each node only observes a single predecessor. This observation ring is specified by:

$$\forall s, s' \ : \ s \neq s' \ :: \ s.C[s'] \prec (pred.s).C[s']$$

A node $Mutex(s)$ only observes its predecessor directly. For the other nodes, it observes their remote clocks through the images provided by its predecessor. For this transformation, thanks to the transitivity of observations, the properties of the solution are preserved. We have again:

$$\forall s, s' :: \ s.C[s'] \prec s'.C[s']$$

At an implementation level, this observation topology leads to message exchanges according to a ring topology.

Finally, a node may also observe simultaneously the entire array of clocks of its predecessor:

$$\forall s \ : \ (s.C[s'] : s' \neq s) \prec \ ((pred.s).C[s'] : s' \neq s)$$

5 Conclusion

In this paper, we have introduced the observation relation as an abstraction of the properties of usual underlying networks. After showing how its semantics and syntax could be embedded within the UNITY approach, we have stated general observation properties. Finally, we have considered the description and the validation of a distributed mutual exclusion algorithm. We believe that the observation relation is interesting not only for the description of distributed algorithms since it provides an abstraction of communications, but also for the proof of distributed algorithms since it has interesting properties which should facilitate proof steps.

Among our investigations on observations, let us mention the expression of distributed algorithms. For instance we have designed an observations-based version of Mattern's termination detection algorithm [13]. We are also studying the use of observations as a tool to support the development of distributed algorithms. We are working on the properties which are preserved when an initial algorithm with global variables is transformed into another one where some of these variables are replaced by their images. We plan to validate such transformations within an HOL [9] observation theory and to make them available within a UNITY development environment [5]. A further research aspect concerns the study of the observation of several variables, the crucial point being how the dependence between these variables should be dealt with.

References

1. Gregory R. Andrews and Fred B. Schneider. Concepts and notations for concurrent programming. *Computing Surveys*, 15(1):3–43, March 1983.
2. K. Mani Chandy and Jayadev Misra. *Parallel Program Design: A Foundation.* Addison-Wesley, 1988.
3. M. Charpentier, M. Filali, P. Mauran, G. Padiou, and P. Quéinnec. Répartition par observation dans Unity. Technical Report 96-01-R, IRIT, 27 pages, January 1996.
4. Michel Charpentier. Une sémantique pour Unity avec section observe. Technical report, IRIT, April 1996.
5. Michel Charpentier, Abdellah El Hadri, and Gérard Padiou. A Unity-based algorithm design assistant. In *Workshop on Tools and Algorithms for the Construction and Analysis of Systems*, pages 131–145, Aarhus, Denmark, May 1995. BRICS Notes Series NS-95-2.
6. Vilay K. Garg. *Principles of Distributed Systems.* Kluwer Academic Publisher, 1996.
7. Rob Gerth and Amir Pnueli. Rooting UNITY. In *Proc. fifth Int. Workshop on Software Specification and Design*, pages 11–19, May 1989.
8. David Moshe Goldschlag. *Mechanically Verifiyng Concurrent Programs.* PhD thesis, University of Texas at Austin, May 1992.
9. M.J.C. Gordon and T.F. Melham. *Introduction to HOL: A Theorem Proving Environnement for Higher Order Logic.* Cambridge University Press, 1993.

10. C. A. R. Hoare. *Coomunicating Sequential Processes.* Prentice-Hall International, 1984.
11. Leslie Lamport. Time, clocks and the ordering of events in a distributed system. *Communications of the ACM*, 21(7):558–565, July 1978.
12. Zohar Manna and Amir Pnueli. *The Temporal Logic of Reactive and Concurrent Systems.* Springer-Verlag, 1991.
13. Friedemann Mattern. Algorithms for distributed termination detection. *Distributed Computing*, 2:161–175, 1987.
14. Jayadev Misra. A logic for concurrent programming. Technical report, The University of Texas at Austin, April 1994.
15. G. Ricart and A.K. Agrawala. An optimal algorithm for mutual exclusion in computer networks. *Communications of the ACM*, 24(1):9–17, January 1981.
16. Beverly A. Sanders. Eliminating the substitution axiom from Unity logic. *Formal Aspects of Computing*, 3(2):189–205, 1991.
17. R.T. Udink and J.N. Kok. On the relation between Unity properties and sequence of states. In J.W de Bakker, W.-P. de Roever, and G. Rozenberg, editors, *Semantics: Foundations and Applications,* volume 666 of *Lecture Notes in Computer Science*, pages 594–608, 1993.

Simulating Reliable Links with Unreliable Links in the Presence of Process Crashes*

Anindya Basu[1] Bernadette Charron-Bost[2] Sam Toueg[1]

[1] Department of Computer Science, Upson Hall, Cornell University, USA.
[2] Laboratoire d'Informatique LIX, Ecole Polytechnique, FRANCE.

1 Introduction

We study the effect of link failures on the solvability of problems in distributed systems. In particular, we address the following question: *given a problem that can be solved in a system where the only possible failures are process crashes, is the problem still solvable if links can also fail by losing messages?* The answer depends on several factors, including the synchrony of the system, the model of link failures, the maximum number of process failures, and the nature of the problem to be solved.

In this paper, we focus on asynchronous systems (results concerning synchronous systems will be described in a companion paper). The set of problems solvable in asynchronous systems with process crashes include *Reliable, FIFO,* and *Causal Broadcast*, and their *uniform* counterparts [Bir85, HT94], as well as *Approximate Agreement* [DLP+86], *Renaming* [ABND+90], and *k-set Agreement* [Cha90]. The question is whether such problems remain solvable (and if so, how) if we add link failures.

We consider two models of lossy links: *eventually reliable* and *fair lossy*. Roughly speaking, they have the following properties: with an eventually reliable link, there is a time after which all messages sent are eventually received (messages sent before that time may be lost). Such a link can lose only a finite (but unbounded) number of messages. With a fair lossy link, if an infinite number of messages are sent, an infinite subset of these messages is received. Such a link can lose an infinite number of messages. Clearly, any algorithm that works with fair lossy links also works with eventually reliable links. Thus, to make our results as strong as possible, we assume eventually reliable links when we prove impossibility results, and fair lossy links when we show problems to be solvable.[3]

Since an eventually reliable link can lose only a finite number of messages, it may appear that one can mask these message losses by repeatedly sending copies of each message, or by piggybacking on each message all the messages that were previously sent. Such a scheme is highly inefficient, but it does seem to simulate a reliable link. So it appears that, in principle, any problem that can be solved in a system with process crashes and reliable links, remains solvable in a system with process crashes and eventually reliable links.

* Research partially supported by NSF grant CCR-9402896, DARPA/NASA Ames grant NAG-2-593, Air Force Material Contract F30602-94-C-0224 and ONR contract N00014-92-J-1866.

[3] Eventually reliable or fair lossy links do not lead to permanent network partitioning. Such partitioning renders most interesting problems trivially impossible.

Our first two results concern systems where half (or more) processes may crash. We first show that the intuition described above is flawed. We do so by exhibiting a problem, *Uniform Reliable Broadcast* [HT94], that is solvable with reliable links but not with eventually reliable links. However, not all problems are like Uniform Reliable Broadcast. Our second result characterizes a large class of problems that *remain solvable* even with fair lossy links. Informally, this class consists of all the problems whose specifications refer only to the behavior of correct processes (i.e., processes that do not crash) — these are called *correct-restricted* [Gop92] or *failure-insensitive* [BN92] problems.[4] This class of problems includes Reliable, FIFO, and Causal Broadcast, and correct-restricted versions of Approximate Agreement, Renaming, and k-set Agreement. For such problems, we show how to automatically transform any algorithm that works in a system with process crashes and reliable links into one that works with process crashes and fair lossy links.

Our final result concerns systems where a majority of processes is correct. In this case, we show that any problem that is solvable with process crashes and reliable links is also solvable with process crashes and fair lossy links. We do this by showing that given a system with fair lossy links and a majority of correct processes, one can simulate a system with reliable links.

The problem of tolerating crash and/or link failures has been extensively studied (e.g., [AAF+94, AE86, AGH90, BSW69, FLMS93, GA88, WZ89]). Several papers focus on a single link and on how to mask failures of that link [AAF+94, BSW69]. In contrast, we study lossy links in the context of an entire system: we show that the effect of lossy links depends on the proportion of faulty processes in the system. Other works have also studied tolerating lossy links in the context of an entire network. However, some of them do not consider process crashes [GA88], while others assume that crashed processes recover [AGH90], and yet others focus on the solution of specific problems such as Broadcast [AE86]. In contrast, our results assume permanent process crashes, and more importantly, we study the effect of link failures on the solvability of problems in general. This approach stresses the importance of the notion of *correct-restricted* problems.

The paper is organized as follows. In Section 2, we define our model, including the various types of links that we consider. Sections 3 and 4 consider systems where a majority of processes may crash. We first prove that in general, reliable links cannot be simulated by eventually reliable links (Section 3). We then show that "natural" correct-restricted problems that are solvable with reliable links are also solvable with fair lossy links (Section 4). In Section 5, we consider systems where a majority of processes are correct, and show how to simulate reliable links with fair lossy links. Finally, in Section 6 we state our results more formally using a refinement of the model and the notion of translation. This version of the paper does not contain detailed proofs; these are given in [BCBT96].

---

[4] The complement of this class of problems includes all problems with *uniform* specifications [NT90].

2 Model

We consider asynchronous distributed systems where processes communicate by message passing via a completely connected network, and there are no bounds on relative process speeds or message transmission times.

2.1 Variables and States

We postulate an infinite universal set of *variables* $\mathcal{V}$. Each variable v in $\mathcal{V}$ can be assigned a value from the set of natural numbers $\mathbb{N}$. A *state* s is a mapping $V \rightarrow \mathbb{N}$ for some subset of variables V of $\mathcal{V}$. We say that *state* s *is over variables* V, and write $var(s) = V$. For any v in V, the *value of* v *in state* s is $s(v)$. The set of all states is denoted S.

2.2 Processes

Let $P = \{p_1, \cdots, p_n\}$ be an indexed non-empty set of n processes. Each process p_i in P is formally defined by a set of *states* $\mathcal{Q}_i$, a set of *initial states* $\mathcal{Q}_i^0 \subseteq \mathcal{Q}_i$, a set of *actions* $\mathcal{A}_i$, a *transition relation* $\mathcal{T}_i$ on $\mathcal{Q}_i \times \mathcal{A}_i$, and a *state transition function* $\delta_i : \mathcal{Q}_i \times \mathcal{A}_i \rightarrow \mathcal{Q}_i$.

The set $\mathcal{Q}_i$ is a set of states over some finite (non-empty) set of variables $V_i \subset \mathcal{V}$. We say that V_i is the *sets of variables of process* p_i. We assume that the sets of variables of distinct processes are disjoint.

The set $\mathcal{A}_i$ is the set of actions that p_i can execute. There are three types of actions: *send*, *receive*, and *internal*. To define the send and receive actions, we postulate a set $\mathcal{M}(P)$ of all the possible messages that processes in P can send. We assume that each message $m \in \mathcal{M}(P)$ has a header with three fields, *sender*$(m) \in P$, *dest*$(m) \in P$, and *tag*(m), an integer used to differentiate messages.

The sets of send and receive actions of $\mathcal{A}_i$, denoted *Send*$(\mathcal{A}_i)$ and *Receive*$(\mathcal{A}_i)$, respectively, are defined as: $Send(\mathcal{A}_i) = \{send(m, p_j) \mid m \in \mathcal{M}(P),\ sender(m) = p_i,\ dest(m) = p_j\}$, and $Receive(\mathcal{A}_i) = \{receive(m) \mid m \in \mathcal{M}(P),\ dest(m) = p_i\} \cup \{receive(\perp)\}$. Action *send*$(m, p_j)$ models the sending of message m to p_j. Action *receive*(m) models the receipt of message m, and action *receive*$(\perp)$ models the failure of p_i's attempt to receive a message (because no message was sent to p_i yet, or the messages sent to p_i are "in transit", or they were "lost", etc.). The transition relation $\mathcal{T}_i$ on $\mathcal{Q}_i \times \mathcal{A}_i$ specifies which actions p_i can execute from any given state: $(s, a) \in \mathcal{T}_i$ iff p_i in state $s \in \mathcal{Q}_i$ can execute action $a \in \mathcal{A}_i$. To model the fact that it is not possible for a process to block because it does not have an action to execute, we assume that for every state $s \in \mathcal{Q}_i$ there exists at least one action $a \in \mathcal{A}_i$ such that $(s, a) \in \mathcal{T}_i$. To model the fact that a process can try to receive a message, but cannot select which message to receive, we assume if $(s, a) \in \mathcal{T}_i$ and $a \in Receive(\mathcal{A}_i)$, then for all $a' \in Receive(\mathcal{A}_i)$, $(s, a') \in \mathcal{T}_i$.

The state transition function $\delta_i : \mathcal{Q}_i \times \mathcal{A}_i \rightarrow \mathcal{Q}_i$ specifies what the state of p_i is after it executes an action. More precisely, if p_i is in state $s \in \mathcal{Q}_i$ and executes action $a \in \mathcal{A}_i$, then p_i goes into state $s' = \delta_i(s, a)$.

Finally, we find it convenient to assume that in every execution, messages are "unique" (this will be made more precise in Section 2.6). To enforce this, we assume that p_i increments a message counter each time it sends a message, and that each message is tagged with the current value of this counter. More precisely, we make the following assumptions on V_i, δ_i and $\mathcal{T}_i$. The set of variables V_i of p_i has a variable msg_cntr_i. If $s' = \delta_i(s, send(m, p_j))$ and $s(msg\_cntr_i) = k$, then $s'(msg\_cntr_i) = k+1$. Moreover, if $(s, send(m, p_j))$ is in $\mathcal{T}_i$, then $tag(m) = s(msg\_cntr_i)$.

2.3 Events and Histories

An *event of process* $p_i \in P$ is a tuple $e = (p_i, a_i, l)$ where $a_i \in \mathcal{A}_i$ and $l \in \mathbb{N}$. We say that action a_i *is associated with* event e.

A *local history of process* $p_i \in P$, denoted $H[i]$, is a finite or an infinite sequence $s_i^0\ e_i^1\ s_i^1\ e_i^2\ s_i^2 \cdots$ of alternating states and events such that: (a) if $H[i]$ is finite, it terminates with a state, (b) $s_i^0 \in \mathcal{Q}_i^0$, (c) for all $k \geq 1$, $s_i^k \in \mathcal{Q}_i$ and $e_i^k = (p_i, a_i^k, k)$, and (d) for all $k \geq 0$, $(s_i^k, a_i^{k+1}) \in \mathcal{T}_i$ and $s_i^{k+1} = \delta_i(s_i^k, a_i^{k+1})$. The *state history of a local history* $H[i]$, denoted $\overline{H}[i]$, is the sequence of states in $H[i]$, namely $s_i^0\ s_i^1\ s_i^2 \cdots$. A *history* H *of* P is a vector of local histories $< H[1], H[2], \cdots, H[n] >$. The *state history of* H *of* P, denoted $\overline{H}$, is the vector $< \overline{H}[1], \overline{H}[2], \cdots, \overline{H}[n] >$. Vector $\overline{H}$ is also called a *state trace*, or simply a *trace*.

Process p_i is *correct in history* H if $H[i]$ is infinite; otherwise we say that p_i *crashes in history* H. The set of all correct processes in history H is denoted by *correct*(H).

2.4 Event Ordering

We relate events that occur in a history using the *happens-before* (henceforth abbreviated as *before*) relation defined in [Lam78]. The before relation $\prec_H$ over events of a history H is the smallest transitive relation such that: (1) if e and e' are different events in the same local history and e occurs before e' in that local history, then $e \prec_H e'$; (2) if, for some $m \in \mathcal{M}(P)$, $e = (p_i, send(m, p_j), k)$ and $e' = (p_j, receive(m), l)$ are events in H, then $e \prec_H e'$. We write $e \preceq_H e'$ if $e \prec_H e'$ or $e = e'$.

2.5 Systems of *P*

Let P be a set of processes. We define $\mathcal{H}(P)$ to be the set of all histories H of P such that $\prec_H$ is a strict partial order. Let H be any history in $\mathcal{H}(P)$ and H' be any down-set of H (i.e., H' is a vector such that, for every $p_i \in P$, $H'[i]$ is a prefix of $H[i]$, and if $H'[i]$ is finite it terminates with a state). Then, H' is also a history in $\mathcal{H}(P)$.

A system $\mathcal{S}(P)$ *of* P is a subset of $\mathcal{H}(P)$. We denote by $\overline{\mathcal{S}(P)}$ the set of traces in $\mathcal{S}(P)$, i.e., the set $\{\overline{H} \mid H \in \mathcal{S}(P)\}$.

2.6 Link Properties

Let P be a set of processes. As we saw in Section 2.2, each process p_i tags each message that it sends with a counter that is incremented after each sending. This ensures that in every history $H \in \mathcal{H}(P)$, messages are unique: if $(p_i, send(m, p_j), k)$ and

$(p_s, send(m', p_t), k')$ are distinct events in H, then $m \neq m'$ (either $sender(m) \neq sender(m')$ or $tag(m) \neq tag(m')$).

We say that p_i *sends* m *to* p_j *in* H if event $(p_i, send(m, p_j), k)$ is in $H[i]$ for some k. Similarly, p_j *receives* m *from* p_i *in* H if event $(p_j, receive(m), l)$ with $sender(m) = p_i$ is in $H[j]$ for some l.

Reliable Links A reliable link does not create, duplicate, or lose messages. Formally, the link from p_i to p_j is *reliable in history* H *of* P if H satisfies:

L1: (*No Creation*) For all $m \in \mathcal{M}(P)$, if p_j receives m from p_i, then p_i sends m to p_j.
L2: (*No Duplication*) For all $m \in \mathcal{M}(P)$, p_j receives m from p_i at most once.
L3: (*No Loss*) For all $m \in \mathcal{M}(P)$, if p_i sends m to p_j, and p_j executes receive actions infinitely often,[5] then p_j receives m from p_i.

Implementing reliable links in a (non-blocking) asynchronous system requires infinite storage for buffering messages — finite buffers can overflow and thus cause message losses. Note that in our model every process has infinite storage.

Lossy Links A lossy link can lose messages in transit. We consider two types of such links. The link from p_i to p_j is *eventually reliable in history* H *of* P if H satisfies L1, L2 and:

L4: (*Finite Loss*) If p_j executes receive actions infinitely often, then the number of messages sent by p_i to p_j that are not received by p_j is finite.

The link from p_i to p_j is *fair lossy in history* H *of* P if H satisfies L1 and L2 and:

L5: (*Fair Loss*) If p_i sends an infinite number of messages to p_j, and p_j executes receive actions infinitely often, then p_j receives an infinite number of messages from p_i.

A reliable link is also eventually reliable, and an eventually reliable link is also fair lossy. A reliable link does not "lose" messages, an eventually reliable link can lose only a finite number of messages, and a fair lossy link can lose an infinite number of messages.

2.7 Systems of P with Reliable and Lossy Links

The *system of* P *with at most* t *process crashes and reliable links*, denoted $\mathcal{S}^t_R(P)$, is the set of all histories $H \in \mathcal{H}(P)$ such that at most t processes crash in H (i.e., at most t local histories $H[i]$ of H are finite) and all links are reliable in H (i.e., for all $p_i, p_j \in P$, the link from p_i to p_j is reliable in H). The *system of* P *with at most* t *process crashes and eventually reliable links*, denoted $\mathcal{S}^t_{ER}(P)$, and the *system of* P *with at most* t *process crashes and fair lossy links*, denoted $\mathcal{S}^t_{FL}(P)$, are similarly defined. Note that for all t, $\mathcal{S}^t_R(P) \neq \emptyset$ and $\mathcal{S}^t_R(P) \subseteq \mathcal{S}^t_{ER}(P) \subseteq \mathcal{S}^t_{FL}(P) \subseteq \mathcal{H}(P)$.

[5] This implies that p_j is correct in H.

2.8 Problem Specifications, Solving a Problem

A problem specification is often given in the form of requirements on sets of traces. To see this, consider a problem like *Consensus*. Roughly speaking, a system $\mathcal{S}(P)$ of P solves Consensus, if $\overline{\mathcal{S}(P)}$ satisfies the following conditions: (a) in every trace $\overline{H} \in \overline{\mathcal{S}(P)}$, each process has some propose and decision variables that satisfy some agreement and validity requirement (e.g., correct processes agree on the value of their decision variables, a decision value must be a proposed one, etc.), and (b) $\overline{\mathcal{S}(P)}$ must have two traces $\overline{H}_0$ and $\overline{H}_1$ such that the initial value of all the propose variables is 0 in $\overline{H}_0$, and 1 in $\overline{H}_1$. Informally, the specification of Consensus is the set of all $\overline{\mathcal{S}(P)}$ for all P, that satisfy (a) and (b). In other words, it is the set of all sets of traces that satisfy (a) and (b).

To formally define a problem specification, we first need to define the set of all traces. Recall that S is the set of all states. Let *Seq*(S) be the set of all non-empty finite and infinite sequences over S such that all the states in a sequence have the same set of variables (i.e., for each σ in *Seq*(S), and any two states s and s' in σ, $var(s) = var(s')$). If $\sigma \in Seq(S)$, $var(\sigma)$ denotes the set of variables of any state in σ. The set of all traces, denoted *Vec*(S), is $\bigcup_{k \in \mathbb{N}} \{< \sigma_1, \sigma_2, \cdots, \sigma_k > \mid \forall i, j, 1 \leq i, j \leq k, \sigma_i \in Seq(S) \text{ and } var(\sigma_i) \cap var(\sigma_j) = \emptyset\}$.

Two traces $\overline{H}$ and $\overline{H'}$ in *Vec*(S) are *compatible* if they have the same dimension, say k, and for all i, $1 \leq i \leq k$, $var(\overline{H}[i]) = var(\overline{H'}[i])$. A set of traces is *proper* if it is non-empty and all its elements are compatible. The set of all proper sets of traces is $\Sigma^* = \{\overline{\mathcal{S}} \mid \overline{\mathcal{S}} \subseteq Vec(S) \text{ and } \overline{\mathcal{S}} \text{ is proper}\}$. A *specification* Σ is a subset of Σ^*.

Let Σ be a problem specification, P be a set of processes, and $\mathcal{S}(P)$ be a system of P. We say that $\mathcal{S}(P)$ *solves (a problem with specification)* Σ, if $\overline{\mathcal{S}(P)} \in \Sigma$.

2.9 Closure under Non-Trivial Reduction

The specifications of most problems satisfy a natural closure property that we now describe. Let P be a set of processes, and $\mathcal{S} = \mathcal{S}(P)$ and $\mathcal{S}' = \mathcal{S}'(P)$ be two systems of P. Suppose that $\mathcal{S}$ solves some problem specification Σ. Is it reasonable to require that if $\mathcal{S}' \subseteq \mathcal{S}$ then $\mathcal{S}'$ solves Σ? To understand this issue, consider a specific example: let Σ be the specification of Consensus (sketched in the previous section).

Since $\mathcal{S}$ solves Σ, then $\overline{\mathcal{S}}$ satisfies condition (a) of Σ, namely, every trace $\overline{H} \in \overline{\mathcal{S}}$ satisfies agreement and validity. If $\mathcal{S}' \subseteq \mathcal{S}$, it is obvious that $\overline{\mathcal{S}'}$ also satisfies condition (a). But the set $\overline{\mathcal{S}'} \subseteq \overline{\mathcal{S}}$ may not satisfy condition (b): for example, *every* trace $\overline{H} \in \overline{\mathcal{S}'}$ may start with all the propose variables equal to 0. In this case, $\mathcal{S}'$ does not solve Σ.[6] On the other hand, if $\overline{\mathcal{S}'}$ satisfies condition (b), then $\mathcal{S}'$ indeed solves Σ. This motivates the following definitions and assumption.

The *initial state of a trace* $\overline{H}$, denoted *init*$(\overline{H})$, is the vector $< s_1^0, s_2^0, \cdots, s_k^0 >$, where k is the dimension of $\overline{H}$, and for all i, $1 \leq i \leq k$, s_i^0 is the first state in $\overline{H}[i]$. For any $\overline{\mathcal{S}} \in \Sigma^*$, we define $init(\overline{\mathcal{S}}) = \{init(\overline{H}) \mid \overline{H} \in \overline{\mathcal{S}}\}$, the set of all initial states of all traces in $\overline{\mathcal{S}}$.

[6] This is not fortuitous: we do not want to allow a system to trivially "solve" Consensus by just avoiding certain initial states.

For all $\overline{\mathcal{S}}$ and $\overline{\mathcal{S}'}$ in Σ^*, we say that $\overline{\mathcal{S}'}$ *is a non-trivial reduction of* $\overline{\mathcal{S}}$ if $\overline{\mathcal{S}'} \subseteq \overline{\mathcal{S}}$ and $init(\overline{\mathcal{S}'}) = init(\overline{\mathcal{S}})$. A specification Σ is *closed under non-trivial reduction* if $\overline{\mathcal{S}} \in \Sigma$ and $\overline{\mathcal{S}'}$ is a non-trivial reduction of $\overline{\mathcal{S}}$ implies $\overline{\mathcal{S}'} \in \Sigma$. Henceforth, we consider only such specifications.

2.10 Correct-Restricted Specifications

Intuitively, a specification is correct-restricted if it refers only to the states of correct processes (those with infinite traces) [Gop92, BN92]. Formally, let $\overline{H}$ and $\overline{H'}$ be any two traces in $Vec(S)$ with the same dimension, say k. Traces $\overline{H}$ and $\overline{H'}$ are *correct-equivalent*, denoted $\overline{H} \overset{c}{\sim} \overline{H'}$, if for all i, $1 \leq i \leq k$, if $\overline{H}[i]$ or $\overline{H'}[i]$ is infinite then $\overline{H'}[i] = \overline{H}[i]$. For any $\overline{\mathcal{S}}$ and $\overline{\mathcal{S}'}$ in Σ^*, we say that $\overline{\mathcal{S}'}$ is *a correct-restricted extension* of $\overline{\mathcal{S}}$, denoted $\overline{\mathcal{S}'} \geq_c \overline{\mathcal{S}}$, if $\overline{\mathcal{S}'} \supseteq \overline{\mathcal{S}}$ and $\forall \overline{H'} \in \overline{\mathcal{S}'}$, $\exists \overline{H} \in \overline{\mathcal{S}} : \overline{H} \overset{c}{\sim} \overline{H'}$. In other words, $\overline{\mathcal{S}'}$ is obtained from $\overline{\mathcal{S}}$ by adding some traces that are correct-equivalent to those in $\overline{\mathcal{S}}$. Finally, we say that a specification Σ is *correct-restricted* if for all $\overline{\mathcal{S}}, \overline{\mathcal{S}'} \in \Sigma^* : \overline{\mathcal{S}'} \geq_c \overline{\mathcal{S}}$ and $\overline{\mathcal{S}} \in \Sigma$ implies $\overline{\mathcal{S}'} \in \Sigma$.

Reliable Broadcast (RB) and *Consensus* are examples of problems with a correct-restricted specification. Their uniform counterparts (e.g., URB in Section 3) are *not* correct-restricted (their specifications refer to the behavior of faulty processes) [HT94].

3 Reliable is Strictly Stronger than Eventually Reliable

Since an eventually reliable link can lose only a finite number of messages, it may appear that one can mask these message losses by repeatedly sending copies of each message, or by piggybacking on each message all the messages that were previously sent. Such a scheme is certainly inefficient,[7] but it does seem to simulate a reliable link (akin to a data link protocol that uses retransmissions to simulate a reliable link over a lossy one). So it may appear that any problem that is solvable with reliable links is also solvable with eventually reliable links. We now show that this intuition is incorrect: in systems where a majority of processes may crash, there are natural problems that can be solved with reliable links but not with eventually reliable links.

One such problem is *Uniform Reliable Broadcast* (or simply *URB*) [HT94]. Informally, URB is defined in terms of two primitives, *broadcast* and *deliver*, that must satisfy three properties:

- *Validity:* If a correct process broadcasts a message m, then it eventually delivers m.
- *Uniform agreement:* If a process (whether correct or faulty) delivers a message m, then all correct processes eventually deliver m.
- *Integrity:* For any message m, every correct process delivers m at most once, and only if m was previously broadcast by its sender.

A simple algorithm given in [HT94] solves URB with reliable links and any number of process crashes, and a standard partitioning argument shows that URB cannot be solved with eventually reliable links if a majority of processes may crash.

---

[7] Indeed it may require the sending of an infinite number of message copies, or, alternatively, the sending of messages of infinite length.

Theorem 1. *1. For $0 \leq t < n$, there is a set P of n processes such that $\mathcal{S}_R^t(P)$ solves URB.*
2. For $2 \leq n \leq 2t$, there is no set P of n processes such that $\mathcal{S}_{ER}^t(P)$ solves URB.

The above theorem implies that one cannot simulate reliable links with eventually reliable links when a majority of processes may crash. The precise statement of this impossibility result is given in Section 6.5 (Theorem 13), after the formal definition of simulation is given.

4 Solving Correct-Restricted Problems with Fair Lossy Links

The previous result does *not* mean that *all* problems that are solvable with reliable links are unsolvable with eventually reliable links. In fact, (most) correct-restricted problems that are solvable with reliable links are also solvable with fair lossy links, and thus with eventually reliable links. To prove this, we first introduce a new type of link that is weaker than a reliable link but stronger than an eventually reliable link — this intermediate link type is called *weakly reliable* (Section 4.1). We then show that any set of processes that solves a correct-restricted problem with reliable links also solves it with weakly reliable links (Section 4.2). Finally, we show how to simulate weakly reliable links with fair lossy links (Section 4.3). Note that weakly reliable links are introduced for technical reasons only — they may not model any "real" links.

4.1 Weakly Reliable Links: An Intermediate Model

Let P be a set of processes. The link from p_i to p_j is *weakly reliable in history H of P* if H satisfies **L1**, **L2** and:

L6: (*No Visible Loss*) For all $m \in \mathcal{M}(P)$, if p_i sends m to p_j before some event e_l of some correct process p_l (according to $\prec_H$), and p_j executes receive actions infinitely often, then p_j receives m from p_i.

Roughly speaking, **L6** states that if the sending of a message m by p_i to p_j is "visible" to a correct process (because it is in the "causal past" of that process), then m is not lost: if p_j executes receive actions infinitely often, then it eventually receives m.

The *system of P with at most t process crashes and weakly reliable links*, denoted $\mathcal{S}_{WR}^t(P)$, is the set of all histories $H \in \mathcal{H}(P)$ such that at most t processes crash in H and all links are weakly reliable in H. Note that $\mathcal{S}_R^t(P) \subseteq \mathcal{S}_{WR}^t(P) \subseteq \mathcal{S}_{ER}^t(P)$.

4.2 Solving Correct-Restricted Problems with Weakly Reliable Links

Any set of processes that solves a correct-restricted problem with reliable links also solves it with weakly reliable links. To show this formally, we first prove:

Lemma 2. *For any set of processes P and any t, if H is a history in $\mathcal{S}_{WR}^t(P)$ then there is a history H' in $\mathcal{S}_R^t(P)$ such that $\overline{H'} \overset{c}{\sim} \overline{H}$ and* $\text{init}(\overline{H'}) = \text{init}(\overline{H})$.

Proof: Let $H \in \mathcal{S}^t_{WR}(P)$. We construct H' from H by removing from H all the events that are not "visible" to correct processes in H (and deleting all the states that follow removed events). To do so, we first define $\vartheta(H) = \{ e \mid \exists p_l \in \mathit{correct}(H), \exists e' \text{ in } H[l] : e \prec_H e' \}$. Intuitively, this is the set of all events that are "visible" to (i.e., in the "causal past" of) correct processes in H. Note that by transitivity of $\prec_H$, if $e' \in \vartheta(H)$ and $e \prec_H e'$ then $e \in \vartheta(H)$. We then construct H', the down-set of H that contains only the events in $\vartheta(H)$, as follows.

For each $H[i] = s_i^0\, e_i^1\, s_i^1\, e_i^2 \cdots s_i^{k-1}\, e_i^k\, s_i^k\, \cdots$:

1. If $H[i]$ is infinite, we define $H'[i] = H[i]$.
2. If $H[i]$ is finite, we define $H'[i] = s_i^0\, e_i^1\, s_i^1\, e_i^2 \cdots e_i^k\, s_i^k$ where k is the maximum index such that $e_i^k \in \vartheta(H)$. If $H[i]$ has no event in $\vartheta(H)$, then $H'[i] = s_i^0$.

From this construction it is clear that H' is a down-set of H, $\mathit{correct}(H') = \mathit{correct}(H)$, and the set of all events in H' is $\vartheta(H)$. Furthermore, $\overline{H'} \overset{c}{\sim} \overline{H}$ and $\mathit{init}(\overline{H'}) = \mathit{init}(\overline{H})$.

To show $H' \in \mathcal{S}^t_R(P)$, note first that $H' \in \mathcal{H}(P)$ (because it is the down-set of a history in $\mathcal{H}(P)$), and that at most t processes crash in H' (because $\mathit{correct}(H') = \mathit{correct}(H)$ and, since $H \in \mathcal{S}^t_{WR}(P)$, at most t processes crash in H). It remains to show that all links are reliable in H', i.e., for every two processes p_i and p_j, properties **L1**, **L2**, and **L3** hold in H'. We first note that since $H \in \mathcal{S}^t_{WR}(P)$, it satisfies **L1**, **L2**, and **L6**.

L2: Since H satisfies **L2**, for all $m \in \mathcal{M}(P)$, p_j receives m from p_i at most once in H. Since H' is a down-set of H, p_j receives m from p_i at most once in H'.

L1: Suppose p_j receives m from p_i in H', and let e_j be the corresponding receive event. Since H' is a down-set of H, p_j receives m from p_i in H. Since H satisfies **L1**, p_i sends m to p_j in H; let e_i be the corresponding event. We have $e_i \prec_H e_j$. Since e_j is in H', $e_j \in \vartheta(H)$, and so $e_i \in \vartheta(H)$. Thus, e_i is also in H'. In other words, p_i sends m to p_j in H', as we needed to show.

L3: Suppose p_i sends m to p_j in H', and p_j executes receive actions infinitely often in H'. We must show that p_j receives m from p_i in H'. First note that p_j executes receive actions infinitely often, and hence is correct, in *both* H' and H; moreover, $H'[j] = H[j]$. Let e_i be the event corresponding to p_i sending m to p_j in H'. By construction of H', $e_i \in \vartheta(H)$. Thus, there is an event e_l that occurs at some correct process p_l in H, such that $e_i \prec_H e_l$. Since H satisfies **L6**, and p_j executes receive actions infinitely often in H, p_j receives m from p_i in H, and therefore in H'. □

Theorem 3. *Let Σ be any correct-restricted problem specification. For any set of processes P and any t, $\mathcal{S}^t_R(P)$ solves Σ if and only if $\mathcal{S}^t_{WR}(P)$ solves Σ.*

Proof: Let Σ be any correct-restricted specification, and $\mathcal{S} = \mathcal{S}^t_{WR}(P)$ and $\mathcal{S}' = \mathcal{S}^t_R(P)$. Suppose $\mathcal{S}$ solves Σ. Note that $\overline{\mathcal{S}'} \subseteq \overline{\mathcal{S}}$, and, by Lemma 2, $\mathit{init}(\overline{\mathcal{S}'}) = \mathit{init}(\overline{\mathcal{S}})$. Thus, $\overline{\mathcal{S}'}$ is a non-trivial reduction of $\overline{\mathcal{S}}$. Since $\mathcal{S}$ solves Σ, $\mathcal{S}'$ also solves Σ.

Conversely, suppose $\mathcal{S}'$ solves Σ. We must show that $\mathcal{S}$ solves Σ. We know that $\overline{\mathcal{S}} \supseteq \overline{\mathcal{S}'}$. By Lemma 2, $\forall \overline{H} \in \overline{\mathcal{S}}, \exists \overline{H'} \in \overline{\mathcal{S}'} : \overline{H'} \overset{c}{\sim} \overline{H}$. Thus, $\overline{\mathcal{S}}$ is a correct-restricted extension of $\overline{\mathcal{S}'}$. Since $\mathcal{S}'$ solves Σ and Σ is correct-restricted, $\mathcal{S}$ also solves Σ. □

4.3 Simulating Weakly Reliable Links with Fair Lossy Links

Fair lossy links can be used to simulate weakly reliable links. To show this, we describe two procedures, *wr_send*(m, p_j) and *wr_recv*(m), that satisfy the properties of weakly reliable links when executed in any system with fair lossy links. We only give an informal description of this simulation and its proof here (a more formal treatment is postponed to later sections). Since our focus is on solvability (rather than efficiency), we describe the simplest *wr_send*(m, p_j) and *wr_recv*(m) simulation procedures that are sufficient to carry our result. These primitives are inefficient, indeed they assume infinite storage, and infinite message sizes.

Roughly speaking, **L6** stipulates that if a process p_i sends a message m to a process p_j (that executes receive actions infinitely often) and this sending is in the "causal past" of some correct process p_l, then p_j eventually receives m. We can achieve this property with fair lossy links as follows: process p_l maintains a list of all the messages that were sent in its causal past, and this list eventually includes the message m above. In addition, p_l sends this list to every process infinitely often, and in particular to process $p_j = dest(m)$. Since p_l is correct, property **L5** (Fair Loss) of the links from p_l to p_j ensures that p_j eventually receives (a list that contains) m from p_l.

The *wr_send*(m, p_j) and *wr_recv*(m) procedures (for process p_i) given in Figure 1 are based on the simple idea described above. Every process p_i maintains a queue *Prev_Sends*$_i$ that contains all the messages that were *wr_sent* in its "causal past". In order to *wr_send* a message m to p_j, process p_i simply appends m to its queue *Prev_Sends*$_i$. Process p_i also concurrently executes a separate task (called Send Diffusion) that sends *Prev_Sends*$_i$ infinitely often to all other processes.

To *wr_recv* a message, p_i first executes a receive. If it receives some queue of messages *Prev_Sends* from some other process, p_i appends *Prev_Sends* to *Prev_Sends*$_i$. The *wr_recv* procedure now returns the first message in *Prev_Sends*$_i$ with destination p_i that p_i has not yet *wr_recv*d (it returns the null message $\perp$ otherwise).

We now sketch an informal proof that the *wr_send* and *wr_recv* procedures satisfy the properties of weakly reliable links, namely **L1**, **L2**, and **L6**.

Lemma 4. *For every process p_i, the queue* Prev_Sends$_i$ *is non-decreasing.*

Lemma 5. *For every process p_i, a message m is in* Prev_Sends$_i$ *only if* sender(m) wr_send*s m.*

Proof: The underlying links are fair lossy and thus do not create messages (property **L1**). The result is now clear from the way *Prev_Sends*$_i$ is maintained, and can be obtained by a tedious induction that is omitted here. □

Lemma 6. *For every process p_i that executes* wr_recv *infinitely often, if m is in* Prev_Sends$_i$ *and* dest$(m) = p_i$*, then p_i* wr_recv*s m.*

Proof: Every time p_i executes *wr_recv*, it *wr_recv*s the *first* message in the queue *Prev_Sends*$_i$ with destination p_i that it has not yet *wr_recv*d (if such a message exists). It is now clear that once m is in the queue *Prev_Sends*$_i$, p_i will eventually *wr_recv* m. □

Simulation code for process p_i:

```
Variables
    Prev_Sends_i : a queue of messages, initially empty

Procedure wr_send(m, p_j)          {simulating a send over a Weakly Reliable link}
    append m to Prev_Sends_i
end Procedure

Procedure wr_recv(m)               {simulating a receive over a Weakly Reliable link}
    receive(Prev_Sends)
    if Prev_Sends ≠⊥ then append Prev_Sends to Prev_Sends_i
    if Prev_Sends_i has a message m' such that
        dest(m') = p_i and p_i has not yet executed wr_recv(m')
    then m := first such message in Prev_Sends_i
    else m :=⊥
end Procedure

Send Diffusion Task                {concurrent task to repeatedly broadcast Prev_Sends_i}
    Do forever
        forall j ≠ i do send(Prev_Sends_i, p_j)
```

Fig. 1. Simulating Weakly Reliable links with Fair Lossy links

Lemma 7. *If p_i* wr_sends *m to p_j before some event e of a correct process p_l, then m is eventually in* Prev_Sends$_l$.

Proof: If $p_i = p_l$, p_i appends m to *Prev_Sends$_i$* during the execution of *wr_send*(m, p_j). If not (i.e., $p_i \neq p_l$), then since p_i *wr_sends* m to p_j before event e of p_l, by the definition of the before relation, there must exist some messages $m_0, m_1, \cdots, m_{k-1}$ and processes $p_i = p_{i_0}, p_{i_1}, \cdots, p_{i_{k-1}}, p_{i_k} = p_l$ such that:

0. p_{i_0} *wr_sends* m_0 to p_{i_1}, and
1. either $m = m_0$, or p_{i_0} *wr_sends* m to p_j before p_{i_0} *wr_sends* m_0 to p_{i_1}, and
2. p_{i_j} *wr_sends* m_j to $p_{i_{j+1}}$ before $p_{i_{j+1}}$ *wr_recvs* m_j, for $0 \leq j \leq k-1$, and
3. p_{i_j} *wr_recvs* m_{j-1} before p_{i_j} *wr_sends* m_j to $p_{i_{j+1}}$, for $1 \leq j \leq k-1$, and
4. either p_{i_k} *wr_recvs* m_{k-1} at the same time as event e of p_{i_k} occurs (i.e., the two events are the same), or it *wr_recvs* m_{k-1} before e.

Let p and q be any two processes, and m any message such that p *wr_sends* m to q, and q *wr_recvs* m. From Figure 1, it is easy to see that the queue *Prev_Sends$_p$* of p immediately after p *wr_sends* m (note that this *Prev_Sends$_p$* already contains m) is contained in the queue *Prev_Sends$_q$* of q immediately after q *wr_recvs* m. From this observation, Lemma 4, and (1)-(4) above, we conclude that:

1. *Prev_Sends$_{i_0}$* immediately after p_{i_0} *wr_sends* m_0 to p_{i_1} contains m, and

2. *Prev_Sends*$_{i_j}$ immediately after p_{i_j} *wr_sends* m_j to $p_{i_{j+1}}$ is contained in *Prev_Sends*$_{i_{j+1}}$ immediately after $p_{i_{j+1}}$ *wr_recvs* m_j, for $0 \leq j \leq k-1$, and
3. *Prev_Sends*$_{i_j}$ immediately after p_{i_j} *wr_recvs* m_{j-1} is contained in *Prev_Sends*$_{i_j}$ immediately after p_{i_j} *wr_sends* m_j to $p_{i_{j+1}}$, for $1 \leq j \leq k-1$, and
4. *Prev_Sends*$_{i_k}$ immediately after p_{i_k} *wr_recvs* m_{k-1} is is contained in *Prev_Sends*$_{i_k}$ immediately after the event e.

Since m is in *Prev_Sends*$_i$ immediately after p_i *wr_sends* m, by chaining the above facts we conclude that m is contained in the queue *Prev_Sends*$_{i_k}$ of process $p_{i_k} = p_l$ immediately after the event e. □

We can now show that the simulation procedures *wr_send* and *wr_recv* satisfy the three properties of weakly reliable links:

L1: Suppose p_i *wr_recvs* m from p_j. From the code of the *wr_recv* procedure it is clear that m is in *Prev_Sends*$_i$ and *dest*$(m) = p_i$. From Lemma 5, we conclude that p_j *wr_sends* m to p_i.

L2: Obvious from the *wr_recv* procedure.

L6: Suppose p_i *wr_sends* m to p_j before some event e of a correct process p_l, and p_j executes *wr_recv* actions infinitely often. We need to show that p_j *wr_recvs* m. Since p_j executes *wr_recv* actions infinitely often, from Figure 1 it is clear that p_j must execute receive actions infinitely often. By Lemma 7, m is eventually in *Prev_Sends*$_l$. By Lemma 4, m remains in *Prev_Sends*$_l$ forever. Since p_l is correct it sends its (current) queue *Prev_Sends*$_l$ infinitely often to p_j (in the Send Diffusion Task), and these are the only messages that it sends to p_j. Only a finite number of these queues do not contain m. Since the link from p_l to p_j satisfies property **L5**, p_j receives an infinite number of the queues that p_l sends to p_j. Thus, p_j eventually receives a queue that contains m, and from this receipt onwards, m is in *Prev_Sends*$_j$. By Lemma 6, p_j *wr_recvs* m. □

This completes our informal proof that the *wr_send* and *wr_recv* procedures in Figure 1 simulate weakly reliable links using fair lossy links. By Theorem 3, if a correct-restricted problem is solvable with reliable links, then it is also solvable with weakly reliable links. Combining these two results, we conclude that for correct-restricted problems, fair lossy links are "as good as" reliable links in terms of problem solvability. A more precise statement of this claim is postponed to Section 6.6.

5 Simulating Reliable Links with Fair Lossy Links when $n > 2t$

Fair lossy links can be used to simulate reliable links, provided $n > 2t$ (i.e., a majority of processes is correct). To show this, we describe two procedures, *rel_send*(m, p_j) and *rel_recv*(m), that simulate the properties of reliable links when the underlying links are fair lossy and $n > 2t$. The simulation procedures that we give are simple but quite inefficient (they require infinite storage and infinite message sizes).

Every process p_i maintains a queue of messages *Prev_Sends*$_i$ that it uses to store all messages that were *rel_sent* in its "causal past". In addition, every process p_i executes a Diffusion Task after every internal action as well as when it returns from a *rel_send* or a *rel_recv* procedure. This task broadcasts *Prev_Sends*$_i$ and receives the *Prev_Sends* queues from all other processes (if any have been broadcast). If any *Prev_Sends* queue

Simulation code for process p_i:

```
Variables
    Prev_Sends_i : a queue of messages, initially empty
    Proc_Ack_i : a set of processes, initially empty

Procedure rel_send(m, p_j)                {simulating a send over a reliable link}
    append m to Prev_Sends_i
    Proc_Ack_i := {p_i}
    while |Proc_Ack_i| < t + 1 do     {till m is echoed by at least t + 1 distinct processes}
        forall j ≠ i do send(Prev_Sends_i, p_j)
        receive(Prev_Sends)
        if Prev_Sends ≠⊥ then
            append Prev_Sends to Prev_Sends_i
            let p_k := sender(Prev_Sends)
            if m in Prev_Sends then Proc_Ack_i := Proc_Ack_i ∪ {p_k}
end Procedure

Procedure rel_recv(m)                   {simulating a receive over a reliable link}
    if Prev_Sends_i has a message m' such that
        dest(m') = p_i and p_i has not yet executed rel_recv(m')
    then m := first such message in Prev_Sends_i
    else m :=⊥
end Procedure

Diffusion Task        {after every internal action and every rel_send and rel_recv procedure}
    forall j ≠ i do send(Prev_Sends_i, p_j)
    receive(Prev_Sends)
    if Prev_Sends ≠⊥ then append Prev_Sends to Prev_Sends_i
```

Fig. 2. Simulating Reliable links with Fair Lossy links when $n > 2t$

is received, it is appended to the *Prev_Sends*$_i$ queue. Note that if a process is correct, this task is executed infinitely often.

Process p_i *rel_sends* a message m to p_j by first appending m to the queue *Prev_Sends*$_i$. It then repeatedly broadcasts *Prev_Sends*$_i$ (that contains m) till at least $t + 1$ processes echo m by sending their own *Prev_Sends* queues (containing m) to p_i. This means that at least one correct process, say p_l, has m in its *Prev_Sends*$_l$ queue and therefore will send its *Prev_Sends*$_l$ queue containing m to p_j infinitely often. Process p_i now returns from the *rel_send*. Note that even if p_i crashes after it returns from the *rel_send*(m, p_j) procedure, property **L5** ensures that p_j will eventually receive a queue containing m from p_l (and therefore *rel_recv* m) if it executes receive actions infinitely often.

In order to *rel_recv* a message, p_i executes the *rel_recv* procedure which returns the first message in *Prev_Sends*$_i$ with destination p_i that p_i has not yet *rel_recv*d (it returns the null message $\perp$ otherwise).

Lemma 8. *For every process p_i, the queue* Prev_Sends$_i$ *is non-decreasing.*

Lemma 9. *For every process p_i, a message m is in* Prev_Sends$_i$ *only if* sender(m) rel_send*s m.*

Lemma 10. *For every process p_i that executes* rel_recv *infinitely often, if m is in* Prev_Sends$_i$ *and* dest(m) $= p_i$*, then p_i* rel_recv*s m.*

Theorem 11. *The simulation procedures* rel_send *and* rel_recv *satisfy the three properties* **L1**, **L2**, *and* **L3** *of reliable links.*

Proof: L1 and **L2**: Similar to the proof for **L1** and **L2** in Section 4.3.
L3: Suppose p_i invokes the *rel_send* procedure to send a message m to p_j, and p_j executes *rel_recv* actions infinitely often. If p_j *rel_recv*s message m, **L3** is satisfied. Now suppose that p_j does not *rel_recv* m. We will show that p_i crashes while executing the *rel_send*(m, p_j) procedure, i.e., before returning from that procedure. In this case, we pretend that p_i crashes just before invoking *rel_recv*: we can do so because the invocation of *rel_send*(m, p_j) has no "visible" effect, as p_j never receives m. This simulates a reliable link where p_i crashes before sending m — and **L3** is also satisfied.

Assume, for contradiction, that p_i does not crash while executing the *rel_send*(m, p_j) procedure. Then either p_i returns from the *rel_send* procedure, or blocks forever in the *rel_send* procedure.

If p_i returns from the *rel_send* procedure, $|Proc\_Ack_i| \geq t + 1$. This implies that at least $t + 1$ distinct processes are in the set *Proc_Ack*$_i$. From Figure 2, a process p_k is put in the set *Proc_Ack*$_i$ only if $k = i$ or m is in some queue *Prev_Sends* that is received by p_i from p_k. In either case, by property **L1** of fair lossy links, if $p_k \in$ *Proc_Ack*$_i$ then m is in *Prev_Sends*$_k$.

Since *Proc_Ack*$_i$ contains at least $t + 1$ distinct processes and $n > 2t$, at least one of these processes, say p_l, must be correct. Thus m is in *Prev_Sends*$_l$ for some correct process p_l. By Lemma 8, m remains in *Prev_Sends*$_l$ forever. Note that p_l sends *Prev_Sends*$_l$ to p_j infinitely often. Only a finite number of these queues do not contain m. Since the link from p_l to p_j satisfies property **L5**, p_j receives an infinite number of these queues. Thus, p_j eventually receives a queue that contains m. From this receipt onwards, m is in *Prev_Sends*$_j$. Since $dest(m) = p_j$, by Lemma 10, p_j *rel_recv*s m — a contradiction.

Now assume that p_i blocks forever in the *rel_send*(m, p_j) procedure. In this case, p_i sends *Prev_Sends*$_i$ (which contains m) infinitely often. Since $n > 2t$, there are at least $t + 1$ correct processes. Each such correct process would have received *Prev_Sends*$_i$ (containing m) either during an execution of the Diffusion Task or during an execution of the *rel_send* procedure. It would then have echoed m either during an execution of the Diffusion Task or during an execution of the *rel_send* procedure. Thus, p_i would have received at least $t + 1$ distinct echoes of m and could not have blocked forever in the *rel_send*(m, p_j) procedure. This implies that p_i must have crashed during the *rel_send*(m, p_j) procedure, as we needed to show. □

6 Simulation and Translation: Model and Results

In the previous sections we have informally proved that: (1) in general, eventually reliable links cannot simulate reliable links, (2) when $n > t$, fair lossy links can simulate weakly reliable links, and (3) when $n > 2t$, fair lossy links can simulate reliable links. To state these results more precisely, we first refine our model and define the notion of translation [NT90].

6.1 Augmentation

The state of a process that simulates another one has two components: the *simulated variables* (all the variables of the simulated process) and the *simulation variables* (some book keeping variables that are used to carry out the simulation). These two sets of variables are disjoint. So if p' simulates p, the variables of p' include those of p. To formalize this, we introduce the following definitions.

We say that state s' *augments state* s, $s' \geq_a s$, if the set of variables of s' includes all the variables of s, and the variables of s have the same value in s and s'. Formally, $s' \geq_a s$ if $var(s') \supseteq var(s)$ and for all $v \in var(s)$, $s(v) = s(v')$.

Let s_1 and s_2 be two states over disjoint sets of variables V_1 and V_2. Then $s = (s_1, s_2)$ denotes the state over $V_1 \cup V_2$, such that $\forall v \in V_1, s(v) = s_1(v)$ and $\forall v \in V_2, s(v) = s_2(v)$. Note that if $s' \geq_a s$ then $s' = (s, r)$ for some state r over $var(s') \setminus var(s)$.

We extend the notion of augmentation to sequences of states, and then to vectors of sequences of states (i.e., to traces), in the natural way. For any two sequences of states σ and σ' in $Seq(S)$, we say that σ' *augments* σ, and write $\sigma' \geq_a \sigma$, if they have the same length, and every element s' in σ' augments the corresponding element s in σ. Similarly, for any two traces $\overline{H'}$ and $\overline{H}$ in $Vec(S)$, we say that $\overline{H'}$ augments $\overline{H}$, and write $\overline{H'} \geq_a \overline{H}$, if they have the same dimension, say k, and for all $1 \leq i \leq k$, $\overline{H'}[i] \geq_a \overline{H}[i]$.

6.2 Stuttering

When a process simulates another one, it may execute several actions to simulate a single action of the simulated process. Thus, a simulation "stretches" the trace of the simulated process: a segment $s_1\ s_2$ of a trace can be stretched into some "stuttering" version $s_1\ \cdots\ s_1\ s_2\ \cdots\ s_2$ [Lam83]. For any two sequences of states σ and σ', we say that σ' *is a stuttering of* σ, and write $\sigma' \geq_s \sigma$, if (a) either both σ' and σ are infinite or they are both finite, and (b) σ' can be obtained from σ by repeated applications of the following operation: for any state s in σ, replace s by any non-empty finite sequence of the form $s\ \cdots\ s$.

6.3 Specifications Closed under Stuttering and Augmentation

As we saw, simulation leads to both stuttering *and* augmentation: the trace of the simulating process is a stuttered and augmented version of the trace of the simulated process. For any two sequences of states σ and σ' in $Seq(S)$, we write $\sigma' \geq_{sa} \sigma$ if there is a sequence $\sigma_0 \in Seq(S)$ such that $\sigma' \geq_a \sigma_0$ and $\sigma_0 \geq_s \sigma$. Similarly, for any two traces $\overline{H'}$ and $\overline{H}$ in $Vec(S)$, we write $\overline{H'} \geq_{sa} \overline{H}$ if they have the same dimension, say k, and for all $1 \leq i \leq k$, $\overline{H'}[i] \geq_{sa} \overline{H}[i]$. Finally, for all $\overline{\mathcal{S}}, \overline{\mathcal{S}'} \in \Sigma^*$, we say that $\overline{\mathcal{S}'}$ *is a stuttered and augmented version of* $\overline{\mathcal{S}}$, and write $\overline{\mathcal{S}'} \geq_{sa} \overline{\mathcal{S}}$, if there is a mapping τ from $\overline{\mathcal{S}'}$ onto $\overline{\mathcal{S}}$ such that $\forall \overline{H'} \in \overline{\mathcal{S}'}, \overline{H'} \geq_{sa} \tau(\overline{H'})$. Note that $\geq_{sa}$ is a transitive relation.

We focus on problem specifications that are insensitive to stuttering (i.e., state repetitions) and augmentation (i.e., state extensions). Formally, a specification Σ *is closed under stuttering and augmentation* if:

$$\forall \overline{\mathcal{S}}, \overline{\mathcal{S}'} \in \Sigma^* \ : \ \overline{\mathcal{S}'} \geq_{sa} \overline{\mathcal{S}} \text{ and } \overline{\mathcal{S}} \in \Sigma \text{ implies } \overline{\mathcal{S}'} \in \Sigma.$$

Many problems, including Consensus and URB, have specifications in this class.

6.4 Simulation and Translation

Let $\mathcal{S}$ and $\mathcal{S}'$ be any two systems. Intuitively, $\mathcal{S}'$ simulates $\mathcal{S}$ if the traces in $\overline{\mathcal{S}'}$ are stuttered and augmented versions of a subset of the traces in $\overline{\mathcal{S}}$ that has the same initial states as $\overline{\mathcal{S}}$. In other words, $\mathcal{S}'$ simulates $\mathcal{S}$ if $\overline{\mathcal{S}'}$ is a stuttered and augmented version of a non-trivial reduction $\overline{\mathcal{S}_0}$ of $\overline{\mathcal{S}}$. Formally, $\mathcal{S}'$ *simulates* $\mathcal{S}$ iff:

$$\exists \mathcal{S}_0 \subseteq \mathcal{S} : \mathit{init}(\overline{\mathcal{S}_0}) = \mathit{init}(\overline{\mathcal{S}}) \text{ and } \overline{\mathcal{S}'} \geq_{sa} \overline{\mathcal{S}_0}.$$

Note that $\overline{H'} \geq_{sa} \overline{H}$ implies $\mathit{correct}(H') = \mathit{correct}(H)$. Thus, a simulation does not crash any process or mask any process failures.

Observation 12. *Let Σ be any specification closed under stuttering and augmentation. If $\mathcal{S}'$ simulates $\mathcal{S}$ and $\mathcal{S}$ solves Σ, then $\mathcal{S}'$ also solves Σ.*

Proof: Since $\mathcal{S}'$ simulates $\mathcal{S}$, $\overline{\mathcal{S}'} \geq_{sa} \overline{\mathcal{S}_0}$ for some non-trivial reduction $\overline{\mathcal{S}_0}$ of $\overline{\mathcal{S}}$. Thus, $\overline{\mathcal{S}} \in \Sigma$ implies that $\overline{\mathcal{S}_0} \in \Sigma$. Since $\overline{\mathcal{S}'} \geq_{sa} \overline{\mathcal{S}_0}$, and Σ is closed under stuttering and augmentation, $\overline{\mathcal{S}'} \in \Sigma$. □

For any set P of processes, we use the notation $\mathcal{S}^t_X(P)$ where $X \in \{R, WR, ER, FL\}$ to denote any one of the systems $\mathcal{S}^t_R(P)$, $\mathcal{S}^t_{WR}(P)$, $\mathcal{S}^t_{ER}(P)$, and $\mathcal{S}^t_{FL}(P)$. A *translation from X links to Y links for systems with n processes and at most t crashes*, denoted $X \overset{n,t}{\rightarrow} Y$, is a translation function $\mathcal{T}$ that maps any set P of n processes into a set $P' = \mathcal{T}(P)$ of n processes such that $\mathcal{S}^t_Y(P')$ simulates $\mathcal{S}^t_X(P)$.

6.5 Impossibility of Translation $R \overset{n,t}{\rightarrow} ER$ for $n \leq 2t$

Consider a system with at least two processes where a majority of processes may crash (i.e., $2 \leq n \leq 2t$). In Section 3, we saw that URB can be solved with reliable links but not with eventually reliable links. This implies that reliable links cannot be simulated with eventually reliable links. More precisely:

Theorem 13. *For $2 \leq n \leq 2t$, there is no translation $R \overset{n,t}{\rightarrow} ER$.*

Proof: For contradiction, suppose there is a translation $R \overset{n,t}{\rightarrow} ER$ for some n and t, $2 \leq n \leq 2t$. Let $\mathcal{T}$ be the translation function of $R \overset{n,t}{\rightarrow} ER$. As noted earlier, URB is closed under stuttering and augmentation. By Theorem 1(1), for any $n > 0$, there is a set P of n processes such that $\mathcal{S}^t_R(P)$ solves URB. Let $P' = \mathcal{T}(P)$. By the definition of translation, $\mathcal{S}^t_{ER}(P')$ simulates $\mathcal{S}^t_R(P)$. By Observation 12, $\mathcal{S}^t_{ER}(P')$ also solves URB — a contradiction to Theorem 1(2). □

6.6 Translation $WR \stackrel{n,t}{\rightarrow} FL$

In Section 4.3, we informally showed that two procedures, *wr_send* and *wr_recv*, can be used to simulate weakly reliable links using fair lossy links. Based on these procedures, we can define a translation $\mathcal{T} = WR \stackrel{n,t}{\rightarrow} FL$ that maps any set of n processes P into a set of n processes $P' = \mathcal{T}(P)$ such that $\mathcal{S}^t_{FL}(P')$ simulates $\mathcal{S}^t_{WR}(P)$. Roughly speaking, P' is obtained from P by replacing the send and receive actions of processes in P, with the actions of the *wr_send* and *wr_recv* procedures, respectively. A precise description of the mapping from P to P' that defines the translation $\mathcal{T}$, together with a proof of correctness, is given in [BCBT96]. We can now state our main result:

Theorem 14. *Let Σ be any specification that is correct-restricted, and closed under stuttering and augmentation. Let $\mathcal{T}$ be the translation referred to above. For any set of processes P, if $\mathcal{S}^t_R(P)$ solves Σ then $\mathcal{S}^t_{FL}(P')$ solves Σ where $P' = \mathcal{T}(P)$.*

Proof: Suppose $\mathcal{S}^t_R(P)$ solves Σ. Since Σ is correct-restricted, Theorem 3 implies that $\mathcal{S}^t_{WR}(P)$ also solves Σ. Let $P' = \mathcal{T}(P)$. By the definition of $WR \stackrel{n,t}{\rightarrow} FL$, $\mathcal{S}^t_{FL}(P')$ simulates $\mathcal{S}^t_{WR}(P)$. Since $\mathcal{S}^t_{WR}(P)$ solves Σ, and Σ is closed under stuttering and augmentation, Observation 12 implies that $\mathcal{S}^t_{FL}(P')$ solves Σ. □

6.7 Translation $R \stackrel{n,t}{\rightarrow} FL$ for $n > 2t$

In Section 5, we informally proved that procedures *rel_send* and *rel_recv* can be used to simulate reliable links using fair lossy links when a majority of processes are correct. These procedures are the basis of a formal translation $\mathcal{T} = R \stackrel{n,t}{\rightarrow} FL$ for any $n > 2t$. Roughly speaking, P' is obtained from P by replacing the send and receive actions of P by the actions of the *rel_send* and *rel_recv* procedures, respectively. The formal description of the mapping from P to P' and its correctness proof is similar to that for the translation $WR \stackrel{n,t}{\rightarrow} FL$ and is also given in [BCBT96].

Theorem 15. *Let Σ be any specification closed under stuttering and augmentation. Let $\mathcal{T}$ be the translation referred to above. For any set of $n > 2t$ processes P, if $\mathcal{S}^t_R(P)$ solves Σ then $\mathcal{S}^t_{FL}(P')$ solves Σ where $P' = \mathcal{T}(P)$.*

Acknowledgments

We are grateful to David Cooper and Vassos Hadzilacos for valuable discussions that helped us significantly improve the presentation of the results.

References

[AAF+94] Y. Afek, H. Attiya, A. D. Fekete, M. Fischer, N. Lynch, Y. Mansour, D. Wang, and L. Zuck. Reliable communication over unreliable channels. *Journal of the ACM*, 41(6):1267–1297, 1994.

[ABND+90] H. Attiya, A. Bar-Noy, D. Dolev, D. Peleg, and R. Reischuk. Renaming in an asynchronous environment. *Journal of the ACM*, 37(3):524–548, 1990.

[AE86] B. Awerbuch and S. Even. Reliable broadcast protocols in unreliable networks. *Networks: An International Journal*, 16, 1986.

[AGH90] B. Awerbuch, O. Goldreich, and A. Herzberg. A quantitative approach to dynamic networks. In *Proceedings of the 9th ACM Symposium on Principles of Distributed Computing*, pages 189–204, Québec City, Québec, Canada, 1990.

[BCBT96] A. Basu, B. Charron-Bost, and S. Toueg. Simulating reliable links with unreliable links in the presence of process crashes. Technical report, Cornell University, Dept. of Computer Science, Cornell University, Ithaca, NY 14853, 1996.

[Bir85] K. Birman. Replication and fault-tolerance in the ISIS system. In *Proceedings of the 10th ACM Symposium on Operating Systems Principles*, pages 79–86, Orcas Island, WA USA, 1985.

[BN92] R. Bazzi and G. Neiger. Simulating crash failures with many faulty processors. In A. Segal and S. Zaks, editors, *Proceedings of the Sixth International Workshop on Distributed Algorithms*, volume 647 of *Lecture Notes on Computer Science*, pages 166–184. Springer-Verlag, 1992.

[BSW69] K. A. Bartlett, R. A. Scantlebury, and P. T. Wilkinson. A note on reliable full-duplex transmission over half-duplex links. *Comm. of the ACM*, 12(5):260–261, 1969.

[Cha90] S. Chaudhuri. Agreement is harder than consensus: Set consensus problems in totally asynchronous systems. In *Proceedings of the 9th ACM Symposium on Principles of Distributed Computing*, pages 311–324, Québec City, Québec, Canada, August 1990.

[DLP+86] D. Dolev, N. A. Lynch, S. S. Pinter, E. W. Stark, and W. E. Weihl. Reaching approximate agreement in the presence of faults. *Journal of the ACM*, 33(3):499–516, July 1986.

[FLMS93] A. D. Fekete, N. Lynch, Y. Mansour, and J. Spinelli. The impossibility of implementing reliable communication in the face of crashes. *Journal of the ACM*, 40(5):1087–1107, 1993.

[GA88] E. Gafni and Y. Afek. End-to-end communication in unreliable networks. In *Proceedings of the Seventh ACM Symposium on Principles of Distributed Computing*, pages 131–148, Toronto, Ontario, Canada, August 1988.

[Gop92] A. Gopal. *Fault-Tolerant Broadcasts and Multicasts: The Problem of Inconsistency and Contamination*. PhD thesis, Cornell University, January 1992.

[HT94] V. Hadzilacos and S. Toueg. A modular approach to fault-tolerant broadcasts and related problems. Technical Report TR 94-1425, Cornell University, Dept. of Computer Science, Cornell University, Ithaca, NY 14853, May 1994.

[Lam78] L. Lamport. Time, clocks, and the ordering of events in a distributed system. *Communications of the ACM*, 21(7):558–565, July 1978.

[Lam83] L. Lamport. What good is temporal logic? In R. E. A. Mason, editor, *Information Processing 83: proceedings of the IFIP Ninth World Congress*, pages 657–668. IFIP, North-Holland, September 1983.

[NT90] G. Neiger and S. Toueg. Automatically increasing the fault-tolerance of distributed algorithms. *Journal of Algorithms*, 11(3):374–419, 1990.

[WZ89] D. Wang and L. Zuck. Tight bounds for the sequence transmission problem. In *Proceedings of the 8th ACM Symposium on Principles of Distributed Computing*, pages 73–83, August 1989.

A Cyclic Distributed Garbage Collector for Network Objects

Helena Rodrigues* and Richard Jones

Computing Laboratory, University of Kent, Canterbury, Kent CT2 7NF, UK
Tel: +44 1227 764000 x7754, Fax +44 1227 762811
email: {hccdr,R.E.Jones}@ukc.ac.uk

Abstract. This paper presents an algorithm for distributed garbage collection and outlines its implementation within the Network Objects system. The algorithm is based on a *reference listing* scheme, which is augmented by *partial tracing* in order to collect distributed garbage cycles. Processes may be dynamically organised into groups, according to appropriate heuristics, to reclaim distributed garbage cycles. The algorithm places no overhead on local collectors and suspends local mutators only briefly. Partial tracing of the distributed graph involves only objects thought to be part of a garbage cycle: no collaboration with other processes is required. The algorithm offers considerable flexibility, allowing expediency and fault-tolerance to be traded against completeness.

Keywords: distributed systems, garbage collection, algorithms, termination detection, fault tolerance

1 Introduction

With the continued growth of interest in distributed systems, designers of languages for distributed systems are turning their attention to garbage collection [24, 21, 16, 14, 15, 3, 18, 19, 17, 9, 22], motivated by the complexity of memory management and the desire for transparent object management. The goals of an ideal distributed garbage collector are

safety: only garbage should be reclaimed;
completeness: all objects that are garbage at the start of a garbage collection cycle should be reclaimed by its end. In particular, it should be possible to reclaim distributed cycles of garbage;
concurrency: distributed garbage collection should not require the suspension of mutator or local collector processes; distinct distributed garbage collection processes should be able to run concurrently;
efficiency: garbage should be reclaimed promptly;
expediency: wherever possible, garbage should be reclaimed despite the unavailability of parts of the system;

* Work supported by JNICT grant (CIENCIA/BD/2773/93-IA) through the *PRAXIS XXI* Program (Portugal).

scalability: distributed garbage collection algorithms should scale to networks of many processes;

fault tolerance: the memory management system should be robust against message delay, loss or replication, or process failure.

Inevitably compromises must be made between these goals. For example, scalability, fault-tolerance and efficiency may only be achievable at the expense of completeness, and concurrency introduces synchronisation overheads. Unfortunately, many solutions in the literature have never been implemented so there is a lack of empirical data for the performance of distributed garbage collection algorithms to guide the choice of compromises.

Distributed garbage collection algorithms generally follow one of two strategies: tracing or reference counting. Tracing algorithms visit all 'live' objects [12, 7]; global tracing requires the cooperation of all processes before it can collect any garbage. This technique does not scale, is not efficient and requires global synchronisation. In contrast, distributed reference counting algorithms have the advantages for large-scale systems of fine interleaving with mutators, and locality of reference (and hence low communication costs). Although standard reference counting algorithms are vulnerable to out-of-order delivery of reference count manipulation messages, leading to premature reclamation of live objects, many distributed schemes have been proposed to handle or avoid such race conditions [2, 11, 20, 23, 3, 18].

On the other hand, distributed reference counting algorithms cannot collect cycles of garbage. Collecting interprocess garbage cycles is an important issue of our work: we claim that it is fairly common for objects in distributed systems to have cyclic connections. For example, in client-server systems, objects that communicate with each other remotely are likely to hold references to each other, and often this communication is bidirectional [27]. Many distributed systems are typically long running (e.g. distributed databases), so floating garbage is particularly undesirable as even small amounts of uncollected garbage may accumulate over time to cause significant memory loss [19]. Although interprocess cycles of garbage could be broken by explicitly deleting references, this would lead to exactly the scenario that garbage collection is supposed to replace: error-prone manual memory management.

Systems that use distributed reference counting as their primary distributed memory management policy must reclaim cycles by using a complementary tracing scheme [14, 16, 13, 17], or by migrating objects until an entire garbage cyclic structure is eventually held within a single process where it can be collected by the local collector [24, 19]. However, migration is communication-expensive and existing complementary tracing solutions either require global synchronisation and the cooperation of all processes in the system [14], place additional overhead on the local collector and application [17], rely on cooperation from the local collector to propagate necessary information [16], or are not fault-tolerant [16, 17].

This paper presents an algorithm and outlines its implementation for the Network Objects system [4]. Our algorithm is based on a modification of distributed reference counting — *reference listing* [3] — augmented by *partial tracing* in order to collect distributed garbage cycles [13]. We use heuristics to form groups of processes dynamically that cooperate to perform partial traces of subgraphs suspected of being garbage.

Our distributed algorithm is designed not to compromise our primary goals of efficient reclamation of local and distributed acyclic garbage, low synchronisation overheads, avoidance of global synchronisation, and fault-tolerance. To these ends, we trade some degree of completeness and efficiency in collecting distributed cycles. However, eventually distributed garbage cycles will be reclaimed. In brief, our aim is to match rates of collection against rates of allocation of data structures. Objects that are only reachable from local processes have very high allocation rates, and therefore must be collected most rapidly. The rate of creation of references to remote objects that are not part of distributed cycles is much lower, and the rate of creation of distributed garbage cycles is lower still and hence should have the lowest priority for reclamation.

The paper is organised as follows. Section 2 briefly describes the overall design of the Network Objects system and introduces terminology. Section 3 describes how partial tracing works in the absence of failures. Section 4 describes termination and how the collectors are synchronised. Section 5 describes how process failures are handled. We discuss related work in Section 6, and conclude and present some points for future work in Section 7.

2 Terminology and Network Objects Overview

Our algorithm is designed for use with the Network Objects system, a distributed object-based programming system for Modula-3 [4]. A distributed system is considered to consist of a collection of *processes*, organised into a network, that communicate by exchange of *messages*. Each process can be identified unambiguously, and we identify processes by upper-case letters, e.g. A, B, ..., and objects by lower-case letters suffixed by the identifier of the process to which they belong, e.g. xA, xB, ...

From the garbage collector's point of view, *mutator* processes perform computations independently of other mutators in the system (although they may periodically exchange messages) and allocate objects in local heaps. The state of the distributed computation is represented by a *distributed graph* of objects. Objects may contain references to objects in the same or another process. Each process also contains a set of *local roots* that are always accessible to the local mutator. Objects that are reachable by following from a root a path of references held in other objects are said to be *live*. Other objects are said to be *unreachable* or *dead*, and they constitute the *garbage* to be reclaimed by a *collector*. A reference to an object held on the same process is said to be *local*; a reference to an object held on a remote process is said to be an *external* or *remote*. A collector that operates solely within a local heap is called a *local collector*.

Only *network objects* may be shared by processes. The process accessing a network object for which it holds a reference is called the *client*, and the process containing the network object is called its *owner*. Clients and owners may run on different processes within the distributed system. Network objects cannot migrate from one process to another.

A client cannot directly access the data fields of a network object for which it holds a reference, but can only invoke the object's methods. A reference in the client process points to a ***surrogate*** object (see the dashed circle in figure 1), whose methods perform remote procedure calls to the owner, where the corresponding method of the *concrete* object is invoked. A process may hold at most one surrogate for a given concrete object, in which case all references in the process to that object point to the surrogate.

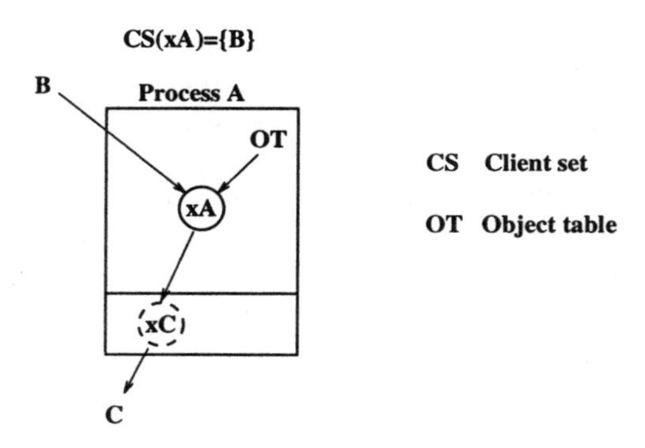

Fig. 1. Surrogates, object table and client sets

Each process maintains an *object table* (see figure 1). The object table of the owner of a concrete object xA contains a pointer to xA as long as any other process holds a surrogate for it. A process's object table also holds entries for any surrogates held (not represented in the figure).

The heap of a Modula-3 process is managed by garbage collection. Local collections are based on tracing from local roots — the stack, registers, global variables and also the object table. We shall refer to those public network objects that are referenced from other processes through the table as *OT roots*. The object table is considered a root by the local collector in order to preserve objects reachable only from other processes.

Object table entries are managed by the distributed memory manager. The Network Objects system uses a variation of reference counting called *reference listing* to detect distributed acyclic garbage [3]. Rather than maintain a count in each concrete object of the number surrogates for it, each object maintains a *client set*[2] of the names of all those processes that hold a surrogate for it.

For the purpose of the algorithm there are two operations on references that are important in the system: transmission of a reference to another process and deletion of a remote reference.

[2] In Network Objects terminology this set is called the *dirty set*.

References to a network object may be *marshalled* from one process to another either as arguments or results of methods. A network object is marshalled by transmitting its *wireRep* — a unique identifier for the owner process plus the index of the object at the owner. If the process receiving the reference is not the owner of the object, then the process must create a local surrogate. In order for a process to marshall a wireRep to another process, the sender process needs either to be the owner of the object or to have a surrogate for that object. This operation must preserve a key invariant: whenever there is a surrogate for object xA at client B, then $B \in xA.clientSet$.

Suppose process A marshalls a network object xA to process B, as an argument or result of a remote method invocation. A may be the owner of xA, or it may be a client that has a surrogate for xA. In either case:

1. A sends to B the wireRep of xA and waits for an acknowledgement.
2. Before B creates a surrogate for xA, it sends a *dirty call* to xA's owner.
3. Assuming no communication failure, the owner receives the call, adds B to xA's client set and then sends an acknowledgement back to B.
4. When the acknowledgement is received, B creates the surrogate and then returns an acknowledgement back to A. If B already has a surrogate for xA, the surrogate creation step is skipped.

Surrogates unreachable from their local root set are reclaimed by local collectors. Whenever a surrogate is reclaimed, the client sends a *clean call* to the owner of the object to inform it that the client should be removed from its client set. When an object's client set becomes empty, the reference to the object is removed from the object table so that the object can be reclaimed subsequently by its owner's local collector.

Race conditions between dirty and clean calls must be avoided. If a clean call were to arrive before a dirty call, removing the last entry from the client set, an object may be reclaimed prematurely. To prevent this scenario arising, an object's client set is kept non-empty while its wireRep is being transmitted. This scheme also handles messages in transit. Each clean or dirty call is also labelled with a sequence number; these increase with each new call sent from the client. This scheme makes the transmission and deletion of references tolerant to delayed, lost or duplicated messages.

Processes that terminate, whether normally or abnormally, cannot be expected to notify the owners of all network objects for which they have surrogates. The Network Objects collector detects termination by having each process periodically ping the clients that have surrogates for its objects. If the ping is not acknowledged in time, the client is assumed to have terminated, and is removed from all client sets at that owner. A more detailed description of these operations, with a proof of their correctness, is described in [3].

3 Three-Phase Partial Tracing

Our algorithm is based on the premise that distributed garbage cycles exist but are less common than acyclic distributed structures. Consequently, distributed cyclic garbage must be reclaimed but its reclamation may be performed more slowly than that of acyclic or local data. It is important that collectors — whether local or distributed — should not unduly disrupt mutator activity. Local data is reclaimed by Modula-3's *Mostly Copying collector* (slightly modified) [1], and distributed acyclic structures are managed by the Network Objects reference listing collector [3]. We augment these mechanisms with an incremental three-phase partial trace to reclaim distributed garbage cycles. Our implementation does not halt local collectors at all, and suspends mutators only briefly. The local collectors reclaim garbage independently and expediently in each process. The partial trace merely identifies garbage cycles without reclaiming them. Consequently, both local and partial tracing collector can operate independently and concurrently.

Our algorithm operates in three phases. The first, *mark-red*, phase identifies a distributed subgraph that may be garbage: subsequent efforts of the partial trace are confined to this subgraph alone. This phase is also used to form a group of processors that will collaborate to collect cycles. The second, *scan*, phase determines whether members of this subgraph are actually garbage, before the final, *sweep*, phase makes those garbage objects found available for reclamation by local collectors. A new partial trace may be initiated by any process not currently part of a trace. There are several reasons for choosing to initiate such an activity: the process may be idle, a local collection may have reclaimed insufficient space, or the process may not have contributed to a distributed collection for a long time.

The distributed collector requires that each object has a *colour* — red or green — and that initially all objects are green. Network objects also have a *red set* of process names, akin to their client set.

3.1 Mark-Red Phase

Partial tracing is initiated at *suspect* objects: surrogates suspected of belonging to a distributed garbage cycle. We observe that any distributed garbage cycle must contain some surrogate. Suspects should be chosen with care both to maximise the amount of garbage reclaimed and to minimise redundant computation or communication. At present, we consider a surrogate to be suspect if it is not referenced locally, other than through the object table. This information is provided by the local collector — any surrogate that has not been marked is suspect.

This heuristic is actually very simplistic and may lead to undesirable wasted and repeated work. For example, it may repeatedly identify a surrogate as suspect even though it is reachable from a remote root. Only measurements of real implementations will show if this is indeed a serious problem. However, our algorithm should be seen as a framework: any better heuristic could be used. In Section 6 we show how more sophisticated heuristics improve the algorithms discrimination and hence its efficiency.

The mark-red phase paints the transitive referential closure of suspect surrogates red. Any network object receiving a mark-red request also inserts the name of the sending process into their red set to indicate that this client is a member of the suspect subgraph (*cf.* client sets)[3].

The second purpose of the mark-red phase is to identify dynamically groups of processes that will collaborate to reclaim distributed cyclic garbage. A group is simply the set of processes visited by mark-red. Group collection is desirable for fault-tolerance, decentralisation, flexibility and efficiency. Fault-tolerance and efficiency are achieved by requiring the cooperation of only those processes forming the group: progress can be made even if other processes in the system fail. Groups lead to decentralisation and flexibility as well. Decentralisation is achieved by partitioning the network into groups, with multiple groups simultaneously but independently active for garbage collection. Communication is only necessary between members of the group. Flexibility is achieved by the choice of processes forming each group. This can be done statically by prior negotiation or dynamically by mark-red. In the second case, heuristics based on geography, process identity, distance from the suspect originating the collection, or time constraints can be used.

An interesting feature of this design is that it does not need to visit the complete transitive referential closure of suspect surrogates. The purpose of this phase is simply to determine the scope of subsequent phases and to construct red sets. Early termination of the mark-red phase trades *conservatism* (tolerance of floating garbage) for a number of benefits: expediency, bounds on the size of the graph traced (and hence on the cost of the trace), execution of the mark-red process concurrently with mutators without need for synchronisation, and cheap termination of the phase. Section 4 explains how termination of mark-red is detected.

We allow multiple partial traces, initiated by different processes, to operate concurrently, but for now we do not permit groups (hence partial tracings) to overlap. Although inter-group references are permissable, mark-red is not propagated to processes that are members of other groups. One reason for this restriction is to prevent a mark-red phase from interfering with a scan phase, or vice-versa. A second reason is to allow simpler control over the size of any group: merging groups would add considerable complexity.

The example in figure 2 illustrates a mark-red process. The figure contains a garbage cycle ($yA \rightarrow yB \rightarrow yD \rightarrow xC \rightarrow yA$). Process A has initiated a partial trace; yB is a suspect because it is not reachable from a local root (other than through the object table). The mark-red process paints the suspect's transitive closure red, and constructs the red sets. In the figure, the red set of an object xX is denoted by $RS(xX)$; clear circles represent green objects and shaded ones red objects. Note that objects xD and yC are not garbage although they have been painted red: their liveness will be detected by the scan phase.

---

[3] Notice that cooperation from the acyclic collector and the mutator would be required if, instead, mark-red removed references from client sets or copies of client sets (see [13]). Red sets avoid this need for cooperation as well as allowing the algorithm to identify which processes have sent mark-red requests.

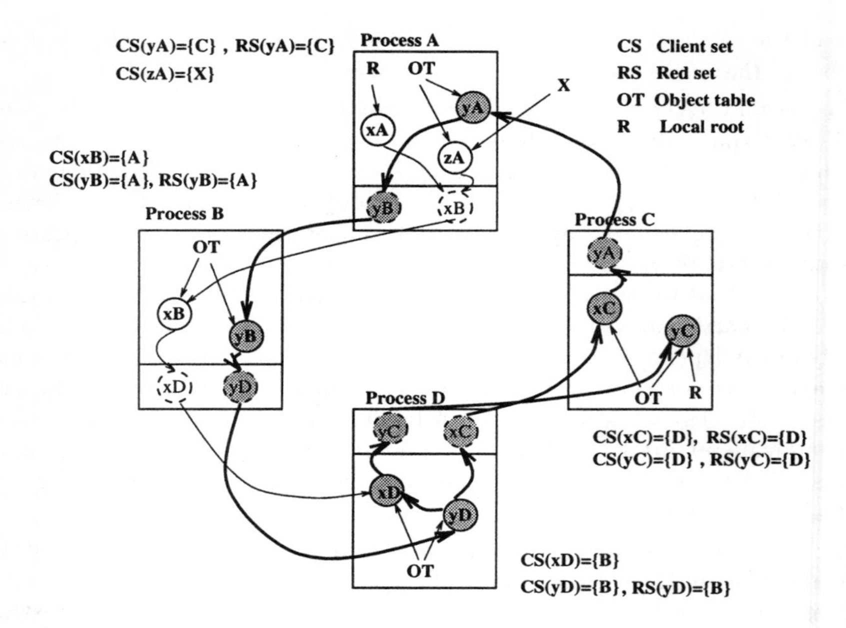

Fig. 2. Mark-Red Phase

3.2 Scan Phase

At the end of the mark-red phase, a group of processes has been formed. Members of this group will cooperate for the scan-phase. The aim of this phase is to determine whether any member of the red subgraph is reachable from outside that subgraph. The phase is executed concurrently on each process in the group. The first step is to compare the client and red sets of each red concrete object. If a red object does not have a red set (e.g. xD in figure 2), or if the difference between its client and its red sets is non-empty, the object must have a client outside the suspect red graph. In this case the object is painted green to indicate that it is live. All red objects reachable from local roots or from green concrete objects are now repainted green by a *local-scan* process. If a red surrogate is repainted green, a *scan-request* is sent to the corresponding concrete object. If this object was red, it is repainted green, along with its descendents. The scan phase terminates when the group contains no green objects holding references to red children in the group.

Continuing our example, each process calculates the difference between client and red sets for each red concrete object it holds. For instance, xD in process D has no red set so xD is painted green and becomes a root for the local-scan. Figure 3 shows the result of the scan phase: the live objects xD and yC have been repainted green.

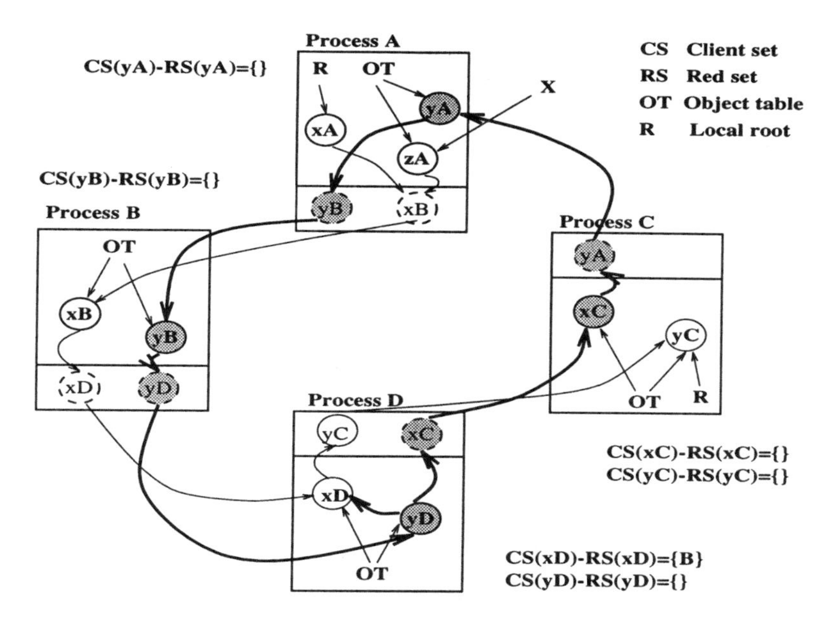

Fig. 3. Scan Phase

3.3 Sweep Phase

At the end of the scan phase, all live objects are green[4]. Any remaining red objects must be part of inaccessible cycles, and can thus be safely reclaimed. The sweep phase is executed in each process independently. Each object table is swept for references to red objects. If any are found, any local references held in the object, including references to surrogates, are deleted, thus breaking the cycle. The red descendents of these objects are now available for reclamation by the local collector. Reclamation of a surrogate causes the Networks Objects reference listing system to send a delete message to the owner of the corresponding concrete object: when its client set becomes empty, that object will also be reclaimed.

4 Synchronisation and Termination

4.1 Mark-Red Phase

By accepting that the red subgraph may include only a subset of the set of garbage objects — i.e. a conservative approximation — we gain a number of benefits including the removal of any need for synchronisation with mutators and cheaper termination.

[4] Note that the converse, *i.e.* that all green objects are live, is not necessarily true.

The solution we have currently adopted is based on that presented by Derbyshire [7]. A mark-red process is launched by the process initiating the collection. Subsequent mark-red requests export further mark-red processes. Terminating processes return an acknowledgement, identifying their process and those visited by any mark-red processes that they have exported. A process terminates when it has finished colouring its local subgraph and it has received acknowledgements for all the mark-red requests it has sent. As soon as the initiating process has received acknowledgements from all the mark-red processes that it has exported, the mark-red phase is complete and the membership of the group is known. Members of the group are then instructed by the initiating process to start the scan phase and informed of their co-members.

4.2 Scan Phase

The scan phase terminates when all members of the group have completed their scan processes and no messages are in transit. In contrast to the mark-red phase, the scan phase must be complete with respect to the red subgraph, since it must ensure that all live red objects are repainted green. As for other concurrent marking schemes (e.g. Dijkstra *et al.* [8]), this requires synchronisation between mutator and collector. Termination detection is also expensive as local-scans must be able to detect any change to the connectivity of the graph made by a mutator.

A local mutator may only change this connectivity by overwriting references to objects. Such writes can be detected by a *write barrier* [26]. We have adapted the Mostly Parallel garbage collection algorithm for the scan phase [5]. This technique uses operating system support to detect those objects modified by mutators (actually pages that have been updated within a given interval). When the local-scan process has visited all objects reachable from its starting points, the mutator is halted while the local-scan retraces the graph from any modified objects, as well as from the roots. Because most of the scanning work has already been done, it is expected that this retrace will terminate promptly (the underlying assumption is that the rate of allocation of network objects, and of objects reachable from those network objects, is low).

On termination of a local-scan, any red concrete object o and its descendents must be isolated from the green subgraph held in that process. Thus a red object cannot become reachable through actions of the local mutator. However its reachability can still be changed if:

1. a remote method is invoked on o;
2. a new surrogate in some other process is created for o;
3. another object in the same process receives a reference to o.

Notice that this scenario could only occur if the red surrogate were still alive. Although such mutator activity could be handled by the Mostly Parallel scan, this would be implementationally expensive. Instead, mutator messages are trapped by a *dirty-barrier*, a 'snapshot-at-the-beginning' barrier [26]. If a client invokes a remote method on a red surrogate, or copies the wireRep held by a red surrogate to another process, before the client's local-scan has terminated (including the receipt of acknowledgements for all the scan-requests that it has made), a scan-request is sent to the corresponding concrete object to arrive before the mutator message[5]. How this scan-request is handled depends on the state of the owner process. If the owner's local-scan has not terminated, the concrete object is repainted green and becomes an additional root for the local-scan; the scan-request is acknowledged immediately. If, on the other hand, the local-scan has terminated, the local mutator is halted and the descendents of this concrete object are repainted green as well by an *atomic-scan* process. We claim that this does not cause excessive delay as it is likely that many of its descendents will already have been repainted green. In this case, the scan-request is not acknowledged until all descendents (in the group) of the red surrogate have been painted green, if necessary by further scan-requests to other processes.

Global scan phase termination is again detected by a distributed termination detection algorithm based on Derbyshire's algorithm. A local-scan process notifies all other members of its group as soon as it has finished colouring green all objects reachable from its local roots and local green concrete objects and has received the acknowledgements for all the scan-requests it has generated (*cf.* mark-red phase). However, this account does not take the mutator actions described above into consideration. Correct termination of Derbyshire's algorithm requires that each scan-request (and subsequent scan) has a local-scan responsible for it — scan-requests from atomic-scan processes1 generated by mutator activity breach this invariant. However, trapping mutator messages with the snapshot-at-the-beginning barrier preserves the invariant. If the owner's local-scan has not terminated, it takes over the responsibility for scanning the descendents of the object. If the local-scan has terminated, the scan-request is not acknowledged until all the descendents have been scanned; the local-scan in the client process cannot terminate until it has received this acknowledgement. Notice that the mutator operation cannot have been made from a red surrogate in a process which has completed its local-scan. If the local-scan had been completed, the red surrogate would have been unreachable from the client's local roots: the action must therefore have been caused by a prior external mutator action. But in this case the surrogate would have been repainted green by an atomic-scan process. Thus the barrier suffices to ensure that any scan-request has a local-scan process ultimately responsible for it.

As before, each process informs other members of the group as soon as it has received acknowledgements from all these scan-requests. Termination of the scan phase is complete once a process has received notification of termination from each member of the group.

---

[5] In our implementation, we send both messages in the same remote procedure call.

5 Fault-Tolerance

5.1 Network Objects

Our algorithm is built on top of the reference listing mechanism provided by the Network Objects distributed memory manager, albeit slightly modified. The Network Objects collector is resilient to communication failures or delays, and to process failures.

Network Objects uses reference lists (actually reference *sets*) rather than reference counts. Furthermore, any given client process holds at most one reference — the surrogate – to any given concrete object. Communication failures are detected by a system of acknowledgements. However, a process that sends a message but does not receive an acknowledgement cannot know whether that message was received or not: it does not know whether the message or its acknowledgement was lost. Its only course of action is to resend the message. Unlike reference counting, reference listing is resilient to duplication of messages; the Network Objects dirty and clean operations are idempotent.

As we showed in Section 2, the dirty call mechanism also prevents out-of-order delivery of reference count messages from causing the premature reclamation of objects.

An owner of a network object can also detect the termination of any client process. Any client that has terminated is removed from the dirty set of the corresponding concrete object. Reference listing therefore allows objects to be reclaimed even if the client terminates without making a clean call. However, communication delay may be misinterreted as process failure, in which case an object may be prematurely reclaimed. Such an error will be detected when and if an attempt is made to use the the surrogate after the restoration of communication.

Proof of the safety and liveness of the Network Objects reference listing system is beyond the scope of this paper, but may be found in [3]. In the rest of this section, we shall informally discuss two aspects of the correctness of our partial tracing algorithm: safety and liveness.

5.2 Safety

First we shall establish the necessary conditions for our algorithm to be unsafe and show that these cannot arise. The safety requirement for our algorithm is that no live objects are reclaimed.

First we note that the system of acknowledgements ensures that marking requests are guaranteed to be delivered to their destination unless either client or owner process fail before the message is safely delivered and acknowledged. Although it is possible that messages might be duplicated, marking is an idempotent operation (*cf.* reference listing, above).

For the safety requirement to be breached, that is, for an object o to be incorrectly reclaimed in the sweep phase, the conditions below must hold at the end of the scan phase.

1. o must be red.
2. All members of o's client set must be members of the group.
   If this condition does not hold, then o would have been repainted green at the start of the scan phase, when o's dirty set was compared with the group membership.
3. All members of o's red set must be members of the group.
   Likewise, if this condition does not hold, then o would have been repainted green at the start of the scan phase, when o's red set was compared with the group membership.
4. All processes on at least one path of references from a root, possibly in another process, to o are still active.
   If this condition does not hold, then o is no longer reachable. Further more, the Network Objects reference listing mechanism would detect the failed process and would have deleted it from all client sets. Eventually this mechanism would reclaim o too.

From this we can conclude that o would only be prematurely reclaimed if no scan-request has traversed a path of references from a root to o, even though all processes on that path are still active. The system of acknowledgements implies that there is at least one scan-request that has not been acknowledged, and hence that there is at least one responsible local-scan process that has not terminated. But this means that the scan phase has not yet terminated.

5.3 Termination

We have argued informally above that our algorithm cannot incorrectly reclaim live objects through communication or process failures. We now show informally that each phase of the partial trace must terminate in each (non-terminating) process, and how this is achieved. First, we note that a mark-red or local-scan process will not terminate until it has received acknowledgements for all the mark-red (scan) requests that it has made. We note further that communication failures are handled by acknowledgements and the idempotency of marking requests and acknowledgements.

However, if a process with a group fails, the current phase may not terminate unless that failure is detected. If a process fails, its clients know that they need not wait for an acknowledgement from the failed process of any outstanding requests. Fortunately, the Network Objects system provides that owners detect failures of clients, if necessary reclaiming the corresponding concrete objects. We therefore provide the same system of timeouts to client processes that the Network Objects provide for owners. Thus a mark-red or local-scan process may terminate as soon as it has received acknowledgements of requests from all processes that are still active.

This leaves the scenario that a client of an object in a process p may fail, leaving p partially detached from the rest of the group. This process and its descendents may reach two undesirable states. First, any phase of the partial trace may not terminate in p before the instruction to move to the next phase arrives (through another concrete object owned by p). In this case, p must abandon this partial trace by deleting all its red sets and restoring all objects to green.

The second, and at first sight more intransigent problem, is that p may be fully detached from the rest of the group, and thus never receive the instruction to move to the next phase of the trace. Worse, not only it cannot proceed to the next phase, but the rule that a process may only collaborate in one group at a time means that it cannot participate in any group in the future! We resolve this problem by requiring that all garbage collection messages are stamped with the identity of the initiating process, *Init*. If a process has terminated a phase but the client to which it is to send the acknowledgement has failed, then responsibility for the request is inherited by *Init*, i.e. the acknowledgement is sent directly to *Init*.

The *Init* process treats such 'unexpected' acknowledgements according to when they arrive. If the acknowledgement of a request arrives before the end of the phase for which the request was issued, it is treated like any other acknowledgement: the sending process will continue to collaborate as a member of the group. If such an acknowledgement arrives at *Init* out of phase, then the sending process is effectively suspended from group membership: no further instruction to proceed to the next phase will be issued to it. Instead, at the end of the sweep phase, it will be instructed by *Init* to abandon the work that it has done, before passing this instruction on to the owners of any surrogates it holds.

If the initiating process itself fails, then the objects originally suspected of being garbage certainly are. The reference listing mechanism will propagate knowledge of *Init*'s death to all other members of the group.

6 Related Work

Distributed reference counting can be augmented in various ways to collect distributed garbage cycles. Some systems, such as Juul and Jul [14], periodically invoke global marking to collect distributed garbage cycles. With this technique, the whole graph must be traced before any cyclic garbage can be collected. Even though some degree of concurrency is allowed, this technique cannot make progress if a single process has crashed, even if that process does not own any part of the distributed garbage cycle. This algorithm is complete, but it needs global cooperation and synchronisation, and thus does not scale.

Maeda *et al.* [17] present a solution also based on earlier work by Jones and Lins using partial tracing with weighted reference counting [13]. Weighted reference counting is resilient to race conditions, but cannot recover from process failure or message loss. As suggested by Jones and Lins, they use secondary reference counts as auxiliary structures. Thus they need a weight-barrier to maintain consistency, incurring further synchronisation costs.

Maheshwari and Liskov [19] describe a simple and efficient way of using object migration to allow collection of distributed garbage cycles, that limits the volume of the migration necessary. The *Distance Heuristic*[6] estimates the length

[6] This idea was also discussed by Fuchs [10].

of the shortest path from any root to each object. The estimate of the distance of a cyclic distributed garbage object keeps increasing without bound; that of a live object does not. This heuristic allows the identification of objects belonging to a garbage cycle, with a high probability of being correct. These objects are migrated directly to a selected destination process to avoid multiple migrations. However, this solution requires support for object migration (not present in Network Objects). Moreover, migrating an object is a communication-intensive operation, not only because of its inherent overhead but also because of the time necessary to prepare an object for migration and to install it in the target process [25]. Thus, this algorithm would be inefficient in the presence of large objects.

The total overhead performed by our algorithm depends on how frequently it is run. A very simplistic heuristic may led to wasted and repeated work. However, even with a simplistic heuristic, a probability of being garbage can be assigned to each suspect object that has survived a partial tracing. For example, we could take a round-robin approach by tracing only from the suspect that was least recently traced. Better still, the Distance Heuristic should increase the chance of our algorithm tracing only garbage subgraphs. The more accurate this heuristic is, the more likely that its use would:

1. decrease the number of times a partial trace is run;
2. limit the mark-red trace to just garbage objects;
3. reduce the number of messages for the scan phase to the best case (in which no atomic-scans would be generated);
4. avoid suspension of mutators to handle scan-requests.

Lang *et al.* [16] also presented an algorithm for marking within groups of processes. Their algorithm uses standard reference counting, and so inherits its inability to tolerate message failures. It relies on the cooperation from the local collector to propagate necessary information. Firstly, they negotiate to form a group. A initial marking marks all concrete objects within the group depending on whether they are referenced from inside or from outside the group. Marks of the concrete objects are propagated locally towards surrogates. Finally marks of surrogates are propagated towards the concrete objects within the group to which they refer. When there is no more information to propagate, any dead cycles can be removed.

This algorithm is difficult to evaluate because of the lack of detail presented. However, the main differences between this and our algorithm is that we trace only those subgraphs suspected of being garbage and that we use heuristics to form groups opportunistically. In contrast, Lang's method is based on Christopher's algorithm [6]. Consequently it repeatedly scans the heap until it is sure that it has terminated. This is much more inefficient than simply marking nodes red. For example, concrete objects referenced from outside the suspect subgraph are considered as roots by the scan phase, even if they are only referenced inside the group. In the example of figures 2 and 3 our algorithm would need a total of 6 messages (5 for mark-red phase and 1 for scan phase), against a total of 10 messages (7 for the initial marking and 3 for the global propagation) for Lang's algorithm. Objects may also have to repeat traces on behalf of other objects (i.e. a trace from a 'soft' concrete object may have to be repeated if the object is hardened). Their 'stabilisation loop' may also require repeated traces. Finally, failures cause the groups to be completely reorganised, and a new group garbage collection restarted almost from scratch.

7 Conclusions and Future Work

This paper has presented a solution for collecting distributed garbage cycles. Although designed for the Network Objects system, it is applicable to other systems. Our algorithm is based on a *reference listing* scheme [3], augmented by partial tracing in order to collect distributed garbage cycles [13]. Groups of processes are formed dynamically to collect cyclic garbage. Processes within a group cooperate to perform a partial trace of only those subgraphs suspected of being garbage.

Our memory management system is highly *concurrent*: mutators, local collectors, the acyclic reference collector and distributed cycle collectors operate mostly in parallel. Local collectors are never delayed, and mutators are only halted by a distributed partial tracing either to complete a local-scan or, after that, to handle incoming messages to garbage-suspect objects; if all suspects are truly garbage, the latter event will never occur.

Our system reclaims garbage *efficiently*: local and acyclic collectors are not hindered. The efficiency of the distributed partial tracing can be increased by restricting the size of groups, thereby trading *completeness* for promptness. Appropriate choice of groups ensures completeness: eventually all cyclic garbage is reclaimable. The use of the acyclic collector and groups also permits some *scalability*, although our strategy could be defeated by pathological configurations in which a single garbage cycle spans a large number of processes.

Finally, our distributed collector is *fault-tolerant*: it is resilient to message delay, loss and duplication, and to process failure. *Expediency* is achieved by the use of groups.

Our algorithm is presently being implemented and measured. In particular, some choices for cooperation with the mutator require further study and depend mainly on experimental results and measurements. We are also interested in heuristics for suspect identification and group formation. Aspects of the concurrency and termination should be supported by formal proof and formal analysis of costs.

The management of overlapping groups is a further area for study. Currently, we prevent groups from overlapping. The distributed partial traces never deadlock, but continue their work without the cooperation of any process refusing a mark-red request. This can lead to wasted work, mainly if subgraphs span groups. We are currently exploring avenues to improve this, based on merging of groups and partitioning of work. We believe that this may also improve the algorithm's handling of failures.

References

1. Joel F. Bartlett. Compacting garbage collection with ambiguous roots. Technical Report 88/2, DEC Western Research Laboratory, Palo Alto, California, February 1988. Also in Lisp Pointers 1, 6 (April–June 1988), pp. 2–12.
2. David I. Bevan. Distributed garbage collection using reference counting. In *PARLE Parallel Architectures and Languages Europe*, pages 176–187. Springer Verlag, LNCS 259, June 1987.
3. Andrew Birrel, David Evers, Greg Nelson, Susan Owicki, and Edward Wobber. Distributed garbage collection for network objects. Technical report SRC 116, Digital - Systems Research Center, 1993.
4. Andrew Birrel, David Evers, Greg Nelson, Susan Owicki, and Edward Wobber. Network objects. Technical report SRC 115, Digital - Systems Research Center, 1994.
5. Hans-Juergen Boehm, Alan J. Demers, and Scott Shenker. Mostly parallel garbage collection. *ACM SIGPLAN Notices*, 26(6):157–164, 1991.
6. T.W. Christopher. Reference count garbage collection. *Software Practice and Experience*, 14(6):503–507, June 1984.
7. Margaret H. Derbyshire. Mark scan garbage collection on a distributed architecture. *Lisp and Symbolic Computation*, 3(2):135 – 170, April 1990.
8. Edsgar W. Dijkstra, Leslie Lamport, A.J. Martin, C.S. Scholten, and E.F.M. Steffens. On-the-fly garbage collection: An exercise in cooperation. *Communications of the ACM*, 21(11):965–975, November 1978.
9. Paulo Ferreira and Marc Shapiro. Larchant: Persistence by reachability in distributed shared memory through garbage collection. In *Proceedings of the 16th International Conference on Distributed Computing Systems (ICDCS), Hong Kong*, May 27-30 1996.
10. Matthew Fuchs. Garbage collection on an open network. In *International Workshop IWMM95*, pages 251–265, Berlin, 1995. Springer Verlag, LNCS 986.
11. Benjamin Goldberg. Generational reference counting: A reduced-communication distributed storage reclamation scheme. In *Proceedings of SIGPLAN'89 Conference on Programming Languages Design and Implementation*, pages 313–321. ACM Press, June 1989.
12. Paul R. Hudak and R.M. Keller. Garbage collection and task deletion in distributed applicative processing systems. In *Conference Record of the 1982 ACM Symposium on Lisp and Functional Programming, Pittsburgh, Pa.*, pages 168–78, August 1982.
13. Richard E. Jones and Rafael D. Lins. Cyclic weighted reference counting without delay. In Arndt Bode, Mike Reeve, and Gottfried Wolf, editors, *PARLE'93 Parallel Architectures and Languages Europe, Munich*, Berlin, June 1993. Springer-Verlag. LNCS 694.
14. Neils-Christian Juul and Eric Jul. Comprehensive and robust garbage collection in a distributed system. In *Proceedings of International Workshop on Memory Management, St. Malo, France*, volume LNCS 637, DIKU, University of Copenhagen, Denmark, September 16–18 1992. Springer Verlag.
15. Rivka Ladin and Barbara Liskov. Garbage collection of a distributed heap. In *International Conference on Distributed Computing Systems, Yokahama*, June 1992.
16. Bernard Lang, Christian Quenniac, and José Piquer. Garbage collecting the world. In *ACM Symposium on Principles of Programming, Albuquerque*, pages 39–50, January 1992.

17. Munenori Maeda, Hiroki Konaka, Yutaka Ishikawa, Takashi TomoKiyo, Atsushi Hori, and Jorg Nolte. On-the-fly global garbage collection based on partly mark-sweep. In *Proceedings of International Workshop on Memory Management, Kinross, UK*, Tsukuba Research Center, Japan, September 27–29 1995. Springer Verlag. LNCS 986.
18. U. Maheshwari and B. Liskov. Fault-tolerant distributed garbage collection in a client-server objected-oriented database. In *Proceedings of the third International Conference on Parallel and Distributed Information Systems*, pages 239–248, September 1994.
19. U. Maheshwari and B. Liskov. Collecting cyclic distributed garbage by controlled migration. In *Proceedings of the Principles of Distributed Computing*, 1995.
20. J. Piquer. Indirect reference counting: A distributed garbage collection algorithm. In Aarts *et al.*, editor, *PARLE'91 Parallel Architectures and Languages Europe*, Berlin, 1991. Springer Verlag, LNCS 505.
21. David Plainfossé and Marc Shapiro. Experience with fault-tolerant garbage collection in a distributed Lisp system. In *Proceedings of International Workshop on Memory Management, St. Malo, France*, INRIA, France, September 16–18 1992. Springer Verlag. LNCS 637.
22. David Plainfossé and Marc Shapiro. A survey of distributed garbage collection techniques. In *Proceedings of International Workshop on Memory Management, Kinross, UK*, INRIA, France, September 27–29 1995. Springer Verlag. LNCS 986.
23. M. Shapiro, P. Dickman, and D. Plainfossé. Robust, distributed references and acyclic garbage collection. In *Proceedings of the Symposium on Principles od Distributed Computing*, 1992.
24. Marc Shapiro, O. Gruber, and David Plainfossé. A garbage detection protocol for a realistic distributed object-support system. Rapports de Recherche 1320, INRIA-Rocquencourt, November 1990. Also in ECOOP/OOPSLA'90 Workshop on Garbage Collection.
25. N. G. Shivaratri, P. Krueger, and M. Singhal. Load Distributing for Locally Distributed Systems. *Computer*, 25(12):33–44, December 1992.
26. Paul R. Wilson. Garbage collection and memory hierarchy. In *Proceedings of International Workshop on Memory Management, St. Malo, France*, volume LNCS 637, University of Texas, USA, September 16–18 1992. Springer Verlag.
27. Paul R. Wilson. Distr. gc general discussion for faq. gclist (gclist@iecc.com), March 1996.

Incremental, Distributed Orphan Detection and Actor Garbage Collection Using Graph Partitioning and Euler Cycles

Peter Dickman

Department of Computing Science, University of Glasgow, Glasgow, UK

Abstract. A new algorithm is presented for incremental, distributed, concurrent garbage collection in systems with Actors. The algorithm also serves to detect orphan computations at low cost, as a side-effect of garbage collection, and permits the accurate elimination of unnecessary work without prejudicing the integrity of applications. Unlike all previous related algorithms, the new technique efficiently constructs a graph representation of the reachability relation within which Euler cycles can be used to determine the garbage objects. The new algorithm uses $O(N+E)$ space and $O(N+E)$ time, in the worst-case, to collect a graph of N objects and E references; this is comparable to one previously known algorithm and superior to all others (which require $O(N^2)$ time and $O(N+E)$ space in the worst-case). The new algorithm also avoids an uneven space utilisation problem exhibited by the only other $O(N+E)$ time algorithm, making it more suitable for use in non-shared-memory distributed systems.

1 Introduction

Garbage Collection (GC) is the automatic management of dynamically allocated storage (most commonly heap memory) in a computer system and has been studied for almost four decades. Automating memory management eliminates two of the major memory management errors made by programmers: premature freeing of memory containing data structures which are still required, and failure to release memory containing data structures which are no longer in use. The former error may lead to arbitrary behaviour in the program; the latter causes space leaks which reduce performance and may eventually exhaust the address space. As well as reducing the scope for errors, garbage collection both reduces the volume of code, because the programmer doesn't need to provide additional code to free memory, and permits more complex interactions between components, as manual memory management is difficult if the code is not layered. Garbage collection does not, however, guarantee perfect utilisation of memory; the programmer might, for example, construct an ever-expanding data structure that fills the address space. Furthermore, there are costs involved in garbage collection which, although comparable with the cost of manual memory management, are non-trivial and which might exceed the cost of doing no recycling of memory in small, short-lived applications.

Most modern programming languages incorporate a garbage collector in the run-time system, with the simplification of programs and reduction in programmer error rates more than offsetting the small additional costs involved. In addition, the development of high-quality incremental and generational garbage collectors has eliminated the disruptive pauses in applications caused by naïve stop-the-world garbage collectors in some early systems.

Over the last decade, new algorithms have been proposed which extend garbage collection to a new class of languages and systems, those using active objects. The difference between traditional collectors and the new "Actor Garbage Collectors" is explained in the next section, following the brief introduction to traditional GC algorithms which concludes this section. Section 2 also presents the known algorithms for Actor GC, to inform the later comparisons. In §3 a new algorithm for Actor garbage collection is presented, which eliminates the drawbacks of the existing algorithms. Incremental and distributed versions of the algorithm are then given in §4 and §5 respectively. The complexity of the algorithm is discussed in §6 and it is compared with the known algorithms. Our on-going implementation and measurement effort is briefly introduced in §7. Finally, the use of Actor GC algorithms for orphan detection is motivated in §8.

[18] provides an excellent overview of traditional uniprocessor GC algorithms and gives references to the various sorts of algorithm described in the remainder of this section; [10] surveys distributed GC techniques. Traditionally, whether an object is garbage (available for recycling) or live (potentially usable by the application) is defined in terms of reachability from a defined 'root' in the graph of objects and references. This root would classically be the stack in a simple programming system.

Their are two principle families of algorithms for garbage collection. Reference counting and reference listing algorithms directly determine when an object becomes garbage, by detecting the removal of the last reference to the object. Such algorithms are straightforward and effective but suffer one major drawback, they cannot detect and eliminate cyclic structures. Because of their low cost and avoidance of global information, variants of reference counting and reference listing algorithms are often used in distributed systems, as partial collectors forming one component of a hybrid GC. This hybridisation reduces the usage of the more expensive but complete algorithms which are also merged into the hybrid.

The only known algorithms which can detect all garbage are tracing algorithms, they walk the object graph from the root(s) determining which objects are live. The live objects are retained and all other objects are presumed to be garbage and recycled. This process may involve copying all the live objects and destroying all of the original objects, or it can be achieved by eliminating all objects which have not been detected to be live, although this requires more detailed knowledge of the collection of all objects in the system. Many distributed versions of tracing algorithms have been proposed, but all require many messages to be exchanged and most suffer from a lack of fault-tolerance.

The existing Actor GC algorithms, and the new algorithm presented in §3, are all tracing algorithms, i.e. they detect all garbage. In Actor systems, however, the definitions of garbage and the root object are modified, with the result that the algorithms can also eliminate orphan computations (as described in $8).

2 Actor Garbage Collection Algorithms

Actor systems [1] offer a different computational model from that normally associated with the best-known object-based programming languages, such as C++. In an Actor system each object contains a thread of control and a message queue, as well as the encapsulated behaviour and state including references to other Actors. Actors communicate by sending asynchronous messages (which may contain references), and the processing of messages by the embedded thread within an Actor may cause the Actor to change its subsequent behaviour. This contrasts with the synchronous invocation model of most object languages. The algorithms proposed here do not, however, depend on the asynchronous nature of communication, or the embedding of a single thread in each Actor; they are equally applicable to a wide-range of concurrent object models.

2.1 Problem Statement

The key distinction, for the purposes of GC, between Actor systems and the invocation based systems is that Actors contain a thread of control at all times: in effect, these objects are always reachable from a stack, namely that associated with the embedded thread. The term "Actor" is used loosely throughout this paper, to describe any system in which threads of control, or loci of execution, are directly associated with objects in this way. The proposed algorithm is usable in pure Actor systems and in hybrids in which only some objects have an embedded thread, i.e. systems which mix Actor entities with traditional objects.

Traditionally, the definition of 'the root of reachability' used by GC algorithms includes the stacks associated with every thread in the system. If this is done in an Actor system, every object would have to be retained as live, which is clearly inappropriate. Consider an object, A, which: is not referenced by any other object, has an empty message queue, and contains a thread of control that is blocked awaiting receipt of a message. No other object can send this object a message, since no references to the object are held. Furthermore, the blocking of the thread means that it will never spontaneously resume processing. Such an object serves no useful purpose and can be eliminated from the system. Early Actor systems therefore defined the root of reachability to include only those Actors in which the thread was engaged in computation, or where the message queue was not empty; reachability from the root was then used to determine which other Actors should be retained. The term *active* will be used henceforth to indicate an Actor which is engaged in computation, the converse, *passive*, indicating an object that will never engage in further computation unless it receives a message.

In the late 1980's the group led by Dennis Kafura at the Virginia Polytechnic Institute produced algorithms which allowed the root set to be further reduced. They noted that some active Actors can also be eliminated without adversely affecting the application. Consider two Actors, A and B, each of which holds a reference to the other and no other references whatsoever. Assume further that there are no other references to A or B in the system, and that A and B are active, endlessly passing messages back and forth to each other. The definition of root given above would include these objects in the root set, and hence they would never become garbage. However, they serve no useful purpose and could be removed without affecting the outcome of the overall computation.

Conversely, omitting all active Actors from the root set could affect the outcome of the application. Consider an unreferenced active Actor A, which holds a reference to a live object, B. At some future time A might send a message to B which contains a self-reference, B would then hold a reference to A, and since B is live A must also be live. At the time of the GC, however, A would not be reachable from the root and hence would be discarded (and therefore never send a message to B). The behaviour of the application might thus depend on whether and/or when the garbage collector executed.

There is one important feature of this problem which must be emphasised. If an active Actor could, for example, write to a file or send a message to a terminal, it must be retained. This means that the concept of "reference" used when discussing Actor GC algorithms is wider than sometimes understood. There are, in effect, references implicit in certain code fragments and these must be detected and followed when necessary. In a pure Actor system such references are explicit, since the files, terminal etc. are themselves modelled by Actors which are members of the root set. In hybrid systems rather more care is needed to ensure that all "references" are identified.

2.2 The Kafura Algorithms

Kafura's group proposed two algorithms [8, 7, 9, 15] which do not assume that active Actors should be retained. Special roots, such as terminals and the root of the filing system, are identified. The algorithms make many passes over the object graph, colouring the Actors in order to determine which active Actors could influence the outcome of the application. In effect, an active Actor which can reach (possibly indirectly) a non-garbage object must itself be non-garbage, and this definition is applied recursively, the fixed point determining which Actors are garbage and which are not.

These algorithms define the live objects to be: those which are reachable from the roots, and those objects which are active, or could become so in the future, that can potentially reach a live object. This has become the standard definition of live and garbage objects in the field of Actor GC.

2.3 The Backpointer Algorithm

In [3] the author proposed two new algorithms for Actor GC. The Backpointer algorithm had the same outcomes as Kafura's Push-Pull and Is-Black algorithms, but was considerably more efficient. The key factor in achieving the performance improvement was the caching of reachability information in the form of back-pointers added to the object graph.

Consider a graph of N objects (nodes) and E references (edges). Whereas, in the worst case, the incremental variants of Kafura's algorithms require $O(N+E)$ space and $O(N^2)$ (for Is-Black) or $O(N \times E)$ (for Push-Pull) time, with up to N passes over the graph; the Backpointer algorithm makes 2 passes over the object graph, requiring $O(N + E)$ space and $O(N + E)$ time.

The main drawback of the Backpointer algorithm is that it distributes the space it uses unevenly across a distributed system, and this limits its usefulness in some architectures. The partition merging algorithm presented in this paper is a refinement of my FCM (Fixed-Cost-Merge) algorithm [3]. The crucial advantage of the new algorithm is its avoidance of the uneven space distribution exhibited by the Backpointer algorithm.

2.4 Related Work

Other work in the field has concentrated on the implementation of Kafura's algorithms, giving careful consideration to the initiation and termination of the collection [13, 14], studying the use of the algorithms in real-time systems [16], and further investigating the use of the algorithms in distributed systems [11, 12]. The early work by Halstead [6] differs from the other algorithms in that it could not detect cyclic garbage, and hence is now of little interest.

3 The Partition Merging Algorithm

Due to space limitations, this paper concentrates on the core mechanism of the new algorithm, which is fundamentally different from its predecessors, requiring only a single pass over the initial object graph and achieving a uniform space distribution. The techniques used to initiate and terminate the algorithm are briefly sketched in §5.1. This section presents the new algorithm, the specific issues which arise when making the algorithm incremental and utilising it in a distributed system are addressed in the following two sections.

3.1 Defining Liveness: Roots, Non-Roots and Potential Roots

As noted above, an Actor is considered to be live if it is either reachable from the roots or if it both (a) is or might subsequently become active and (b) could potentially initiate the sending of messages which eventually lead to messages being received by a live object.

The standard Actor GC approach is taken, with all active Actors being potential roots (whether or not they are roots is implicitly determined by the GC algorithm), a clearly specified root set, and the remaining objects being non-roots. An active Actor is defined to be live if any object reachable from it is live, since it could send a messages which is forwarded along the path from the active Actor to the live object and hence change the live object. Furthermore, if an active Actor is live, everything reachable from it is live.

Suppose that we could construct data structures which precisely detailed the sets of objects that are currently reachable from each active Actor. The definition of liveness given in the preceding paragraph implies that every object in the reachability set of an active Actor (including the active Actor itself) has the same status, either all are garbage or none. Also, if two of these sets overlap, they must have the status, either both sets contain garbage or neither does.

It is these observations that underpin the new algorithm, which is described in sections §3.2 and §3.3.

3.2 Reachability Partitions for Active Objects

Every object in the graph is extended with mark bits and a pointer. At the start of garbage collection the mark bits are cleared and the pointers are made into self-references. The first main phase of the PMA (partition merging algorithm) then iterates over the active Actors, for each in turn it constructs a data structure, using the additional pointer added to each object header.

The principle action of the PMA is to recursively follow the reference graph, setting the mark bits to avoid infinite looping. Whenever a reference is found, for example from an object A to another object B, the reachability partitions for A and B are merged, as illustrated in figure 1.

If one or both of the objects still has a self-reference as its header pointer, the merging is achieved by exchanging the header pointers. If, however, both objects have non-trivial header pointers, a new data structure, called an X-node, is constructed. An X-node simply holds two references, and these are set to be copies of the header pointers; the header pointers are then replaced with references to the new X-node.

Invariants There are two crucial invariants which are maintained by the merging of the partitions.

The first is that the in-degree and out-degree of nodes in the constructed graph data structure are equal. Initially every object has a self-referencing header pointer, thus in-degree $= 1 =$ out-degree for every node. The two forms of merge operation, direct linking and the introduction of an X-node, both maintain the invariant, with X-nodes having in-degree $= 2 =$ out-degree. This property is used later, to justify the existence of Euler cycles in the construct.

The second invariant is that all objects which are linked in a partition have the same GC status, i.e. either they are all live, or they are all garbage. Initially this is certainly true, since each partition contains a single object. Two partitions are merged only if a member of one directly references a member of another; since the references are investigated only through reachability from an active Actor,

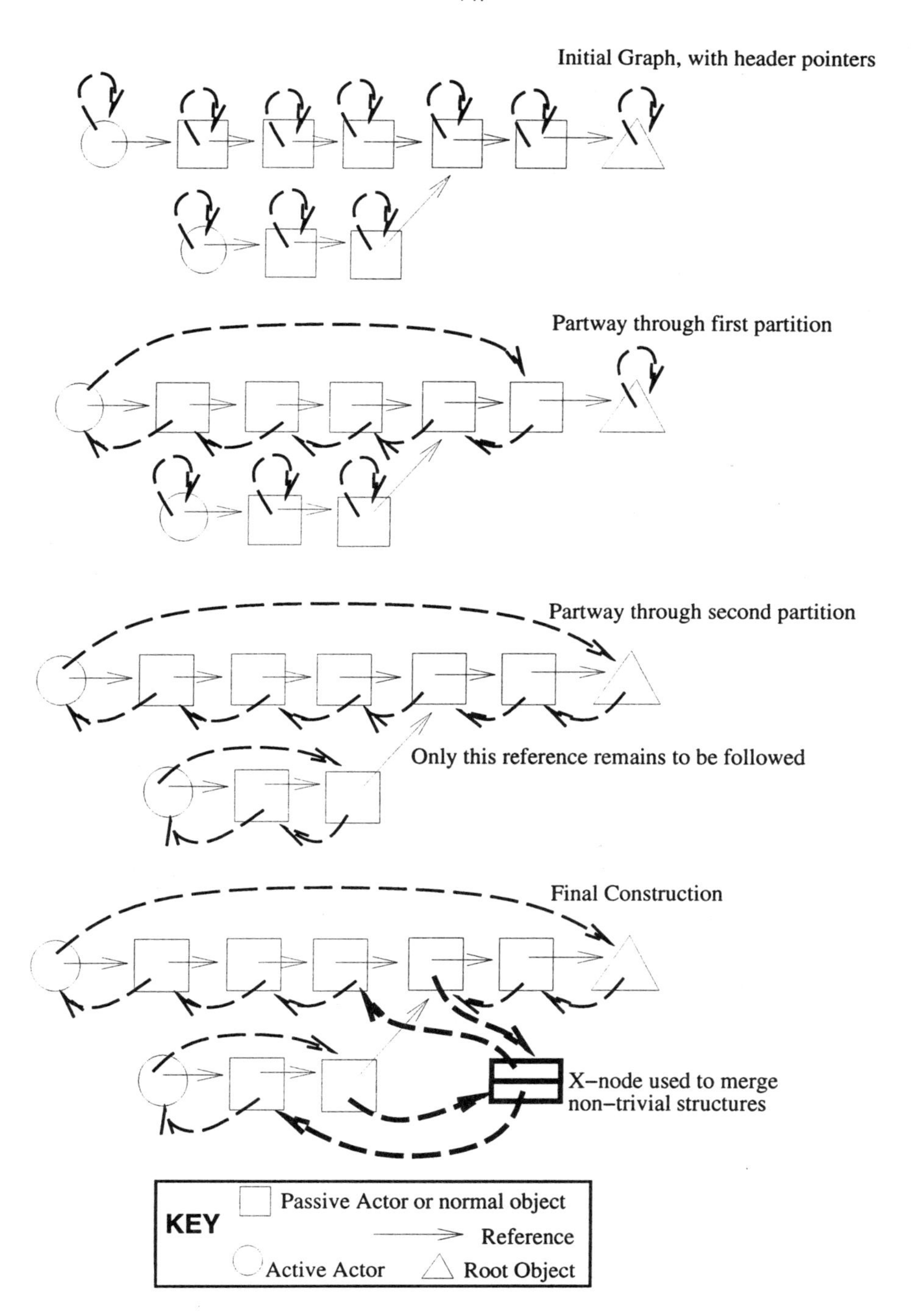

Fig. 1. Partition merging to construct linked reachability sets for Euler cycle traversal

it follows that merging only occurs if an active Actor can reach members of both partitions, which must therefore have the same GC status.

It is also the case that, after the traversal and merging are complete, each connected partition contains a collection of objects which are the union of overlapping active Actor reachability sets[1]. In other words, the partitions precisely divide the objects into sets which have common GC status.

3.3 Classification of Partitions using Euler cycles

An Euler cycle in a connected directed graph is a cycle that traverses each edge of the graph exactly once. There is an Euler cycle for an arbitrary graph, G, if and only if the graph is connected and the in-degree is equal to the out-degree at every vertex in the graph. Thus the merged partitions constructed above have Euler cycles and, since the in-degree is 1 at each object vertex (as against X-nodes) the Euler cycles will visit every object precisely once. (see e.g. [2] for a more formal treatment of Euler cycles).

After completion of the merging of partitions, the second phase of the Partition Merging Algorithm is executed. All objects reachable from the root sets are now coloured to indicate that they are live and then, starting from each active object, an Euler cycle traversing algorithm is used to traverse each of the merged partitions in turn. If a live object is found in a partition the entire partition is retained, if no live object is found the entire partition is garbage and can be discarded.

The Euler cycle traversal algorithm is a recursive walk, following the header pointers until an X-node is encountered. If the X-node has been visited previously this segment of the walk is completed; if the X-node is being encountered for the first time it is marked and each of the embedded pointers is followed in turn. This recursive algorithm is, however, implemented iteratively using an additional pointer in each X-node to construct a list of outstanding work. This permits the traversal to be completed in $O(N + E)$ time without risking stack overflow.

Once all partitions have been traversed the garbage collection is completed.

3.4 Constant Time Merging of Partitions

At first sight, the use of the header pointers and X-nodes may appear over-complex. This approach has the significant advantage of allowing two arbitrary partitions to be merged in constant time. If instead, partition membership was represented by, for example, numbers, the merging of partitions could lead to non-linear costs in certain worst-case scenarios [4].

The X-nodes are crucial to this approach: if the merging was performed using a simpler pointer exchange it might behave incorrectly. Linking two disjoint cycles can be achieved by exchanging two pointers, one from each cycle. If, however, this same approach is used on a single cycle the result can be two separate cycles. This is illustrated in figure 2. Again, non-linear costs might be

[1] The proof of this is omitted here due to space constraints, see [4] for details.

incurred if attempts were made to detect that the objects being linked were already part of a single merged partition.

The mechanism proposed here, although not proven to be optimal, offers a low-cost constant-time mechanism for merging the partitions.

4 Incrementality

Incremental garbage collection algorithms run in parallel with the mutator (application) code, thereby reducing jitter in the application's perceived performance. The standard means of making a GC algorithm incremental is to use a read or write barrier. When the GC executes, it constructs a logical barrier that prevents the mutator from accessing objects. This is often achieved using virtual memory mechanisms to make objects appear to have been paged out. When the GC algorithm finishes processing an object it moves the object across the barrier, making it accessible to the application again.

Objects which have crossed the barrier may hold references to objects which are inaccessible, and if the mutator follows such a reference the barrier will be touched. When this happens, either the mutator blocks or, more commonly, the GC algorithm is immediately invoked (for example by the page fault handler) to process the object which is being referenced. As soon as that processing is complete, the object can be moved across the barrier thereby permitting the mutator to proceed.

4.1 Scanning References

In traditional tracing collectors, such as mark and sweep, only a single pass over the application visible state of an object is required, scanning the references. Further passes may be made but they will not access the references, instead being scans of GC-specific fields in the headers of all objects. Thus, once a single scan of the object's references is complete the mutator can be allowed to access the object and, if it wishes, the mutator can modify the reference set held in that object.

Similar techniques can be used to make an Actor GC algorithm incremental. However, since the Kafura algorithms make many passes through the object graph, and the Backpointer algorithm makes two passes, it is necessary to retain a snapshot of the reference set associated with an object immediately prior to it being transferred across the barrier[2].

It may appear that the partition merging algorithm also makes two passes over the graph, one to merge the partitions and the second to compute reachability from the root set. However, these two passes can be combined into a single pass: in effect the root set members are treated as active Actors with the added act of colouring the objects reachable from them.

[2] Note that these will be the same as the references held when GC was initiated, as the mutator cannot modify an object prior to it crossing the barrier.

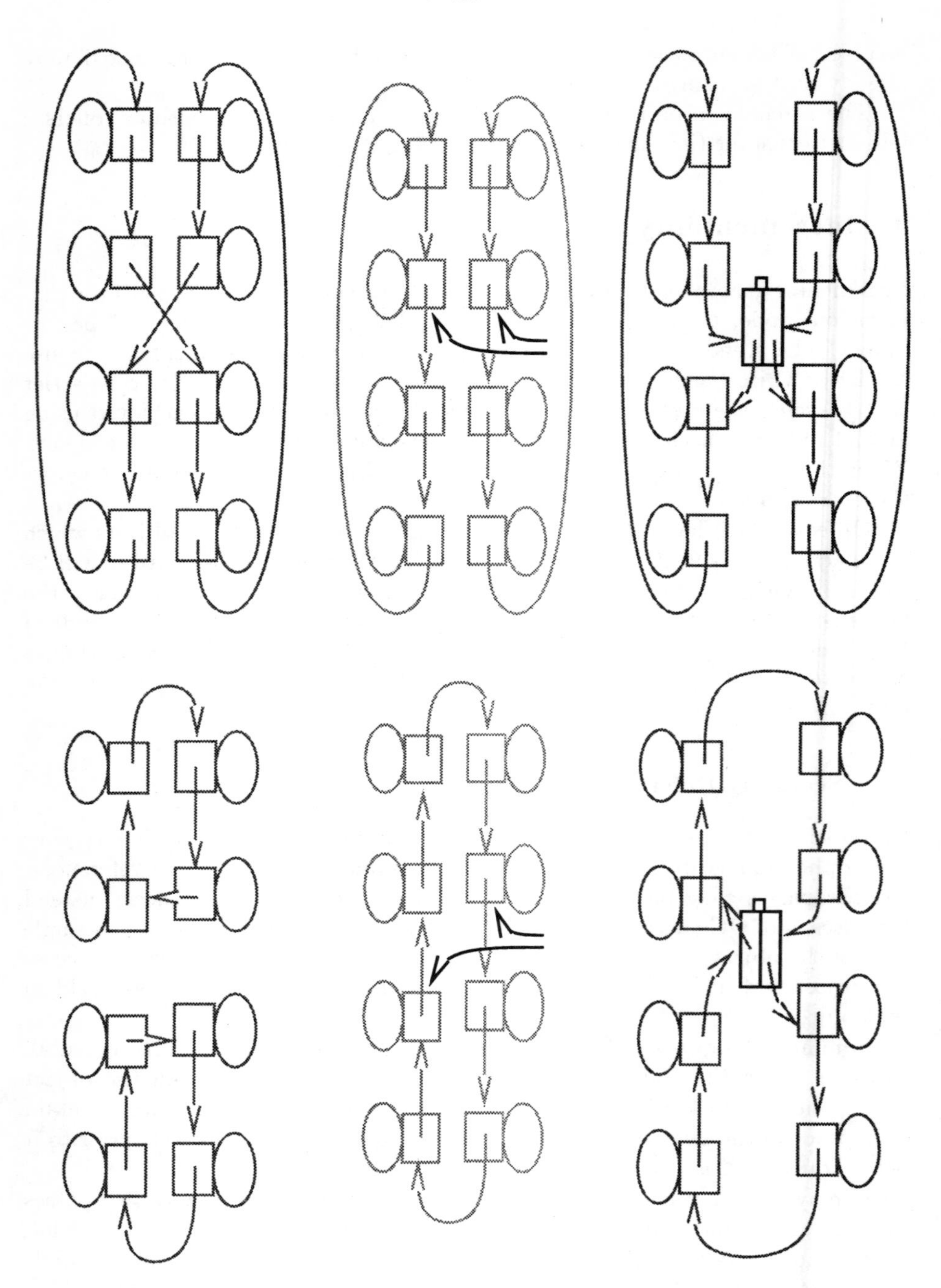

Fig. 2. Merging two cycles (above) and a single cycle (below), by pointer exchange (left) and X-node inclusion (right), due to a link between the indicated nodes (centre).

4.2 The Outstanding Work List and Concurrency

The two phases of the algorithm both require iteration over the active Actors. Furthermore, the early processing of objects to allow the mutator to access them will cause some objects to be marked even though the recursive descent from those objects is incomplete. This means that additional work items will have to be generated when objects are prematurely exposed to the mutator. There are several possible means of implementing the outstanding work list, using stacks or lists. One simple technique is to add an additional pointer and action bit to every object header. The pointers are used to chain together objects and where the pointer is non-nil, the action bit is used to indicate whether colouring is required in addition to merging. The initial work-item list will contain the root set and all active Actors.

The use of outstanding work lists in the barrier-based incremental version of the algorithm also offers the opportunity for concurrency in the GC algorithm. If appropriate locking is added, more than one thread can execute the GC code in parallel, alongside a multi-threaded mutator. This potential for concurrency is of particular significance for Actor systems aimed at parallel computers, especially highly-parallel shared-memory machines.

4.3 New Objects

During incremental GC, the mutator may construct new objects. There are two possible approaches that can be taken, usually called "allocate-black", if the new items are always retained by the on-going GC, and "allocate-white" which requires further processing of the new objects by the GC. The second approach incurs additional work, but has the advantage of recovering new objects which rapidly become garbage during the on-going GC.

It is imperative in Actor GC that new objects are not blindly retained. A new object might have been created by a garbage active Actor and subsequently been activated, having been given a reference to its creator. If the garbage Actor is removed but the new object is retained a dangling pointer will have been created.

In the incremental partition merging algorithm, new objects are given a distinct colour and a pointer to their creator is embedded in their header (occupying one of the two header pointer slots already discussed, one of which is not needed for new objects). These pointers are followed after the traversal of partitions is completed, and the new object is given the same GC status (live or garbage) as its creator.

5 Distribution

Actor programming models were designed for use in distributed systems as well as single address space uniprocessor applications. There are several problems specific to distributed garbage collection that must be addressed by any Actor GC algorithm.

5.1 Algorithm Initiation and Termination

Determining when to initiate garbage collection in a distributed system is non-trivial, but simple schemes are generally effective. The major problems instead concern the means used to reliably inform other nodes in the system that a GC has begun. The barrier based approach to incrementality presupposes that the barrier cleanly and consistently divides the entire object graph into two, with no messages crossing the barrier. In a distributed system this requires that the barrier form a consistent cut. Fortunately the techniques proposed for use with Kafura's algorithms, e.g. in [13], are also applicable to the algorithm presented here.

A means of detecting distributed termination is also required, to determine when the merging of partitions is complete, when all of the Euler cycle traversals have been performed, and when the status of the new objects has been resolved. For simplicity here we assume that a standard (costly) standalone termination algorithm is used; again, however, techniques used with Kafura's algorithms will also work here.

5.2 Distributing the Outstanding Work List

The outstanding work list naturally partitions across the components of a distributed system and Puaut's work [11] appears to generalise to the new algorithm. Messages must be sent across the distributed system to indicate that processing of a remotely referenced object is required. Since the loss of such a message might lead to the deletion of non-garbage objects it is essential that at-least-once message passing is used, fortunately the operations are idempotent so precisely-once message passing (which is, in general, not possible) is not needed.

5.3 The Apportionment of Working Space

The main benefit of the new algorithm concerns the distribution of the working space across a distributed system. There are two main uses of space in the algorithm: additional fields in object headers and the X-nodes.

The space consumed in object headers is distributed across the system in a similar fashion to the objects themselves and is allocated when objects are instantiated. This guarantees that these additional pointers will neither form hot-spots in the system nor require additional storage during a GC, when memory is likely to be in short supply.

An X-node can only be introduced when a reference between objects is traversed during partition merging. If the X-node is stored on the same system component as the object containing the reference, the potential X-node space distribution will match the distribution of references across the system and again this avoids hot spots. In this case, however, additional space is allocated during the GC, but the complexity arguments in §6.2 show that this algorithm is comparable with all three of its predecessors in this respect.

Although space is required to implement the new algorithm, the memory is distributed across the system in proportion to the distribution of objects and references. This consonance eliminates the main drawback of the Backpointer algorithm and means that the new algorithm can be used in systems containing widely and heavily referenced objects, e.g. name servers.

6 Algorithm Complexity

Given a range of algorithms with similar effects, any choice between them is usually made on the grounds of either performance or ease of implementation. The new algorithm is unusual in that it is both straightforward to implement and more efficient than its predecessors. Throughout this section, we assume that there are N objects and E references in the graph, with $N < E \ll N^2$. We further assume that the number of new objects, C, introduced during a GC is $O(N)$.

6.1 Time complexity

The Kafura algorithms make multiple passes through the garbage objects and as a consequence are not time-efficient. The worst-case time complexity of the Push-Pull algorithm is $O(N \times E)$ while that of the Is-Black algorithm is $O(N^2)$. The Backpointer algorithm makes two passes, with time-complexity $O(N + E)$.

The new algorithm requires an initialisation phase which visits every object, costing $O(N)$, then merges partitions which in the worst case involves visiting every object and following every reference, at a cost of $O(N+E)$. An Euler cycle of length L can be traversed in time $O(L)$, and the sum of the component cycle lengths must be less than $N + E$, therefore the worst case cost of the partition traversals is again $O(N + E)$. Finally, the new object resolution costs $O(\mathrm{C})$, if there are C new objects; this is assumed to be $O(N)$. The use of the barrier for incrementality introduces a small amount of additional work, noting the new work tasks, but again this is at most $O(N)$. The worst-case time-complexity of the new algorithm is, thus, $O(N + E)$.

6.2 Space complexity

The incremental forms of the Kafura algorithms require, in the worst case, snapshots of the object references as well as various data structures or header fields for managing the outstanding work. The net effect is that these algorithms can be implemented using no better than $O(N + E)$ space. The Backpointer algorithm also requires space for snapshots of the references, as well as additional space in which to construct the backpointers upon which its time-efficiency depends. It follows that the worst-case space complexity of the Backpointer algorithm is also $O(N + E)$.

The new algorithm adds two pointers and a few bits to the object headers, and will in addition require X-nodes to be constructed. An X-node has fixed size and is only needed when two multi-object partitions are merged. There are N single-object partitions initially, and there can be at most E mergings. The worst-case space complexity of the new algorithm is thus $O(N+E)$.

6.3 Constant Factors

All four incremental algorithms exhibit similar worst-case space-complexity, $O(N+E)$, and the worst-case time-complexity distinguishes the $O(N^2)$ and $O(N \times E)$ Kafura algorithms from my $O(N+E)$ algorithms. Selecting the appropriate algorithm for a given system will therefore depend upon the constant factors and whether or not the worst case graph structures for a given algorithm arise in that system. These factors are currently being investigated, as described in the next section.

7 Implementation

At present two implementation efforts involving Actor GC algorithms are underway in Glasgow. The first is the construction of a highly instrumented testbed and animation system which will allow the algorithms to be executed against artificial workloads. The aim being to determine the constant factors that apply in the complexity arguments above.

The second effort is the construction of an Actor-like programming system over the DRASTIC distributed systems platform [5]. The DRASTIC platform is a highly-instrumented scalable object-based system which, in addition to the standard object and reference management facilities required of any distributed object support platform, supports software evolution through a large-grained encapsulation mechanism. A simple Actor-like system is being implemented with selectable, instrumented garbage collection facilities, to allow us to investigate the behaviour of the algorithms under loads generated by more realistic applications.

7.1 Preliminary Results

The instrumented testbed is coded in Modula-3 and, although incomplete, can be used to compare the performance of a variety of GC algorithms on constructed graphs. The main components of the testbed are: a suite of GC algorithms, all implemented in a similar straightforward style including a sweep phase to recover garbage; a simulated heap structure; and a graph constructor which builds objects and references within the heap. Graph structures can be loaded into, and retrieved from, the simulated heap for checking and to allow the same structures to be reused. Additional tools have been produced which generate deterministic or random graphs containing specified patterns. The experiments summarised here were all conducted on lightly loaded DEC Alpha workstations running OSF-1 and involved executing the GC algorithms as singly-threaded single-address-space stop-the-world collectors over a memory-resident simulated heap. This avoided paging and networking costs and permitted the specific cost of the GC algorithm to be determined. Future tests using the DRASTIC platform, will investigate the overall effect on application performance.

The testbed incorporates the new algorithm presented here, Kafura's Is-Black and Push-Pull algorithms and two further mark-and-sweep algorithms. The first additional algorithm is a mark-and-sweep algorithm which retains only objects reachable from the roots. The second is a mark-and-sweep which retains all objects reachable from the roots or from active objects. Thus one mark-and-sweep retains too little, the other too much, and neither is acceptable as an Actor GC algorithm. These additional algorithms do, however, provide a useful indication of reasonable performance.

On small graphs within small heaps, all of the algorithms executed in under 1 millisecond (the granularity of our timings). As the heap size was increased the costs increased linearly, dominated by the whole-heap sweeps. All of the algorithms were comparable for small graphs, with the new algorithm and Kafura's Push-Pull taking time similar to the second mark-and-sweep, while Is-Black was roughly twice as slow.

As the graphs gained in complexity, however, the new algorithm remained comparable (timings consistently within 15% of each other) with the more extensive of the mark and sweep algorithms, whereas the Kafura algorithms, as expected, scaled less well. On a simple graph with about 200 live and 300 garbage nodes which was rich in short linked lists, the new algorithm and mark-and-sweep algorithms required half of the time taken by Push-Pull and only a quarter of the time required by Is-Black. On larger graphs containing pathological structures neither of Kafura's algorithms terminated within an hour, whereas the mark and sweep algorithms and the new algorithms completed within a second.

These results are as yet tentative, with much work remaining in tuning the algorithms and conducting more detailed and extensive measurements. It is clear, however, that the new algorithm outperforms the previously known ones on non-trivial graphs and is no worse in the smallest cases we can sensibly measure.

8 Orphan Detection

In distributed systems it is often the case that one component of the system is performing a computation on behalf of a different component. For example, in client-server architectures clients send requests which the servers then endeavour to satisfy. If, however, the client crashes while the server is computing, there is often little or no point in continuing the computation. There are some situations, however, where the server must continue processing to ensure that its internal data structures remain consistent.

The usual mechanisms for detecting such "orphan" computations involve the use of timeouts, and in congested systems these may not perform well and may even abort valid on-going activities. Worse still, application and system programmers can make mistakes in determining which orphans can be aborted and which should continue in order to maintain validity of the system. Transaction mechanisms can also be used to resolve such situations, but at a high cost in terms of lost work, due to unnecessary rollbacks.

The Actor GC algorithms offer a much improved solution to this problem, automatically and correctly detecting which on-going computations (active Actors) must be retained and which can be safely aborted. The new algorithm, being the first Actor GC algorithm that is both effective and efficient, is suitable for use with a wide range of application architectures and offers the opportunity for accurate orphan detection at low cost, as a side-effect of garbage collection.

The difficulties inherent in using these techniques have been clearly indicated, notably the need to accurately determine when I/O is possible, but in pure object models, where these problems can be overcome, there is a clear potential benefit.

9 Concluding Remarks

This paper has presented a new algorithm for the incremental, distributed garbage collection of Actors. The emphasis has been on the key ideas underlying the algorithm, with only brief discussion of the techniques used to make it incremental and distributed. In this section various possible extensions to the algorithm and implementation are reported, before the paper is concluded with a summary of the significant features of the work.

9.1 Future Work

Many optimisations are possible to the basic algorithm presented in this paper. Generational Actor GC is possible, though the techniques are not quite as effective, because any potentially active young object holding a reference to an old object must be retained. The partition merging can be optimised in a variety of ways, for example by annotating object headers during the merging to avoid repeated merging of the same object into the reachability set of any one active object. The current emphasis is, however, on the performance evaluation of the algorithms, and on studying applications to determine the frequency with which pathological data structures arise[3].

There are two major items of related work planned. One is to improve the fault-tolerance of the distributed version of the algorithm, which currently requires at-least-once message passing and separate distributed termination algorithms. The second is to hybridise the algorithm, combining it with a new variant of weighted reference counting [17, 3], to reduce the need for expensive full garbage collections.

[3] For example: a time-cost pathological data structure for Kafura's Is-Black algorithm is a singly linked list with active head and live tail. An example of a space-cost pathological data structure for the new algorithm is a heavily indexed matrix structure composed of cells and pointers with active column and row headers and live embedded data structures.

9.2 Summary

In this paper a new algorithm for the incremental, distributed, concurrent garbage collection of Actor-like systems has been presented. Unlike all previously known algorithms for this problem, the new algorithm is both time and space efficient and avoids hot-spots in its working space. The algorithm depends on a new approach to determining whether an active Actor can influence the live objects in the system, using a fixed-cost graph merging mechanism and Euler cycle traversal to determine which on-going computations can be terminated without prejudicing the applications integrity. As such, it also offers an efficient algorithm for orphan detection in distributed systems.

Acknowledgements

The work described here has been carried out intermittently over a six year period. It was begun at the University of Cambridge Computer Laboratory, funded by a UK EPSRC PhD studentship; it continued while the author was an ERCIM Fellow working at INRIA, INESC and GMD and a visiting researcher at DIKU; I am now based at the University of Glasgow and my recent work on Actor GC has been partly funded through the EPSRC AIKMS "DRASTIC" project, grant number GR/J99285. I would like to thank Huw Evans and several anonymous reviewers for their helpful suggestions concerning the presentation of this work.

References

1. Gul A. Agha: Actors: A Model of Concurrent Computation in Distributed Systems. MIT Press (Series in Artificial Intelligence) 1986
2. Béla Bollobás: Graph Theory: An Introductory Course. Springer-Verlag (Graduate Texts in Mathematics #63) 1979
3. Peter Dickman: Distributed Object Management in a Non-Small Graph of Autonomous Networks with Few Failures. PhD Thesis, University of Cambridge Computer Laboratory, 1992
4. Peter Dickman: Efficient, Incremental, Distributed Orphan Detection and Actor Garbage Collection. In preparation (extended version of this paper), to be a University of Glasgow, Department of Computing Science Technical Report
5. Peter Dickman and Huw Evans: Material about the DRASTIC project is available from http://www.dcs.gla.ac.uk/~drastic/
6. R. H. Halstead Jr: Multiple-processor implementation of message-passing systems. MIT LCS TR 198, April 1978
7. Dennis Kafura, Doug Washabaugh and Jeff Nelson: Garbage Collection of Actors. Proceedings of OOPSLA/ECOOP '90 pp 126–134 (ACM SIGPLAN Notices Vol. 25 #10, October 1990)
8. Dennis Kafura, Manibrata Mukherji and Doug Washabaugh: Concurrent and Distributed Garbage Collection of Active Objects. IEEE Transactions on Parallel and Distributed Systems, Vol. 6 #4, April 1995

9. Jeff Nelson: Automatic, Incremental, On-The-Fly Garbage Collection of Actors. MSc Thesis, Department of Computing Science, Virginia Polytechnic Institute, 1989
10. David Plainfossé and Marc Shapiro: A Survey of Distributed Garbage Collection Techniques. IWMM'95 pp 211–250, published as LNCS Vol.986
11. Isabelle Puaut: Distributed Garbage Collection of Active Objects with no Global Synchronisation. IWMM'92 pp 148–164, published as LNCS Vol.637
12. Isabelle Puaut: Gestion d'objets actifs dans les systèmes distribués: problématique et mise en œuvre. PhD Thesis, Université de Rennes I, 1993
13. Nalini Venkatasubramanian, Gul Agha and Carolyn Talcott: Scalable Distributed garbage Collection for Systems of Active Objects. IWMM'92 pp 134–147, published as LNCS Vol.637
14. Nalini Venkatasubramanian: Hierarchical Memory Management in Scalable Parallel Systems. MSc Thesis, University of Illinois at Urbana-Champaign, 1991
15. Doug Washabaugh: Real-time garbage collection of Actors in a Distributed System. MSc Thesis, Department of Computing Science, Virginia Polytechnic Institute, 1989
16. Doug Washabaugh and Dennis Kafura: Real-time garbage collection of Actors. Proceedings of the 11th Real-Time Systems Symposium, pp 21–30, December 1990
17. Paul Watson and Ian Watson: An efficient garbage collection scheme for parallel computer architectures. PARLE'87 pp 423–443, appears as LNCS Vol.259
18. Paul R. Wilson: Uniprocessor Garbage Collection Techniques. IWMM'92 pp 1–42, published as LNCS Vol.637

A Framework for the Analysis of Non-Deterministic Clock Synchronisation Algorithms

Pedro Fonseca[1,2*] and Zoubir Mammeri[1,3]

[1] CRIN/ENSEM, 2 av. de la Forêt d'Haye
F-54516 VANDOEUVRE-LES-NANCY, France
[2] Universidade de Aveiro, Portugal
[3] ENSAM, Chalons en Champagne, France
e-mail: pf@ua.pt,mammeri@loria.fr

Abstract. In recent years, non-deterministic clock synchronisation algorithms (NDCSA) have appeared as an attractive alternative to deterministic ones. NDCSA offer a precision that can be made as small as desired. The price to pay is a probability of success that is less than one. We propose an uniform analysis of NDCSA. Our approach strives at decomposing and identifying the factors that affect the performance of these algorithms. Our aim is to determine simple local conditions that guarantee that the desired precision will be attained with the desired probability.
The results are then used to estimate the communication burden imposed by the synchronisation algorithm and provide us with some guidelines to compare the different NDCSA.

1 Introduction

In recent years, non-deterministic clock synchronisation algorithms (NDCSA) have appeared as an attractive alternative to deterministic ones. Interest on NDCSA increased after the paper by Lundelius and Lynch [7] in 1984. There, they prove that, in a system composed of n sites, where the message delay Γ is bounded by $\Gamma_{\max}$ and $\Gamma_{\min}$ such that $\Gamma_{\min} \leq \Gamma \leq \Gamma_{\max}$, no deterministic clock synchronisation algorithm can attain a precision better than $\zeta(1 - 1/n)$, where $\zeta = \Gamma_{\max} - \Gamma_{\min}$. This result expresses two characteristics of deterministic clock synchronisation algorithms. First, the attainable precision has a non-null lower bound. Second, there must be an upper bound for the message delivery delay ($\Gamma_{\max}$ must be finite).

NDCSA strive at circumvent this limitation, by offering a precision that can be made as small as desired. The price to pay is a probability of success that is less than one. In general, this probability of success can be made as close to one

* Pedro Fonseca was partially supported by Portuguese Government grant PRAXIS XXI BD-3729-94

as desired by sending a sufficient number of messages (provided there is enough time and bandwidth to do so).

When trying to compare NDCSA, we are often faced with the problem of identifying the causes of different performances. Many NDCSA that have been published are strongly connected to a given architecture or system model. If a new algorithm performs better than its predecessors, is this due to its novel characteristics? Or is the system where the algorithm executes specially suited for it? Would another algorithm perform as well in the same system? If there is no common ground for comparison, the list of similar questions can be endless. We need then some means to establish the comparison between the different NDCSA.

We propose an uniform analysis of NDCSA. Our approach strives at decomposing and identifying the factors that affect the performance of these algorithms. Our method is similar to the one presented in [10]; we have formalised the approach and extended it to other algorithms.

We study the mechanisms how one site constructs a local view of the system in order to resynchronise its clock, either by reading the other site's clocks, or by detecting an event to restart its local clock. In other words, how it associates a clock value (local or remote) to an instant in time. We do not address the issue of the usage that can be made of such information, in order to compute corrections that yield optimal precision and accuracy or some degree of fault-tolerance. Another aspect of our work is that we assume that the delays are described by random variables, for which we know some of the statistical properties (mean, variance, distribution, ...). In these aspects, our work is complementary to [2].

Section 2 specifies the envisaged system and defines some terms; the basis for the analysis is presented in Sect. 3. Sections 4 and 5 present the application of our method to published non-deterministic clock synchronisation algorithms and in Sect. 6 the results are applied to a comparative analysis of these algorithms.

2 The Target System

The system to be considered consists of n sites, that communicate by means of messages. There is no shared or common memory: all communication is done through the messages. No condition is imposed on message delay other than it is non-null.

Each site p has a *physical clock*, $C_p(t)$. This physical clock gives an approximation of physical time t, such that, given two instants t_1 and t_2 ($t_1 < t_2$), $(1-\rho)(t_2-t_1) \leq C_p(t_2) - C_p(t_1) \leq (1+\rho)(t_2-t_1)$. ρ is the *drift rate* of the physical clock.The physical clock is a read-only device, and the site has no means of changing its value.

The aim of a clock synchronisation algorithm is to provide every site p with a *synchronised clock*, $SC_p(t)$. Synchronised clocks are local approximations of global time and their value is computed based on the local physical clock and on the information received from other sites. For the moment, we consider only algorithms that work by rounds; R is the duration of one round.

The closeness of synchronised clock values in two different sites p and q is measured by the *precision*. There are commonly two definitions for the precision of clock synchronisation. In the first, the precision is the upper bound on the difference between the values displayed by two different clocks at the same instant t, and we represent it by $\hat{\delta}$.

Definition 1. $\hat{\delta}$ is the *precision* of clock synchronisation if and only if

$$\forall_{p,q=1..n}, \forall_t, \quad |CS_p(t) - CS_q(t)| \leq \hat{\delta}$$

In another alternative definition, the precision δ is the upper bound on the interval between any two (real-time) instants two different synchronised clocks display the same value.

Definition 2. δ is the *precision* of clock synchronisation if and only if

$$\forall_{p,q=1..n}, \forall_{t_a,t_b}, \quad SC_p(t_a) = SC_q(t_b) \Rightarrow |t_a - t_b| \leq \delta$$

We adopt definition 1 as the definition of precision.

3 Guaranteeing Precision

NDCSA offer the advantage of a better precision than deterministic clock synchronisation algorithms, but this has a price: there is a non-null probability that the system will fail to synchronise. This probability of failure can be made as small as desired by sending a sufficient number of messages. Guaranteeing a small probability of failure with a tight synchronisation (a small value of $\hat{\delta}$) may require a large number of messages. In order to use NDCSA, we need to estimate the number of messages required to synchronise the system to a precision $\hat{\delta}^{\mathrm{des}}$, with a probability of success $\mathrm{P}_{\mathrm{des}}$.

Our aim is to determine the conditions that guarantee that $\mathbf{P}[\hat{\delta} \leq \hat{\delta}^{\mathrm{des}}] \geq \mathrm{P}_{\mathrm{des}}$. The approach will consist on decomposing the algorithm execution in smaller tasks, and deriving the conditions that are imposed on each of these smaller tasks, in order to guarantee the specified precision and probability of success. We decompose the overall execution of the algorithm in three levels: system, site and step level. In the system, there are u $(1 \leq u \leq n)$ sites that run the synchronisation algorithm. In each of these u sites, the algorithm is performed in v steps $(v \geq 1)$ during each synchronisation round. In this section, we will derive the conditions that apply to the step-level that guarantee the desired conditions at the system level.

3.1 Site-Level Guarantee

Each site executes the NDCSA to read the clocks of other sites or to detect a resynchronisation event (the moment for starting the clock with a new value), following the paradigm presented by [11].

Assumption 3. *Each site executes the NDCSA to perform one of the following: i) read the clocks of other sites or ii) detect a resynchronisation event (the moment for starting the clock with a new value). It will not perform both.*

Assumption 3 is valid for all NDCSA that have been published so far. If one algorithm executes both, the error will be the sum of the errors in each of the two actions: read the remote clock and detect an event.

To guarantee a precision $\hat{\delta}^{\mathrm{des}}$ at any time t, the precision right after the execution of the synchronisation algorithm, $\hat{\pi}$, must be less than $\hat{\delta}^{\mathrm{des}} - 2\rho R$, in order to guarantee a precision $\hat{\delta}^{\mathrm{des}}$ until the following execution, R seconds later. If we allow each site p to make an error $\epsilon_p^{\mathrm{site}}$ on the value it computes, the difference between two clock values will not be larger than $2\epsilon_p^{\mathrm{site}}$. Therefore, the maximum error tolerated at site level is $\epsilon^{\mathrm{site\_tol}} = \hat{\pi}/2 = \hat{\delta}^{\mathrm{des}}/2 - \rho R$.

$$\mathrm{Max}\{\epsilon_{\mathrm{p}}^{\mathrm{site}}\} \leq \epsilon^{\mathrm{site\_tol}} \Rightarrow \hat{\delta} \leq \hat{\delta}^{\mathrm{des}} \tag{1}$$

$$\mathbf{P}[\mathrm{Max}\{\epsilon_{\mathrm{p}}^{\mathrm{site}}\} \leq \epsilon^{\mathrm{site\_tol}}] \leq \mathbf{P}[\hat{\delta} \leq \hat{\delta}^{\mathrm{des}}] \tag{2}$$

Assumption 4. *All site errors $\epsilon_p^{\mathrm{site}}$, $(p = 1..u)$ are identically distributed independent random variables. They are represented by a random variable ϵ^{site}.*

Given Assumpt. 4, we have

$$\left(\mathbf{P}[\epsilon^{\mathrm{site}} \leq \epsilon^{\mathrm{site\_tol}}]\right)^u \leq \mathbf{P}[\hat{\delta} \leq \hat{\delta}^{\mathrm{des}}] \tag{3}$$

In real life, errors will not always be independent and identically distributed (i.i.d.) random variables. When errors are not i.i.d., this can be taken as a worst-case situation. When errors are not identical, we can take the largest error to be the error in every site. Error independence will be a worst-case situation if the correlation between errors in different sites is positive. Sending synchronisation messages that are shared by the different sites in a ring is an example of a system that will tend to have such a behaviour. This means also that the synchronisation tasks in each site do not compete with each other to succeed.

3.2 Step-Level Guarantee

Each site executes the NDCSA in v steps $(v \geq 1)$ in one synchronisation round. The step is the basic unit of the NDCSA, and it consists on sending r messages that will be used by one or more receivers. The error associated with the i-th step on site p, $\epsilon_{p,i}^{\mathrm{step}}$, is the error in the result produced by the receiver (if there is only one) or the maximum of these errors (if there is more than on receiver). We assume that it is possible to compute a bound ϵ^{tol} on $\epsilon_{p,i}^{\mathrm{step}}$ so that if the error in every step is less than ϵ^{tol}, the error at each site will be less than $\epsilon^{\mathrm{site\_tol}}$.

Assumption 5.

$$\exists_{\epsilon^{\mathrm{tol}}>0} : \forall_{p=1..u}, \forall_{i=1..v}, \quad \epsilon_{p,i}^{\mathrm{step}} \leq \epsilon^{\mathrm{tol}} \Rightarrow \epsilon_p^{\mathrm{site}} \leq \epsilon^{\mathrm{site\_tol}}$$

We extend Assumpt. 4 to the errors in the algorithm steps:

Assumption 6. *The error in the different steps of the algorithm* $\epsilon_{p,i}^{\mathrm{step}}$, *(*$p = 1..u, i = 1..v$*) are identically distributed independent random variables. They are represented by a random variable* ϵ^{step}.

Again, the considerations concerning Assumpt. 4 apply to Assumpt. 6. Assumptions 5 and 6 imply that

$$\left(\mathbf{P}[\epsilon^{\mathrm{step}} \leq \epsilon^{\mathrm{tol}}]\right)^{v} \leq \mathbf{P}[\epsilon^{\mathrm{site}} \leq \epsilon^{\mathrm{site\_tol}}] \tag{4}$$

From (3) and (4), we have

$$\left(1 - \mathbf{P}[\epsilon^{\mathrm{step}} > \epsilon^{\mathrm{tol}}]\right)^{uv} \leq \mathbf{P}[\hat{\delta} \leq \hat{\delta}^{\mathrm{des}}] \tag{5}$$

$\mathbf{P}[\epsilon^{\mathrm{step}} > \epsilon^{\mathrm{tol}}]$, the probability that the error in one step is greater than ϵ^{tol} may be difficult to compute. But, in most cases, we can compute an upper bound for this probability, f^{err}, which is also a function of r, the number of messages sent in each step.

Assumption 7.

$$\exists_{f^{\mathrm{err}}} : f^{\mathrm{err}} \geq \mathbf{P}[\epsilon^{\mathrm{step}} > \epsilon^{\mathrm{tol}}]$$

From (5) and Assumpt. 7:

$$(1 - f^{\mathrm{err}})^{uv} \leq \mathbf{P}[\hat{\delta} \leq \hat{\delta}^{\mathrm{des}}] \tag{6}$$

We can now state the conditions on the step of the NDCSA guaranteeing that the desired precision will be attained with the desired probability.

Theorem 8.

$$f^{\mathrm{err}} \leq \frac{1 - \mathrm{P_{des}}}{uv} \Rightarrow \mathbf{P}[\hat{\delta} \leq \hat{\delta}^{\mathrm{des}}] \geq \mathrm{P_{des}} \tag{7}$$

Proof. To prove theorem 8 we introduce the following

Lemma 9. $\forall_{x \in [0,1[}, \forall_{n \in \mathbb{N}}, (1 - x)^n \geq 1 - nx$

$$\begin{aligned} \mathbf{P}[\hat{\delta} \leq \hat{\delta}^{\mathrm{des}}] &\geq (1 - \mathbf{P}[\epsilon^{\mathrm{step}} > \epsilon^{\mathrm{tol}}])^{uv} \quad (5) \\ &\geq 1 - uv\mathbf{P}[\epsilon^{\mathrm{step}} > \epsilon^{\mathrm{tol}}] \quad (\text{Lemma } 9) \\ &\geq 1 - uvf^{\mathrm{err}} \quad (\text{Ass. } 7) \end{aligned}$$

Stating $1 - uvf^{\mathrm{err}} \geq \mathrm{P_{des}}$ completes the proof of theorem 8. □

The result in (7) allows relating the probability that the system will synchronise to a precision $\hat{\delta}^{\mathrm{des}}$ (a global property of the system) with the probability that the error in one step is larger than a given value (a local property). This can be used to establish the conditions that one single step of the algorithm must fulfill in order to guarantee the desired precision with the desired probability.

The results were established under three assumptions. The first relates to the independence of errors. The validity of this assumption will depend on the underlying assumptions about the behaviour of the system errors, as discussed in Sect. 3.1. The two others suppose that some parameter or function can be computed. These are:

1. the value of the upper bound on the step error, ϵ^{tol}, as a function of $\hat{\delta}^{\mathrm{des}}$;
2. the function f^{err}, the upper bound on the probability that the error in one step will be greater than ϵ^{tol}.

To apply our method to concrete cases, these two parameters have to be specified. That is the object of the next two sections.

4 Upper Bound on the Step Error

The computation of ϵ^{tol} will depend on the mechanism of the NDCSA: whether it detects a resynchronisation event or it reads the remote clocks.

4.1 Detection of a Resynchronisation Event

The error made by the NDCSA will cause the moments chosen by the sites to spawn an interval of size δ, where $(1-\rho)\hat{\delta} \leq \delta \leq (1+\rho)\hat{\delta}$. The value adopted in every site for its new clock is the same, so the scattering of the resynchronisation instants is the only source of error. To guarantee $\hat{\pi} \leq \hat{\delta}^{\mathrm{des}} - 2\rho R$, we must have $\epsilon^{\mathrm{site\_tol}} \leq \hat{\delta}^{\mathrm{des}}/2 * (1-\rho) - \rho R$.

To our knowledge, the only algorithm that uses non-deterministic techniques to detect a resynchronisation event is Le-Lann's ([6]). In this algorithm, each site executes the algorithm once, so $\epsilon^{\mathrm{tol}} = \epsilon^{\mathrm{site\_tol}}$.

4.2 Reading Remote Clocks

We will consider two cases: a Master-Slave (MS) structure and a distributed structure. In the MS structure, each Slave site reads the clock of the Master, and adopts the estimate as the correct value. The algorithm is executed in the Master $(u = 1)$. In a distributed structure the algorithm is executed in all sites $(u = n)$.

The number of steps in each site will depend on the type of network. In a broadcast network, each site needs one single step to send the r messages to all receivers: $v = 1$. In a point-to-point network, each site will need one step for each receiver: $v = n - 1$.

For the MS structure, the error for one step is the error in the Slave's estimate. To keep the difference between the values of two Slave clocks within $\hat{\delta}^{\text{des}}$ of each other, we must have $\epsilon^{\text{tol}} = \hat{\delta}^{\text{des}}/2 - \rho R$.

In a decentralised structure, each site reads the clock of every other site. The site will then compute the new value or correction for its clock. This is done by means of a Convergence Function (CF). For the CF average, median and mid-point, the precision function $\hat{\pi}(\hat{\delta}, \varepsilon)$ is always of the form $\varepsilon + f_{\text{ftc}}(k)$ ([11]), where k is the fault-tolerance degree, f_{ftc} a term that reflects how much the precision degrades with fault tolerance and ε is the maximum difference between any two corresponding values in two correct sites. In our case, $\varepsilon = 2\epsilon^{\text{tol}}$. If we want to guarantee that the precision is always less than $\hat{\delta}^{\text{des}}$ at any time, we must have $\hat{\pi}(\hat{\delta}, \varepsilon) + 2\rho R \leq \hat{\delta}^{\text{des}}$, which leads to $\epsilon^{\text{tol}} \leq \hat{\delta}^{\text{des}}/2 - \rho R - f_{\text{ftc}}(k)/2$.

The NDCSA that read the clocks of other processors are those of Arvind [1], Cristian [3], both algorithms of Olson and Shin [8], and the algorithm of Rangarajan and Tripathi [10].

5 Upper Bound on the Probability of Error

The upper bound on the probability of error, f^{err}, depends on the action the NDCSA performs. We find that the algorithms published so far fall into one of two categories: those that are based on a averaging process, and those based on an interval of possible values. The algorithms of Le-Lann, Arvind, Rangarajan and Tripathi and the averaged based algorithm of Olson and Shin fall into the first category; the algorithm of Cristian and the interval-based algorithm of Olson and Shin into the second.

5.1 Averaging-Based Algorithms

These NDCSA produce a result W based on the average of s samples of a r.v. X. The number of samples s is related to the number of messages r sent in each step: $s = Cr$, where C is the *cooperation constant*, that depends on the algorithm. It reflects how many sets of r samples are used for each of the computations performed; its value is n for Le-Lann's algorithm and 1 for all the others.

The delay suffered by a message when crossing a connection can be decomposed in several terms: Send time (Γ_{send}), Access time (Γ_{acc}), Propagation time (Γ_{prop}), and Receive time (Γ_{rec}), that account respectively for the time to prepare the message, access the medium, propagate through the communication channel and being processed by the receiver [5]. All these terms are random variables and the total delay when crossing a connection can be represented by the r.v. $\Gamma_{\text{cross}} = \Gamma_{\text{send}} + \Gamma_{\text{acc}} + \Gamma_{\text{prop}} + \Gamma_{\text{rec}}$. If the time a message is sent is relevant for the synchronisation algorithm, all these terms contribute to the randomness (and therefore to the error) of the result. In this case, X is the sum of all delays suffered by a message during transit. If D is an upper bound on the number of connections crossed by one message, then $\mathbf{Var}[X] \leq D\mathbf{Var}[\Gamma_{\text{cross}}]$.

In a broadcast network, the difference between the times one message arrives at different nodes depends only on $\Gamma_{\text{brd}} = \Gamma_{\text{prop}} + \Gamma_{\text{rec}}$. This term has usually a randomness much smaller that Γ_{cross}. If a synchronisation algorithm relies only on the arrival times of such messages, the error in the result is much smaller and $\mathbf{Var}[X] = \mathbf{Var}[\Gamma_{\text{brd}}]$. The algorithm of Le-Lann is based on this principle.

The error will depend on the variance of W, $\mathbf{Var}[W] = \mathbf{Var}[X]/(Cr) \leq (D\mathbf{Var}[\Gamma_{\text{alg}}])/(Cr)$, where Γ_{alg} is the relevant delay for the algorithm (Γ_{brd} or Γ_{cross}) and D and C are as defined above. Table 1 presents the values of these parameters. To compute f^{err}, we will assume that the number of samples used in the averaging process is sufficient to approximate the result to a normal r.v. (another alternative would be the Tchebychev's inequality, which imposes no a priori assumptions). Under this assumption we have

$$\begin{aligned}\mathbf{P}[|W - \mathbf{E}[W]|] > \epsilon^{\text{tol}}] &= \text{erfc}\left(\frac{\epsilon^{\text{tol}}}{\sqrt{2}\sigma_{\text{W}}}\right)\\ &\leq \sqrt{\frac{2D}{\pi r C}}\frac{\sigma_{\Gamma_{\text{alg}}}}{\epsilon^{\text{tol}}}\exp\left(-\frac{rC\epsilon^{\text{tol2}}}{2D\sigma^2_{\Gamma_{\text{alg}}}}\right)\\ &\equiv f^{\text{err\_avg}}\left(\sqrt{\frac{rC}{2D}}\frac{\epsilon^{\text{tol}}}{\sigma_{\Gamma_{\text{alg}}}}\right) \end{aligned} \tag{8}$$

where erfc is the complementary error function of a normal random variable, $\sigma_{\mathcal{X}} = \sqrt{\mathbf{Var}[\mathcal{X}]}$ and $f^{\text{err\_avg}}(\lambda) = \frac{1}{\sqrt{\pi}}\frac{e^{-\lambda^2}}{\lambda}$ is the error function for the averaging algorithms.

Table 1. Parameters of averaging algorithms

Algorithm	B	D	Γ
Arvind	1	D	Γ_{cross}
Le-Lann	n	1	Γ_{brd}
Olson and Shin (avg)	1	n	Γ_{cross}
Rangarajan and Tripathi	1	$\ln(n)$	Γ_{cross}

5.2 Interval Based Algorithms

These NDCSA rely on the computation of an interval where the remote clock value lies by means of sending a request message and receiving a response. The estimate for the remote clock is the mid point of the interval of possible values and the error half the width of the interval. The i-th delay suffered by a message can be represented by a r.v. $d_i = \min_i + \alpha_i$, where $\min_i$ is the minimum value for d_i and α_i the variable part of d_i. The width of the interval is (neglecting the terms in ρ), the sum of the variable parts of the delays. The probability that one message allows reading the remote clock with an error less than ϵ^{tol} is $\mathrm{F_A}(2\epsilon^{\mathrm{tol}})$, where A is the sum of the variable parts of the delays suffered by the message and $\mathrm{F_A}$ the distribution function (d.f.) of A.

In Cristian's algorithm, one step of the algorithm corresponds to repeating this process r times, until one reading has an error less than ϵ^{tol}. If each attempt to read the remote clock is independent of all others, $f^{\mathrm{err}} = (1 - \mathrm{F_A}(2\epsilon^{\mathrm{tol}}))^{\mathrm{r}}$. Note that A can be a function of n. In Olson and Shin's interval algorithm, after all the r messages are sent, the interval where the remote clock lies is the intersection of the intervals corresponding to each message. In this case, the d.f. and the probability density function (p.d.f.) of the interval bounds are relatively easy to compute. But the p.d.f. of the interval width (from which the error depends) is the convolution of the p.d.f. of the interval bounds, which leaves us with a result that is difficult to deal with analytically. If we note that the width of the smallest interval is larger or equal to the width of the intersection of all intervals, we can use the error corresponding to the smallest interval as an upper bound for the error of the intersection. Therefore, we need that at least one of the intervals has an error which is less than ϵ^{tol}, and the error function f^{err} is identical to Cristian's algorithm's. Table 2 presents the error function f^{err} for all the NDCSA.

Table 2. Error functions

Algorithm	f^{err}
Le-Lann	$f^{\mathrm{err\_avg}}\left(\frac{\sqrt{n}\epsilon^{\mathrm{tol}}}{\sigma_{\Gamma_{\mathrm{brd}}}}\right)$
Arvind	$f^{\mathrm{err\_avg}}\left(\frac{\epsilon^{\mathrm{tol}}}{\sqrt{D}\sigma_{\Gamma_{\mathrm{cross}}}}\right)$
OS-avg	$f^{\mathrm{err\_avg}}\left(\frac{\epsilon^{\mathrm{tol}}}{\sqrt{n}\sigma_{\Gamma_{\mathrm{cross}}}}\right)$
RT	$f^{\mathrm{err\_avg}}\left(\frac{\epsilon^{\mathrm{tol}}}{\sqrt{\ln(n)}\sigma_{\Gamma_{\mathrm{cross}}}}\right)$
Cristian	$(1 - \mathrm{F_A}(2\epsilon^{\mathrm{tol}}))^{\mathrm{r}}$
OS- int	$(1 - \mathrm{F_A}(2\epsilon^{\mathrm{tol}}))^{\mathrm{r}}$

6 Estimating Message Complexity

In this section, we will apply the results from previous sections to compute estimates on the communication burden imposed by the synchronisation algorithm. The parameters to estimate this burden are the number of messages sent during each step of the algorithm, r, and the number of *frames* in the system, m. One frame is the unit of data sent from one site to another, and it may contain one or several messages.

The framework we propose can be used in two ways. We can use the result in (6) to numerically compute the values of r that comply with the specified precision and probability of success; that is presented in Sect. 6.2. Another solution is to use (7) to establish, in an approximative way, the major trends for the values of r and m; this is done in Sect. 6.1.

6.1 Analytical Approximation

The total number of frames is related to the number of messages by $m \leq \frac{uvD}{Q} r$, where uv accounts for the total number of steps, D is an upper bound on the number of connections crossed by one message and Q is the number of messages in each frame. Q is n for Olson and Shin's algorithms, $\ln(n)$ for Rangarajan and Tripathi's and 1 for all others.

We present the results concerning the number of messages and the number of frames in table 3. We consider two situations: a MS and a distributed structure. Some of the NDCSA were developed specifically for a distributed structure, so applying them to a centralised structure would be meaningless.

Table 3. Message complexity with growing n

	r		m	
	MS	Dist.	MS	Dist.
Le-Lann	—	$O\left(\frac{\ln(n)}{n}\right)$	—	$O(\ln(n))$
Arvind	$O(D\ln(\sqrt{D}n))$	$O(D\ln(\sqrt{D}n))$	$O(nD^2\ln(\sqrt{D}n))$	$O(n^2D^2\ln(\sqrt{D}n))$
OS-avg	—	$O(n\ln(n))$	—	$O(n^2\ln(n))$
RT	—	$O(\ln^2(n))$	—	$O(n\ln^3(n))$
Cristian	$O\left(\frac{\ln(n)}{-\ln(1-p_A)}\right)$	$O\left(\frac{\ln(n)}{-\ln(1-p_A)}\right)$	$O\left(\frac{nD\ln(n)}{-\ln(1-p_A)}\right)$	$O\left(\frac{n^2D\ln(n)}{-\ln(1-p_A)}\right)$
OS-int	—	$O\left(\frac{\ln(n)}{-\ln(1-p_A)}\right)$	—	$O\left(\frac{n\ln(n)}{-\ln(1-p_A)}\right)$

Note: $p_A \equiv F_A(2\epsilon^{\text{tol}})$

6.2 Numerical Results

In this section, we present the results obtained by numerical inversion of (6) to yield the required value of r to comply with a given specification of P_{des}, $\hat{\delta}^{\text{des}}$ and ϵ^{tol} for a system of n sites with a given delay variance, $\mathbf{Var}[\Gamma]$. Figures 1 to 4 display the dependency of r on the probability of failure ($r(\mathrm{P}_{\text{fail}})$, $\mathrm{P}_{\text{fail}} = 1 - \mathrm{P}_{\text{des}}$) and on the allowed error, expressed as normalised error, $\epsilon^{\text{norm}} = \epsilon^{\text{tol}}/\mathbf{Var}[\Gamma]$ for two algorithms: Arvind's and Le-Lann's. Arvind's algorithm is one of the simplest and has been used as a base for two other algorithms; Le-Lann's algorithm presents, in some aspects, a performance remarkably better than other NDCSA.

In all the cases, the values of n are 2, 8, 32 and 128. Both algorithms are executed with a distributed structure on a broadcast network.

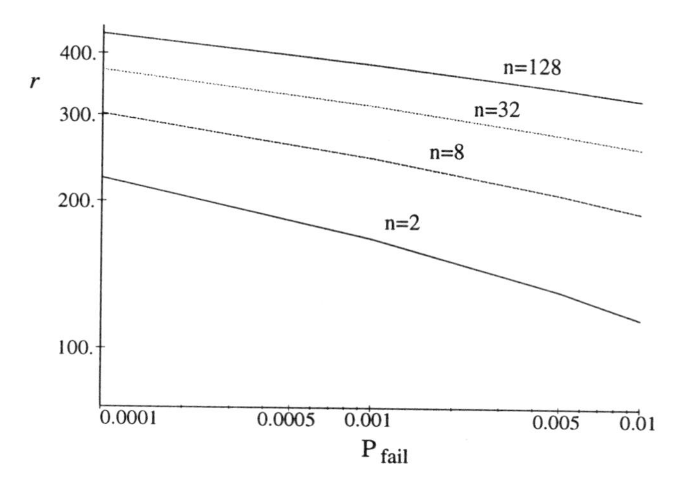

Fig. 1. Arvind's algorithm: $r(\mathrm{P_{fail}})$, $\epsilon^{\mathrm{norm}} = .2$

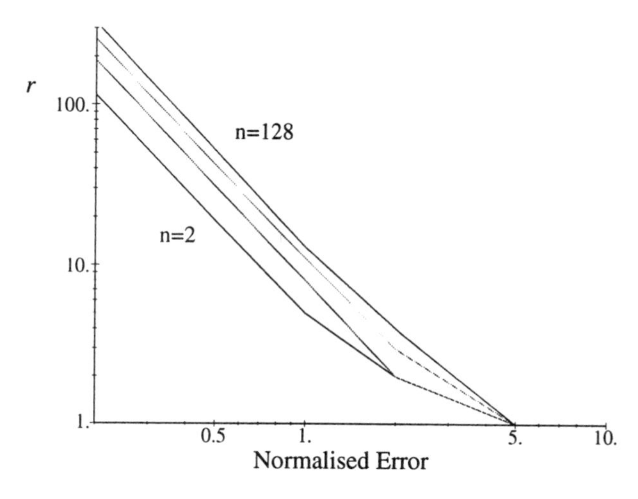

Fig. 2. Arvind's algorithm: $r(\epsilon^{\mathrm{norm}})$, $\mathrm{P_{des}} = 0.99$

Le-Lann's algorithm has a performance with growing n inverse to Arvind's: then n grows, r diminishes. We can see also how trying to guarantee synchronisation with small error and large probability of success can yield very large values of r. For instance, for Arvind's algorithm, $\mathrm{P_{des}} = 0.99$, $\epsilon^{\mathrm{tol}} = 0.2\mathbf{Var}[\Gamma]$

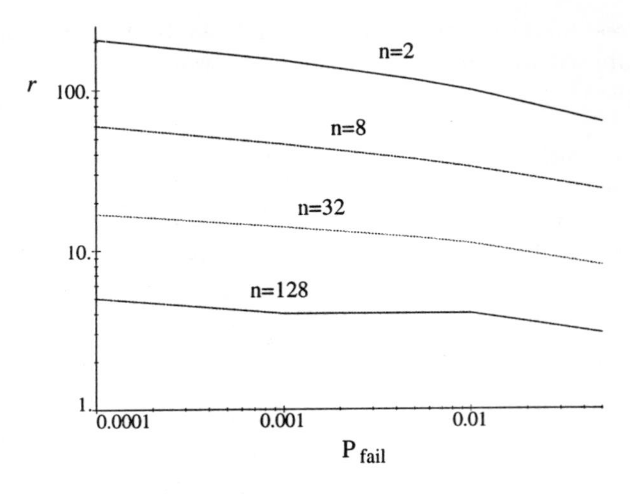

Fig. 3. Le-Lann's algorithm: $r(\mathrm{P_{fail}})$, $\epsilon^{\mathrm{norm}} = .2$

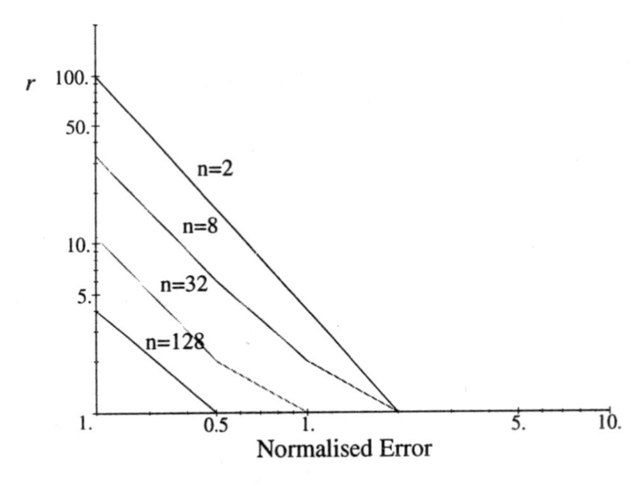

Fig. 4. Le-Lann's algorithm: $r(\epsilon^{\mathrm{norm}})$, $\mathrm{P_{des}} = 0.99$

and $n = 8$, the number of messages r is more than 100, which corresponds to more than 800 messages in each synchronisation round.

Apart the difference concerning the number of sites, both algorithms present similar performance with respect to changing $\mathrm{P_{des}}$ and ϵ^{tol} in identical circumstances.

6.3 Comparative Analysis

The results obtained provide us with some guidelines to compare the different NDCSA. Using simple remote clock reading algorithms like Cristian's and

Arvind's with no adaptions in a distributed structure (where all sites read all other site's clocks) may lead to an "explosion" in the number of messages and frames. When the system is not fully connected ($D > 1$), every message must cross several connections and the effect on increasing the communication burden is double-fold. On one side, the probability that one single result is correct goes down. With every extra connection crossed, the variance of the result grows: this leads to an increase in the number of messages, r, in order to prevent the probability of error from growing. On the other side, each message must be carried by several frames: the ratio m/r grows too. For situations where connectivity is not complete, and therefore messages have to be relayed, the algorithms of Olson and Shin and the algorithm of Rangarajan and Tripathi present a better performance. When comparing these two, we find again that better connectivity leads to fewer messages. Rangarajan and Tripathi's algorithm requires that one site can access every other site by relaying one message at most $\ln(n)$ times; Olson and Shin's only requires a Hamiltonian cycle. In fact, these two are specialisations of a more general algorithm (Arvind's) to specific architectures, and they do not radically change the algorithm's performance.

Reducing the error in each step is obtained also by means of large sampling populations. Le-Lann's algorithm exploits this. In this algorithm, each site uses all received values to compute one single statistic, and each value sent by one site is used by all sites (including itself). This means that the size of the sample is nr ($C = n$). By contrast, in the other averaging algorithms, the communication is done in an one-to-one basis: one value sent by one site is used for one single result in the receiver. The size of the sample for each estimate of the result is r ($C = 1$). This grants Le-Lann's algorithm with a dependence on growing n that is much smaller than the others. On the other hand, the results for Le-Lann's algorithm should be taken with some care, as the are based on the hypothesis that Γ_{brd} does not change with n.

The variance of Γ_{alg} (the delay that is relevant for the averaging algorithms) is also an important quality factor. In algorithms where there is no reading of remote clocks, the time information is obtained by the arrival of the messages only. In this case, the scattering of the results obtained by different sites is measured by $\mathbf{Var}[\Gamma_{\text{brd}}] = \mathbf{Var}[\Gamma_{\text{prop}}] + \mathbf{Var}[\Gamma_{\text{rec}}]$. When the value of the sender clock must be used in the computations, the scattering is measured by $\mathbf{Var}[\Gamma_{\text{cross}}] = \mathbf{Var}[\Gamma_{\text{send}}] + \mathbf{Var}[\Gamma_{\text{acc}}] + \mathbf{Var}[\Gamma_{\text{prop}}] + \mathbf{Var}[\Gamma_{\text{rec}}]$. For most cases, and specially in broadcast media, the major source of uncertainty in message delay is due to Γ_{acc}, the time to access the medium. So algorithms that rely only on the arrival time of one message, such as Le-Lann's, and not on the send time, will need a smaller number of messages to guarantee the same probability of error. But these strategies are associated to broadcast media, which limits their application.

6.4 Guidelines for Reducing the Number of Messages

NDCSA, by their own nature, require a larger number of messages than deterministic algorithms, and this can be a obstacle to their use. Some guidelines for keeping the required number of messages low can be extracted from the results above.

Reduce distance between processors Each connection crossed increases the number of messages necessary to yield the desired probability. In a non fully connected network, synchronisation messages should be routed through the shortest path.

Centralise If fault tolerance provided by Master Slave structures is sufficient, these can be used to keep the number of messages low.

Strive for broadcast If one message is broadcast to several sites, the total number of messages will be smaller for the same probability of failure.

Reduce uncertainty Algorithms that rely only on the arrival time of messages present a smaller uncertainty than algorithms that rely on transmitting clock values. A smaller uncertainty means a smaller number of messages for the same probability of error.

Cooperate If all values received are used to compute one single statistic, the size of the sample is larger and the probability that the result is correct increases. For the same probability of error, a smaller number of messages can be employed

7 Conclusion

We have presented a general framework to the comparative analysis of non-deterministic clock synchronisation algorithms. This allowed establishing a common ground to base the analysis of several published algorithms and identify the reasons for the different performances. We addressed the issue of how a site constructs a local view of the system (how it relates clock values to instants in time) from the representation of the delays as random variables. Our aim is to propose a simple and general model. Making a simple model usually requires approximations that degrade accuracy, but it allows expressing the effect of several parameters more clearly.

The method is general and can be applied to any NDCSA as long as the underlying assumptions remain valid. In this paper, we have used some approximations that are rather crude in other to present results that are simple and analytically tractable. For a specific case, when the delay distribution is known, for instance, or by using numerical methods, better approximations can be employed, in order to provide more accurate results. The results are based on the i.i.d. assumption as a worst case situation. This is based on several assumptions regarding the system where the algorithm executes.

So far, the method is directed towards algorithms that work in "rounds". Recently proposed "continuous" non-deterministic clock synchronisation algorithms, such as the one in [9], require our method to be adapted.

Some assumptions regarding the tolerated errors can be relaxed. For instance, the relation between the event $A = \mathbf{P}[\epsilon_p^{\text{site}} \leq \epsilon^{\text{site\_tol}}]$ and $B = \mathbf{P}[\epsilon_{p,i}^{\text{step}} \leq \epsilon^{\text{tol}}$ for all $i]$ is deterministic. We are looking at guaranteeing $\mathbf{P}[A] \geq \mathbf{P}[A \cap B] = \mathbf{P}[B]\mathbf{P}[A|B]$. The conditions are established so that $B \Rightarrow A$ is always true: $\mathbf{P}[A|B] = 1$. We can envisage a scheme where the conditions imposed on one step can be relaxed, by allowing a larger error. This larger error may not always guarantee $\mathbf{P}[A|B] = 1$. But the increase in $\mathbf{P}[B]$ may compensate the decrease in $\mathbf{P}[A|B]$ and the overall effect would be an increase in $\mathbf{P}[A]$. This approach could also provide some insight into which convergence functions are more suitable (if any) to NDCSA.

Another open issue is how to use the information gathered during the phase of message interchange to compute a new value for the clock correction, and how it affects the algorithm's performance. As Cristian pointed out in [4], the rationale employed for deterministic clock synchronisation may not always be valid for NDCSA.

References

1. K. Arvind. Probabilistic clock synchronization in distributed systems. *IEEE Trans. Parallel and Distributed Systems*, 5(5):474–487, May 1994.
2. Hagit Attiya, Amir Herzberg, and Sergio Rajsbaum. Optimal clock synchronization under different delay assumptions. *SIAM J. Comput.*, 25(2):369–389, April 1996.
3. F. Cristian. Probabilistic clock synchronization. *Distributed Computing*, 3:146–158, 1989.
4. Flaviu Cristian and Christof Fetzer. Probabilistic internal clock synchronization. In *Proc. 13th Symposium on Reliable Distributed Systems*, pages 22–31. IEEE, 1994.
5. Hermann Kopetz and Wolfgang Schwabl. Global time in distributed real-time systems. Technical Report 15/89, Technische Universitat Wien, October 1989.
6. G. Le-Lann. Synchronisation statistique d'horloges physiques: étude algorithmique. Document HORA, INRIA, September 1990.
7. Jennifer Lundelius and Nancy Lynch. An upper and lower bound for clock synchronization. *Information and Control*, 62:190–204, 1984.
8. A. Olson and K. G. Shin. Probabilistic clock synchronization in large distributed systems. In *Proc. Distributed Computing Systems*, pages 290–297, May 1991.
9. Alan Olson, Kang G. Shin, and Bruno J. Jambor. Fault-tolerant clock synchronization for distributed systems using continuous synchronization messages. In *Proc. 25th International Symposium on Fault Tolerant Computing*, pages 154–163, Pasadena, California, June 1995. IEEE.
10. S. Rangarajan and S. K. Tripathi. Efficient synchronization of clocks in a distributed system. In *Proceedings of Real-Time Systems Symposium*, pages 22–31, 1991.
11. Fred B. Schneider. Understanding protocols for Byzantine clock synchronization. Technical Report 87-859, Department of Computer Science, Cornell University, Ithaca, New York, August 1987.

A Notation

Symbol	Description
n	number of sites
p,q	sites
$C_p(t)$	physical clock of site p
ρ	drift rate of physical clocks
$SC_p(t)$	synchronised clock of site p
u	number of sites executing the algorithm
v	number of steps in each site
r	number of messages sent in each step
R	duration of a synchronisation round
C	cooperation constant
D	upper bound on the number of connections crossed by one message
W	result computed by one site
$\hat{\delta}$,δ	precision of clock synchronisation
$\hat{\delta}^{\text{des}}$	desired value for the precision
P_{des}	desired probability for clock synchronisation
$\hat{\pi}$	precision after synchronisation
ϵ_p^{site}	error in site p
ϵ^{site}	site error (r.v.)
$\epsilon^{\text{site\_tol}}$	error tolerated at site level
$\epsilon_{p,i}^{\text{step}}$	error on the i-th step on site p
ϵ^{step}	error on one step (r.v.)
ϵ^{tol}	error allowed in one step
ϵ^{norm}	normalised error
f^{err}	upper bound on the probability of error
$f^{\text{err\_avg}}$	f^{err} for averaging algorithms
Γ	communication delay
$\Gamma_{\min}$, $\Gamma_{\max}$,	lower and upper bounds for the communication delay
Γ_{brd}	delay for broadcasting a message
Γ_{cross}	delay for sending a message across one connection
F_X	distribution function of r.v. X

Optimal Time Broadcasting in Faulty Star Networks

Aohan Mei, Feng Bao, Yukihiro Hamada, and Yoshihide Igarashi

Department of Computer Science
Gunma University, Kiryu, 376 Japan
Email: {mei, bao, hamada, igarashi}@comp.cs.gunma-u.ac.jp

Abstract. This paper investigates fault-tolerant broadcasting in star networks. We propose a non-adaptive single-port broadcasting scheme in the n-star network such that it tolerates $n-2$ faults in the worst case and completes the broadcasting in $O(n \log n)$ time. The existence of such a broadcasting scheme was not known before. A new technique, called *diffusing-and-disseminating*, is introduced to design our broadcasting scheme. This technique is useful to improve the efficiency of broadcasting in star networks. We analyze the reliability of the broadcasting scheme in the case where faults are randomly distributed in the n-star network. The broadcasting scheme in the n-star network can tolerate $(n!)^{\alpha}$ random faults with a high probability, where α is any constant less than 1.

1 Introduction

Broadcasting is one of the fundamental tasks in network communications. It is the process of disseminating a message from the source node to all other nodes in the network. It can be accomplished in such a way that each node repeatedly receives and forwards messages. For the past decade a lot of studies on broadcasting in networks have been done [9],[13].

In large networks, some nodes and/or links may be faulty. However, multiple copies of the message can be disseminated through disjoint paths. The fault tolerance of broadcasting can be achieved by processing this type of message multiplicities. We say that the broadcasting is successful if the correct message from the source node can reach all the healthy nodes in the network within a certain limit of time. Apparently, the maximum number of faults that can be tolerated in this way is less than the connectivity of the network in the worst case. Various models of fault-tolerant broadcasting and communication in networks have been proposed. We can contrast the features of these models. *Adaptive* versus *non-adaptive*, *single-port* versus *multi-port*, and *full-duplex* versus *half-duplex* are interesting contrasts [10],[13]. For some models it is assumed that each node has some information about the set of faults [8],[14]. For other models it is assumed that each node does not have any information about faults. For a model of the latter type each healthy node receives a message from an adjacent node and sends the message to an adjacent node no matter where faulty nodes and/or links exist. Broadcasting schemes in hypercubes discussed in [5] and [15] are for a model of the latter type.

Star networks were proposed as attractive interconnection networks [1], and they have been much studied [2],[3],[6],[7],[16],[17]. Previous theoretical work related to this paper is as follows. The n-star network consists of n $(n-1)$-star networks and additional $n!/2$ links. The connectivity and the diameter of the n-star network are $n-1$ and $\lfloor 3(n-1)/2 \rfloor$, respectively [1],[2]. For any pair of distinct nodes u and v in the n-star network, there are $n-1$ node-disjoint paths from u to v with lengths at most $dist(u,v)+4$, where $dist(u,v)$ denotes the distance between u and v [6],[16]. The diameter of the n-star network with at least 1 and at most $n-2$ faulty nodes and/or links is $diam+1$, where $diam$ is the diameter of the healthy n-star network [16]. In [12] a single-port broadcasting scheme for the n-star with no faults was proposed. Its running time is at most $n \log n$ and optimal in the sense that for any constant $c < 1$, there does not exist any single-port broadcasting scheme with running time $cn \log n$. A problem of finding a single-port broadcasting scheme in faulty star networks was also presented in [12]. In [11] a simple fault-tolerant scheme for the n-star was presented. It tolerates at most $O(\sqrt{n \log n})$ faults and completes the broadcasting in $O(n \log n)$ time. For the multi-port model, a fault-tolerant broadcasting using routing was introduced in [4], and its running time is $O(n^{\frac{3}{2}} \log n)$. In [8] a multi-port broadcasting using an additional assumption that all nodes of the star network know the location of every fault in advance was designed. It takes the minimum broadcasting time and the minimum number of message transmissions [8].

In this paper we consider single-port broadcasting in faulty star networks under the assumption that the source node is healthy and each node does not have any information about faults. We give a single-port broadcasting scheme that can tolerate $n-2$ faults. We show that for any constant $\epsilon > 0$, broadcasting by our scheme terminates within $(1+\epsilon)n \log n$ time for a sufficiently large n. The running time of our fault-tolerant broadcasting scheme is optimal for the asymptotic order and almost optimal for the constant factor of the order. That is, the performance of the broadcasting scheme is asymptotically good.

The broadcasting process by our scheme can be divided into two stages. The first stage is called the diffusing stage, and the second stage is called the disseminating stage. In the diffusing stage the message from the source node is transmitted along $n-1$ disjoint paths in the n-star network. This stage can be implemented in the single-port manner in $O(n \log n)$ time. This stage will be explained in Section 4. The disseminating stage is essentially the same as the broadcasting process commonly used in networks (e.g., broadcasting in hypercubes [15]). It can be also implemented in the single-port manner, and its running time is $O(n \log n)$ time. This stage will be explained in Section 3. In Section 5 we give a probabilistic analysis of the reliability of the broadcasting by our scheme in the n-star network with faults randomly distributed. We show that the broadcasting is successful with a probability higher than $1 - 1/n!$ if there are at most $(n!)^{\alpha}$ faulty nodes randomly distributed in the n-star network and α is any constant less than 1.

2 Preliminaries

Let $a_1a_2\cdots a_n$ be a permutation of n symbols $1, 2, \cdots, n$. For each i $(2 \le i \le n)$ and a permutation $a_1a_2\cdots a_n$, a generator g_i is defined as $g_i(a_1\cdots a_ia_{i+1}\cdots a_n) = a_i\cdots a_{i-1}a_1a_{i+1}\cdots a_n$. An undirected graph $G = (V, E)$ is called the n-star graph (denoted by S_n) if $V = \{a_1a_2\cdots a_n \mid a_1a_2\cdots a_n$ is a permutation of $1, 2, \cdots, n\}$ and $E = \{(u, v) \mid u, v \in V$ and $v = g_i(u)$ for some $i\}$. The n-star graph is also called the n-star network. For $2 \le i \le n$, edges of S_n specified by g_i are said to be of dimension i. We often denote edge (u, v) by g_i^u if $g_i(u) = v$. It is immediate that S_n has $n!$ nodes and it is $(n-1)$-regular. We can choose $(n-1)!$ permutations with an identical last symbol from $n!$ permutations. Hence, we can decompose the n-star network into n node-disjoint $(n-1)$-star networks. From this property we can say that star networks are *hierarchical*. As an example, we show the 4-star network in Figure 1, where $\#a$, $\#b$, $\#c$ and $\#d$ are connected to $\#a$, $\#b$, $\#c$ and $\#d$, respectively.

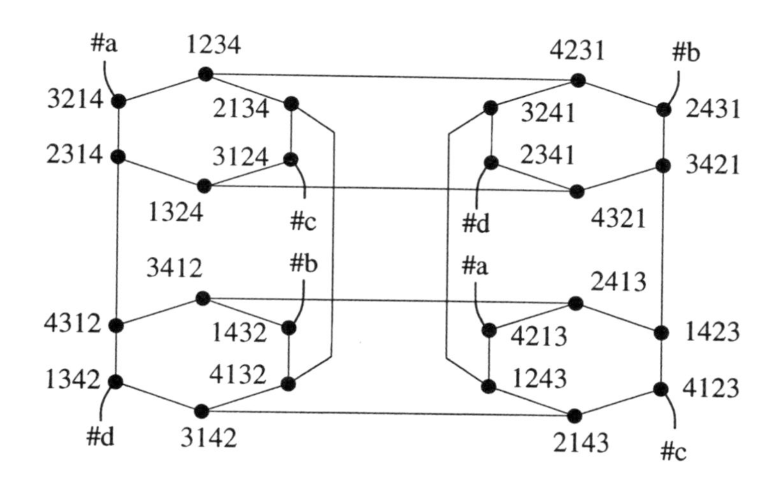

Fig. 1. The 4-star network.

Let $a_1, a_2, \cdots, a_m$ be m distinct symbols chosen from $\{1, 2, \cdots, n\}$. There are $m!$ permutations of $a_1, a_2, \cdots, a_m$. For each permutation $a_1a_2\cdots a_m$, we assign an integer between 0 and $m! - 1$ in the lexicographic order. This integer is called the rank of $a_1a_2\cdots a_m$ and denoted by $r(a_1a_2\cdots a_m)$.

In Section 4, we propose a fault-tolerant broadcasting scheme that exploits smaller sub-star networks S_i's $(2 \le i < n)$ recursively. For the analysis of the scheme, we specify sub-star networks in the following way. Each sub-star network S_i of S_n is specified by $n - i$ symbols and the set of generators $\{g_2, g_3, \cdots, g_i\}$. In other words, for each i, S_n can be decomposed into $(n!)/(i!)$ disjoint S_i's by partitioning the node set so that nodes belong to the same sub-star network of the decomposition if and only if $n-i$ symbols from the rightmost of all the nodes of the sub-star network are an identical sequence. Note that each sub-star network S_i can be also decomposed into disjoint smaller sub-star networks. For simplicity, we use $1*n$ to denote a permutation with the first and last symbols being 1 and n, respectively. This notation can be extended to denote other permutations. For example, $1*n-1n$ denotes a permutation with the first, the $(n-1)$st and the nth symbols being 1, $n-1$ and n, respectively.

We assume that each node represents a processor and each link represents a bidirectional communication line connecting two nodes at the extremes of the link. All the nodes in a network are synchronized with a global clock. Broadcasting time is measured as the number of steps to complete the broadcasting. In each step every node can send a message to at most one neighbor node and can receive a message from at most one neighbor node. Such a model is called a single-port network. Communication from node $b_1b_2\cdots b_n$ to node $a_1a_2\cdots a_n$ can be done in the same fashion as communication from the identity permutation $12\cdots n$ to node $f(a_1)f(a_2)\cdots f(a_n)$, where $f(b_k)=k$ for each $k (1\leq k\leq n)$. Hence, it is sufficient to consider broadcasting only from the source node $12\cdots n$. All logarithms in this paper are to the base 2.

3 An Information Disseminating Scheme

In this section we give a natural broadcasting scheme in S_n. The principle of the broadcasting scheme is the same as that of broadcasting in hypercubes given in [5], [15]. It is described as the following procedure. The procedure is executed at each node u of S_n concurrently.

```
procedure    dissem(n, t)
  (* it is executed at each node u *)
  repeat t times
    for i := 2 to n do
      if u held the message before the current step then
         u sends the message along dimension i
```

For the implementation of procedure *dissem*(n,t), it is not necessarily that each node knows whether its incident nodes and links are healthy and which is the source node of broadcasting.

Theorem 1. *If there are at most $n-2$ faulty nodes and/or links in S_n, then every node receives the message from the source node by executing procedure* dissem$(n, diam+1)$ *at each node in S_n.*

Proof. For any pair of distinct nodes u and v in S_n, there are $n-1$ node-disjoint paths from u to v with lengths at most $\min\{dist(u,v)+4, diam+1\}$ [6], [16]. Hence, the theorem holds.

Corollary 2. *If there are at most $n-2$ faulty nodes and/or links in S_n, then the message from the source node can reach every node within distance d by executing procedure* dissem$(n, d+4)$ *at each node in S_n.* □

Theorem 3. *For any $t < \lfloor n/2 \rfloor$, procedure* dissem(n,t) *cannot complete the broadcasting even if there are no faults in S_n.*

Proof. Without loss of generality we may assume that the source node is $12\cdots n$. Consider a node $\alpha = a_1a_2\cdots a_n$ such that for each i ($1 \leq i \leq \lfloor n/2 \rfloor$), $a_i > \lfloor n/2 \rfloor$. Obviously there exists such a node. We show that the message from the source node cannot reach node α by executing *dissem*(n,t) if $t < \lfloor n/2 \rfloor$.

We call each loop of the **repeat** statement of *dissem*(n,t) a round. Then *dissem*(n,t) consists of t rounds, and each round consists of $n-1$ steps. For each node $u = u_1u_2\cdots u_n$, let $l(u) = |\{u_i : 1 \leq i \leq \lfloor n/2 \rfloor \text{ and } u_i > \lfloor n/2 \rfloor\}|$. Then apparently $l(12\cdots n) = 0$ and $l(\alpha) = l(a_1a_2\cdots a_n) = \lfloor n/2 \rfloor$. If the message held in node u moves to node v during a round of *dissem*(n,t), then $l(v) \leq l(u)+1$. Hence, if the message from the source node reaches node v by *dissem*(n,t) then $l(v) \leq t$. We therefore need at least $\lfloor n/2 \rfloor$ rounds to complete the broadcasting by *dissem*(n,t).

From Theorem 3.1 and Theorem 3.3 we can say that procedure *dissem* can broadcast a message from the source node throughout the network S_n in $3n^2/2 + O(n)$ steps if there are at most $n-2$ faults, but cannot complete broadcasting in S_n in $\lfloor n/2 \rfloor (n-1)$ steps even if there are no faults.

4 An Efficient Fault-Tolerant Broadcasting Scheme

The broadcasting scheme proposed in [12] uses the hierarchical structure of star networks. Roughly speaking, the scheme consists of $n-2$ rounds. Suppose that the source node $12\cdots n$ has initially a message which will be broadcast in S_n. The source node sends the message to n sub-star networks S_{n-1}'s so that at least one node of each S_{n-1} receives the message. Recursively continue this process. Then at least one node of each sub-star network S_{n-i+1} in the ith round ($i \geq 2$) has the message. Each node with the message in each S_{n-i+1} sends the message to $n-i+1$ sub-star networks S_{n-i} in the $(i+1)$st round so that at least one node in each S_{n-i} receives the message. Clearly, all the nodes in S_n receive the message from the source node after repeating the round $n-2$ times. This scheme, however, may fail if there is a fault in S_n.

Let us recall the definition of function r given in Section 2. For example, function r assigns a rank between 0 and 5 in lexicographic order to each permutation of three distinct symbols. For permutations of integers 1, 3, and 5, $r(135) = 0$, $r(153) = 1$, $r(315) = 2$, $r(351) = 3$, $r(513) = 4$, and $r(531) = 5$. For a pair of permutations p_1 and p_2 of the same length (they are not necessarily permutations of the same set of integers), if $r(p_1) \neq r(p_2)$ then there is at least one position where the symbols of p_1 and p_2 are distinct. Note that p_1 and p_2

are distinct if $r(p_1) \neq r(p_2)$ and that p_1 and p_2 are not necessarily identical even if $r(p_1) = r(p_2)$.

Let d be the minimum positive integer such that $d! \geq n$. Let $p = p_1 p_2 \cdots p_n$ be an arbitrary permutation of $1, 2, \cdots, n$, and let $p[i,j]$ denote $p_i p_{i+1} \cdots p_j$, where $1 \leq i \leq j \leq n$. We partition the label of each permutation p into three intervals. The first interval is just the first symbol $p[1,1]$ and called the *head*. The second interval is $p[2, 3d+1]$ and called the *identifier district*. The identifier district is divided into three blocks, $p[2, d+1]$, $p[d+2, 2d+1]$, and $p[2d+2, 3d+1]$. The last interval is $p[3d+2, n]$. (See Figure 2.)

The broadcasting process by our scheme is divided into two stages, called the *diffusing* stage and the *disseminating* stage. In the *diffusing* stage, the message from the source node $s = 12 \cdots n$ is transmitted along $n-1$ internally disjoint channels. Each channel contains at least one node in every $(3d+1)$-sub-star network. Note that there are totally $\frac{n!}{(3d+1)!}$ such sub-star networks. In other words, after the *diffusing* stage, at least one node in each $(3d+1)$-sub-star network holds the message if there exist at most $n-2$ faults in S_n. Hence, after the *diffusing* stage, for every node u in S_n, there exists a node v holding the message within distance $\lfloor \frac{9d}{2} \rfloor$ from u. Then, $dissem(n, \lfloor \frac{9d}{2} \rfloor + 4)$ is executed in the *disseminating* stage. By Corollary 3.3, every node can receive the message in this way if there exist at most $n-2$ faults in S_n.

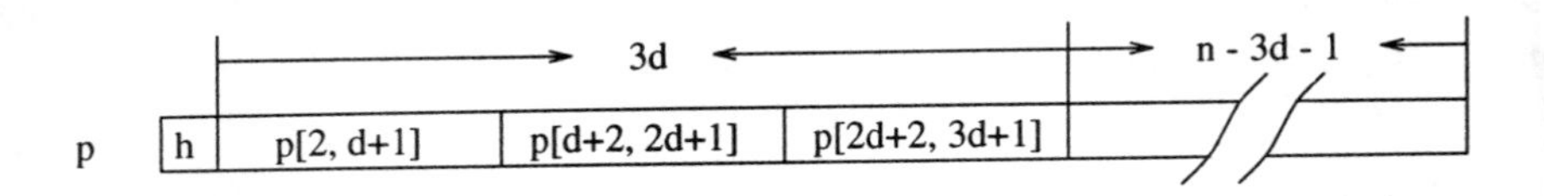

Fig. 2. The identifier district of p.

The *diffusing* stage consists of the *pre-stage* and the *recursive stage*. During the *pre-stage*, node $s = 12 \cdots n$ sends its message to its $n-2$ neighbors $g_2(s) = 213 \cdots n$, $g_3(s) = 3214 \cdots n$, $\cdots$, $g_{n-1}(s) = (n-1)2 \cdots (n-2)1n$. Then the $n-1$ nodes with the message (including s) send the message along dimension n. Then, $n-1$ nodes $t_1 = n * 1$, $t_2 = n * 2$, $\cdots$, $t_{n-1} = n * (n-1)$ receive the message. Next for each j $(1 \leq j \leq n-1)$, the message is transmitted from t_j to w_j along an appropriate route, where w_j satisfies the conditions that $w_j[3d+2, n] = t_j[3d+2, n]$ and $r(w_j[2, d+1]) = r(w_j[d+2, 2d+1]) = r(w_j[2d+2, 3d+1]) = j$. For each j $(1 \leq j \leq n-1)$) we can specify such a route from t_j to w_j by choosing appropriate nodes whose labels are obtained by some changes of symbols in $t_j[2, d+1]$, $t_j[d+2, 2d+1]$, and $t_j[2d+2, 3d+1]$. For any pair of distinct i and j $(1 \leq i, j \leq n-1)$ the route from t_i to w_i and the route from t_j to w_j are node-disjoint since the last symbol of the label of each node on the former route is i while the last symbol of each node on the latter route is j.

The *recursive* stage follows the *pre-stage*. The *recursive* stage is consistent with the hierarchical structure of S_n. For clear explanation, we now assume that no faults exist in S_n. Suppose that there exist $n-1$ nodes holding the message in an m-sub-star network in the recursion stage. Let these $n-1$ nodes be $x_1, x_2, \cdots, x_{n-1}$, where for each j $(1 \leq j \leq n-1)$, $r(x_j[2, d+1]) = r(x_j[d+2, 2d+1]) = r(x_j[2d+2, 3d+1]) = j$. Then for each j $(1 \leq j \leq n-1)$, the message is transmitted from x_j to m nodes that are in different $(m-1)$-sub-star networks. That is, for each j $(1 \leq j \leq n-1)$, x_j transmits the message to a node, say y_j in each of the m S_{m-1}'s, where y_j does not necessarily satisfy $r(y_j[2, d+1]) = r(y_j[d+2, 2d+1]) = r(y_j[2d+2, 3d+1]) = j$. However, at least two of $r(y_j[2, d+1])$, $r(y_j[d+2, 2d+1])$ and $r(y_j[2d+2, 3d+1])$ are equal to j. If one of these three ranks is not equal to the others, then we can choose a route from y_j to a node in the same S_{m-1}, say z_j which is obtained by some changes of symbols in the corresponding intervals so that $r(z_j[2, d+1]) = r(z_j[d+2, 2d+1]) = r(z_j[2d+2, 3d+1]) = j$. Then we move to the next round of the recursive stage. The recursive stage stops when it reaches $(3d+1)$-sub-star networks. The principle of the recursive stage is depicted in Figure 3. More detailed description about the recursive stage is given in Subsection 4.1.

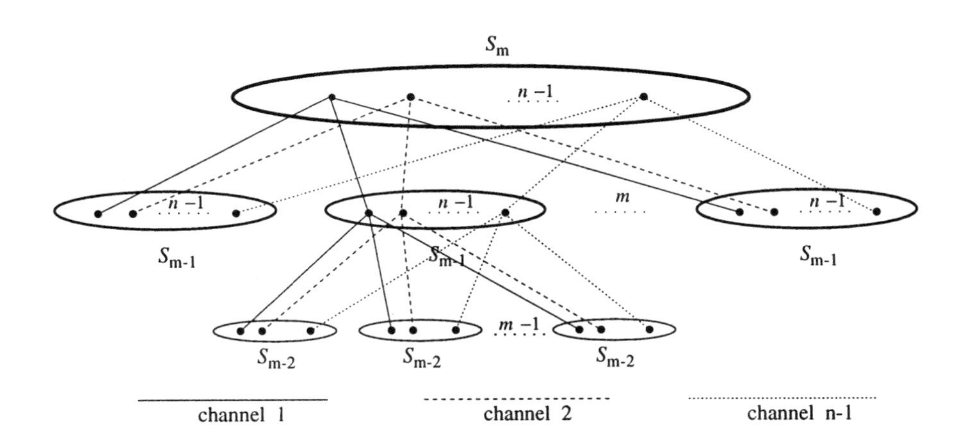

Fig. 3. Message routes from S_m to S_{m-1} in the recursive stage.

4.1 Construction of $n-1$ Node-Disjoint i-Level Channels

Definition 4. For a node p in S_n, the identifier (id for short) of p is $r(p[i, i+d])$ if $r(p[i, i+d]) = r(p[j, j+d])$ for some $i \neq j$ $(i, j = 1, d+1$ or $2d+1)$, and otherwise it is 0.

From Definition 1 we have the next lemma.

Lemma 5. *For any pair of permutations* p_1 *and* p_2 *of length* n*, they are distinct if their id's are not identical.* □

We now explain the process of the recursive stage in detail. For each j $(1 \leq j \leq n-1)$, w_j has the message at the beginning of the recursive stage. We therefore may consider that each w_j is a message source at the beginning of the recursive stage. Hereafter, we call them message sources. Remember that $r(w_j[2, d+1]) = r(w_j[d+2, 2d+1]) = r(w_j[2d+2, 3d+1]) = j$. The recursive stage is divided into $n-3d-1$ rounds. For each i $(1 \leq i \leq n-3d-1)$ and each j $(1 \leq j \leq n-1)$, let $W_i(w_j)$ be the set of nodes that hold the message from w_j and are ready to broadcast the message at the beginning of round i. Initially, let $W_1(w_j) = \{w_j\}$ and $W_i(w_j) = \phi$ for $i > 1$. Its contents will be renewed at the beginning of each round i. For each i $(1 \leq i \leq n-3d-1)$ and each j $(1 \leq j \leq n-1)$, during round i each node in $W_i(w_j)$ executes the following operations:

(1) Each node $p \in W_i(w_j)$ sends its message to $g_2(p), g_3(p), \cdots, g_{3d+1}(p)$ sequentially. Then node p broadcasts its message in a binary jumping way. That is, p sends its message to $g_{3d+1+1}(p), g_{3d+1+2}(p), \cdots, g_{3d+1+2^k}(p)$ sequentially, where k is the maximum integer such that $3d+1+2^k \leq n-i$.

When all of $g_2(p), g_3(p), \cdots, g_{3d+1}(p)$ have received the message, each node $u \in \{g_2(p), g_3(p), \cdots, g_{3d+1}(p)\}$ executes the following operations:

(2) Along g_{n-i+1}, u sends the message received from p. Then after $g_{n-i+1}(u)$ receives the message, it transmits the message to node u' through a path such that $r(u'[2, d+1]) = r(u'[d+2, 2d+1]) = r(u'[2d+2, 3d+1]) = j$. Let $Route_p^{(i,j)}(u)$ be the set of nodes on the path from u to u'. Node u' is added to $W_{i+1}(w_j)$.

Let $3d+1 < dim \leq 3d+1+2^k$. Each node v in the set of nodes (including $g_{3d+1+1}(p), g_{3d+1+2}, \cdots, g_{3d+1+2^k}(p)$) that received the message along an edge g^v_{dim} broadcasts the message in a binary jumping way as described in (3) below:

(3) When p sends a message along $g^p_{3d+1+2^{l-1}}$, v first receives a message from g^v_{dim} (it is not necessarily that v receives the message directly from p). Then v sends its message to $g_{dim+2^l}(v), g_{dim+2^{l+1}}(v), \cdots, g_{dim+2^q}(v)$ sequentially, where q is the maximum integer such that $dim+2^q \leq n-i$. When the binary jumping transmissions finished, p and all the nodes that received the message along a dimension among $g_{3d+2}, g_{3d+3}, \cdots, g_{n-i}$, send the message along g_{n-i+1}. The set of the final $n-3d-i$ nodes that have received the message along g_{n-i+1} is denoted by $BJ_i(p)$. All the nodes in $BJ_i(p)$ are added to $W_{i+1}(w_j)$, too.

The operations listed (1), (2) and (3) above are called Rule (1), Rule (2) and Rule (3), respectively. A detailed description about this scheme will be given in Subsection 4.2. From the scheme given above we have the next lemma.

Lemma 6. *Let $w_j = *j$ $(1 \leq j < n)$ be a message source in S_n at the beginning of the recursive stage. For each i $(1 \leq i < n-3d)$ and for an arbitrary node $p \in W_i(w_j)$, $r(p[2,d+1]) = r(p[d+2,2d+1]) = r(p[2d+2,3d+1]) = j$.* □

Definition 7. Let $p \in W_i(w_j)$, $H = \{g_2(p), g_3(p), \cdots, g_{3d+1}(p)\}$ and $Y_i(w_j,p) = \bigcup_{u \in H} Route_p^{(i,j)}(u)$. Let $X_i(w_j,p)$ be the set of nodes that have received the message directly or indirectly from node p but not in $Y_i(w_j,p)$ during round i. We define $U_i(w_j) = \bigcup_{p \in W_i(w_j)} Y_i(w_j,p)$ and $V_i(w_j) = \bigcup_{p \in W_i(w_j)} X_i(w_j,p)$.

Definition 8. Let w_j be a message source in S_n at the beginning of the recursive stage. For $1 \leq i < n-3d$, the *i-level channel* rooted at w_j, denoted by $C_i(w_j)$, is defined as $\bigcup_{k=1}^{i} (U_k(w_j) \cup V_k(w_j))$.

The next lemma is immediate from the definition of the *i-level channel.*

Lemma 9. *For $1 \leq i < n-3d$, at least one node in each S_{n-i} obtained by fixing the last i symbols of the labels of nodes of S_n is contained in the i-level channel rooted at w_j, $C_i(w_j)$.* □

Lemma 10. *Let $w_j = *j$ $(1 \leq j < n)$ be a message source in S_n at the beginning of the recursive stage satisfying $r(w_j[2,d+1]) = r(w_j[d+2,2d+1]) = r(w_j[2d+2,3d+1]) = j$. Then for each i $(1 \leq i < n-3d)$ there exists an i-level channel rooted at w_j, $C_i(w_j)$, such that the id of every node in $C_i(w_j)$ is j.*

Proof. By the definition of the i-level channel rooted at w_j, $C_i(w_j) = \bigcup_{k=1}^{i} (U_k(w_j) \cup V_k(w_j))$. From Lemma 4.2, for each node $p \in W_i(w_j)$, $r(p[2,d+1]) = r(p[d+2,2d+1]) = r(p[2d+2,3d+1]) = j$. From Rule 3 of the recursive stage, the identifier district of each node in $V_k(w_j)$ $(1 \leq k < n-3d)$ remains unchanged. Hence, the *id*'s of all the nodes in $\bigcup_{k=1}^{n-3d-1} V_k(w_j)$ are equal to j. On the other hand, from Rule 1 and Rule 2 of the recursive stage, at most one of $u[2,d+1]$, $u[d+2,2d+1]$ and $u[2d+2,3d+1]$ of each node $u \in U_k(w_j)$ $(1 \leq k < n-3d)$ has changed. Hence, the *id*'s of all the nodes in $\bigcup_{k=1}^{n-3d-1} U_k(w_j)$ are equal to j.

From Lemma 4.1 and Lemma 4.4, the following theorem is immediate.

Theorem 11. *Let $w_1, w_2, \cdots, w_{n-1}$ be $n-1$ message sources in S_n. Then for each i $(1 \leq i < n-3d)$, there exist $n-1$ node-disjoint i-level channels $C_i(w_1)$, $C_i(w_2), \cdots, C_i(w_{n-1})$.* □

Lemma 12. *For $1 \leq i < n-3d$, round i of the recursive stage takes $3d + \max\{\log(n-3d-i)+1, \lfloor \frac{3d}{2} \rfloor + 1\}$ steps.*

Proof. The process of round i contains two parts. One is to broadcast the message to $U_i(w_j)$, and the other is to broadcast the message to $V_i(w_j)$. At the beginning of round i, each node $p \in W_i(w_j)$ sends its message to its $3d$ neighbors $g_2(p), g_3(p), \cdots, g_{3d+1}(p)$ sequentially. This process takes $3d$ steps. Then node p sends the message to $g_{3d+1+1}(p), g_{3d+1+2}(p), \cdots, g_{3d+1+2^k}(p)$ sequentially. Nodes $g_2(p), g_3(p), \cdots, g_{3d+1}(p)$ that received the message from p, send concurrently the message along g_{n-i+1}. Since the distance between u and u' specified in Rule 2 of the recursive stage is at most $1 + \lfloor \frac{3}{2}d \rfloor$, the transmission of the message from u to u' takes at most $1 + \lfloor \frac{3}{2}d \rfloor$ steps. On the other hand, broadcasting to $V_i(w_j)$ takes $3d + \log(n - 3d - i) + 1$ steps. Hence, the lemma holds.

Theorem 13. *Let s be the source node in S_n. For an arbitrary sub-star network S_{3d+1} obtained by fixing the last $n - 3d - 1$ symbols of the labels of nodes in S_n, s can send the message to $n - 1$ distinct nodes in the S_{3d+1} through $n - 1$ node-disjoint paths within $n + \lfloor \frac{9d}{2} \rfloor - 1 + \sum_{i=1}^{n-3d-1} (3d + \max\{\log(n - 3d - i) + 1, \lfloor \frac{3d}{2} \rfloor + 1\})$ steps.*

Proof. Initially, s sends the message to $n - 2$ neighbors $g_2(s), g_3(s), \cdots, g_{n-1}(s)$ sequentially. Then the $n - 1$ nodes with the message (including s) send the message along dimension n. Then $n - 1$ nodes $*1, *2, \cdots, *n - 1$ receive the message. Furthermore, each node $*j$ $(1 \leq j < n)$ transmits the message to node w_j that satisfies $r(w_j[2, d+1]) = r(w_j[d+2, 2d+1]) = r(w_j[2d+2, 3d+1]) = j$. Clearly, these $n - 1$ paths are node-disjoint. Then we may regard such node w_j as a message source, and let these nodes be $w_1, w_2, \cdots, w_{n-1}$. The total number of steps of these operations is at most $n + \lfloor \frac{9d}{2} \rfloor - 1$.

From Lemma 4.3, $C_{n-3d-1}(w_j)$ $(1 \leq j < n)$ will be used to send the message of w_j to all S_{3d+1}'s obtained by fixing the last $n - 3d - 1$ symbols of the node labels. That is, at least one node in each S_{3d+1} has received the message from w_j. From Theorem 4.5, arbitrary two $C_{n-3d-1}(w_{j_1})$ and $C_{n-3d-1}(w_{j_2})$ are node-disjoint if $j_1 \neq j_2$. From Lemma 4.6, the total number of steps is $n + \lfloor \frac{9d}{2} \rfloor - 1 + \sum_{i=1}^{n-3d-1} (3d + \max\{\log(n - 3d - i) + 1, \lfloor \frac{3d}{2} \rfloor + 1\})$. Hence, the theorem holds.

4.2 The Diffusing-and-Disseminating Scheme

We assume that there exist at most f faults in S_n, where $f < n - 1$. Our broadcasting scheme, called the *diffusing-and-disseminating*, is described as follows:

In the diffusing stage, a node with the message often has a block in its *identifier district* such that the rank of the block is different from the rank of the corresponding block of the message source. Such a block is called a destroyed block.

procedure *diffusing-and-disseminating*
(* for each node u *)
if u is the source node **then** /* the pre-stage */
  for $i := 2$ **to** $n - 1$ **do**

```
        send the message along g_i^u;
  if u has the message then send the message along g_n^u;
  if u received a message from g_n^u
    then u decides a unique node v satisfying r(v[2,d+1]) = r(v[d+2,2d+1])
          = r(v[2d+2,3d+1]) = u[n,n] and sends a message to v by calling
          route(u,v)
  for i := 0 to n - 3d - 2 do /* the recursive stage */
  begin
     if u has the message then
     begin
        for j := 2 to 3d+1 do
            send the message along g_j^u
        call binary-jump(3d+1, n-i)
     end;
     if u received the message from g_j^u (1 < j < 3d+2)
       then send the message along g_{n-i}^u;
     if u received a message from g_{n-i}^u
       then u finds the destroyed block and sends a message to a unique node
            v without any destroyed block by calling route(u,v)
  end;
  call dissem(n, ⌊9d/2⌋ + 4). /* the disseminating stage */
```

Remark: Node u can calculate the node v mentioned in the description of procedure *diffusing-and-disseminating*. From v, node u creates a packet $\langle message, k, v_1 v_2 \cdots v_d\rangle$, where k is 0, 1, or 2, and initiates a subroutine $route(u, v)$. The detail of this part will be given in a full version of this paper.

The following procedure $binary\text{-}jump(n_1, n_2)$ is to distribute the message from a node in S_n to its neighbors in a binary jumping way. The node with the message at the beginning of each round i is regarded as a message source in round i.

```
  procedure    binary-jump(n1, n2)
   (* for each node u *)
  if u is a message source then X := n1;
  for i := 1 to ⌈log(n2 - n1 + 1)⌉ do
     if u has the message then
     begin
        dim := X + 2^(i-1);
        if dim ≤ min{n - 1, n2} then
            send the message along g_dim^u
     end
     else
        if the message was first received by u then
           X := the dimension through which the message came;
  if u has the message then send the message along g_{n2}^u.
```

Lemma 14. *Let* $u = h_1 d_2 \cdots d_{n_1} a_{n_1+1} \cdots a_{n_2} *$ *be a message source. After the* *execution of the* **for** *loop of* binary-jump(n_1, n_2) *at* u, *there exists a set of* $n_2 - n_1$ *nodes* $\{a_{n_1+1}d_2 \cdots d_{n_1}*, a_{n_1+2}d_2 \cdots d_{n_1}*, \cdots, a_{n_2}d_2 \cdots d_{n_1}*\}$ *such that each node in the set has received the message.*

Proof. We prove the lemma by induction on i, the number of iteration **for** loop. For $i = 1$, the message source sends the message along $g^u_{n_1+1}$, and then the node that receives the message has a_{n_1+1} in the first position of its label.

Suppose that the lemma holds true up to the ith iteration. Then after the ith iteration, the nodes that received the message have $a_{n_1+1}, \cdots, a_{n_1+2^i-1}$ in the first positions of their labels. We first assume that $2^{i+1} \leq n_2$. Then at the $(i+1)$st iteration the message source u sends the message to a node with $a_{n_1+2^i}$ at the first position of its label. Since node $a_{n_1+l}*$ $(1 \leq l < 2^i)$ received the message from dimension $n_1 + l$, the node sends the message to a node with $a_{n_1+l+2^i}$ at the first position of its label. If $2^i \leq n_2$ and $2^{i+1} > n_2$, then there exists a number c $(1 \leq c < 2^i)$ such that $2^i + c = n_2$.

The following procedure *route*(u, v) can transmit a message from u to a node with no destroyed block in its *id* district.

```
procedure    route(u,v)
 (*for each node u *)
begin
   for i:= 1 to ⌊3d/2⌋ do
      if u created or received a packet ⟨message, k, v1v2 ··· vd⟩ in the last
       step /* k = 0,1 or 2 */
      then if u[kd + 2, (k + 1)d + 1] ≠ v1v2 ··· vd
         then
         begin
            if u[1,1] ∈ {v1, v2, ··· , vd}
            then Let j be the subscript such that u[1,1] = vj
            else Let j be the smallest subscript such that
             u[j + kd + 1, j + kd + 1] ≠ vj;
            u sends the packet ⟨message, k, v1v2 ··· vd⟩ along (j + kd + 1)
         end
end.
```

Lemma 15. *For an arbitrary node* u *in* S_n, route(u, v) *can transmit a message from* u *to a node* v *with* $r(v[2, d+1]) = r(v[d+2, 2d+1]) = r(v[2d+2, 3d+1])$ *within* $\lfloor \frac{3d}{2} \rfloor$ *steps if exactly one of* $u[j+1, j+d]$ $(j = 1, d+1,$ *or* $2d+1)$ *is a destroyed block.*

Proof. These two nodes u and u' are in the same S_{d+1} obtained by fixing the last $n - d - 1$ symbols of the node labels in S_n. Hence, the maximum number of steps needed to transmit the message is at most the diameter of S_{d+1}.

Theorem 16. *Procedure* diffusing-and-disseminating *can broadcast a message from the source node to all other nodes in* S_n *within* $(1 + \varepsilon)n \log n$ *steps if there exist at most* $n - 2$ *faulty nodes and/or links in the network, where* ε *is any positive constant less than* 1.

Proof. For a sufficiently large n, $(\varepsilon_1 \log n)! \geq n$. Assume $d = \varepsilon_1 \log n$. The total number of steps $T(n)$ needed by ***diffusing-and-disseminating*** is evaluated as follows:

$$\begin{aligned}
T(n) &= n + \frac{9d}{2} - 1 + (\frac{9d}{2} + 4)(n-1) \\
&\quad + \sum_{i=1}^{n-3d-1} (3d + \max\{\log(n-3d-i) + 1, \frac{3d}{2} + 1\}) \\
&\leq n + \frac{9d}{2} - 1 + (\frac{9d}{2} + 4)(n-1) + (\frac{9d}{2} + 1)(n - 3d - 1) \\
&\quad + \sum_{i=1}^{n-3d-1} (\log(n-3d-i) + 1) \\
&= \varepsilon n \log n + \log(n - 3d - 1)! + O(n) \\
&= (1 + \varepsilon) n \log n + O(n).
\end{aligned}$$

5 Analysis of Broadcasting in S_n with Random Faults

Since the connectivity of S_n is $n-1$, any broadcasting scheme for S_n cannot tolerate more than $n-2$ faults in the worst case. However, if faulty places are randomly distributed in a network, then the worst case rarely occurs. Even if there exist many more than $n-2$ faults in S_n, broadcasting may succeed with a high probability. In this section, we show that for any constant $\alpha < 1$, if there are no more than $(n!)^\alpha$ faulty nodes randomly distributed in S_n, then broadcasting by our scheme succeeds with a probability higher than $1 - 1/n!$.

We assume that there are at most f faulty nodes and no faulty links. We also assume that for each i $(1 \leq i \leq f)$ all configurations of S_n with i faulty nodes are equally probable and that the probability of a configuration with i faults is p_i, where $\sum_{i=1}^{f} p_i = 1$. We denote a configuration of S_n with i faulty nodes by $conf(n,i)$. The number of different $conf(n,i)$'s is $\binom{n!}{i}$. We denote $CONF(n,i)(fail) = \{conf(n,i) : \text{broadcasting fails in } conf(n,i)\}$. Then the following equality holds:

$$Prob\{\text{broadcasting succeeds}\} = 1 - \sum_{i=1}^{f} |CONF(n,i)(fail)| \binom{n!}{i}^{-1} p_i. \quad (1)$$

Let s be the source node and v be an arbitrary node in S_n. Denote $CONF(n,i)(fail, v) = \{conf(n,i) : \text{the message from } s \text{ cannot reach } v \text{ by our broadcasting scheme in } conf(n,i)\}$.

Then we have the next inequality.

$$|CONF(n,i)(fail)| < \sum_{v \in V(S_n) - \{s\}} |CONF(n,i)(fail, v)|. \quad (2)$$

The exact evaluation of $|CONF(n,i)(fail,v)|$ is complicated, but we can derive an upper bound on $|CONF(n,i)(fail,v)|$.

Let us review how the message is transmitted from the source node s to a node v. In the *diffusing* stage, the message is sent to $n-1$ nodes, say $s_1, s_2, \cdots, s_{n-1}$, through $n-1$ disjoint channels. If any of $s_1, s_2, \cdots, s_{n-1}$ receives the message then the *diffusing* is said to be successful, and otherwise, it is said to be unsuccessful. Since the length of the path used to send the message from s to v in each channel is not greater than $n \log n$, we have the following inequality.

$$|\{conf(n,i): \textit{ diffusing} \text{ stage fails}\}| < (n \log n)^{n-1} \binom{n! - (n-1)}{i - (n-1)}. \quad (3)$$

We next consider the *disseminating* (*dissem* for short) stage. In the *disseminating* stage, for each j $(1 \le j \le n-1)$, the message is sent from s_j to v through $n-1$ node-disjoint paths. For the difficulty of deriving an accurate estimation of the failure probability, we loosen the definition of a failure in the *disseminating* stage as follows: We say that the *disseminating* stage fails if there exists a s_j such that sending the message from s_j to v fails. Then, we have the following inequality.

$$|\{conf(n,i): \textit{dissem} \text{ stage fails}\}| < \sum_{j=1}^{n-1} |\{conf(n,i): \textit{ dissem} \text{ from } s_j \text{ to } v \text{ fails}\}|. \quad (4)$$

Since each of the $n-1$ node-disjoint paths from s_j to v is not longer than $\log n$, we have the following two inequalities.

$$|\{conf(n,i): \textit{ disseminating} \text{ from } s_j \text{ to } v \text{ fails }\}| < (\log n)^{n-1} \binom{n! - (n-1)}{i - (n-1)}. \quad (5)$$

$$|CONF(n,i)(fail,v)| < (n \log n)^{n-1} \binom{n! - (n-1)}{i - (n-1)} + n(\log n)^{n-1} \binom{n! - (n-1)}{i - (n-1)}. \quad (6)$$

From (1), (2) and (6), we obtain the following inequality,

$$Prob\{\text{broadcasting succeeds}\} >$$
$$1 - \sum_{i=1}^{f} 2n!(n \log n)^{n-1} \binom{n! - (n-1)}{i - (n-1)} \binom{n!}{i}^{-1} p_i. \tag{7}$$

Theorem 17. *For any constant $\alpha < 1$, if all links are healthy and there are no more than $(n!)^\alpha$ faulty nodes randomly distributed in S_n, then broadcasting by our scheme succeeds with a probability higher than $1 - 1/n!$.*

Proof. Let $f = (n!)^\alpha$. From inequality (7) and the following inequality

$$\binom{n! - (n-1)}{i - (n-1)} \binom{n!}{i}^{-1} = \frac{i(i-1)\cdots(i-n+2)}{n!(n!-1)\cdots(n!-n+2)} < \left(\frac{i}{n!}\right)^{n-1}.$$

Hence,

$$Prob\{\text{broadcasting success}\} > 1 - \sum_{i=1}^{f} p_i \left(\frac{n^2 i \log n}{n!}\right)^{n-1} >$$
$$1 - \left(\frac{(n!)^\alpha n^2 \log n}{n!}\right)^{n-1} > 1 - \frac{1}{n!}.$$

The lower bound shown in the proof of Theorem 5.1 on the probability of successful broadcasting is not tight. In our analysis, some failure configurations are counted more than once in our analysis, and some configurations that are actually not failure ones, are also counted as failure configurations in our analysis. For a sufficiently large n, $(n!)^\alpha$ is much larger than $n-2$, and the lower bound $1-\frac{1}{n!}$ on the probability of successful broadcasting is very close to 1. We therefore can say that our probabilistic analysis is satisfactory.

We can similarly analyze the reliability of broadcasting in the case where all nodes are healthy and there exist failed links randomly distributed in S_n. We can also analyze the reliability of broadcasting in a similar way in the case where both failed nodes and failed links exist, if an appropriate probabilistic distribution of such faults in S_n is given.

6 Conclusion

We showed that our broadcasting scheme tolerates up to $n-2$ faults in S_n and that its running time is $O(n \log n)$. This running time is optimal for the asymptotic order and almost optimal for the constant factor of the order. We conjecture that the scheme might be optimal even for the constant factor of the order, too. This problem is theoretically interesting, and worthy for the further investigation. Another interesting problem is to consider the situation where faults are Byzantine type. It is obviously that at most $t < (n-1)/2$ Byzantine faulty nodes can be tolerated in worst case. Is there any single-port broadcasting scheme with running time $O(n \log n)$ for this?

References

1. S. B. Akers, D. Harel, and B. Krishnamurthy, "The star graph: An attractive alternative to the n-cube." In *Proc. Int. Conf. Parallel Processing,* pp. 393–400, 1987.
2. S. B. Akers and B. Krishnamurthy, "On group graphs and their fault tolerance." *IEEE Trans. Computers,* Vol. C-36, pp. 885–888, 1987.
3. S. B. Akers and B. Krishnamurthy, "A group-theoretic model for symmetric interconnection networks." *IEEE Trans. Computers,* Vol. 38, pp. 555–566, 1989.
4. N. Bagherzadeh, N. Nassif, and S. Latifi, "A routing and broadcasting scheme on faulty star graphs." *IEEE Trans. Computers,* Vol. 42, pp. 1398–1403, 1993.
5. S. Carlsson, Y. Igarashi, K. Kanai, A. Lingas, K. Miura, and O. Petersson, "Information disseminating schemes for fault tolerance in hypercubes." *IEICE Trans. Fundamentals,* Vol. E75-A, pp. 255–260, 1992.
6. K. Day and A. Tripathi, "A comparative study of topological properties of hypercubes and star graphs." *IEEE Trans. Parallel and Distributed Systems,* Vol. 5, pp. 31–38, 1994.
7. P. Fragopoulou and S. G. Akl, "Optimal communication algorithms on star graphs using spanning tree constructions." *Journal of Parallel and Distributed Computing,* Vol. 24, pp. 55–71, 1995.
8. L. Gargano, A. A. Rescigno, and U. Vaccaro, "Optimal communication in faulty star networks." Manuscript, 1995.
9. S. M. Hedetniemi, S. T. Hedetniemi, and A. L. Liestman, "A survey of gossiping and broadcasting in communication networks." *Networks,* Vol. 18, pp. 319–349, 1988.
10. A. L. Liestman, "Fault-tolerant broadcast graphs." *Networks,* Vol. 15, pp. 159–171, 1985.
11. A. Mei, Y. Igarashi, and N. Shimizu, "Efficient broadcasting on faulty star networks." In *Proc. 4th International School and Symposium: Formal Techniques in Real-Time and Fault-Tolerant Systems*, Uppsala, Sweden, 1996, to appear.
12. V. E. Mendia and D. Sarkar, "Optimal broadcasting on the star graph." *IEEE Trans. Parallel and Distributed Systems,* Vol. 3, pp. 389–396, 1992.
13. A. Pelc, "Fault-tolerant broadcasting and gossiping in communication networks." Manuscript, 1995.
14. D. Peleg, "A note on optimal time broadcast in faulty hypercubes." *Journal of Parallel and Distributed Computing,* Vol. 26, pp. 132–135, 1995.
15. P. Ramanathan and K. G. Shin, "Reliable broadcast in hypercube multicomputers." *IEEE Trans. Computers,* Vol. 37, pp. 1654–1657, 1988.
16. Y. Rouskov and P. K. Srimani, "Fault diameter of star graphs." *Information Processing Letters,* Vol. 48, pp. 243–251, 1993.
17. S. Sur and P. K. Srimani, "A fault tolerant routing algorithm in star graph interconnection networks." In *Proc. Int. Conf. Parallel Processing,* Vol. III, pp. 267–270, 1991.

A Lower Bound for Linear Interval Routing

T. Eilam, S. Moran and S. Zaks

Department of Computer Science
Technion, Haifa, Israel

Abstract. Linear Interval Routing is a space-efficient routing method for point-to-point communication networks. It is a restricted variant of Interval Routing where the routing range associated with every link is represented by an interval with no wrap-around. A common way to measure the efficiency of such routing methods is in terms of the maximal length of a path a message traverses. For Interval Routing the upper bound and lower bound on this quantity are $2D$ and $1.75D - 1$, respectively, where D is the diameter of the network. We prove a lower bound of $\Omega(D^2)$ on the length of a path a message traverses under Linear Interval Routing. We further extend the result by showing a connection between the efficiency of Linear Interval Routing and the bi_diameter of the network.

1 Introduction

In a communication network, where communication between nodes is accomplished by sending and receiving messages, a routing algorithm is employed to ensure that every message will reach its destination. An *optimal* routing method routes every message to its destination in the shortest way. Usually optimal routing is achieved by keeping in each node a table with n entries and such that the i-th entry in the table determines the edge to be traversed by a message destined to node i. For large networks it is practical to consider routing methods in which less than $O(n)$ space is used in each node. Such a routing scheme is *Interval Routing* which was introduced in [SK82], and was discussed together with other compact routing methods in [LT83]. Under Interval Routing the nodes of the network are labeled with unique integers from the set $\{0, ..., n-1\}$, where n is the number of nodes in the network, and at each node each of its adjacent edges is labeled with one interval of $\{0, ..., n-1\}$. Cyclic intervals, (i.e., intervals that wrap-around over the end of the name segment to its beginning), are allowed. Under such Interval Labeling Scheme (ILS), at each node i, messages destined to node j are sent on the unique edge that is labeled with an interval that contains j. A *valid* ILS of a graph is an ILS under which every message will eventually arrive at its destination. It was shown in [SK85] that every graph admits a valid ILS, under which every message will traverse a path of length at most $2D$, where D is the diameter of the graph. In [R91] a lower bound of $1.5D + 0.5$ was proved for the maximal length of a path a message traverses under Interval Routing, and this bound was recently improved ([TL94]) to $1.75D - 1$.

Linear Interval Routing ([BLT91]) is a restricted variant of Interval Routing which uses only linear intervals (i.e., wraps-around are not allowed). Its advantage over Interval Routing is that it can be used for routing in dynamic networks, where insertion and deletion of nodes can occur. In fact it is the simplest form of Prefix Routing, introduced in [BLT90] as a space-efficient routing method for dynamic networks. In [BLT90] it was shown that there is a valid PLS (Prefix Labeling Schema) for every dynamically growing network and that insertions of links or nodes require an adaptation cost of $O(1)$. It was noted in [BLT91] that not every graph has a valid Linear Interval Labeling Scheme (LILS). A complete characterization of the graphs that admit a valid LILS was recently presented in [FG94], together with an algorithm that generates a valid LILS in case one exists. There are no known non-trivial upper or lower bounds for the longest path traversed by a message under Linear Interval Routing. The algorithm of [FG94] does not imply a (non-trivial) upper bound (such as the $2D$ bound for Interval Routing); actually, a graph with an arbitrary number of nodes n and diameter $D = 2$ is shown in [FG94], for which there exists a labeling scheme with longest path of $O(1)$, but the algorithm generates an LILS under which the maximal length a message traverses is $n - 1$.

In this paper we prove that, under this measure of the longest path of a single message, the restriction to linear intervals is very costly. We prove a lower bound of $\Omega(D^2)$ on the maximal length of a path a message traverses under Linear Interval Routing. We show a connection between the bi_diameter of a graph (defined in Sec. 4) and a lower bound on the efficiency of an LILS for it, then we analyze the relation between the bi_diameter and the diameter of graphs, and as a consequence we conclude that, using our technique, a lower bound of $\Omega(D^2)$ is the best one can achieve.

In Sec. 2 we present the model together with a precise description of the routing methods under discussion. The lower bound is proved in Sec. 3. In Sec. 4 we generalize the result and show the connection to the bi_diameter. Conclusions and open questions are discussed in Sec. 5.

2 Preliminaries

We assume a point-to-point asynchronous communication network where processors communicate with their neighbors by exchanging messages along the communication links, which are bidirectional. The network topology is modeled by a symmetric digraph $G = (V, E)$, $|V| = n$, where the set V of nodes corresponds to the processors, and the set E of edges corresponds to their communication links. For an edge $e = (u, v) \in E$, the edge (v, u) is termed the ***opposite edge*** of the edge e.

Each message has an header that includes its destination. As a message arrives at each intermediate node, if it is the target then the message is processed, otherwise the edge on which it will continue is determined by a local routing decision function.

The routing methods in discussion involve a labeling of the nodes and edges of the graph. Denote $N = \{0, ..., n-1\}$. Each node is labeled with a unique integer in N, termed *node number*, and each edge (in each direction) is labeled with an *interval of* N, defined as follows.

Definition 1. An *interval of* N is one of the following:

1. A *linear interval* $\langle p, q \rangle = \{p, p+1, ...q\}$, where $p, q \in N$ and $p \leq q$.
2. A *wrap-around interval* $\langle p, q \rangle = \{p, p+1, ..., n-1, 0, ..., q\}$, where $p, q \in N$ and $p > q$.
3. The *null interval* $\langle\rangle$ is the empty set.

We say that a node $u \in V$ is *contained* in an interval $\langle p, q \rangle$ if $u \in \langle p, q \rangle$. Note that the null interval does not contain any node.

Definition 2. An *Interval Labeling Scheme* (ILS) $L_G = (L_V, L_E)$ of a graph $G = (V, E)$ is defined by:

1. A one-to-one function $L_V : V \rightarrow N$ that labels the vertices of V.
2. An edge labeling function $L_E : E \rightarrow I$, where I is the set of intervals of N, that satisfies the following properties for every $u \in V$:

   union property the union of all the intervals corresponding to the edges outgoing u is equal to N or $N - \{L_V(u)\}$.

   disjunction property the intervals corresponding to any two directed edges outgoing u are disjoint.

   (In other words, the non-empty intervals associated with all the edges outgoing any vertex u form a partition of N or of $N - \{L_V(u)\}$.)

   For $e = (u, v)$ we will write $L_E(u, v)$ instead of $L_E((u, v))$.

For simplicity, from now on, given an ILS and a node u, we will not distinguish between the node u and its node number $L_V(u)$.

Example 1. Figure 1 shows a graph and an ILS of it. The edge $(4, 0)$ is labeled with the linear interval $\langle 0, 1 \rangle$ the edge $(2, 3)$ is labeled with the wrap-around interval $\langle 3, 0 \rangle$ and the edge $(1, 2)$ is labeled with the null interval $\langle\rangle$. Note that the union property and the disjunction property are satisfied at each node.

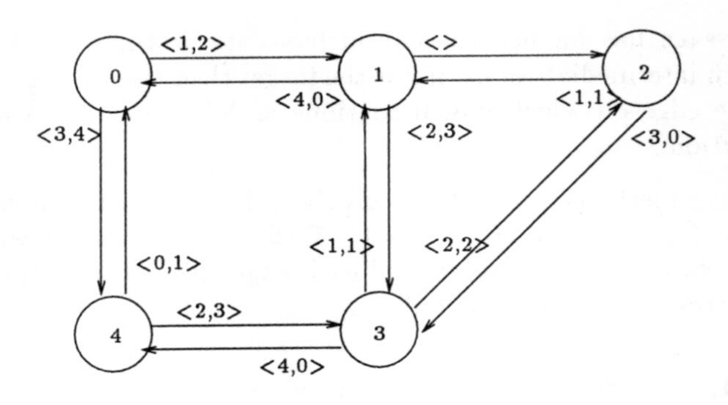

Fig. 1. Example of Interval Routing.

Two special cases of Interval Labeling Scheme are now presented; in *linear interval labeling scheme* no wrap-around intervals are allowed, and in *strict-linear interval labeling scheme* we also require that no interval on an edge outgoing a vertex includes it. Formally:

Definition 3. A *linear* interval labeling scheme (LILS) $L_G = (L_V, L_E)$ of a graph $G = (V, E)$ is an interval labeling scheme in which the interval $L_E(e)$ is either linear or null, for every $e \in E$.

Definition 4. A *strict-linear* interval labeling scheme (strict-LILS) $L_G = (L_V, L_E)$ of a graph $G = (V, E)$ is a linear interval labeling scheme in which $u \notin L_E(u, w)$ for every node $u \in V$ and every edge $(u, w) \in E$.

By a *labeling scheme* we mean ILS or any of its two variants. Given a graph with any labeling scheme, the routing of message is performed as follows. If node u has a message destined to node v, $v \neq u$, then u will send the message on the *unique* edge (u, w) such that $v \in L_E(u, w)$.

Example 2. Refer to Fig. 1. Assume that node 2 has a message to send to node 0. It will send the message via the edge $(2, 3)$ (since 0 is contained in the interval $\langle 3, 0\rangle = L_E(2, 3)$), the message will then be sent to node 4 ($0 \in \langle 4, 0\rangle = L_E(3, 4)$) and then to node 0 ($0 \in \langle 0, 1\rangle = L_E(4, 0)$).

For a graph G, let $d_G(u, v)$ denote the (shortest) distance between the vertices u and v of G. The *diameter* of G, denoted D_G, is defined as

$$D_G = \max_{u,v \in V} d_G(u, v).$$

We measure the efficiency of a labeling scheme L_G of G by means of the longest path a message will traverse if the routing on G is performed according to L_G. Given a labeling scheme L_G of G, and two nodes u and v, $Path_G(u,v,L_G) = \langle u = x_0, x_1, ..., x_n = v \rangle$ denotes the path a message, destined to v, will traverse starting from u, under the labeling scheme L_G, and $Dist_G(u,v,L_G)$ denote its length (number of edges). We define

$$Dist_G(L_G) = \sup_{u,v \in V} Dist_G(u,v,L_G).$$

Obviously, if the labeling scheme is arbitrary, routing cannot ensure that every message will eventually arrive to its destination; though a message from u to v cannot be stuck at any node, it still can cycle forever without getting to v. We thus introduce the following definition.

Definition 5. A *valid* labeling scheme of a graph G is a labeling scheme L_G for which $Dist_G(L_G) < \infty$.

From the definitions it is clear that if the routing is done according to a valid labeling scheme, each message will follow a simple path (in which all vertices are distinct), which implies that $Dist_G(L_G) < n$, and thus in the definition of $Dist_G(L_G)$ we could replace sup with max when considering valid labeling schemes.

Example 3. The ILS presented in Fig. 1 is valid.

Often there is more than one labeling for a graph G. The issue in question is: How good can we do with each of the routing methods in discussion? To express this question in formal terms we denote by $OPT_{ILS}(G)$ the minimum $Dist_G(L_G)$ over all ILSs of G. Formally,

$$OPT_{ILS}(G) = \min_{L_G} Dist_G(L_G).$$

$OPT_{LILS}(G)$ and $OPT_{strict-LILS}(G)$ are defined in a similar way considering the LILSs, and strict-LILSs, of G, respectively.

For Interval Routing, upper and lower bounds for $OPT_{ILS}(G)$ are known. An upper bound of $2D$ was presented in [SK85], where D is the diameter of the network, and a lower bound of $1.5D + 0.5$ was presented in [R91] and was later improved to $1.75D - 1$ in [TL94].

In our proof we use the following characterization of graphs that admit a valid LILS.

Definition 6. ([FG94]) A lithium graph is a connected graph $G = (V, E)$, $V = V_0 \cup V_1 \cup V_2 \cup V_3$, the sets V_i are disjoint, $|V_1|, |V_2|, |V_3| \geq 2$ and there is a ***unique*** edge connecting V_0 with each of the V_i's (and no edge between the V_i's, for $i = 1, 2, 3$).

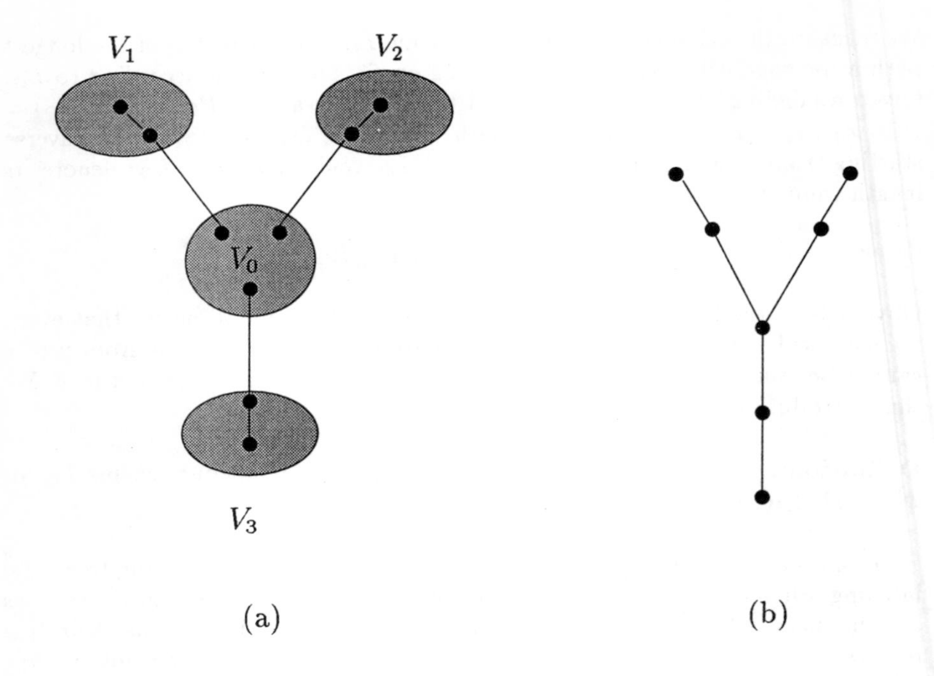

Fig. 2. (a) The general form of a lithium graph. (b) The simplest lithium graph.

Example 4. ([FG94]) Figure 2 shows the general form of a lithium graph and the simplest lithium graph (known as the Y graph).

Theorem 7. *([FG94]) A graph G has a valid LILS iff G is* not *a lithium graph.*

In [FG94], an algorithm that generates LILS for graphs that are not lithium graphs, was presented, but the algorithm does not imply a non-trivial upper bound for $OPT_{LILS}(G)$. Actually, for a graph G with an arbitrary number of nodes n and diameter $D = 2$, it could generate an LILS, L_G, such that $Dist_G(L_G) = n - 1$. Note that this leaves the upper bound question still open.

3 An $\Omega(D^2)$ Lower Bound

Following is the statement of our lower bound result.

Theorem 8. *For every constant D, There is a graph G with diameter $D_G > D$, that has a valid LILS, and $OPT_{LILS}(G) \geq D_G^2/18$.*

Following are two general observations that we use in the proof of Theorem 8.

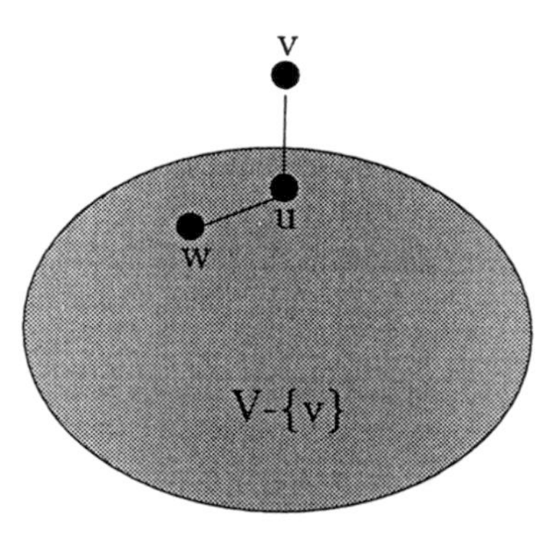

Fig. 3. For any LILS, the interval label of the edge (u, w) cannot contain both 0 and $n-1$.

Observation 9. *Let L_G be a strict-LILS of a graph with n vertices G. For any edge $e = (u, v)$ in G, the interval label $L_E(e)$ of e cannot contain both 0 and $n-1$.*

Proof. If $L_E(e)$ contains both 0 and $n-1$, then $L_E(e) = \langle 0, n-1 \rangle$, thus $u \in L_E(e)$, a contradiction. □

Observation 10. *Consider a graph $G = (V, E)$ with n vertices and any valid LILS of it, L_G. Let $(u, v) \in E$ such that degree$(v) = 1$ (i.e., v has only u as a neighbor). If $L_V(u) \neq 0, n-1$ and $L_V(v) \neq 0, n-1$, then $L_E(u, w) \neq \langle 0, n-1 \rangle$, where w is any neighbor of u. (See Fig. 3.)*

Proof. It is clear that if L_G is a *valid* LILS then any node $x \neq u, v$ satisfies: $L_V(x) \notin L_E(u, v)$. Any different edge e, outgoing u, will satisfy $v \notin L_E(e)$, and thus $L_E(e) \neq \langle 0, n-1 \rangle$. □

Following is the proof of Theorem 8.

Sketch of Proof. We prove the theorem in two steps. In step 1 we use Observation 9 to prove it for strict-LILS and in step 2 we use Observation 10 to extend the result to any LILS.

Step 1: strict-LILS.

We construct graphs G_k (for simplicity we assume that k is even), as follows (G_k, for $k = 4$, is depicted in Fig. 4, where each undirected edge between u and v stands for a directed edge (u, v) and its opposite edge (v, u)). The graph $G_k = (V_k, E_k)$ has three isomorphic *wings*, W_1, W_2 and W_3. Each wing W_i

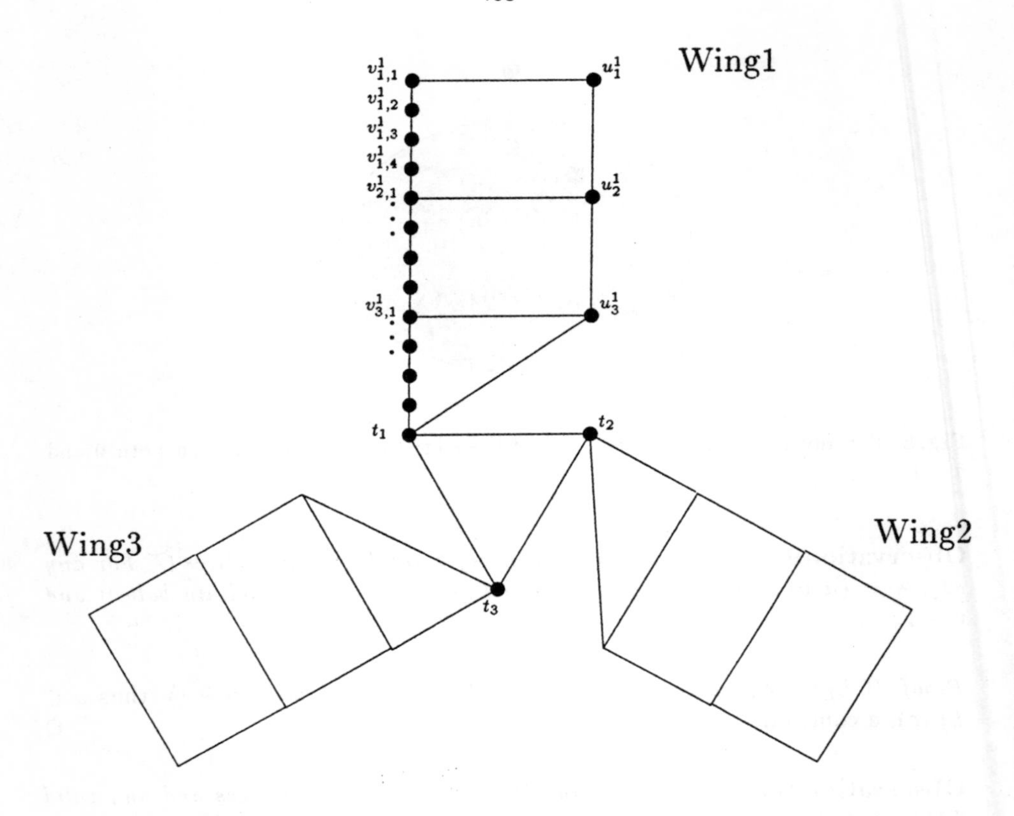

Fig. 4. The graphs G_k for $k = 4$. There are three isomorphic wings. Only Wing 1 is detailed.

($i = 1, 2, 3$) contains the vertices $\{u^i_x | 1 \leq x \leq k-1\} \cup \{v^i_{x,y} | 1 \leq x \leq k-1, 1 \leq y \leq k\} \cup \{t_i\}$ and the following sets of edges. The *inward* edges: $\{(u^i_x, u^i_{x+1}) | 1 \leq x \leq k-2\} \cup \{(v^i_{x,y}, v^i_{x,y+1}) | 1 \leq x, y \leq k-1\} \cup \{(v^i_{x,k}, v^i_{x+1,1}) | 1 \leq x \leq k-2\} \cup \{(v^i_{k-1,k}, t_i)\} \cup \{(u^i_{k-1}, t_i)\}$, the *outward* edges which are the opposites of the edges above, and the *horizontal* edges: $\{(u^i_x, v^i_{x,1}) | 1 \leq x \leq k-1\}$ and their opposites. In addition, we have in G_k the following set of edges (which are not part of any wing): $\{(t_1, t_2), (t_2, t_3), (t_3, t_1)\}$ and their opposites. Note that the graph G_k is 2-edge-connected, it contains $3k^2$ vertices and, by Theorem 7, it has a valid LILS.

Proposition 11. *The diameter D_{G_k} of the graph G_k satisfies $D_{G_K} \leq 3k$.*

Sketch of Proof. Let s and t be any two vertices on G_k. It is easy to see that there is a path of length of at most $3k$ between s and t. The path starts from s, goes (in the shortest way) to the nearest u-node, continues on u-nodes (plus possibly two of the t_is) until it reaches the nearest u-node which is the nearest to t, and then continues in the shortest way to t. The proposition follows. □

Let L be any valid strict-LILS of G_k. We will denote by min and max the nodes u and v in G_k such that $L_V(u)$ is minimal and $L_V(v)$ is maximal (clearly, $min = 0$ and $max = 3k^2 - 1$).

There are three wings in G_k, and therefore in L at least one wing does not contain any of the vertices min and max. We assume w.o.l.g that the wing W_1 does not contain the vertices min and max. Let $P_{min} = Path_G(u_1^1, min, L) = \langle u_1^1 = x_0, x_1, ..., x_{l-1}, x_l = min\rangle$ and $P_{max} = Path_G(u_1^1, max, L) = \langle u_1^1 = y_0, y_1, ..., y_{m-1}, y_m = max\rangle$.

Proposition 12. *The* directed *paths P_{min} and P_{max} are edge-disjoint.*

Proof. $min \in L_E(e)$ for any edge e in the path P_{min}, and $max \in L_E(e)$ for any edge e in the path P_{max}, thus by Observation 9, P_{min} and P_{max} must be edge-disjoint. □

The following proposition is stronger than what we need for the proof; it also shows the general form of the paths P_{min} and P_{max}.

Proposition 13. *No outward edges of Wing* 1 *appear in either of the paths P_{min} or P_{max}.*

Sketch of Proof. An outward edge in Wing 1 has one of three forms: (u_{x+1}^1, u_x^1), $(v_{x,y+1}^1, v_{x,y}^1)$ or $(v_{x+1,1}^1, v_{x,k}^1)$. We will prove the proposition for an edge of the first form; the proof for the other two cases is similar. Consider any edge $e = (u_{i+1}^1, u_i^1)$, $1 \le i \le k-2$. If e appears in the path P_{min} then, by the structure of the graph G_k, it is clear that one of the vertices u_i^1 or $v_{i,1}^1$ appears in P_{min} before u_{i+1}^1. If it is u_i^1, then the node u_i^1 appears twice in P_{min}, which implies a cycle, in contradiction with the assumption that L is valid. If it is $v_{i,1}^1$, then consider the suffix of the path P_{min} from the point that the node u_{i+1}^1 (followed by the node u_i^1) appears in it. Recall that the node min is not one of the nodes in the wing W_1. By the assumption that L is a *valid* strict-LILS, we conclude that the path P_{min} will eventually meet t_1. Again, from the structure of the graph G_k, it is obvious that in order to reach t_1 from the node u_i^1, one of the nodes u_{i+1}^1 or $v_{i,1}^1$ should be met; in either case, this will be the second time the node appears in P_{min}, which implies a cycle, in contradiction with the assumption that L is valid. The proof that the same holds for the path P_{max} is similar. □

The general form of the paths P_{min} and P_{max} is now clear. Each of the paths goes from u_1^1 straight down towards the center triangle with possible changes of sides from a u-node to a v-node and vice-versa. Each square in wing W_1 is traversed via the short right path of length 1, or via the long left path of length k.

Proposition 14. *At least one of the paths P_{min} or P_{max} has length of at least $k^2/2$.*

Sketch of Proof. Each one of P_{min} and P_{max} contains either the edge (u_i^1, u_{i+1}^1) or the path (of length k) $\langle v_{i,1}^1, v_{i,2}^1, ..., v_{i,k}^1, v_{i+1,1}^1 \rangle$ for every $1 \leq i \leq k-1$. If the path P_{min} (P_{max}) contains the edge (u_i^1, u_{i+1}^1) for some i, $1 \leq i \leq k-1$, then, by Proposition 12, P_{max} (P_{min}) does not contain the same edge and thus it must contain the path $\langle v_{i,1}^1, v_{i,2}^1, ..., v_{i,k}^1, v_{i+1,1}^1 \rangle$. We conclude that either P_{min} or P_{max} must contain at least $k/2$ such paths each of length k, and thus have length of at least $k^2/2$. □

Proposition 14 implies that $OPT_{strict-LILS}(G_k) \geq k^2/2$, and by Proposition 11 we have $OPT_{strict-LILS}(G_k) \geq k^2/2 = (3k)^2/18 \geq D_{G_k}^2/18$.

Step 2: general LILS.
Assume now that L is a general LILS. Here we use the graphs G'_k, $k > 1$, (k is even), that are constructed from the graphs G_k in the following way. For each vertex u in G_k, add a *leaf* which is a unique vertex u' and the edges (u, u') and (u', u) . The resulting graphs, G'_k are demonstrated in Fig. 5. Note that, by Theorem 7, the graphs G'_k have a valid LILS.

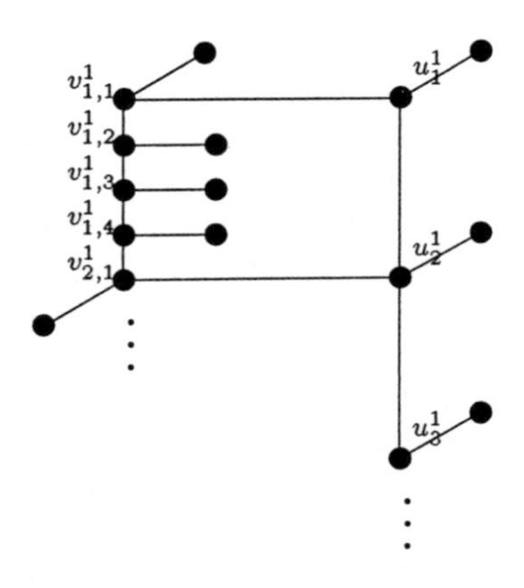

Fig. 5. A piece of the graph G'_k, $k = 4$.

It can be shown that the diameter of G'_k is bounded by $3k$ [1]. Note that, by Observation 10, Proposition 12 holds also for the graphs G'_k. It is clear that Proposition 13 and Proposition 14 hold also for the graphs G'_k (since, by Observation 10, the new edges are not part of the paths P_{min} and P_{max}). We conclude that under any LILS there will be a path in G'_k, that a message will have to follow, of length at least $D^2_{G'_k}/18$, and thus the theorem holds. □

Remark. Step 2 of the proof could also be proved on the original graphs G_k, but with a smaller constant (and a more detailed proof).

4 The Lower Bound as a Function of the Bi_diameter

We show a connection between the bi_diameter of a large family of graphs and a lower bound on the efficiency of their LILS. In order to simplify the definitions, we consider undirected graphs, which represent symmetric digraphs where every undirected edge is considered as a pair of directed anti-parallel edges. We define bi_diameter as follows.

Definition 15. Let $G = (V, E)$ be an undirected graph. For any $u, v \in V$, the *bi_distance* between u and v, denoted $bi\_dist_G(u, v)$, is the length of the shortest edge-simple cycle in G that includes the vertices u and v. If there is no such cycle then $bi\_dist_G(u, v) = \infty$. The *bi_diameter* of G is $bi\_diam(G) = \max_{u,v \in V} bi\_dist_G(u, v)$.

(The notion of bi_diameter was mentioned in [LSX95] in connection with the study of de Bruijn graphs, but their definition of bi_diameter is related to vertex-connectivity while ours is related to edge-connectivity.)

Definition 16. A *leaf* is a vertex u with only one adjacent vertex (i.e., degree(u)=1). For a graph G, the *reduced graph* $R(G)$ is the graph obtained from G by removing its leaves (and their corresponding edges). A *2-edge-connected graph with leaves* is a connected graph G such that $R(G)$ is 2-edge-connected. Note that a 2-edge-connected graph is also 2-edge-connected with leaves.

Definition 17. An *end-component* $G_i = (V_i, E_i)$ of a connected graph $G = (V, E)$ is a non-trivial (connected) subgraph of G that contains a unique vertex $t_i \in V_i$ such that any path in G between two vertices $u \in V_i$ and $v \in V - V_i$ passes through the vertex t_i. The vertex t_i is the *separation vertex* of G_i.

[1] It is clear that the diameter $D_{G'_k}$ of the graphs G'_k satisfies $D_{G'_k} \leq 3k + 2$. If we change a little the construction and do not attach a vertex to the three furthest vertices: $v^1_{1,(k+1)/2}, v^2_{1,(k+1)/2}$ and $v^3_{1,(k+1)/2}$ we achieve $D_{G'_k} \leq 3k$ and have the same proof. For simplicity we do not mention this detail in the construction above.

An end-component is ***non-trivial*** if it contains at least one vertex other than the separation vertex which is not a leaf. Two non-trivial end-components are *disjoint* if they have at most one vertex in common (such vertex, if exists, must be their separation vertex).

Definition 18. A *petal graph* is a connected undirected graph with at least three mutually disjoint non-trivial end-components.

Example 5. Figure 6 (a) is a demonstration of the general form of a petal graph; Figure 6 (b) shows two simple petal graphs (the upper of which is also a lithium graph). The graphs G_k and the extended graphs G'_k of Sec. 3 are also examples of petal graphs.

Note that the only graphs which are not petal graphs are graphs G for which the superstructure (defined in [E79]) of $R(G)$ is a simple path. Also note that a lithium graph is a special case of a petal graph.

Theorem 19. *Let G be a petal graph and let G_1, G_2 and G_3 be any three non-trivial mutually disjoint end-components of G such that $bi\_diam(R(G_1)) \geq bi\_diam(R(G_2)) \geq bi\_diam(R(G_3))$. Then*

$$OPT_{LILS}(G) = \Omega(bi\_diam(R(G_3))).$$

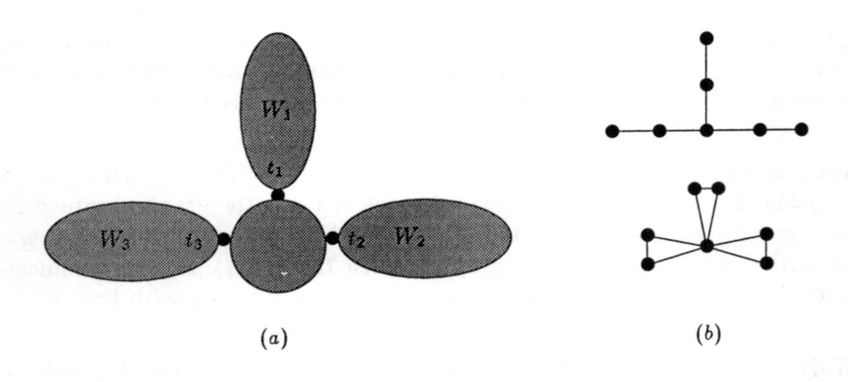

Fig. 6. a. The general structure of a petal graph. b. Two simple petal graphs.

Note that if a petal graph G has three non-trivial mutually disjoint end-components G_1, G_2 and G_3 that are not 2-edge-connected graphs with leaves then $bi\_diam(R(G_3))$ is ∞. Indeed, such a graph is a lithium graph. Thus Theorem 19 implies the following corollary, which is the *only if* part of Theorem 7.

Corollary 20. *A lithium graph does not have a valid LILS.*

We give a brief description of the idea of the proof of Theorem 19. Consider any LILS of G, L_G. For any three non-trivial mutually disjoint end-components of G, G_1, G_2 and G_3, one of them, say G_3, does not contains min and max. For a vertex u in G_3 we consider three paths; the sub-paths of the paths $Path_G(u, min, L_G)$ and $Path_G(u, max, L_G)$ from u to t_3 (the separation node of G_3), and the path $Path_G(t_3, u, L_G)$. We show that these paths form a 2-edge-connected subgraph of G_3 or a 2-edge-connected subgraph of G_3 with one weak bridge (i.e., a bridge that one of its endpoints is a leaf). A lower bound of $\Omega(bi\_dist_G(u, t_3))$ on the length of a path a message traverses follows. we then show that $bi\_diam(G_3) \leq 2[max_{u \in G_3}(bi\_dist_G(u, t_3))]$.

Since we are interested in a lower bound in terms of the diameter of the network, the connection between the diameter and the bi_diameter is important. The graphs G_k of Sec. 3 are an example of graphs whose bi_diameter is proportional to the square of their diameter. (Specifically, $bi\_diam(G_k) = 2/9 \cdot D^2_{G_k} + O(D_{G_k})$.) It follows that for any constant D, there is a graph G with diameter $D_G \geq D$ such that $bi\_diam(G) \geq C \cdot D^2_G$, where C is a constant. On the other hand we have recently proved:

Theorem 21. *([EMZ]) For any 2-edge-connected graph G with diameter D_G,*

$$bi\_diam(G) = O(D^2_G).$$

(This result was independently proved by [D].)

Theorem 21 shows that we could not hope to improve the lower bound using this technique, i.e., by finding petal graphs G with $bi\_diam(G) = D^n_G$, $n > 2$, since such graphs do not exist.

Note that the lower bound of $\Omega(D^2_{G_k})$ of Sec. 3 is a consequence of Theorem 19 and of the fact that $bi\_diam(G_k) = \Theta(D^2_{G_k})$. However, by using the specific structure of the graphs G'_k in the proof of Theorem 8, we obtain a lower bound which is larger by a multiplicative constant than the one derived from the proof of Theorem 19 (which is omitted from this paper).

5 Conclusions and Open Questions

In this paper we studied the performance of Linear Interval Routing in terms of the longest path traversed by a message, relative to the diameter D of the network. There are two advantages of Linear Interval Routing over Interval Routing. The first one is its simplicity in implementation. This is important especially in fast networks, where the decision function is often the bottleneck of the communication. The second and main advantage is that it can be used for routing in dynamic networks. Until now it was not clear what the efficiency of Linear Interval Routing is. Whereas an upper bound of $2D$ and a lower bound of $1.75D-1$ were known for Interval Routing, there is still no non-trivial upper bound for Linear Interval Routing, while our result shows a lower bound of $\Omega(D^2)$. We thus conclude that by restricting ourselves to linear intervals, we might need to pay a significant price in time. Moreover, in Sec. 4 we revealed a connection between the bi_diameter of a graph (which may be as large as the square of its diameter) and a lower bound on the efficiency of an LILS for it. We introduced the family of petal graphs (which includes all lithium graphs), and presented a lower bound result for these graphs. The lower bound achieved by this result for lithium graphs is ∞, which implies the impossibility result of [FG94]. While our lower bound technique can be extended to graphs which are not petal graphs, it is not clear whether it can be used to obtain optimal lower bounds to all graphs. Other related open question are:

1. Improve the lower bound (it follows from our discussion that $\Omega(D^2)$ is the best lower bound that can be achieved by our technique; however different techniques could possibly do better).
2. Find any non-trivial upper bound for the same problem (all known techniques for generating LILS provide an $O(|V|)$ upper bound).

References

[BLT90] Bakker, E.M., van Leeuwen, J., Tan, R.B.: Prefix Routing Schemas in Dynamic Networks. Tech. Rep. RUU-CS-90-10, Dept. of Computer science, Utrecht University, Utrecht (1990).

[BLT91] Bakker, E.M., van Leeuwen, J., Tan, R.B.: Linear Interval Routing. Algorithms Review **2** (1991) 45–61.

[D] Dinitz, Y., Private communication.

[E79] Even, S.: Graph Algorithms. Computer Science Press, Inc., (1979).

[EMZ] Eilam, T., Moran, S., Zaks, S.: Bi-Diameter in 2-Edge-Connected Graphs. In preparation.

[FG94] Fraigniaud, P., Gavoile, C.: Interval Routing Schemes. Research Rep. 94-04, LIPS-ENS Lyon (1994).

[LSX95] Li, Q., Sotteau, D., Xu, J.: 2-Diameter of De Bruijn Networks. Rapport de Recherche 950, Universite de Paris Sud, centre d'Orsay, Laboratoire de Recherche en Informatique, 91405 Orsay, France (1995).

[LT83] van Leeuwen, J., Tan, R.B.: Routing with Compact Routing Tables. Tech. Rep. RUU-CS-83-16, Dept. of Computer Science, Utrecht University (1983). Also as: Computer Networks with Compact Routing Tables. In: G. Rozenberg and A. Salomaa (Eds.) The book of L, Springer- Verlag, Berlin (1986) 298–307.

[LT86] van Leeuwen, J., Tan, R.B.: Computer Network with Compact Routing Tables. In: G. Rozenberg and A. Salomaa (Eds.), The Book of L., Springer-Verlag, Berlin (1986) 259–273.

[R91] Ružička, P.: A Note on the Efficiency of an Interval Routing Algorithm. The Computer Journal **34** (1991) 475–476.

[SK82] Santoro, N., Khatib, R.: Routing Without Routing Tables. Tech. Rep. SCS-TR-6, School of Computer Science, Carleton University (1982). Also as: Labeling and Implicit Routing in Networks. The Computer Journal **28** (1) (1985) 5–8.

[SK85] Santoro, N., Khatib, R.: Labeling and Implicit Routing in Networks. The Computer Journal,**28** (1) (1985) 5–8.

[TL94] Tse, S. S.H., Lau, F. C.M.: A Lower Bound for Interval Routing in General Networks. Tech. Rep. TR-94-09, Department of Computer Science, University of Hong Kong (1994).

Topological Routing Schemes *

Giorgio Gambosi and Paola Vocca

Dipartimento di Matematica,
Università di Roma "Tor Vergata",
Via della Ricerca Scientifica, I-00133 Rome, Italy.
{gambosi,vocca}@utovrm.it

Abstract. In this paper, the possibility of using topological and metrical properties to efficiently route messages in a distributed system is evaluated.
In particular, classical interval routing schemes are extended to the case when sets in a suitable topological (or metrical) space are associated to network nodes and incident links, while predicates defined among such sets are referred in the definition of the routing functions.
In the paper we show that such an approach is strictly more powerful than conventional interval and linear interval routing schemes, and present some applications of the technique to some specific classes of graphs.

Keywords: Distributed systems, compact routing tables, interval routing, shortest paths.

1 Introduction

Routing messages efficiently between processors is a fundamental task in any distributed computing system. The network of processors is modeled as an (undirected) connected graph $G = (V, E)$, where V is a set of n processors and E is a set of pairwise communication links: it is then important to route each message along a shortest path from its source to the destination.

While this can be easily done by referring, at each node v, to a complete routing table which specifies, for each destination u, the set of links incident to v which are on shortest paths between u and v, in the general case such tables are too space consuming for large networks and it is necessary to make use of routing schemes with smaller tables.

Research activities aim at identifying classes of network topologies over which routing schemes using a smaller amount of information can be defined. In the classical ILS (*Interval Labeling Scheme*) routing scheme, for example, ([14], [16]) node labels belong to the set $\{1, \ldots, n\}$, assumed cyclically ordered, while link labels are pairs of node labels representing disjoint intervals on $\{1, \ldots, n\}$.

* Work partially supported by the E.U. ESPRIT project No.20244 (ALCOM-IT) and by the Italian MURST 40% project "Efficienza di Algoritmi e Progetto di Strutture Informative"

To transmit a message m from node v_i to node v_j, m is broadcast by v_i on a link $e = (v_i, v_k)$ such that the label of v_j belongs to the interval associated to e. With this approach, one always obtains an optimal memory occupation, while the problem is to choose node and link labels in such a way that messages are routed along shortest paths.

In [14], [15], [16] it is shown that ILS can be applied to optimally route on specific network topologies. Moreover, ILS has also been used as a basic building block for routing schemes based on network decomposition and clustering ([1], [2], [9], [10], [11], [12]).

In [15], the approach has been extended to allow that more than 1 interval is associated to each node, while in [4] intervals in multidimensional spaces are considered.

In this paper, we extend the approach to the use of labelings which associate suitable sets in some topological or metrical space both to nodes and to incident links and to the definition of routing functions referring to predicates defined among such sets. This approach is clearly an extension of ILS and we show that, indeed, it is a strict extension of such an approach. Moreover, we show that routing schemes referring to the new definition can be used to optimally route messages on some classes of graphs over which no interval routing scheme has been defined.

The paper is organized as follows: in Sect. 2, we describe the topological labeling scheme and show its properties; in Sect. 3, we introduce a topological labeling scheme defined on the Euclidean d-space, the *Hyper-Rectangle Labeling Scheme*, and in Sect. 4 we analyze the main differences between this scheme and the conventional Interval Labeling Scheme; in Sects. 5 and 6, we present some applications of the proposed technique to some specific classes of graphs; finally, in Sect. 7, conclusions and future research are described.

2 Topological Labeling Scheme

A computer communication network can be described as a connected graph $G = (V, E)$ where V is the set of n processor elements and E represents the set of communication links.

We assume that links are bidirectional and all processors are connected to the network, therefore the graph is undirected and connected.

Given a vertex $x \in V$, we denote by $\delta(x)$ the degree of x and by $out(x)$ the set of edges incident on x, that is $out(x) = \{(x, y) \in E\}$.

Most of the well known routing schemes are based on interval labeling schemes, which encode edges with intervals of $\{1, \dots, n\}$, either linear or cyclic, and vertices with points of $\{1, \dots, n\}$ [14, 3, 8, 16, 7, 6]. In [4] the approach has been extended to multidimensional intervals.

In this paper, we move in the direction of generalizing the above approach by defining labeling schemes which assigns to both vertices and edges opens of suitable topological spaces.

Given two enumerable structures $\mathcal{D}_1$ and $\mathcal{D}_2$ on a topological space $\mathcal{S}$, a *Topological Labeling Scheme* $\mathcal{L}$ is composed by two mappings $\mathcal{L}_1 : V \mapsto \mathcal{D}_1$ and $\mathcal{L}_2 : V \times E \mapsto \mathcal{D}_2$, where $\mathcal{L}_1$ is a total function and $\mathcal{L}_2$ is a partial function defined for $x \in V$ and $e \in out(x)$.

A *routing scheme* $\mathcal{R} = \{\mathcal{R}_x, \forall x \in V\}$ is a collection of *routing functions* $\mathcal{R}_x : V \mapsto 2^{out(x)}$ such that any message from x to y is routed through edge $e \in out(x)$ if and only if $e \in \mathcal{R}_x(y)$.

Given a topological labeling scheme $\mathcal{L}$ on $\mathcal{S}$, a routing scheme based on $\mathcal{L}$ is any collection $\mathcal{R}$ such that:

$$e \in \mathcal{R}_x(y) \Leftrightarrow \mathbf{P}(\mathcal{L}_1(y), \mathcal{L}_2(x,e))$$

where $\mathbf{P}$ is a suitable predicate defined on $\mathcal{D}_1$ and $\mathcal{D}_2$.

According to the choice of $\mathcal{S}$, $\mathcal{D}_1$, $\mathcal{D}_2$, and $\mathbf{P}$, we can obtain most of the well known routing schemes.

For example, let $\mathcal{S} = \{1, \ldots, n\}$, $\mathcal{L}_1 : V \mapsto \mathcal{D}_1 = \{\{1\}, \ldots, \{n\}\}$ and $\mathcal{L}_2 : V \times E \mapsto \mathcal{D}_2 = \{[i,j] | 1 \leq i,j \leq n\}$. Moreover, let

$$\mathbf{P}(\mathcal{L}_1(y), \mathcal{L}_2(x,e)) = \begin{cases} True & \text{if } \mathcal{L}_1(y) \subseteq \mathcal{L}_2(x,e) \\ False & \text{otherwise.} \end{cases}$$

then the resulting routing scheme is the classical Interval Routing Scheme.

We assume that any routing function $\mathcal{R}_x$ is computable in optimal $\Theta(\delta(x))$ time. A routing scheme is *valid* if e is on a path from x to y, is *optimal* if e is on a shortest path from x to y, and is *overall optimal* [5] if *all* shortest paths are represented in the scheme.

Examples of topological spaces are rectangles, spheres, regular polygons in the Euclidean d-space. In these cases, possible routing functions are:

$$\mathcal{R}_y(x) = e \Leftrightarrow \begin{cases} \mathcal{L}_1(y) \subseteq \mathcal{L}_2(x,e) \\ \mathcal{L}_1(y) \cap \mathcal{L}_2(x,e) \neq \emptyset \\ \mathcal{L}_1(y) \cap \mathcal{L}_2(x,e) \neq \emptyset \text{ and with minimal measure among all} \\ \qquad (\mathcal{L}_1(y) \cap \mathcal{L}_2(x,e')) \text{ for } e' \in out(x) \\ \ldots \end{cases}$$

The last routing function requires that the underlying topological space is endowed with a metric.

It is interesting to underscore that the proposed generalization is twofold. From one side we associate to vertices and edges opens of a topological space which better allows to represent different network topologies. From the other, we generalize the routing functions by introducing a general predicate instead of the inclusion predicate used in the Interval Routing Scheme. In the following sections, we will show that the generalizations introduced allow to derive optimal routing schemes for some networks which do not admit optimal Interval Routing Schemes.

3 Hyper-Rectangle Labeling Scheme

In this paper, we start the analysis of Topological Labeling Schemes restricting the study to topologies where the topological space $\mathcal{S}$ is the Euclidean d-space and the open sets of the topological structures $\mathcal{D}_1$ and $\mathcal{D}_2$ are hyper-rectangles defined as follows:

Definition 1. A *hyper-rectangle* I in the Euclidean d-space is the cartesian product of 1-dimensional intervals, $I = J_1 \times J_2 \times \ldots \times J_d$ where each $J_i = [a_i, b_i]$ is a set of points r such that:

$$\begin{cases} a_i \leq r \leq b_i & \text{if } a_i \leq b_i \quad \text{linear interval} \\ a_i \leq r \text{ or } r \leq b_i & \text{if } b_i \leq a_i \quad \text{cyclic interval} \end{cases} .$$

The labeling scheme $\mathcal{L}$ we consider assigns d-dimensional hyper-rectangles to both vertices and edges, *Hyper-Rectangle Labeling Scheme* (H^d). If all 1-dimensional intervals are linear then we have a *Linear Hyper-Rectangle Labeling Scheme* (LH^d).

One natural extension to H^d (LH^d) schemes is to consider a multi-labeling scheme which associates to each link up to k hyper-rectangles.

Let us denote as $k(x,e)$ the number of hyper-rectangles $I^1_{x,e}, \ldots, I^{k(x,e)}_{x,e}$ associated to an edge $e \in out(x)$, and let $I_{x,e} = \bigcup_{i=1}^{k(x,e)} I^i_{x,e}$, then to route a message from x to y the routing function sends the message through the link that satisfies the following condition:

$$e \in \mathcal{R}_x(y) \Leftrightarrow \mathbf{P}(\mathcal{L}_1(y), I_{x,e}) \ .$$

Definition 2. Given a Hyper-Rectangle Labeling Scheme $\mathcal{L} \in \mathrm{H}^d$ (LH^d) on a network $G = (V, E)$, the *compactness* k of $\mathcal{L}$ is defined as:

$$k = max_{x \in V} max_{e \in out(x)} k(x,e) \ .$$

The *compactness* of G is the minimum k over all optimal labeling schemes on G.

Hence, it is possible to classify communication networks according to the compactness degree of the optimal d-dimensional Hyper Rectangle Labeling Scheme.

Definition 3. For any positive integer k we define:

- k-$\mathbf{H}^d_*$: the class of graphs having an optimal routing scheme on a d-dimensional Hyper-Rectangle Labeling Scheme $\mathcal{L}$ of compactness less than or equal to k;
- k-$\mathbf{LH}^d_*$: the class of graphs having an optimal routing scheme on a d-dimensional Linear Hyper-Rectangle Labeling Scheme $\mathcal{L}$ of compactness less than or equal to k;
- **overall** k-$\mathbf{H}^d_*$ (k-$\mathbf{LH}^d_*$): the class of graphs having an overall optimal routing scheme on a d-dimensional (Linear) Hyper-Rectangle Labeling Scheme $\mathcal{L}$ of compactness less than or equal to k.

Among the above defined classes the following relationship holds:

Theorem 4. *For every $k > 0$, we have:*

$$k\text{-}H_*^d \subset (k2^d)\text{-}LH_*^d \ .$$

Proof. Any k-H_*^d can be transformed into an $(k2^d)$-LH_*^d decomposing each d-dimensional hyper-rectangle in 2^d d-dimensional linear hyper-rectangles (for every cyclic interval there are at most d 1-dimensional intervals). □

One interesting feature of the labeling scheme considered is that it allows to easily deal with the *cartesian product* of graphs, an important class of graphs since it includes the topologies of some interconnection networks commonly used in parallel architectures, such as hypercubes, d-dimensional grids and tori.

Definition 5. Let $G_1 = (V_1, E_1), \ldots, G_h = (V_h, E_h)$. The *product graph* $G_1 \times \ldots \times G_h = (V, E)$ is defined as:

$$V = V_1 \times \ldots \times V_h;$$
$$E = \Big\{(\langle v_1, \ldots, v_{i_j}, \ldots, v_h\rangle, \langle v_1, \ldots, v_{i_{j'}}, \ldots, v_h\rangle)| \quad v_i \in V_i, 1 \leq i \leq k, (v_{i_j}, v_{i_{j'}}) \in E_i\Big\}.$$

It is important to underscore that the product of graphs having an optimal Interval Routing Scheme of compactness k, does not in general admit an optimal Interval Routing Scheme of compactness k ([13, 16]). Instead, this is true for our generalization, as shown by the following theorem whose proof can be straightforwardly derived.

Theorem 6. *Let $G_1, \ldots, G_h$ be a set of graphs having an optimal or overall optimal k_i-$H_*^{d_i}$ (k_i-$LH_*^{d_i}$), where $1 \leq i \leq h$, then the product graph $G_1 \times \ldots \times G_h$ has an optimal or overall optimal k-H_*^d (k-LH_*^d) with $k = max(k_1, \ldots, k_h)$ and $d = d_1 + \ldots + d_h$.*

For example, let $G_1 = (V_1, E_1) \in k_1$-$\mathrm{H}_*^{d_1}$ with labeling scheme $(\mathcal{L}_1, \mathcal{L}_2)$ and routing function represented by the predicate $\mathbf{P}= (\mathcal{L}_1(y) \subseteq \mathcal{L}_2(x, e))$, and let $G_2 = (V_2, E_2) \in k_2$-$\mathrm{H}_*^{d_2}$ with labeling scheme $(\mathcal{L}_1', \mathcal{L}_2')$ and routing function represented by the predicate $\mathbf{P}' = (\mathcal{L}_1'(y) \cap \mathcal{L}_2'(x, e))$. Let $\mathcal{L}_1(x) = I_x$ and $\mathcal{L}_2(x, e) = I_{x,e}$ for $x \in V_1$ and $e \in E_1$, and let $\mathcal{L}_1'(x') = I_{x'}'$ and $\mathcal{L}_2'(x', e') = I_{x',e'}'$ for $x' \in V_2$ and $e' \in E_2$. Let $\langle x, y\rangle$ denote the node in $V_1 \times V_2$ corresponding to the pair $x \in V_1$, $y \in V_2$, then the labeling scheme $\mathcal{L}^{G_1 \times G_2} = (\mathcal{L}_1^{G_1 \times G_2}, \mathcal{L}_2^{G_1 \times G_2})$ for $G_1 \times G_2$ is defined as follows:

- $\mathcal{L}_1^{G_1 \times G_2}(\langle x, x'\rangle) = I_x \times I_{x'}$, and
- if $e = (\langle x, x'\rangle, \langle y, x'\rangle)$ then $\mathcal{L}_2^{G_1 \times G_2}(\langle x, x'\rangle, e) = I_{x,e} \times \mathcal{R}^{d_2}$, otherwise, if $e = (\langle x, x'\rangle, \langle x, y'\rangle)$ then $\mathcal{L}_2^{G_1 \times G_2}(\langle x, x'\rangle, e) = \mathcal{R}^{d_1} \times I_{x',e}$ where $\mathcal{R}^d$ is the Euclidean d-space.

If we define the predicate $\mathbf{P}^{G_1 \times G_2}$ as $\mathbf{P} \wedge \mathbf{P}'$, it is easy to see that the routing scheme obtained on $G_1 \times G_2$ is optimal.

4 Hyper-Rectangle vs. Interval Labeling Scheme

In this section we show that Hyper-Rectangle Labeling Scheme associated with a suitable optimal routing function is more powerful than the Interval Routing Scheme. Let us consider the routing scheme $\mathcal{R} = \{\mathcal{R}_x, \forall x \in V\}$ defined as follows:

$$\mathcal{R}_y(x) = e \Leftrightarrow$$
$$\mathcal{L}_1(y) \cap \mathcal{L}_2(x,e) = min\,\{\mathcal{L}_1(y) \cap \mathcal{L}_2(x,e') | \mathcal{L}_1(y) \cap \mathcal{L}_2(x,e') \neq \emptyset \text{ and } e' \in out(x))\} \quad (1)$$

Obviously, $\mathcal{R}_x \in \Theta(\delta(x))$.

Adopting the same notation as in [16, 8, 3], let k-IRS* (k-LIRS*) be the class of graphs having an optimal Interval Routing Scheme (Linear Interval Routing Scheme) of compactness k, we have:

Theorem 7. *k-IRS* $\subset$ k-H^1_** .

Proof. It is trivial to show that every valid Interval Labeling Scheme is also a Hyper-Rectangle Labeling Scheme. Moreover, the inclusion is strict as shown by the labeling of the circular-arc graph of Fig. 1 which admits an 1-H^1_* but not 1-IRS_* ([8]). □

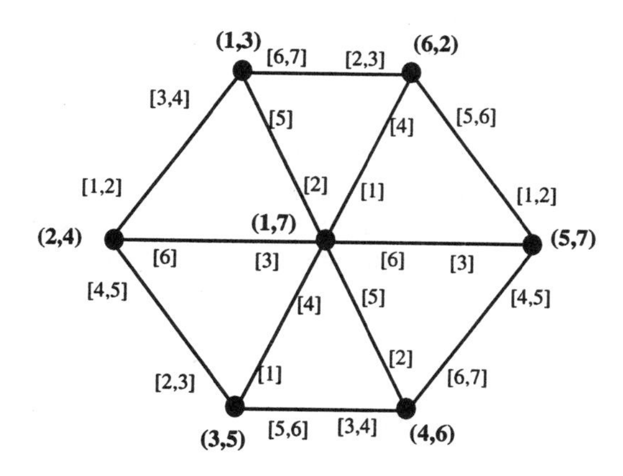

Fig. 1. An optimal Hyper-Rectangle Labeling Scheme.

Analogously, for Linear Routing Schemes we have:

Theorem 8. *k-LIRS* $\subset$ k-LH^1_** .

Proof. Any valid Linear Interval Labeling Scheme is a Linear Hyper-Rectangle Labeling Scheme. Also in this case the inclusion is strict as the Y-graph which does not admit an optimal 1-LIRS ([8]) belongs to 1-LH^1_*. (see Fig. 2). □

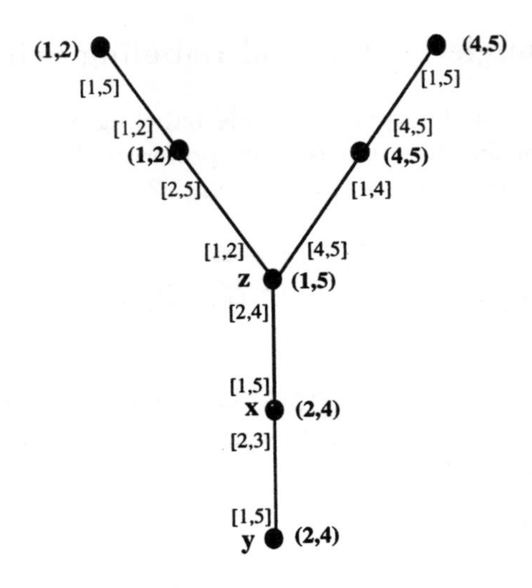

Fig. 2. An optimal Linear Hyper-Rectangle Labeling Scheme.

Notice that, in both the labeling schemes of Figs. 1 and 2, the minimality condition on interval intersection allows to route messages along the shortest path, only.

For example, referring to Fig. 2, the minimality condition allows to correctly route a message from x to y along the edge (x, y), even if both $\mathcal{L}_1(y) \cap \mathcal{L}_2(x, (x, y)) \neq \emptyset$ and $\mathcal{L}_1(y) \cap \mathcal{L}_2(x, (x, z)) \neq \emptyset$.

In Sect. 6 we will generalize the results presented here by showing general Hyper-Rectangle Labeling Schemes for two classes of graphs which do not belong to 1-LIRS* and to 1-IRS*, and by proving that they belong 1-LH^1_* and to 1-H^1_*, respectively.

5 An Overall Optimal Routing Scheme for Bipartite Graphs

In this section we will show how the introduced labeling scheme favorably compares not only with the classical routing schemes as the IRS but also with other generalization such as the Multi-Dimensional Interval Routing Scheme proposed in [4].

In [4] it has been proved that complete bipartite graphs have an overall optimal Multi-Dimensional Interval Routing Scheme with 1-dimensional intervals and compactness 2. In this section, we substantially improve the above result by introducing two labeling schemes which optimally represent all shortest paths for two classes of graphs which properly include complete bipartite graphs with smaller. In particular, for complete bipartite graphs we derive a Hyper-Rectangle Labeling Scheme with compactness smaller than for Multi-Dimensional Interval Routing Scheme.

Moreover, the labeling schemes proposed have an interesting geometric interpretation which allows to better understand the expressive power of our method. The routing scheme used is the one defined in previous section, that is:

$$\mathcal{R}_y(x) = e \Leftrightarrow$$

$$\mathcal{L}_1(y) \cap \mathcal{L}_2(x,e) = min\,\{\mathcal{L}_1(y) \cap \mathcal{L}_2(x,e') | \mathcal{L}_1(y) \cap \mathcal{L}_2(x,e') \neq \emptyset \text{ and } e' \in out(x))\} .$$

Definition 9. A graph $G = (V, E)$ is a *complete k-partite graph* if V can be partitioned into k disjoint sets $V_1, \ldots, V_k$, and the edge set is:

$$E = \{(x_i, x_j) | x_i \in V_i, x_j \in V_j, 1 \leq i, j \leq k, i \neq j\} .$$

For this class of graphs the following theorem holds:

Theorem 10. *Any complete k-partite graph belongs to overall 1-$LH_*^{\lceil \frac{k}{2} \rceil}$* .

Proof. Let us consider the following labeling scheme. The vertex labeling function $\mathcal{L}_1(x_{i,h})$, with $1 \leq i \leq k$ and $1 \leq h \leq |V_i|$, is defined as follows:

$$\mathcal{L}_1(x_{i,h}) = J_{i,h}^1 \times \ldots \times J_{i,h}^{\lceil \frac{k}{2} \rceil} ,$$

where $J_{i,h}^p = \left[a_{i,h}^p, b_{i,h}^p\right]$, $1 \leq p \leq \lceil \frac{k}{2} \rceil$, and

$$\begin{cases} a_{i,h}^p = b_{i,h}^p = 0 & \text{if } p \neq \lceil \frac{i}{2} \rceil \\ a_{i,h}^{\lceil \frac{i}{2} \rceil} = -2(n_i - h + 1) \text{ and } b_{i,h}^{\lceil \frac{i}{2} \rceil} = -2(n_i - h) & \text{if } i \text{ is odd} \\ a_{i,h}^{\lceil \frac{i}{2} \rceil} = 2(h-1) \text{ and } b_{i,h}^{\lceil \frac{i}{2} \rceil} = 2h & \text{if } i \text{ is even} \end{cases} .$$

The edge labeling function $\mathcal{L}_2(x_{i,h}, e)$, where $e = (x_{i,h}, x_{j,k})$, for $1 \leq i, j \leq k$, $i \neq j$, $1 \leq h \leq n_i$ and $1 \leq k \leq n_j$, is as follows:

$$\mathcal{L}_2(x_{i,h}, e) = J_{x_{i,h},e}^1 \times \ldots \times J_{x_{i,h},e}^{\lceil \frac{k}{2} \rceil} ,$$

where $J_{x_{i,h},e}^p = \left[a_{x_{i,h},e}^p, b_{x_{i,h},e}^p\right]$ is so defined:

$$\begin{cases} a_{x_{i,h},e}^{\lceil \frac{i}{2} \rceil} = -2n_i & \text{if } i \text{ is odd} \\ b_{x_{i,h},e}^{\lceil \frac{i}{2} \rceil} = 2n_i & \text{if } i \text{ is even} \\ a_{x_{i,h},e}^{\lceil \frac{j}{2} \rceil} = -2(n_j - k) + 1 & \text{if } j \text{ is odd} \\ b_{x_{i,h},e}^{\lceil \frac{j}{2} \rceil} = 2k - 1 & \text{if } j \text{ is even} \\ a_{x_{i,h},e}^{j} = b_{x_{i,h},e}^{p} = 0 & \text{otherwise} \end{cases} .$$

It is easy to verify that the routing scheme in (1) based on the above defined labeling scheme represents all shortest paths (see Fig. 3 for $k = 3$). □

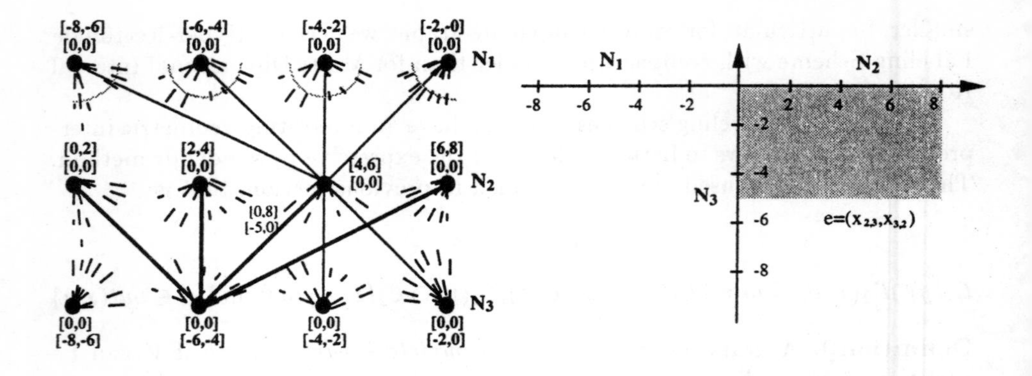

Fig. 3. A valid Hyper-Rectangle Labeling Scheme for complete k-partite.

The other case considered is the class of *k-stratified complete graphs*, defined as follows:

Definition 11. A graph $G = (V, E)$ is a *k-stratified complete graph* if the vertex set V can be partitioned into k disjoint sets $V_1, \ldots, V_k$ and the edge set is:

$$E = \{(x_i, x_{i+1}) | x_i \in V_i, x_{i+1} \in V_{i+1}, 1 \leq i \leq k-1\} \ .$$

For this class of graphs the following theorem holds:

Theorem 12. *Any k-stratified complete graph belongs to overall* $1\text{-}LH_*^{\lceil \frac{k}{2} \rceil}$.

Proof. The vertex labeling function $\mathcal{L}_1$ is defined as in theorem 10. The edge labeling function $\mathcal{L}_2(x_{i,h}, e) = J^1_{x_{i,h},e} \times \ldots \times J^{\lceil \frac{k}{2} \rceil}_{x_{i,h},e}$, is defined as follows.

For each 1-dimensional interval two cases are possible. Let $J^p_{x_{i,h},e} = \left[a^p_{x_{i,h},e}, b^p_{x_{i,h},e}\right]$ we have:

1. $e = (x_{i,h}, x_{i+1,k})$, for $1 \leq i \leq k-1$, $1 \leq h \leq n_i$ and $1 \leq k \leq n_{i+1}$.
   If i is odd:

$$\begin{cases} a^p_{x_{i,h},e} = -2n_i \text{ and } b^p_{x_{i,h},e} = -2k-1 & \text{for } p = \lceil \frac{i}{2} \rceil \\ a^p_{x_{i,h},e} = b^p_{x_{i,h},e} = 0 & \text{for } 1 \leq p < \lceil \frac{i}{2} \rceil \\ a^p_{x_{i,h},e} = -2n_{p-1} \text{ and } b^p_{x_{i,h},e} = -2n_p & \text{for } \lceil \frac{i}{2} \rceil < p \leq \lceil \frac{k}{2} \rceil \end{cases}$$

   If i is even:

$$\begin{cases} a^p_{x_{i,h},e} = 0 \text{ and } b^p_{x_{i,h},e} = 2n_i & \text{for } p = \lceil \frac{i}{2} \rceil \\ a^p_{x_{i,h},e} = -2(n_{i+1} - k) + 1 \text{ and } b^p_{x_{i,h},e} = 2n_{i+2} & \text{for } p = \lceil \frac{i+1}{2} \rceil \\ a^p_{x_{i,h},e} = b^p_{x_{i,h},e} = 0 & \text{for } 1 \le p < \lceil \frac{i}{2} \rceil \\ a^p_{x_{i,h},e} = -2n_{p-1} \text{ and } b^p_{x_{i,h},e} = -2n_p & \text{for } \lceil \frac{i+1}{2} \rceil < p \le \lceil \frac{k}{2} \rceil \end{cases}$$

2. $e = (x_{i,h}, x_{i-1,k})$, for $2 \le i \le k$, $1 \le h \le n_i$ and $1 \le k \le n_{i-1}$.
   If i is odd:

$$\begin{cases} a^p_{x_{i,h},e} = -2n_i \text{ and } b^p_{x_{i,h},e} = 0 & \text{for } p = \lceil \frac{i}{2} \rceil \\ a^p_{x_{i,h},e} = -2n_{i-2} \text{ and } b^p_{x_{i,h},e} = 2k - 1 & \text{for } p = \lceil \frac{i-1}{2} \rceil \\ a^p_{x_{i,h},e} = -2n_{2p-1} \text{ and } b^p_{x_{i,h},e} = 2n_{2p} & \text{for } 1 \le p < \lceil \frac{i-1}{2} \rceil \\ a^p_{x_{i,h},e} = b^p_{x_{i,h},e} = 0 & \text{for } \lceil \frac{i}{2} \rceil < p < \lceil \frac{k}{2} \rceil \end{cases}$$

   If i is even:

$$\begin{cases} a^p_{x_{i,h},e} = -2(n_{i-1} - k) + 1 \text{ and } b^p_{x_{i,h},e} = 2n_i & \text{for } p = \lceil \frac{i}{2} \rceil \\ a^p_{x_{i,h},e} = -2n_{2p-1} \text{ and } b^p_{x_{i,h},e} = -2n_{2p} & \text{for } 1 \le p < \lceil \frac{i}{2} \rceil \\ a^p_{x_{i,h},e} = b^p_{x_{i,h},e} = 0 & \text{for } \frac{i}{2} \rceil < p \le \lceil \frac{k}{2} \rceil \end{cases}$$

Also in this case, it is easy to verify that the routing scheme in (1) based on the above defined labeling scheme represents all shortest paths (see Fig. 4 for $k = 3$). □

6 An Optimal Routing Scheme for Interval and Circular-Arc Graphs

The Hyper-Rectangle Labeling Scheme finds a natural application field in routing on intersection graphs. In particular, we present optimal routing schemes for interval graphs and for circular-arc graphs.

Definition 13. A graph $G = (V, E)$ is an *interval graph* if there exists a collection $\mathcal{I}$ of intervals of the real line such that each vertex $x \in V$ can be represented by an interval $I_x \in \mathcal{I}$, and $(x, y) \in E$ if and only if $I_x \cap I_y \neq \emptyset$.

In [8] it has been proved that any *unit interval graph*, that is an interval graph such that all intervals in $\mathcal{I}$ have the same length, belongs to 1-LIRS*, but, as they show, this is not true for general interval graphs.

The next theorem will underscore the difference between LIRS and LH by showing that the interval graphs can be optimally represented with a Linear Hyper-Rectangle Labeling Scheme.

Theorem 14. *The class of interval graphs belongs to* 1-LH^1_* .

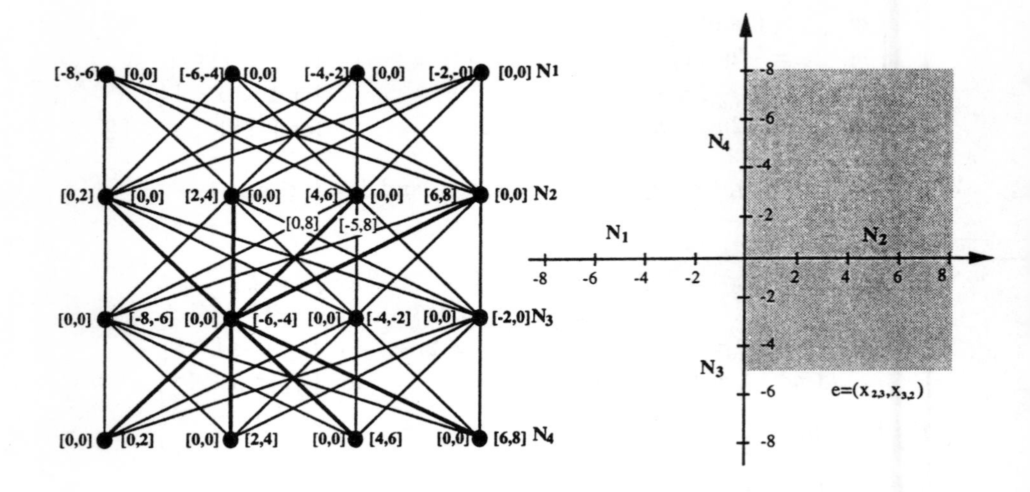

Fig. 4. A Hyper-Rectangle Labeling Scheme for k-stratified complete graph.

Proof. Let $G = (V, E)$ be an interval graph and let $\mathcal{I}$ be a corresponding set of intervals, which, w.l.o.g., we assume normalized with respect to the interval $[1, 2n]$, where $|V| = n$, that is each vertex $x \in V$ is associated to an interval $[\alpha_x, \beta_x] \subset [1, 2n]$.

Let us define the following Linear Hyper-Rectangle Labeling Scheme 1-LH$^1_*$:

$$\mathcal{L}_1 : x \in V \mapsto [\alpha_x, \beta_x] \in \mathcal{I} \ .$$

The partial function $\mathcal{L}_2$ is defined as follows: for each $x \in V$ and for each $e = (x, y) \in out(x)$, let $[\alpha_y, \beta_y]$ the interval associated to y. The following three cases are possible:

1. If $\alpha_y = min \{\alpha_{y_i} | (x, y_i) \in out(x)\}$ then $\mathcal{L}_2(x, e) = [1, \beta_y]$;
2. If $\beta_y = max \{\beta_{y_i} | (x, y_i) \in out(x)\}$ then $\mathcal{L}_2(x, e) = [\alpha_y, 2n]$;
3. Otherwise $\mathcal{L}_2(x, e) = [\alpha_y, \beta_y]$;

Let us now consider the following routing functions:

$$\mathcal{R}_x(y) = e \Leftrightarrow \mathcal{L}_2(x, e) = min \{\mathcal{L}_2(x, e') | \mathcal{L}_1(y) \cap \mathcal{L}_2(x, e') \text{ is maximal for } e' \in out(x)\} \ . \tag{2}$$

Obviously, $\mathcal{R}_x \in \Theta(\delta(x))$. By induction on the length of shortest paths, it is possible to show that the above labeling scheme allows to optimally route any message in the network. □

Fig 5 shows an interval graph and the corresponding labeling scheme. Notice that the labeling scheme can be maintained in constant worst case time per operation under any sequence of update operations (edge insertion or deletion), assuming the resulting graph is still an interval graph.

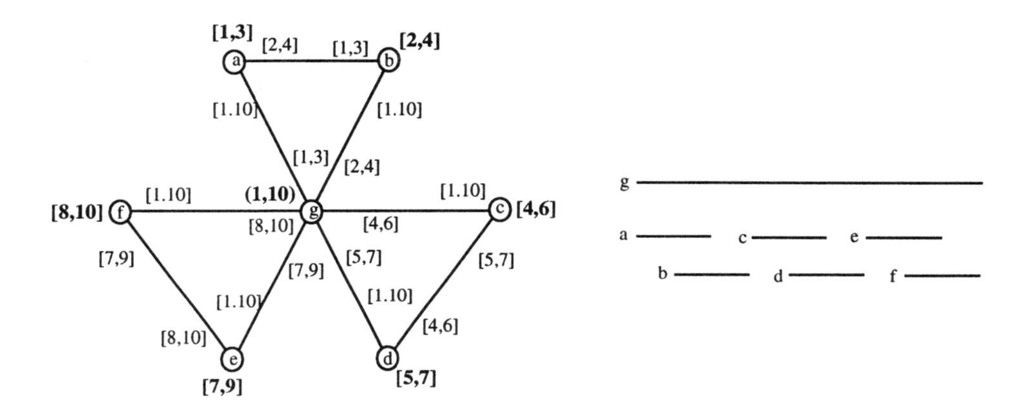

Fig. 5. An optimal Hyper-Rectangle Labeling Scheme for interval graphs.

Analogously to LH with respect to LIRS, it is possible to prove that the Hyper-Rectangle Labeling Scheme is more expressive with respect the Interval Labeling Scheme by showing that the class of circular-arc graphs belongs to 1-H_*^1 but not to 1-IRS*.

Definition 15. A graph $G = (V, E)$ is a *circular-arc graph* if there exists a circle $\mathcal{C}$ such that each vertex $x \in V$ can be represented by an arc $c_x \subseteq \mathcal{C}$, and $(x, y) \in E$ if and only if $c_x \cap c_y \neq \emptyset$.

In [8] has been proved that any *unit circular-arc graph* belongs to 1-IRS*, while this is not true for general circular-arc graphs.

Theorem 16. *The class of circular-arc graphs belongs to* 1-H_*^1 .

Proof. Let $G = (V, E)$ a circular-arc graph. Consider any circular representation of G, on the trigonometric circle. For every vertex x with associated interval $[a_x, b_x]$, let $(1, \alpha_x)$, $(1, \beta_x)$ the polar coordinates of a_x and b_x, respectively. Then, the labeling scheme is defined as follows:

$$\mathcal{L}_1 : x \in V \mapsto [\alpha_x, \beta_x] \ .$$

For each $x \in V$ and for each $e = (x, y) \in out(x)$, let $S(x, e) \subseteq V$ the set of vertices optimally reachable from x through edge e and let $[\alpha_e, \beta_e]$ the minimal arc such that:

$$[\alpha_e, \beta_e] \supseteq \bigcup_{z \in S(x,e)} [\alpha_z, \beta_z] \ .$$

Then $\mathcal{L}_2(x, e) = [\alpha_e, \beta_e]$.

It is easy to show that, due to the structural property of circular-arc graphs, the routing scheme defined in (2) is optimal. □

7 Conclusions

In this paper, the use of topological and metrical properties to efficiently route messages in a distributed system has been introduced and evaluated in some cases.

In particular, we introduced a labeling scheme which associates to both vertices and edges opens of a suitable topological or metrical space. The routing functions are based on predicates defined among such sets and optimally computable.

We showed that the proposed approach favorable compares to the conventional routing schemes.

There are still many open problems with this method. One of them is to study the proposed routing framework with graphs representing relations among geometrical objects, as for example the intersection graphs, which better exploits the topological and metrical capabilities of our scheme.

Moreover it will be interesting to study different topological spaces as for example spheres, regular polygons in the Euclidean d-space and compare their expressive power with respect to the hyper-rectangle labeling scheme.

References

1. B. Awerbuch, A. Bar-Noy, N. Linial, and D. Peleg. Compact Distributed Data Structures for Adaptive Routing. In *Proc. 21st ACM Symp. on Theory of Computing*, pages 479–489, 1989.
2. B. Awerbuch, A. Bar–Noy, N. Linial, and D. Peleg. Improved Routing Strategies with Succinct Tables. *Journal of Algorithms*, 11:307–341, 1990.
3. E. M. Bakker, J. van Leeuwen, and R. B. Tan. Linear Interval Routing Schemes. *Algorithms Review*, 2:45–61, 1991.
4. M. Flammini, G. Gambosi, U. Nanni, and R.B. Tan. Multi-dimensional interval routing schemes. In *Proc. 9th International Workshop on Distributed Algorithms (WDAG'95)*, LNCS. Springer-Verlag, 1995.
5. M. Flammini, G. Gambosi, and S. Salomone. Boolean routing. In *Proc. 7th International Workshop on Distributed Algorithms (WDAG'93)*, volume 725 of *LNCS*. Springer-Verlag, 1993.
6. M. Flammini, G. Gambosi, and S. Salomone. Interval routing schemes. In *Proc. 12th Symp. on Theoretical Aspects of Computer Science (STACS'95)*, volume 900 of *LNCS*. Springer-Verlag, 1995.

7. M. Flammini, J. van Leeuwen, and A. Marchetti Spaccamela. The complexity of interval routing on random graphs. In *Proc. 20th Symposium on Mathematical Foundation of Computer Science (MFCS'95)*, 1995.
8. P. Fraigniaud and C. Gavoille. Interval routing schemes. In *Proc. 13th Annual ACM Symposium on Principles of Distributed Computing*, 1994.
9. G. N. Frederickson and R. Janardan. Designing networks with compact routing tables. *Algorithmica*, 3:171–190, 1988.
10. G. N. Frederickson and R. Janardan. Efficient message routing in planar networks. *SIAM Journal on Computing*, 18:843–857, 1989.
11. G. N. Frederickson and R. Janardan. Space efficient message routing in *c*-decomposable networks. *SIAM Journal on Computing*, 19:164–181, 1990.
12. D. Peleg and E. Upfal. A trade–off between space and efficiency for routing tables. *Journal of the ACM*, 36(3):510–530, 1989.
13. P. Ružička. On efficiency of interval routing algorithms. In *M.P. Chytil, L. Janiga, V. Koubek (Eds.), Mathematical Foundations of Computer Science 1988*, volume 324 of *LNCS*. Springer-Verlag, 1988.
14. N. Santoro and R. Khatib. Labelling and implicit routing in networks. *Computer Journal*, 28(1):5–8, 1985.
15. J. van Leeuwen and R. B. Tan. Computer networks with compact routing tables. In *G. Rozenberg and A. Salomaa (Eds.)* The Book of L, volume 790. Springer-Verlag, 1986.
16. J. van Leeuwen and R. B. Tan. Interval routing. *Computer Journal*, 30:298–307, 1987.

Maintaining a Dynamic Set of Processors in a Distributed System

Satoshi Fujita[1] and Masafumi Yamashita[1]

Faculty of Engineering, Hiroshima University
Kagamiyama 1-4-1, Higashi-Hiroshima, 739, Japan

Abstract. Consider a distributed system consisting of a set V of processors, and assume that every pair of processors can directly communicate with each other. A processor structure is proposed, for implementing a dynamic set $U \subseteq V$ of processors in the distributed system. The dynamic set supports the following three operations: Insert inserts the caller (i.e., the processor executing this operation) in U, Delete removes the caller from U, and Find searches for a processor in U. To evaluate the efficiency of the implementation, an amortized analysis of the message complexity of operations is performed; the amortized number of messages per each operation is $8 + 12\log_2(|V| - 1)$, in the worst case.
The dynamic set is applicable to many important problems, including the load balancing problem, and the proposed processor structure is used to solve the mutual exclusion problem, and to construct a more complex dynamic set of processors like FIFO queue.

Keywords: processor structure, data structure, dynamic set, amortized message complexity, group communication

1 Introduction

Developing algorithms to manipulate dynamic sets is an essential part of designing time efficient algorithms, and therefore, many data structures that can be used to efficiently implement dynamic sets have been proposed.

Dynamic set manipulations are very important in designing efficient distributed algorithms, also. As in sequential algorithms, many distributed problems can be solved efficiently by accommodating key information in a dynamic set to share and manipulate it among the processors. For example, the load balancing problem is reducible (in part) to the problem of searching processors for an idle one; i.e., to the problem of manipulating a "dictionary"[1] that accommodates currently idle processors. If a (FIFO) "queue" can be manipulated efficiently, the distributed mutual exclusion is easily solvable by accommodating requests for entering the critical section in the queue. Since a dynamic set is

[1] A dictionary is a dynamic set supporting three operations, Insert that augments the dynamic set with a specified element, Delete that removed a specified element from the dynamic set, and Search that returns an element in the dynamic set with a specified key, or a special value nil if no such element belongs to the dynamic set.

shared by the processors, it can be manipulated by plural processors, and data in the dynamic set may look to change autonomously (for some processors, or even they may not notice the change). Algorithms to manipulate the dynamic set must be designed so that they work efficiently and correctly under those situations.

In designing distributed algorithms, logical structures of processors, which we simply call *processor structures*, have been used to reduce the communication complexity. For example, a *spanning tree* of the underlying communication network is used to design broadcast and information collection algorithms. If each processor can tell which local ports correspond to tree edges of the spanning tree, both broadcasting and collecting information are efficiently achieved by flowing information along tree edges. Important applications of efficient broadcast and collection algorithms include the leader election problem (e.g., the extrema-finding problem), and the problem of constructing the minimum spanning tree was first discussed in this context [1, 3, 5, 7]. An *in-tree* is another useful structure and is used to solve the mutual exclusion problem [2, 10, 11] and the decentralized object finding problem [4].

In this paper, we propose an idea of implementing dynamic sets by processor structures. To show that our idea is promising, we propose a processor structure that can be used to efficiently implement a dynamic set that supports the following three operations:

- Insert: This operation inserts the processor executing the operation, i.e., caller, in the set as a new element.
- Delete: This operation removes the caller from the set.
- Find: This operation returns the identifier of a processor in the set, or a special symbol *Fail* if the set is empty.

This dynamic set is slightly simpler than dictionary, but can be used to solve many problems, including the load balancing problem.

The proposed implementation of this dynamic set is very efficient; the amortized message complexity per operation is $8 + 12\log_2(|V| - 1)$ in the worst case. The analysis is based on the amortized analysis invented by Ginat *et al* [6].

The main idea of our processor structure is to maintain a dynamic set U of processors in a form of a distributed circular list: Each processor has a local register to store the *next* pointer to the "next" processor. We naturally identify the distributed system with a directed graph G; the node set is the set of processors, and there is a directed edge from u to v iff the next pointer of u points to v. Dynamic set U is maintained in such a way that all processors in U are contained in a unique directed cycle C in G, and that $G - C$ are in-trees, each of whose sink has the next pointer to a processor in C.

Cycle C may contain processors not in U, since Delete is simply implemented by marking the processor executing Delete in most of the cases. Actual removal of marked processors from C will be done later, when either Find or Insert is executed. Insert and Find executed by a processor not in C, say u, follow the directed path starting with u, until it encounters a processor in cycle C, say v, and for Insert, u and v update the next pointers so as to insert u as the next processor of v. The message complexity of those operations therefore mainly depends on the length of the directed path the executing processor traverses. To shorten the length of the traversed path, in the proposed processor structure, we adopt the heuristic of path compression used in the Union-Find algorithm [8].

The processor structure proposed in this paper can be used to solve important problems besides to implement the dynamic set in this paper: To solve the mutual exclusion problem, we circulate a single token along C and regard it as a token ring system. We can use it to implement a more complex dynamic set like a queue.

The paper is organized as follows. In Section 2, we propose a processor structure for implementing the dynamic set defined in this section. In Section 3, we show several applications of our processor structure. Section 4 concludes the paper with future problems.

2 The Processor Structure

2.1 The Model

In what follows, we call a processor a *node*, since we will identify a distributed system with a directed graph with the node set being the set of processors (in the sense we explained in Section 1). Consider a distributed system with a node set $V = \{0, 1, \ldots, n-1\}$. There is a communication link between each pair of nodes in V, so that they can directly communicate with each other. Each link is FIFO in the sense that messages are received in the order they are sent. We assume that the processors and links are reliable.

Let $U \subseteq V$ be a dynamic set of processors that we implement in the distributed system. Set U supports the three operations Insert, Delete and Find defined in Section 1. In the rest of this section, we first present a processor structure for implementing U, then show its correctness, and finally evaluate its efficiency.

2.2 Local Registers

Our processor structure uses some local registers. In this subsection, we introduce them.

Each node $v \in V$ has a local register $\boldsymbol{next}_v$ to store the *next* pointer that points to a node in V. (We use two more local registers $\boldsymbol{token}_v$ and $\boldsymbol{mark}_v$. We explain about them later.) The node pointed by $\boldsymbol{next}_v$ is called the *next* node of v. By associating a directed edge from v to $\boldsymbol{next}_v$ with each ordered pair $(v, \boldsymbol{next}_v)$, we naturally obtain a directed graph G, in which every node has exactly one outgoing edge. Figure 1 illustrates an example of G, in which, for example, node 1 points to node 3 (i.e., $\boldsymbol{next}_1 = 3$) and node 3 points to node 6 (i.e., $\boldsymbol{next}_3 = 6$). Graph G is dynamic in the sense that its edge set dynamically changes, as nodes update their local registers.

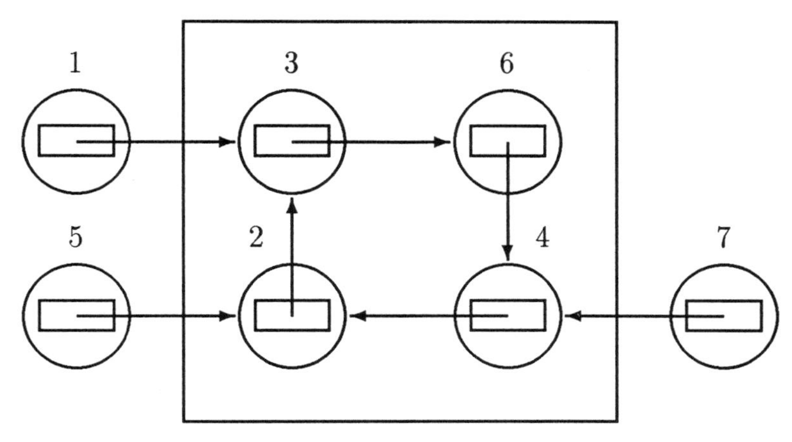

Fig. 1. A graph G representing the relation defined by the next pointers.

In our processor structure, the corresponding graph G is always kept connected. It therefore consists of a unique directed cycle C and a set of in-trees whose sink points to a node in C, since every node in G has exactly one outgoing edge. Further, circular list C is maintained so as to include all nodes in U. That is, we maintain U in C. As you will see, C may contain nodes not in U, since, in many cases, Delete only marks the caller (in C) to distinguish it from the nodes in U and the actual removal process from C is postponed until Find or Insert is executed. A local register (flag) $mark_v$ is used to memorize whether v is marked or not. That $v \in U$ iff $mark_v = 0$ is intended to hold.

When the processor structure is initialized, we appoint a single node in C to be the *anchor*. The anchor works as the arbiter, and breaks a symmetric situation occurred in C, which may cause a deadlock or a starvation. The anchor, however, is not a fixed node, but the role is taken over from one node to another. More clearly, we circulate a token along C, and the node having the token acts as the anchor. A local register (flag) $token_v$ is used to memorize whether or not v has the token. That v is the anchor iff $token_v = 1$ is intended to hold. The anchor is not marked unless $U = \emptyset$. If the current anchor executes Delete, it first sends the token to another node in U, and then marks itself. The anchor is used to check whether $U = \emptyset$ or not.

2.3 Primitive Operations

The processor structure is initialized as follows. We assume that initially $U = V$ holds. Hence initially, $mark_v = 0$ for each $v \in V$. Local register $next_v$ is initialized to $next_v := v + 1 \pmod n$ for each $v \in V$, i.e., the corresponding

```
procedure FindNode        { For initiator v. }
begin
        S := ∅;        { Nodes to be contracted. }
        w := next_v;
        repeat
                Send(inquire, w);
                Receive(M, w);
                if M = skip_me(next_w) then S := S ∪ {w} and w := next_w
        until M = found(next_w, token_w, mark_w);
        return w, next_w, mark_w and S
end
```

Fig. 2. Procedure FindNode.

G is a single directed cycle. The token is initially given to node 0 ($\in V$). Hence, initially, $\boldsymbol{token}_v = 1$ iff $v = 0$.

In the following, we present three procedures Find, Insert and Delete that are implementations of the corresponding operations. Those procedures are assumed to be executed as atomic ones, in the sense that once the execution starts, the processor is dedicated to execute it until it finishes. The only exception is the case in which Delete is initiated by the anchor; the anchor may need to handle messages that before Delete finishes, to avoid possible deadlocks. The procedures are described in a Pascal-like language. Note that comments are enclosed by '{' and '}'.

(A) Procedure FindNode

We first present a procedure FindNode that is used as a central subroutine in the implementations of the three operations. When a node v calls FindNode, it returns $w, next_w, token_w, mark_w$ and S, where w is either the first node in U appeared in the unique directed path from v (if $U \neq \emptyset$), or the anchor node (otherwise). Let $P_v(G)$ denote the directed path in G connecting from v to w. In FindNode, variable S is used to store the set of nodes appeared in $P_v(G)$, excluding v and w, and is used to apply the heuristic of path compression later.

If FindNode returns $mark_w = 1$, w is the anchor since FindNode does not find a node in U, and therefore $U = \emptyset$ since the anchor is marked. FindNode is given in Figure 2.

Each node u on the path traversed, upon receiving a message M' from v, responds as in Figure 3. All nodes u traversed by FindNode block themselves until a further instruction (message) arrives from v, which will be issued in the procedure that calls FindNode as a subroutine. Note that for the simplicity of description, v may send or receive a message to or from v itself. In this case, we assume that v behaves like u (without actually exchanging messages), except that v does not block itself.

```
Case M' of
inquire: begin
            if mark_u = 1 and token_u = 0 then
            { if u ∉ U and u is not the anchor }
                  Send(skip_me(next_u), v) and
                  block the execution of u
                  until it receives message contract from v
            else { if u ∈ U or u is the anchor }
                  Send(found(next_u, token_u, mark_u), v) and
                  block the execution of u
                  until it receives a message from v
      end;
contract(w):  next_u := w;
place_token:  token_u := 1;        { Used in Delete. }
remove_token:  token_u := 0 and next_u := v;        { Used in Insert. }
unblock:  unblock the execution of u, but do nothing else
end
```

Fig. 3. Procedure for a node u upon receiving a message M'.

```
procedure Find        { For initiator v. }
begin
      if mark_v = 0 then return v;
      if token_v = 1 then return Fail;
      Call FindNode;        { It returns w, next_w, token_w, mark_w, and S. }
      next_v := w;        { next_v points to the found node. }
      Send(contract(w), x) for all x ∈ S;        { Path compressed. }
      if mark_w = 0 then return w else return Fail
end
```

Fig. 4. Procedure Find.

(B) Procedure Find

Suppose that a node v wishes to find a node $u \in U$ and calls Procedure Find. If $mark_v = 0$, it returns v itself, since $mark_v = 0$ implies $v \in U$. If $token_v = mark_v = 1$, it returns *Fail*, since $U = \emptyset$ in this case. When v is neither an element in U nor the anchor, Find returns a $u \in U$ as long as $U \neq \emptyset$. If $U = \emptyset$, *Fail* is returned. The node u found is the first node in U appeared in the directed path $P_v(G)$ from v. Find is given in Figure 4.

(C) Procedure Delete

Suppose that a node v wishes to delete itself from U and calls Procedure Delete. If v is not the anchor, i.e., if $token_v = 0$, then the deletion is simply achieved by marking itself. Otherwise, if it is the anchor, Delete first finds a node $u \in U$ by calling FindNode, transfers the token to u, marks itself, and then the path compression is applied. Finally, if FindNode cannot find a node in U and returns the anchor v itself as w, then $U = \{v\}$. The anchor v marks itself and the path compression is applied. As a result, U becomes empty and $token_v = mark_v = 1$

```
procedure Delete        { For initiator v with mark_v = 0. }
begin
    if token_v = 0 then mark_v := 1 and terminate;
    Call FindNode;      { It returns w, next_w, token_w, mark_w, and S. }
    next_v := w;
    if w ≠ v then token_v := 0 and Send(place_token,w);  { Send token. }
    mark_v := 1;   { Mark itself. }
    Send(contract(w), x) for all x ∈ S;       { Path compressed. }
    Send(unblock, w)
end
```

Fig. 5. Procedure Delete.

Interrupt Handler { For the anchor v. }
When the anchor v receives an `inquire` message from a node v' and is waiting for a reply from a node w to the previous `inquire` message to w:

```
begin
    { Flush S. }
    Send(contract(w), x), for all x ∈ S;
    S := ∅;
    Send(unblock, w) and ignore a reply message arriving from w;

    { Acknowledge to v'. }
    Send(found(next_v, token_v, mark_v), v') and block itself
    until it finishes processing the next message from v';

    { Resume execution. }
    Resume the execution of FindNode by resending an inquire
    message to w
end
```

Fig. 6. Interrupt handler for the anchor.

hold. Delete is given in Figure 5.

As mentioned earlier, the execution of a procedure is usually indivisible. The only exception is the execution of FindNode by the anchor, which can occur only in Delete. If the anchor v receives an `inquire` message from a node v' when it executes FindNode and is waiting for a reply from a node w, it first applies the path compression for the nodes in S, then suspends the execution of FindNode, and responds to the `inquire` message. FindNode resumes the control after processing it. Such an interrupt is necessary to guarantee deadlock-freedom. A formal description of this *interrupt handler* is given in Figure 6.

(D) Procedure Insert

Suppose that a node v wishes to insert itself in U and calls procedure Insert. It first looks for a node w in U by calling FindNode. Then it inserts v in the circular list as the next node of w. If $token_w = mark_w = 1$, which implies that $U = \emptyset$,

```
procedure Insert        { For initiator v with mark_v = 1. }
begin
        if token_v = 1 then mark_v := 0
        else
         begin
                Call FindNode;  { It returns w, next_w, token_w, mark_w, and S. }
                mark_v := 0;        { Remove mark. }
                if token_w = mark_w = 1 then
                        Send(remove_token,w), next_v := v
                        and token_v := 1        { Token Received. }
                else Send(remove_token,w) and next_v := next_w;  { v inserted. }
                Send(contract(w), x) for all x ∈ S        { Path compressed. }
         end
end
```

Fig. 7. Procedure Insert.

the anchor is transferred to v from w. A formal description of Insert is given in Figure 7.

2.4 Proving Correctness

Lemma 1. *The circular list C always contains all nodes in U.*

Proof. We show that every node not in C is marked. Every node is not marked and resides in the circular list C at the time of initiation. Suppose that an unmarked node u exists outside of C at some time instant. Then either u was removed from C despite that it was not marked, or its mark was removed despite that it was outside of C. The latter case never occur, since the removal of mark occurs only in Insert and it always inserts the caller as the next node of the node in U that FindNode found. Let us check that the former does not occur, either. This occurs only when a `contract` message is sent to an unmarked node. However, FindNode always returns a set S of marked nodes. Therefore, C includes all nodes in U. □

Lemma 2. *The graph G is weakly connected all the time, i.e., the procedures never generate a new directed cycle besides C.*

Proof. We show that the following two claims hold:

1. At any time instant, C contains the anchor.
2. A `contract` message to a node x asks x for setting its next pointer to a node in C.

The first claim holds, since the anchor is in C at the initiation time, and it is transferred to a node found by FindNode, i.e., to a descendant of the current anchor. To show that the second claim holds, examine the value of parameter w that a `contract` message takes. If it is issued in Find or Insert, w is the node found by FindNode, which belongs to C. Otherwise, if it is issued in Delete by the anchor, w is a node reachable from the anchor, which also belongs to C.

Using the two claims, it is easy to show that the anchor is reachable from every node in G at any time instant. Hence, G is weakly connected. □

Lemma 3. *Deadlocks never occur in the proposed processor structure.*

Proof. Suppose that a deadlock occurs. Then there are a set D of nodes who are executing FindNode, and each of search paths initiated by D encounters a node already blocked by another search path. Since the number of outdegree of every node is 1, it implies that every node in C belongs to one of the search paths (because there is only one cycle in G). Consider the search path containing the anchor v. v must be the initiator of the path and $mark_v = 1$. Hence there is a node $w \in D$ such that the search path initiated by w sent an **inquire** to v but cannot get a reply from v (which causes the deadlock). However, this is a contradiction, since v will execute the interrupt handler which interrupts the search initiated by v and processes the **inquire**. The execution of FindNode called by w will return the data about v and terminate. □

The following theorem can be shown using the above three lemmas, by induction on the set of time instants, at which an operation is performed.

Theorem 4. *The implementations of operations* Find, Insert *and* Delete *are correct.* □

2.5 Performance Analysis of Implementation

In this subsection, we evaluate the performance of the implementation proposed in the previous subsections, by deriving the amortized message complexity per operation in the worst case.

Let $\mathcal{E}$ be a sequence of operations with

- X Insert operations and
- Y Find or Delete operations.

Let N^- denote the total number of nodes deleted from U during $\mathcal{E}$.

We estimate the number of messages issued in $\mathcal{E}$. First, observe that a sequence of communications always starts with an **inquire** message issued in FindNode and at most two more messages will be exchanged to processing it. Suppose that a node v sends an **inquire** to a node w. Then w replies either a **skip_me** or **found** message and blocks itself. Later, v will send one more message (either a **contract**, **place_token**, **remove_token**, or **unblock**) to unblock w. Thus we can count the number of messages issued in $\mathcal{E}$ by counting the number of **inquire** messages, if FindNode is not called by the anchor and no message is discarded without processing it by the interrupt handler in Figure 6.

Suppose that the execution of FindNode by the anchor v is interrupted by an **inquire** message from a node v', when it is waiting for a reply from a node

w. Since v needs to send an **unblock** message to unblock w, two messages are wasted by the interrupt. The search for a node in U by v will be restarted from w by resending an **inquire** message to w. Let us estimate the number interrupts in $\mathcal{E}$. Clearly, it is bounded from above by the number of times that **FindNode** is called in $\mathcal{E}$. Hence at most $2(X+Y)$ messages can be wasted.

We first estimate the number M of **inquire** messages sent to nodes in C. Then

$$M \leq X + Y + N^{-},$$

since all but the last node in each search path return a **skip_me** message and are removed from C. Therefore, the total number Z of messages processed (i.e., received or sent) by nodes in C is at most

$$2(X+Y) + 3(X+Y+N^{-}).$$

Since $N^{-} \leq Y$,

$$Z \leq 5X + 8Y,$$

which is a constant per operation in the amortized sense.

Next, we count the total number Z' of messages handled by nodes in T, letting $T = V - C$. Graph G changes in $\mathcal{E}$, so does T. Note that T always forms a forest.[2] Since T does not contain the anchor, $Z' \leq 3M'$, where M' is the number of **inquire** messages processed by nodes in T in $\mathcal{E}$. In what follows, we estimate M' by the potential method.

An operation in $\mathcal{E}$ modifies the configuration of T, by deleting a node, by inserting a node, and/or by applying the heuristic of path compression. Since no node in T returns a **found** message, the number of **inquire** messages processed by nodes in T in the execution of the ith operation O_i in $\mathcal{E}$ is equal to the length L_i of the path compressed in O_i. Using the terminology in amortized analysis, the *actual cost* c_i of O_i is defined to be L_i. Hence,

$$M' = \sum_i c_i.$$

To apply the potential function method invented in [6], define the potential of T by

$$\Phi(T) = \frac{1}{2} \sum_{v \in T} \log_2 s(v),$$

where $s(v)$ is the size of a node v in T defined to be the number of descendants of v including v itself. Then define the *amortized cost* $\hat{c_i}$ of O_i by

$$\hat{c_i} = c_i - \Phi(T_{i-1}) + \Phi(T_i),$$

[2] Formally, the subgraph of $G = (V, E)$ induced by $T \subseteq V$, i.e., $(T, E \cap (T \times T))$, is a forest.

where T_{i-1} and T_i are the forests T before and after the execution of O_i, respectively. For any sequence $\mathcal{E}$ of m operations, we have

$$\sum_{i=1}^{m} \widehat{c_i} = \sum_{i=1}^{m} (c_i - \Phi_{i-1} + \Phi_i) = \sum_{i=1}^{m} c_i - \Phi_0 + \Phi_m.$$

where Φ_0 is the potential of the initial forest and Φ_m is the potential of the forest after the mth event. Since $\Phi_0 = 0$ (because T_0 is empty) and $\Phi_m \geq 0$ (by definition), we have

$$\sum_{i=1}^{m} c_i \leq \sum_{i=1}^{m} \widehat{c_i}.$$

Now, let us bound the amortized cost $\widehat{c_i}$ of each operation. A key lemma is obtained by Ginat, et al. [6]: The amortized cost of a path compression on a tree is at most $\log_2(|V|-1)$. The execution of operation O_i may call FindNode, and find a node, traversing a search path. Let P_i be the search path. If O_i is Find operation, the path compression along P_i modifies T from T_{i-1} to T_i. We regard that the path compression consists of two phases: In the first phase, the subpath of P_i in T_{i-1} is compressed, and in the second phase, the rest of P_i is compressed. Let T^* be the T immediately after finishing the first phase. We have

$$\widehat{c_i} = c_i - \Phi(T_{i-1}) + \Phi(T^*) - \Phi(T^*) + \Phi(T_i).$$

By the key lemma, $c_i - \Phi(T_{i-1}) + \Phi(T^*) \leq \log_2(|V|-1)$. On the other hand, it is easy to check that $\Phi(T_i) - \Phi(T^*) \leq N_i^- \log_2(|V|-1)$, where N_i^- denotes the number of nodes deleted from C (i.e., added to T) in O_i. Hence, $\widehat{c_i} \leq (1+N_i^-)\log_2(|V|-1)$. The amortized costs of the other two operations can be estimated in the same way, and are bounded by $(1+N_i^-)\log_2(|V|-1)$. Since $\sum_{i=1}^{m} N_i^- = N^-$ and $N^- \leq Y$, we have

$$M' \leq (m+N^-)\log_2(|V|-1) \leq (m+Y)\log_2(|V|-1).$$

We now conclude the following theorem.

Theorem 5. *For any sequence of operations with X* Insert *operations and Y* Find *or* Delete *operations,*

$$(5+6\log_2(|V|-1))X + (8+12\log_2(|V|-1))Y$$

messages are issued, in the worst case. Or equivalently, the amortized message complexity per operation is $8+12\log_2(|V|-1)$, in the worst case. □

3 Applications

3.1 Dynamic Load Balancing

Consider a distributed system, in which tasks are dynamically created, processed and removed on processors. The **dynamic load balancing problem** is the problem of migrating tasks so that the load of processor balances (see, e.g., [9]). Such a migration must be planed and performed by processors autonomously, and not by an instruction of central controller that observes the global configuration of system. As in [9], we classify the degree of load of a processor into three classes; light, medium, and heavy, for the simplicity. If a processor with heavy load can search for a processor with light load efficiently, the load balancing can be solved efficiently. Theimer and Lantz's load balancing algorithm, for example, is an algorithm for finding such a lightly loaded processor, with message complexity $O(|V|)$, in the worst case.

Our dynamic set of processors can be used to this end. We simply accommodate processors with light load in U. If the condition of a processor u changes from medium to light, u executes Insert to insert itself in U. If the condition of u changes from light to medium, it executes Delete to delete itself from U. If a processor u with heavy load wishes to migrate some tasks to a light one, it executes Find to find such a processor. For the implementation we proposed in the last section, the amortized time complexity per operation is $O(\log_2 |V|)$ in the worst case.

3.2 Distributed Mutual Exclusion

The *distributed mutual exclusion* problem is the problem of guaranteeing that at most one processor can be in the critical section at a time. As in the case of dynamic load balancing, we look for a distributed solution.

The problem can be solved using the processor structure proposed in the last section. We initiate the system in such the way that $U = \emptyset$ holds. If a processor wishes to enter the critical section, it executes Insert and waits until it becomes the anchor. The anchor has the right to enter the critical section. When a processor v leaves the critical section, it executes Delete. Since v is the anchor, Delete first sends the token to the next node in C (who is waiting for its turn), and then deletes v from C.

This mutual exclusion algorithm requires only $O(\log_2 |V|)$ messages per entry in the amortized sense. Furthermore, this solution has the following advantages. In many cases, the set of processors that are waiting for their turn to enter the critical section is rather a small dynamic subset U of the set V of all processors. The above solution can be viewed as a token ring system among U, in which U may dynamically change. The token is circulated among a small group U.

3.3 Implementing a FIFO Queue

In Section 2, we propose an implementation of a dynamic set by using a processor structure. The dynamic set implemented supports three simple operations. If we can implement dynamic sets supporting more powerful operations, we can design efficient distributed algorithms using those dynamic sets. A (FIFO) queue of processors can be implemented using the processor structure proposed in Section 2. In this implementation, the queue is implemented in the form of circular list. The anchor represents its *head*. We modify FindNode so that it returns the head. Operation Find then returns the head. Operation Insert first searches for the head and then inserts the caller as the predecessor of the head, i.e., from the tail of the circular list. We do not change the implementation of operation Delete. It passes the token to the next node, and deletes the previous head, i.e., the caller, from C.

4 Concluding Remarks

In this paper, we proposed a processor structure for implementing a dynamic set U of processors supporting three operations; Find, Insert, and Delete. The amortized message complexity per operation is at most $8+12\log_2(|V|-1)$, in the worst case. The dynamic set can be used to many important problems, including the dynamic load balancing problem. The processor structure we proposed to implement the dynamic set can be used to solve the distributed mutual exclusion problem and be used to implement other dynamic sets supporting more powerful operations, like a queue.

An important future problem is to consider an processor structure by which some dynamic set is efficiently implementable on a distributed system in which the communication cost between two processors dynamically changes.

References

1. B. Awerbuch. Optimal distributed algorithms for minimal weight spanning tree, counting, leader election and related problems. In *Proc. 19th STOC*, pages 230–240. ACM, 1987.
2. J. M. Bernabéu-Aubán and M. Ahamad. Applying a path-compression technique to obtain an efficient distributed mutual exclusion algorithm. In *Proc. 3rd WDAG (LNCS 392)*, pages 33–44, 1989.
3. F. Chin and H. F. Ting. An almost linear time and $O(n \log n + e)$ messages distributed algorithm for minimum-weight spanning trees. In *Proc. 26th FOCS*, pages 257–266. IEEE, 1985.
4. R. J. Fowler. The complexity of using forwarding addresses for decentralized object finding. In *Proc. 5th PODC*, pages 108–120. ACM, 1986.
5. R. G. Gallager, P. A. Humblet, and P. M. Spira. A distributed algorithm for minimum-weight spanning tree. In *ACM Transactions on Programming Languages and Systems 5,* 1 66–77, 1983.
6. D. Ginat, D. D. Sleator, and R. E. Tarjan. A tight amortized bound for path traversal. *Information Processing Letters*, 31:3–5, April 1989.
7. G. Singh and A. J. Bernstein. A highly asynchronous minimum spanning tree protocol. *Distributed Computing*, 8:151–161, 1995.

8. R. E. Tarjan. *Data Structures and Network Algorithms.* Society for Industrial and Applied Mathematics, 1983.
9. M. M. Theimer and K. A. Lantz. Finding idle machines in a workstation-based distributed system. In *Proc. 8th ICDCS*, pages 112–122, IEEE, 1988.
10. M. Trehel and M. Naimi. A distributed algorithm for mutual exclusion based on data structures and fault tolerance. In *Proc. 6th Annual Phoenix Conf. on Computers and Communications*, pages 35–39. IEEE, 1987.
11. T.-K. Woo. Huffman trees as a basis for a dynamic mutual exclusion algorithm for distributed systems. In *Proc. 12th IEEE Int. Conf. on Distr. Comp. Sys.*, pages 126–133. IEEE, 1992.

Collective Consistency

Cynthia Dwork, Ching-Tien Ho, Ray Strong

IBM Almaden Research Center

Abstract. We present *collective consistency*, a new paradigm for multi-process coordination arising from our experience building a parallel version of a software package for computer aided engineering. A straightforward checkpointing and reorganization strategy will permit a parallel computation to tolerate failed or slow processes if the failure or tardiness is consistently detected by all participants. To this end, we propose the addition of *autonomous failure detection* and a *consistent reporting* protocol to the collective communication library supporting the industry standard Message Passing Interface. The focus of this paper is the consistent reporting protocol that must provide an approximate solution to the collective consistency problem. We give a family of knowledge-based specifications for a collective consistency protocol and discuss several implementations. We support our implementation choices with an extremely general proof that no trivially dominating solution exists.

1 Introduction

In a large body of parallel mathematical and engineering software computations have a regular structure of alternating communication and computation phases. The communication phase is carried out using the industry standard Message Passing Interface (MPI). In a typical implementation of the MPI, each participant exchanges multiple data packets with the others via a transport layer. A *blocking* (waiting) receive call is issued to the transport layer for all the packets that are to be received from other participants. If one participant fails to send all of its packets to the others, each member blocks (waits forever) pending the arrival of these anticipated packets. A blocked process itself fails to send packets during the subsequent communication phase, causing still other processes to block. Thus, in real computations, when even one process fails *the entire computation becomes blocked pending human intervention.* We do not know of any automatic method currently in use to address this problem.

We develop a method by which such failures can be automatically tolerated by the remaining processes and by which these processes can agree on the identities of failed processes. Our technique allows the processes to agree on a new group membership so that the computation can be reorganized to run on the new group. Of course, such agreement is impossible in a completely asynchronous system: any agreement protocol that guarantees consistency has executions in which some processes must block or take infinitely many steps [DDS, FLP]. However, one goal that *is* both desirable and possible to achieve is the reduction of the window (time period) during which a process failure can cause some or all other processes to block. Specifically, we propose the following approach:

1. Augment the collective calls to the transport layer with a simple failure detector [CT, CHT], so that no process blocks during the collective communication.
2. Run a (preferably simple and brief) *collective consistency* protocol (CCP), during which some participants may block, allowing those processes that return to share a consistent view of the set of non-failed processes. Specifically, a collective consistency protocol ensures that any two processes that do not mutually regard each other as having failed and that do not block will have an identical view of group membership. (A formal definition of the problem appears in Section 2.)
3. Reorganize the application computation to run on the newly agreed-upon group. Continue the computation protocol, ignoring messages from the detected faulty processes.

Our target application is particularly amenable to this approach because each successful completion of a collective communication call presents a natural checkpoint, after which each process can easily recover sufficient information to reorganize and proceed with the computation. Thus when a failure is consistently reported to all surviving participants, the survivors can roll back to the previous check point, reorganize, and roll forward. However, as long as the participants have different views of the set of processes available to participate after reorganization, no further progress is possible. Thus collective consistency is essential to fault tolerance in this setting.

As described informally above and defined formally in Section 2, *collective consistency* is a weak form of consensus in which processes try to reach a common view of group membership under a rather relaxed definition of "commmon." More specifically, a process enters the computation with an *initial view* consisting of a list of suspected failed processes, and, if it returns from the CCP, it does so with an *output view* of suspected faulty processes. The collective consistency requirement is that each process p's output view V_p must agree with the output views of processes that p does not suspect to have failed. In other words, p's ouput view need only agree with the output views of processes not in V_p. Since a symmetric requirement exists for each $q \neq p$, this means that the output views of p and q can differ only if they are mutually suspecting processes.

Thus, a collective consistency protocol partitions the survivors into cliques $C_1, \ldots, C_k$, where the processors in each clique C_i all believe that all processors not in C_i have failed, and each $p_j \in C_i$ has no reason to believe any other process in C_i has (yet) failed. The key difference between collective consistency and consensus is that the definition of collective consistency permits two returning processes, p and q, to have different views under some circumstances. By contrast, in consensus p's output can differ from q's only if at least one of them fails to return. A second difference is that the definition of collective consistency does not require all nonfaulty processes to terminate.

In Section 4 we give a family of knowledge-based specifications for protocols for collective consistency. We prove that certain implementations of these specifications solve the collective consistency problem and we provide two families of such implementations.

Collective consistency is weaker than agreement. Our situation is unlike the typical situation for asynchronous impossiblity proofs: it is not the case that any protocol for collective consistency has executions in which every process takes infinitely many steps. However, we prove the existence of executions E such that *either* every process takes infinitely many steps *or* there exists a particular process p_E such that in E *all* other processes block waiting for p_E. Thus, even in the presence of a single slow or crashed process, there is no non-blocking CCP. Indeed, the lower bound also rules out all hope of designing a single failure tolerant, nontrivial CCP in which, for example, at least one process terminates. This is discussed further in Section 5.

Our target application is computation intensive and certain parts, for example, those involving matrix multiplication, lend themselves easily to parallelization. The work described in this paper arose from our experience in building a prototype parallel implementation of this application, but the results apply to any parallel architecture in which multiple processing elements operate concurrently to perform a single task.

2 Requirements for Collective Consistency of Failure Detection

In this section we define several variants of collective consistency and discuss some design requirements for collective consistency protocols specific to our target applications.

2.1 Definitions of Collective Consistency

The goal of collective consistency is for the participants to reach a consistent view of the set of failed processes. In a collective consistency protocol there is no requirement that a process ever return. Thus, collective consistency can be solved in systems that do not permit solutions to consensus. The question of how to evaluate a collective consistency protocol is a subject for future research. We return to this question in Section 6.

Each process enters a CCP with an *initial view*, which is a list of processes it suspects of failure. (We treat the receipt of the initial view as the first event of the CCP for each process.) In the definition of collective consistency we need not concern ourselves with how these suspicions are generated, but in practice, suspicion is based on the output of a failure detector. During the course of the computation a process may irreversibly decide on an *output view* V – a final list of suspects. In this case we say the process *returns* V. A process that does not decide on an output view is said not to return.

The key requirement of collective consistency essentially says that if p does not regard q as having failed, and if indeed q has not failed and actually returns an output view, then this view is the same as p's:

Collective Consistency: If V_i is the view returned by process i and V_j is the view returned by process j, and if $j \notin V_i$, then $V_i = V_j$.

In other words, since the requirement is symmetric for i and j, if process i does not view process j as having failed, then they must agree on which processes they suspect of failure. We will use the term *weak collective consistency* as a synonym, emphasizing the difference between this property and that of *strong collective consistency*, defined next. Note that strong collective consistency is just consensus without termination.

Strong Collective Consistency: The returned views of all processes are identical.

A *quorum system* is a collection of sets of processes, every pair of which have a non-empty intersection [Gif, PW, NW]. Quorum systems can be used to convert a weak collective consistency protocol into one for strong collective consistency. As explained in the Introduction, a weak collective consistency protocol partitions the survivors into cliques such that, roughly speaking, the members of each clique believe that they, and only they, are alive. To obtain a protocol for strong collective consistency we add the following test: if a process sees that its clique does not contain a quorum then it blocks. To see that the transformation is correct, suppose two processes p and q complete the protocol. Let Q_p be a quorum contained in p's clique, and let Q_q be a quorum contained in q's clique. By definition of a quorum system, there exits a process $r \in Q_p \cap Q_q$. It follows from the definition of weak collective consistency that $V_p = V_r = V_q$. In our target application, each clique continues the computation independently. Under strong collective consistency, at most one clique of processes continues the computation.

In our target application, process q is in the initial view of process p (that is, before the CCP begins p suspects q of having failed), precisely because p did not receive from q an anticipated message containing information essential for p to continue the application computation. While under some circumstances it may be possible for p to obtain this information from other processes, this would be neither simple nor quick. From an engineering point of view, it is easier for all processes in p's group to regard q as having failed and to reorganize the computation accordingly. For this reason we are particularly interested in *monotone* solutions to the collective consistency problem, in which suspicions are never allayed – if p initially suspects q, then nothing during the CCP "convinces" p to stop suspecting q. This motivates the following definition.

Monotonicity: If V_i is the initial view of processor i, and if processor i returns V, then $V_i \subseteq V$.

Finally, we rule out the trivial solution to weak collective consistency in which every terminating process simply views every other as having failed, since this results in a trivial partition (every surviving process forms its own group consisting only of itself) and hence unecessary duplication of work in the target application. To make this precise we need a few definitions.

We model a protocol execution as sequences of events, one sequence for each process. A *cut* is the union of a set of finite prefixes of sequences, one for each process, closed under causality. The *initial cut* is the empty cut. Intuitively, the initial cut "happens before" processes obtain their initial views. This is merely a notational convenience.

Nontriviality: The initial cut c_0 is *multivalent*, that is, there exist cuts d_1 and d_2 containing c_0, such that in d_1 some process p returns an empty output view and in d_2 some (not necessarily different) process q returns a non-empty output view.

Let $\overline{\{i\}}$ denote the set of all processes except i. The nontriviality condition rules out the trivial "solution" to collective consistency in which each process i returns output view $V_i = \overline{\{i\}}$.

A protocol solves (strong) (monotone) collective consistency if it is nontrivial and in all executions the (strong) (monotone) collective consistency condition is satisfied.

2.2 Design Goals

In Section 3 we discuss the relationship between collective consistency and the widely studied group membership problem. The source of the problem, that is, the application in which the problem arises, naturally drives one's view of what constitutes a "good" solution. Much (but by no means all) of the work on group membership arises in the context of replicated databases. As mentioned above, collectively consistency arose in the context of parallel engineering and mathematical software. In this environment there is no danger if the processes split into two groups and each group pursues the computation independently (although of course this results in wasted effort). The same is typically not true in the context of a replicated database.

Our first desideratum is *simplicity*, based on the assumption that a simpler algorithm is more likely to be implemented correctly.

The second desideratum is *speed* (roughly speaking, minimizing the number of protocol steps a process must take before it sends the last message required of it by the protocol), since the system is vulnerable to blocking during execution of the CCP. Intuitively, there appears to be an engineering tradeoff in designing a collective consistency protocol: on the one hand, since failure detection is not performed during the execution of the collective consistency protocol, the more (asynchronous) rounds of communication the greater the chance that a process will block trying to complete a communication round. This would seem to suggest that a collective consistency protocol with few rounds of communication should "perform better" than a protocol requiring many rounds. On the other hand, a more complicated protocol can sometimes permit more processes to return. For example, if a process p requires information from process q, and if a third process r has received this information from q, then r can relay this information to p, even if q does not succeed in doing so. In this case p has had to wait two rounds for the information (one round for it to go from q to r and one round for it to travel from r to p), but p has the information it needs and can return from the protocol.

The advantage to having more processes return is that more processes are then available to complete the application computation. A formalization of this intuition and a thorough analysis of the factors involved in choosing the "right" implementation of collective consistency would be interesting theoretically and extremeley useful in practice. This is an excellent area for future research, especially since this type of tradeoff is not restricted to collective consistency.

3 Related Research

The overall goal of our work is to run a parallel application in the presence of process failure or unacceptable slowness. The study of concurrent execution of work in the presence of failures in a message-passing distributed system was initiated by Bridgeland and Watro [BW]. They consider a system of t asynchronous processes that together must perform n independent units of work. The processes may fail by crashing and each process can perform at most one unit of work during the course of the computation. Bridgeland and Watro provide tight bounds on the number of crash failures that can be tolerated by any solution to the problem. Thus, the goal in the [BW] work is to design a protocol that guarantees that the work will be performed in every execution of the protocol.

Dwork, Halpern, and Waarts designed algorithms for t processors to perform n units of work provided even one of the t processors remains operational [DHW]. They focused on effort-optimal algorithms, in which the sum of the number of messages sent and the number of tasks actually performed (including tasks performed multiple times because a processor fails before being able to report its progress) was $O(n+t)$, which is asymptotically optimal, but the (synchronous) time costs are exhorbitant [DHW]. Faster algorithms were obtained by De Prisco, Mayer, and Yung [DMY]. The approach of Dwork, Halpern, and Waarts yields as a corollary a message-efficient agreement algorithm, improved upon by Galil, Mayer, and Yung [GMY].

Study of another related problem was initiated in a seminal paper by Kanellakis and Shvartsman [KS]. Specifically, they introduce the *Write-All* problem, in which n processes in a shared-memory system cooperate to set all n entries of an n-element array to the value 1. Kanellakis and Shvartsman provide an efficient solution that tolerates up to $n-1$ faults, and show how to use it to derive robust versions of parallel algorithms for a large class of interesting problems. Their original paper was followed by a number of papers that consider the problem in other shared-memory models (see [AW, BR, KS2, KPRS, KPS, MSP]).

There is an extensive literature on the related area of *membership services* (see, for example, [CHT2, DMS, RB] for definitions and results in asynchronous systems). *Group membership* itself is a loosely defined term that generally describes the requirements on (1) the *series* of *views* held by a given process, of the membership of a group (within some universe of interest) as this membership changes over time; and (2) the manner in which the series of views obtained by different processes according to (1) are related. In contrast, collective consistency is a "one-shot deal": (monotone) collective consistency is defined in terms of (inputs and) outputs in a *single* execution of a collective consistency protocol. In particular, in our target application, collective consistency is concerned with the problem of combining initial views of failed processes (provided by a failure detector) to form a collectively consistent output view of failed processes – and hence of group membership (the non-failed processes).

Finally, we note that any algorithm for consensus yields an algorithm for strong monotone collective consistency (and hence all other variants of collective consistency). Very briefly, the reduction is as follows. Run in parallel n independent executions of a single-source agreement protocol to agree on the input views of each process (the ith execution is to agree on the input view of the ith process). Note that it is a property of the single-source agreement problem that if the source remains non-faulty throughout the execution then the value decided upon is the initial value of the source. This is important for monotonicity. Recall that in the collective consistency problem each input view is a set of processes. Take the union of the agreed upon views to be the (common) output view.

Since collective consistency does not require termination, consensus is strictly more powerful than all variants of collective consistency defined in this paper.

In the patent literature, Dwork, Halpern, and Strong describe a solution to the problem of performing work in the presence of failures that is both fast and effort-optimal in the absence of failures, and is correct in all instances [DHS]. A different approach was taken by Whiteside, Freedman, Tasar, and Rothschild: the work is statically divided into a set of tasks, each of which is assigned redundantly to several processing elements. Results produced are subject to majority vote resolution. Beyond voting and the use of timeout to detect failures, there is no attempt to reach agreement among the processes [W].

Our work on collective consistency can be viewed as an adaptation of the general method of Dwork, Halpern, and Strong [DHS] to the environment of parallel computation, in which particular care is taken both to allow for inconsistent failure detection (different processes detect different failures) and to guarantee consistency of response (roughly, differences in detected failures are resolved) at the cost of a small chance of blocking. Intuitively, the chance of blocking is small because the consistency protocol requires few messages; a process would have to fail *during* the consistency protocol to cause blocking.

The collective consistency problem is an example of a coordination problem, like the distributed commit and the consensus problems. Solutions to both of these problems can be and have been used in an *ad hoc* way, under certain assumptions, to solve collective consistency. No solution, including those presented in this paper, is entirely general: each approach works only under certain assumptions about the environment. However, the solutions provided here seem to require the weakest assumptions. We elaborate by characterizing typical assumptions and, in the table below, associating each approach with its required assumptions.

The types of assumptions are

1. type of failure, *e.g.*, processes are restricted to crash failures;
2. number of failures, *e.g.*, $n \geq 3t + 1$;
3. timing, *e.g.*, there is an upper bound on message delivery time;
4. randomness, *e.g.*, each process has access to a private source of random bits; and
5. stable storage.

distributed commit	consensus	collective consistency
1 and 5	(1 or 2) and (3 or 4)	1

Each of the approaches discussed above either in the worst case makes no global decisions (no coordination) and allows all the work to be done by each of the participants [BW, W], or depends on the timing or randomization assumptions we associate with consensus above [DHW, KS]. Strong collective consistency has been studied before, e.g. by [CT, DLS, R]. Rabin called the problem *choice coordination*, and studied it in the context of shared memory with atomic test and set (the atomicity of test and set can be viewed as an additional timing assumption). The relevant results of Chandra and Toueg and Dwork, Lynch, and Stockmeyer discuss conditions under which consensus can be reached if the failure detectors [CT] or the system [DLS] remain sufficiently well behaved for sufficiently long. By contrast, our approach is to take input from failure detectors but make no further use of them during the collective consistency protocol. We know of no literature on monotone collective consistency, strong or weak.

4 Protocols for Monotone Collective Consistency

In this section we present a knowledge-based specification for a family of monotone collective consistency protocols. We give two implementations of this specification. The implementation assumes only crash process failures and that messages are not altered. We also assume (although it is not necessary for correctness) that each message sent eventually reaches its destination unless either the sender or receiver crashes.

In Figure 1 we give a knowledge-based specification for a protocol for monotone collective consistency of failure detection. The operator K_i in the specification is the knowledge operator, indicating that the truth of its argument is known to process i [FHMV].

We use the notation V_i^k to denote the value of V_i formed during iteration $k = 1, 2, \ldots$ by process i (V_i^0 is the initial value). If process i waits forever in iteration ℓ, then $\forall k \geq \ell$ we say V_i^k is *undefined.* The parameter B is an upper bound on the number of times the main loop in the code is iterated. Each value of B yields a different protocol.

In Figure 1, the knowledge operator K_i and its argument ($\phi = \forall j \notin V_i^{k+1}$, either V_j^{k+1} is undefined or $V_j^{k+1} = V_i^{k+1}$) represent an unkown Boolean predicate. An *implementation* of such a knowledge-based specification is an algorithm

Monotone-CC with inputs V_i and B:

```
k = 0;
V_i^0 = V_i;
Repeat for k < B{
        Send V_i^k to all processors;
        ∀j ∉ V_i^k, wait for V_j^k;
        Set V_i^{k+1} to be V_i^k union all V_j^k for j ∉ V_i^k;
        If K_i(∀j ∉ V_i^{k+1}, either V_j^{k+1} is undefined or V_j^{k+1} = V_i^{k+1})
                then return(V_i^{k+1}) and exit;
        else k = k + 1;}
Halt without returning;
```

Fig. 1. Monotone Collective Consistency Protocol; Code for Process i

that replaces $K_i(\phi)$ with a specific predicate. A *sound* implementation replaces $K_i(\phi)$ with a predicate Ψ such that, when Ψ returns True, ϕ holds in the set of possible executions provided by the system in which each participant uses Ψ in place of $K_i(\phi)$. An implementation is said to be *nontrivial* if (1) there is an execution in which each input is empty and after some number of iterations the Ψ at some process returns True and (2) there is an execution in which the input to some process p is nonempty and after some number of iterations the Ψ for p returns True.

Theorem 1. *Any nontrivial sound implementation of Protocol Monotone-CC provides a nontrivial solution to the monotone collective consistency problem.*

Proof. The proof is independent of the bound B on the number of iterations of the main loop, provided $B \geq 1$. In the remainder of the proof we assume $1 \leq k, \ell \leq B$.

Monotonicity is obtained because at all times V_i^k is the union of the initial V_i and other values. Nontriviality follows from monotonicity and the definition of a nontrivial implementation. We now argue collective consistency. Suppose processes p and q return V_p^k and V_q^ℓ respectively, and either $p \notin V_q^\ell$ or $q \notin V_p^k$. Assume without loss of generality that $q \notin V_p^k$. We first show that for all $r \notin V_p^k$, either r does not return or V_r^k is defined and $V_r^k = V_p^k$.

Let $r \notin V_p^k$. Since V_p^c only grows with c, we have that $r \notin V_p^c$ for all $0 \leq c \leq k$; that is, r was never suspected by p. Since p terminates with V_p^k we have that p successfully received V_r^c for all $0 \leq c < k$. If r does not complete the kth iteration of the loop, then it never returns. If r does return then it must complete the kth iteration, and in particular V_r^k is defined. By soundness, if V_r^k is defined, then $V_r^k = V_p^k$.

It follows from the previous argument and the fact that q returns with V_q^ℓ, that $\ell \geq k$ and hence that V_q^k is defined and equal to V_p^k.

Even if process q performs additional iterations, that is, even if $\ell > k$, it will only collect views from processes $r \notin V_q^k = V_p^k$, and all of these satisfy $V_r^k = V_p^k$. A simple induction shows that for all $k' \geq k$, if $V_r^{k'}$ is defined, then $V_r^k \subseteq V_r^{k'} \subseteq \cup_{r' \notin V_r^k} V_{r'}^k = V_p^k = V_r^k$, so $V_r^{k'} = V_r^k$. Thus, process q returns $V_q^\ell = V_p^k$.

One simple nontrivial sound implementation is obtained by letting $\Psi_1 = \forall j \notin V_i^k : (V_j^k = V_i^k)$. That is, Ψ_1 holds in the first iteration (*i.e.*, with $k = 0$) if, for all j not initially suspected of failure by i, j's initial list of suspects agrees with i's initial list of suspects. Letting $B = 1$ (*i.e.*, executing the loop with $k = 0$, returning if possible at the end of this first iteration, and otherwise not returning) Ψ_1 gives an extremely simple and yet nontrivial protocol for monotone collective consistency. We have designed a failure detector and communication transport layer so that if crash failures, significant slowdowns, or occasional message losses occur sufficiently far away in time from the execution of the collective consistency protocol, then failure detection will be uniform among those detecting a given failure. Thus the important window of vulnerability to failure for our simple Ψ is limited to a brief time interval surrounding execution of the collective consistency protocol. In fact, without iteration, the vulnerability window is limited to a brief time before the last failure detector of a machine that has not failed or slowed gives its input to its instance of the consistency protocol.

The *advantage* in iterating (that is, letting $B > 1$), is that it sometimes allows additional processes to terminate in spite of the occurence of failures in the failure window. We elaborate with an example. Since our failure detector uses a timeout, there are circumstances in which failure detection is not uniform. In particular there are executions in which some processes detect a slow process while others do not. Suppose we have a system of only three processes. Consider an execution in which there are no crash failures, but, based on its failure detector, process p suspects that process r has failed while process q does not. Using Ψ_1 with $B = 1$ *no* process returns from Monotone-CC. On the other hand, if we simply change to $B = 2$ then each of p, q, and r returns $\{r\}$ after the second iteration. In the context of our application, this is the desired outcome because p did not receive essential information from r in a timely fashion. The *risk* in iterating is that it increases the chances for a process failure (during execution of the collective consistency protocol) to cause other processes to block.

The predicate specified in Figure 2 provides another family of nontrivial sound implementations of the knowledge based protocol specified in Figure 1. The set W contains those processes j such that someone "trusted" by i (not viewed by i as faulty) "vouches" for j (does not view j as faulty). So in the above scenario this would cause p to wait to hear from r. Using Ψ_2 therefore allows this scenario to be handled even with $B = 1$. Although iteration seems to compensate for a difference in initial views, we conjecture that iteration is not necessary: the same effect can be obtained by a more sophisticated implementation (*e.g.*, using Ψ_2 instead of Ψ_1). Moreover, iteration is always dangerous, as it introduces additional opportunities for blocking.

Predicate Ψ_2:

$W = \{j | \exists m \notin V_i^{k+1} : (j \notin V_m^k)\}$;
$\forall j \in W$, wait for V_j^k;
$\forall m \notin V_i^{k+1}$, set $V_m^{k+1} = V_m^k$ union all V_j^k for $j \notin V_m^k$;
return($\forall m \notin V_i^{k+1} : (V_m^{k+1} = V_i^{k+1})$);

Fig. 2. Code for Predicate Ψ_2

As mentioned in Section 2.1, we can transform a protocol for weak collective consistency into one for strong collective consistency by adding a requirement that the complement of the output set be a member of a previously chosen quorum system. We can also use quorum systems to overcome a weakness of protocols that satisfy the knowledge-based specification of Figure 1: all such protocols block if some process fails at the beginning of the protocol because each other process waits to receive the view of the failed process. Let $\mathcal{Q}$ be a quorum system. In figure 1 replace the line "$\forall j \notin V_i^k$, wait for V_j^k;" with "Wait until $\{j | V_j^k$ received $\} \in \mathcal{Q}$;" and "Set $Q = \{j | V_j^k$ received $\}$;." Replace "$j \notin V_i^k$" in the next line by "$j \in Q$;." Then $\Psi_3 = \forall j \in Q : (V_j^k = V_i^k)$ can be used to provide a sound nontrivial implementation of the resulting specification. In fact it provides a solution for the monotone strong collective consistency problem.

5 A Lower Bound for Collective Consistency Protocols

As discussed above, we could not hope for a collective consistency protocol that tolerates process failure and guarantees termination for all correctly functioning participants; however, we had hoped it would be possible to design a collective consistency protocol that tolerated process failure and guaranteed termination for a nontrivial subset (possibly a minority) of all the correctly functioning participants. The following lower bound result shows that this is not the case: no collective consistency protocol can guarantee the timely (nonblocking) termination of any of the participants in the presence of one slow or failed process. Indeed, we show that, roughly speaking, for every collective consistency protocol either there exists a process p such that in some execution every process must wait to hear from p, or there is an execution in which no process fails but no process ever gives output.

5.1 Definitions and Axioms

We model a protocol execution as sequences of events, one sequence for each process. A *cut* is the union of a set of finite prefixes of sequences, one for each process, closed under causality. The first event at each process is an *input* event, the input being a possibly empty, finite set. Some processes may have later *output* events, at most one per process, the output being a possibly empty, finite set.

Recall that a cut c is *multivalent* if there exist cuts d_1 and d_2, both containing c, such that d_1 has a process that has given as protocol output the empty set and d_2 has a process that has given as protocol output some non-empty set. Note that the two outputs are inconsistent with respect to weak collective consistency. A protocol is *nontrivial* if the empty cut is multivalent. A cut c is *univalent* if there exists a cut d containing c such that in d some process has given output but c is not multivalent. A univalent cut is said to have *projected value* 0 if it is contained in a cut in which the empty set is given as output; otherwise, it is said to have *projected value* 1. As a binary relation on outputs, weak collective consistency is not an equivalence relation; but its projection onto values $\{0, 1\}$ is an equivalence relation. A cut c is *nullvalent* if it is not contained in any cut that has an output event. Note that multivalence, univalence, and nullvalence form a strict trichotomy.

A process p *decides* c if c is multivalent and for all d containing c such that p acts in $d - c$, d is univalent. A cut is *undecided* if it is either multivalent or nullvalent. A process p is *ready* at a cut c if there exists an event e_p at p and a cut d such that $d = c \cup \{e_p\}$. A process p is *permanently blocked* at c if there is no cut d containing c such that p acts in $d - c$. Note that if p is permanently blocked at a multivalent cut c then p trivially decides c.

Our results apply to any system satisfying the axioms of *compatibility of independent actions* and *separability of events*, described below. Note that separability of events is equivalent to there being no cycles in the causal relation between events.

- **Compatibility of Independent Actions:** Let c be a cut, and let d_p and d_q be cuts containing c such that $d_p - c = \{e_p\}$ and $d_q - c = \{e_q\}$, where e_p (respectively, e_q) is a single event occurring at process p (respectively, q). The compatibility axiom states that if $p \neq q$ then $d_p \cup d_q$ is a cut.
- **Separability of Events:** Let e_1 and e_2 be two distinct events contained in a cut c. The separability axiom states that there is a cut d containing one of these events but not the other. That is, there exists a d such that $e_1 \in d \wedge e_2 \notin d$ or $e_2 \in d \wedge e_1 \notin d$.

5.2 The Lower Bound

Lemma 2. *For all cuts c, at most one process both decides and is ready at c.*

Proof. Assume for the sake of contradiction that the lemma is false; specifically, let $p \neq q$ both be ready at and decide a cut c. See Figure 3 for an illustration throughout the proof.

Since p decides c, c must be multivalent. By definition of ready, there are cuts $d_q = c \cup \{e_q\}$ and $d_p = c \cup \{e_p\}$ where e_p (respectively, e_q) is a single event occurring at process p (respectively, process q). Recall that each univalent cut has a unique projected value from $\{0, 1\}$. Because both p and q decide c, d_q and d_p must be univalent with projected values v_q and v_p respectively. By the

Compatibility Axiom, $d = d_p \cup d_q$ is a cut. Cut d is univalent because p decides c, d contains c, and p acts in $d - c$. If d had a projected value different from v_p, then d_p would be multivalent. If d had a projected value different from v_q, then d_q would be multivalent. Thus $v_p = v_q$. Let v represent this projected value.

Since c is multivalent, there must exist a cut c' containing c in which some process has given an output with projected value $v' \neq v$. By the Separability Axiom there is a chain of cuts $c = c_0, c_1, c_2, \ldots, c_m = c'$ where each $c_i = c_{i-1} \cup \{e_i\}$, for $i > 0$, and each e_i is a single event. Note that since $c' = c_m$ has an output with projected value v', if any cut c_i is univalent then it is univalent with projected value v'. Also, note that $c_m \cup \{e_q\}$ is not a cut. This is because $d_q = c_0 \cup \{e_q\}$ is univalent with projected value $v \neq v'$ and $c_m \cup \{e_q\}$ contains both d_q and c_0. Thus, there is a least $i \geq 1$ such that $c_{i-1} \cup \{e_q\}$ is a cut but $c_i \cup \{e_q\}$ is not. Fix this i.

By application of the Compatibility Axiom to cut c_{i-1} and events e_q and e_i, event e_i must occur at process q.

Cut c_{i-1} is multivalent because cut $c_{i-1} \cup \{e_q\}$ is univalent with projected value v and cut $c_m = c_{i-1} \cup_{j=i}^{m} \{e_j\}$ has some process giving output with projected value $v' \neq v$. If either p or q acted in $c_{i-1} - c$, then c_{i-1} would be univalent because both p and q decide c. Thus neither p nor q acts in $c_{i-1} - c$.

Since q decides c and q acts in $c_i - c$, cut c_i is univalent. Since c_m has some process giving output with projected value v' and c_m contains c_i, c_i must be univalent with projected value v'. Thus $d_p \cup c_i$ cannot be a cut because d_p is univalent with projected value $v \neq v'$. But, by $i - 1$ applications of the Compatibility Axiom, using the fact that p does not act in $c_{i-1} - c$, $d_p \cup c_{i-1}$ is a cut. Now by the Compatibility Axiom, e_i must occur at process p.

We now have the desired contradiction: event e_i must occur at both process p and process q, contradicting the assumption that $p \neq q$. This completes the proof of the lemma.

Recall that a cut is undecided if it is either multivalent or nullvalent. The proofs of the following two lemmas are immediate from the definitions.

Lemma 3. *If c is multivalent and p does not decide c, then there exists a cut d containing c such that p acts in $d - c$ and d is undecided.*

Lemma 4. *If c is nullvalent and p is not permanently blocked at c, then there exists a cut d containing c such that p acts in $d - c$ and d is undecided.*

Lemma 5. *If cut c is undecided, then there is an undecided cut d containing c such that, for all processes p,*

- *p decides d, or*
- *p is permanently blocked at d, or*
- *p acts in $d - c$.*

Proof. If p is permanently blocked at a cut x, then p is permanently blocked at every cut containing x. If p decides a cut x and an undecided cut y contains x,

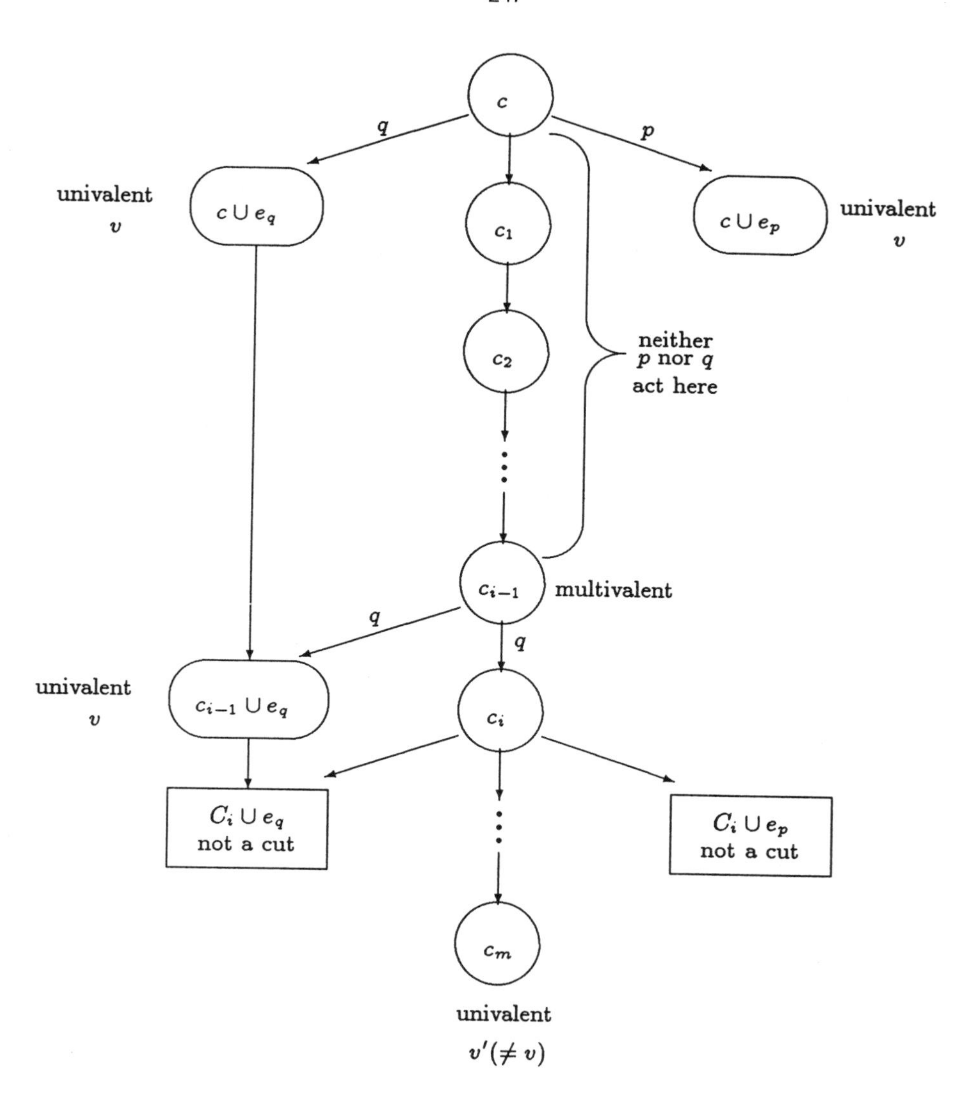

Fig. 3. An illustration of the argument for Lemma 2.

then either y is multivalent and p decides y or y is nullvalent and p is permanently blocked at y. Thus the lemma follows from repeated application of either Lemma 3 or Lemma 4, once for each process.

We now prove that for every nontrivial protocol there is a set W, either empty or containing exactly one process, such that, roughly speaking, (1) if W is empty then the protocol has an execution in which no process fails but no process ever gives output; and (2) if $W = \{p\}$ for some p, then in some execution every process waits for p. Note that the proof is not restricted to protocols for collective consistency, but applies to *any* nontrivial protocol as defined above.

Theorem 6. *For any nontrivial protocol $\mathcal{P}$, there is a set W containing at most one process such that, for all $n \geq 0$, there exists an undecided cut c such that for all $q \notin W$ either q is not ready at c, or in c at least n events have occurred at q. Moreover, these undecided cuts form a containment chain.*

Proof. The proof is by induction on n. Since $\mathcal{P}$ is nontrivial the empty cut is multivalent. Let c_0 be the empty cut (or any undecided cut). Let $D(c)$ be the set of processes that decide or are permanently blocked at cut c. By lemma 2 at most one process in $D(c_0)$ is ready at c_0. In general, let c_i be an undecided cut such that every process not in $D(c_i)$ has had at least i events. By lemma 5 there is a cut c_{i+1} such that every process not in $D(c_{i+1})$ has had at least $i+1$ events. Thus by induction there is a cut c_n such that every process not in $D(c_n)$ has had at least n events. By lemma 2 at most one process in $D(c_n)$ is ready at c_n. Moreover, once a process is ready and decides a cut, it remains ready and decides every containing undecided cut. Thus if p is ready at and decides c_i then p will be ready at and will decide c_{i+1}. So there is at most one process in $W = \{p | (\exists i) p \in D(c_i) \text{ and } p \text{ ready at } c_i\}$.

6 Discussion

In the context of n processes tolerating $c < n$ crash failures and protocols that take input and return output, we call a protocol *non-blocking* if (1) there is a uniform bound on the size (number of protocol relevant events) of a cut in which no process has given output for all executions with at most c crash failures and (2) at each cut of such an execution in which no process has given output, there is at least one process that is not permanently blocked. Our lower bound result shows that if a nontrivial protocol has the property that no undecided cuts contain output events, then it cannot be nonblocking. Since any protocol that achieves at least weak collective consistency has the property that an undecided cut cannot contain an output event, there is no non-blocking protocol that solves weak collective consistency or any stronger version of the problem. In light of this result, we offer the forms of figure 1 with $B = 1$ using either Ψ_1 or Ψ_2 as reasonable choices for protocols that solve weak monotone collective consistency.

As discussed in Section 4, we conjecture that there is a tradeoff between more complicated collective consistency protocols that permit more processes to return and keeping short the window of vulnerability to failure or sudden slowdown, and we have described our intuition for this conjecture. The next logical research step is to design simlation studies to evaluate this conjecture. We believe this tradeoff between protocol sophistication (requiring more rounds of communication) and vulnerability to failures (more rounds during which failures can occur) is not limited to collective consistency protocols; thus any insight gained in learning how to measure protocol effectiveness in the presence of such a tradeoff will have wide applicability.

The open ended nature of the MPI makes it possible to implement our proposed failure detection and collective consistency protocols without any change to the standard. However, our proposal certainly requires major additions and possible revisions to any communication package that supports the MPI in order to provide our fault tolerance services. Although many readers of an early version of this paper were shocked to learn that timeouts were not an option on MPI calls, we know of no such communication package that is currently available and provides any kind of failure detection by timeout.

Acknowledgements

The authors would like to thank Joe Halpern and Yoram Moses for helpful comments, particularly on the material on knowledge based specifications. They also thank Marc Snir for pointing out that our proposed failure detection and collective consistency services could be offered in the context of the current MPI standard without requiring interface modification.

References

[AW] R. J. Anderson and H. Woll, Wait-free Parallel Algorithms for the Union-Find Problem, Proc. 23rd Annual ACM STOC, pages 370-380, 1991.

[BR] J. Buss and P. Ragde, Certified Write-All on a Strongly Asynchronous PRAM, Manuscript, 1990.

[BW] M. F. Bridgeland and R. J. Watro, Fault-Tolerant Decision Making in Totally Asynchronous Distributed Systems, Proc. 6th Ann. ACM Symp. on PODC, pages 52-63, 1987.

[CHT] T. Chandra, V. Hadzilacos, S. Toueg, The weakest failure detector for solving consensus, Proc. 11th ACM Symp. on PODC, pages 147-158, 1992.

[CHT2] T. Chandra, V. Hadzilacos, S. Toueg, On the Impossibility of Group Membership, Proc. 15th ACM Symp. on PODC, 1996.

[CT] T. Chandra and S. Toueg, Unreliable failure detectors for asynchronous systems, Proc. 10th Ann. ACM Symp. on PODC, pages 325-340, 1991, to appear in *J. ACM*.

[DMY] Roberto De Prisco, Alain Mayer, Moti Yung, Time-Optimal Message-Efficient Work Performance in the Presence of Faults, Proc. 13th ACM Symp. on PODC, pages 161-172, Los Angeles, 1994.

[DDS] D. Dolev, C. Dwork, L. Stockmeyer, On the minimal synchronism needed for distributed consensus, *J. ACM* 34:1, pages 77-97, 1987.

[DMS] D. Dolev, D. Malki, R. Strong, A framework for partitionable membership service, Proc. 15th ACM Symp. on PODC, 1996.

[DHS] C. Dwork, J. Halpern, and R. Strong, Fault Tolerant Load Management, application for patent.

[DHW] C. Dwork, J. Halpern, and O. Waarts Accomplishing Work in the Presence of Failures. Proc. 11th Ann. ACM Symp. on PODC, pages 91-102, 1992.

[DLS] C. Dwork, N. Lynch, and L. Stockmeyer, Consensus in the presence of Partial Synchrony, *JACM 35*(2), 1988.

[FHMV] R. Fagin, J. Halpern, Y. Moses, M. Vardi, Knowledge-Based Programs, Proc. 14th Ann. ACM Symp. on PODC, pages 153-163, 1995.

[FLP] M. Fischer, N. Lynch, M. Paterson, Impossibility of distributed consensus with one faulty process. *J. ACM* 32:2, pages 374-382, 1985.

[Gif] D. K. Gifford, Weighted Voting for Replicated Data, Proc. 7th SOSP, pages 150-159, 1979.

[GMY] Zvi Galil, Alain Mayer, Moti Yung, Resolving Message Complexity of Byzantine Agreement and Beyond, private communication, 1995.

[KPRS] Z. Kedem, K. Palem, A. Raghunathan, and P. G. Spirakis. Combining Tentative and Definite Executions for Very Fast Dependable Parallel Computing. Proc. 23rd ACM STOC, pages 381-389, 1991.

[KPS] Z. M. Kedem, K. V. Palem, and P. G. Spirakis, Efficient Robust Parallel Computations. Proc. of 22nd ACM STOC, pages 138-148, 1990.

[KS] P. Kanellakis and A. Shvartsman, Efficient Parallel Algorithms Can Be Made Robust, Proc. 8th Ann. ACM Symp. on PODC, pages 211-219, 1989.

[KS2] P. Kanellakis and A. Shvartsman, Efficient Parallel Algorithms on Restartable Fail-Stop Processes, Proc. 10th ACM Symp. on PODC, pages 23-25, 1991.

[LA] M. Loui and H. Abu-Amara, Memory requirements for agreement among unreliable asynchronous processors, in F. Preparata ed., *Adv. in Computing Research* 4, pages 163-183, JAI Press, 1987.

[MSP] C. Martel, R. Subramonian, and A. Park, Asynchronous PRAMs are (Almost) as Good as Synchronous PRAMs. Proc. 32nd IEEE Symp. on FOCS, pages 590-599, 1991.

[NW] M. Naor and A. Wool, The load, capacity, and availability of quorum systems, Proc. 35th IEEE Symp. on FOCS, pages 214-225, 1994.

[PW] David Peleg and Avishai Wool, Crumbling Walls: A Class of Practical and Efficient Quorum Systems (Extended Abstract), Proc. 14th ACM Symp. on PODC, pages 120-129, 1995.

[R] M. Rabin, The Choice Coordination Problem, *Acta Informatica* 17, pages 121-134, 1982.

[RB] A. Ricciardi and K. Birman, Using process groups to implement failure detection in asynchronous environments, Proc. 10th ACM Symp. on PODC, pages 341-352, 1991.

[W] Arliss Whiteside, Morris Freedman, Omur Tasar, Alexander Rothschild, Operations Controller for a Fault-Tolerant Multiple Computer System, U.S. Patent 4,323,966, 1982.

Planar Quorums

Rida A. Bazzi

School of Computer Science
Florida International University
Miami, Florida 33199 U.S.A.

Abstract. Quorum systems are used to implement many coordination problems in distributed systems such as mutual exclusion, data replication, distributed consensus, and commit protocols. This paper presents a new class of quorum systems based on connected regions in planar graphs. This class has an intuitive geometric nature and is easy to visualize and map to the system topology. We show that for triangulated graphs, the resulting quorum systems are non–dominated, which is a desirable property. We study the performance of these systems in terms of their availability, load, and cost of failures. We formally introduce the concept of cost of failures and argue that it is needed to analyze the message complexity of quorum-based protocols. We show that quorums of triangulated graphs with bounded degree have optimal cost of failure. We study a particular member of this class, the *triangle lattice*. The triangle lattice has small quorum size, optimal load for its size, high availability, and optimal cost of failures. Its parameters are not matched by any other proposed system in the literature. We use percolation theory to study the availability of this system.

Keywords: quorum systems, fault tolerance, load, cost of failures, planar graphs, distributed systems, percolation.

1 Introduction

A quorum system is a collection of sets (quorums) that mutually intersect. Quorum systems have been used to implement mutual exclusion [1, 15], replicated data systems [13], commit protocols [23], and distributed consensus [18]. For example, in a typical implementation of mutual exclusion using a quorum system, processors request access to the critical section from all members of a quorum. A processor can enter its critical section only if it receives permission from all processors in a quorum.[1]

Quorum systems are usually evaluated according to the following criteria.

1. Quorum Size. The size of a quorum determines the number of messages needed to access it. The size of the smallest quorum is usually used as a measure of the number of messages required by protocols using quorum systems.

[1] Additional measures are needed to insure that the implementation is fair and deadlock free.

2. Load. The load of a quorum system measures the share that processors have in handling requests to quorums. Given a probability distribution on quorum accesses, the load on an element is equal to the sum of the access probability of all quorums it belongs to. For a given probability distribution, the quorum system load is the maximum of the loads of all elements. The load of a quorum system is the minimum over all access probability distributions.
3. Availability. The availability of a quorum system measures the probability that the system is usable when failures occur. A quorum system is available if there is a quorum that consists of non-failed processors. The availability of a quorum system is the probability that a quorum system is available given that processors fail according to some probability distribution. The failure probability of a quorum system is the probability that the quorum is not available.
4. Cost of Failures. The cost of failures is used to measure the overhead message-complexity due to failures. If a quorum set has some faulty processors, then it is not usable because it is not guaranteed to share a correct processor with every other quorum. In such a case, a processor attempting to access the quorum with failed processors must find another quorum with no failed processors. A rule that specifies how a replacement quorum is chosen is called a fault-tolerant strategy. Informally, the cost of failures is the additional number of processors that need to be contacted when a failure occurs.

The definition of load above is taken from [20]. The formal definition of cost of failures is a contribution of this work; a similar, but more restricted, concept was informally used by Kumar [12]. We believe that the evaluation of the cost of failures is important to give a better picture of the message complexity of a quorum system in the presence of failures.

In this work, we evaluate quorums according to all of the criteria above and introduce a new general class of quorum systems, the planar quorum systems. We study a particular member of this class and show that it outperforms other proposed quorum systems.

The paper is organized as follows. Sect. 2 gives an overview of related work by other researchers. Sect. 3 summarizes the contributions of the paper. Sect. 4 presents some basic definitions and formally introduces the concept of cost of failures. Sect. 5 introduces the class of planar quorum systems and derives its properties. Sect. 6 defines the triangle system and derives its properties.

2 Related Work

Early work on quorum systems used voting to define quorums [6]. In systems based on voting, each processor has a weighted vote; a quorum consists of any set of processors whose total votes is greater than half the total of all votes. The simple majority system is an example of a voting system [24]. It has the advantage of having high availability, but it suffers from large quorum size and high load. Garcia-Molina and Barbara [5] related quorums to intersecting set systems [3].

They defined *coteries* and introduced the concept of *non-domination.* They showed that non-dominated quorum systems are desirable for high availability and presented an algorithm to construct non-dominated quorums. Maekawa [15] pointed to the connection between quorum systems and finite projective planes and argued that having quorums of almost equal size (balanced systems) is a desirable property. He presented a balanced quorum system in which the size of each quorum is $2\sqrt{n}$ and where each processor belongs to $2\sqrt{n}$ quorum sets, where n is the total number of processors in the system. His system has a low load, a high cost of failures, and a low availability.

Kumar [12] introduced a quorum system based on a hierarchical construction. His system has minimum quorum size of $n^{0.63}$ and a constant cost of failure, which he discusses briefly and informally. Naor and Wool [20] studied the load of many quorum systems. They presented a number of quorum systems with optimal load and high availability. Their best construction is the Paths system with load in the range $[\sqrt{\frac{2}{n}}, 2\sqrt{\frac{2}{n}}]$ and whose availability is $e^{\Gamma(p)\sqrt{n}}$, where $\Gamma(p)$ is a positive function of the failure probability p. The Paths system has an optimal asymptotic load and high availability. In this paper, we improve on both the availability and load of the Paths system. Peleg and Wool [22] presented the CWlog system which has a small quorum size and a relatively high availability. They showed that the CWlog system has a high availability even for small values of n. For some values of n and of the failure probability, the CWlog system has a better availability than the constructions of [12, 20], which have better asymptotic availability.

3 Contributions

In this paper, we present a new class of quorum systems based on planar graphs. Planar quorums consist of minimal connected components that intersect a circuit in a planar graph. Given the geometric nature of the quorums, it is easy to visualize them and map them to an existing network topology. We study the performance of quorum systems in terms of their availability, load, quorum size, and *cost of failures.* We show that, for triangulated graphs, the resulting quorum systems are *non-dominated.* Also, we show that quorums of triangulated graphs with bounded degree have optimal cost of failure and we present an optimal quorum selection strategy.

We study a particular member of this class, the *triangle lattice*, and show that it has an optimal load, small quorum size, high availability, and optimal cost of failures. These combined measures are better than those of any proposed system. Unlike the Paths systems, the triangle lattice is non-dominated, and has better load and availability. In particular, we show using results from percolation theory that if processors fail by crashing with independent failure probability p, then the failure probability of the triangle lattice is *Condorcet.* It follows that the failure probability F_p of the triangle lattice goes to zero as n increases. In particular, if $p = 1/2$ then the failure probability is equal to $1/2$. A direct calculation shows that if $p < 1/4$, $F_p(n) = O(e^{-\frac{4\sqrt{n}}{10}})$. These results are better than the ones given for the Paths system in [20] (see Sect. 2).We show that the load of the triangle system is $\frac{4}{1+\sqrt{8n+1}} \approx \frac{\sqrt{2}}{\sqrt{n}}$ which is exact and better than the results for the load of the Paths system. The proposed system is easily defined in terms of connected components whereas other non-dominated systems with comparable performance are cumbersome to describe [20].

We formally introduce the concept of cost of failures, a contribution of this work. A similar, but more restricted, concept was informally used by other researchers [12]. We argue that the cost of failure is an important parameter that needs to be measured when studying quorum systems.

4 Definitions and System Model

This section introduces the system model, gives some basic definitions and gives a formal definition of the cost of failures.

4.1 Distributed System

A distributed system consists of a set $\mathcal{P} = \{p_1, p_2, \ldots, p_n\}$ of processors. Processors share no memory and can communicate using reliable message passing. Processors can fail by crashing. We assume that failures can be detected by sending a message and using timeout. The failure model is discussed further in Sect. 4.4.

4.2 Coteries and Quorums

Definition 1. A *quorum system* $\mathcal{Q}$ over a set S is a set of subsets (*quorums*) of S such that any two quorums intersect.

In this paper, the set S is $\mathcal{P}$ the set of processors in the system.

Definition 2. A *coterie* $\mathcal{C}$ over a set S is a quorum system over S which is minimal under set inclusion.

Definition 3. A quorum system $\mathcal{Q}$ *dominates* a quorum system $\mathcal{Q}'$ if $\mathcal{Q} \neq \mathcal{Q}'$ and for every $Q' \in \mathcal{Q}'$ there exists $Q \in \mathcal{Q}$ such that $Q \subseteq Q'$.

Definition 4. A quorum system is *non-dominated* if there is no other quorum system that dominates it.

4.3 Availability

Given a quorum system, we are interested in studying its availability when processors fail. A quorum system is available if processors can use it in the presence of failed processors. This can be guaranteed only if one of the quorums has no failed elements.

Definition 5. A quorum system is *available* if one of its quorums has no failed elements

If a quorum system is not available, then we say that it failed and we write *fail*($\mathcal{Q}$). If $\mathcal{Q}$ dominates $\mathcal{Q}'$ then $\mathcal{Q}'$ is available only if $\mathcal{Q}$ is available because every quorum of $\mathcal{Q}'$ contains a quorum of $\mathcal{Q}$. Also, since $\mathcal{Q}$ and $\mathcal{Q}'$ are different, there is a quorum Q_{diff} in $\mathcal{Q}$ that is different from every quorum in $\mathcal{Q}'$. If all processors but those in Q_{diff} fail, $\mathcal{Q}$ is available but $\mathcal{Q}'$ is not. This suggests that we should use quorum systems that are non-dominated because they have higher availability than any quorum system they dominate.

4.4 Failure Model

We assume that processors independently fail by crashing with a failure probability p.[2] We are interested in the availability probability for various values of p. In general, we are interested in systems with higher availability, all other performance parameters being equal.

Definition 6. The failure probability of $\mathcal{Q}$ is $F_p(\mathcal{Q}) = Pr(fail(\mathcal{Q}))$, where Pr denotes the probability function.

Most quorum constructions are given as a function of n, the number of processors. For such constructions, one can study the asymptotic behavior of the failure probability as n increases. It turns out that this asymptotic behavior is sometimes similar to that described by Condorcet Jury Theorem [4] which Peleg and Wool [22] formalized as follows.

Definition 7. A parameterized family of functions $g_p(n) : \mathbf{Z} \mapsto [0,1]$, for $p \in [0,1]$, is Condorcet if and only if $lim_{n\to\infty} g_p(n) = \begin{cases} 0 \text{ if } p < 1/2 \\ 1 \text{ if } p > 1/2 \end{cases}$ and $g_{1/2}(n) = 1/2$ for all n.

To identify the processors that fail, we introduce configurations.

Definition 8. A *configuration* is a mapping $C : \mathcal{P} \mapsto \{\text{true}, \text{false}\}$ such that $C(q) = \text{false}$ if and only if processor q has failed. The set of all configurations is denoted $\mathcal{C}$. The set of faulty processors in C is $faulty(C) = \{q \in \mathcal{P} | C(q) = \text{false}\}$

[2] Models in which processors fail with different or time-varying probabilities are not treated in this paper.

4.5 Strategies and Load

This section presents the formal definitions of strategy and load. It follows the presentation of [20].

A protocol using a quorum system accesses quorums according to some rules. A strategy is a probabilistic rule to choose a quorum. Formally, a strategy is defined as follows.

Definition 9. Let $\mathcal{Q} = \{Q_1, \ldots, Q_m\}$ be a quorum system. A *strategy* $w \in [0,1]^m$ for $\mathcal{Q}$ is a probability distribution over $\mathcal{Q}$.

For every processor $q \in \mathcal{P}$, a strategy w induces a probability that q is accessed. This probability is called the load on q. The *system load* is the load of the *busiest* element induced by the best possible strategy.

Definition 10. Let w be a strategy for a quorum system $\mathcal{Q} = \{Q_1, \ldots, Q_m\}$. For any $q \in \mathcal{P}$, the *load* induced by w on q is $l_w(q) = \Sigma_{Q_j \ni q} w_j$. The *load* induced by w on $\mathcal{Q}$ is

$$\mathcal{L}_w(\mathcal{Q}) = \max_{q \in \mathcal{P}} l_w(q)$$

The *system load* on $\mathcal{Q}$ is

$$\mathcal{L}(\mathcal{Q}) = \min_{w}\{\mathcal{L}_w(\mathcal{Q})\},$$

where the minimum is taken over all strategies.

4.6 Cost of Failures

As defined above, a strategy does not specify how processors choose a quorum when failures occur. When failures occur, processors need to choose an available quorum. The way processors choose an available quorum is called a fault-tolerant strategy. Informally, a *fault-tolerant strategy* is an algorithm that specifies how processors choose an available quorum. Formally, we define a fault-tolerant strategy as follows.

Definition 11. A *fault-tolerant strategy* is a functional $FTS : \mathcal{P} \times \mathcal{C} \mapsto Order$, that maps a configuration and a processor to an order, where $Order : \{1, \ldots, n\} \mapsto \mathcal{P}$ is an ordering of the processors such that:

1. For any C and C' in $\mathcal{C}$, and any $1 \leq k \leq n$, $(\forall i \leq k)(FTS(p,C)(i) = q \wedge C(q) = C'(q)) \Rightarrow (\forall i \leq k)(FTS(p,C)(i) = FTS(p,C')(i))$.
2. If $\mathcal{Q}$ is available, then $Order(\{1, \ldots, n\})$ contains an available quorum.

A fault-tolerant strategy specifies for every processor, and for a given configuration, an order to choose processors. If the quorum system is available, then the chosen processors should contain a live quorum. Informally, this definition requires that the choice of a quorum be deterministic and that processors choose a quorum one processor after another.[3] This definition does not restrict the choice

[3] The requirement that the fault-tolerant strategy be deterministic is made for simplicity. In general, a fault-tolerant strategy can be deterministic or probabilistic.

of a quorum. In fact, if p wants to choose more than one processor at a time, the chosen processors can be ordered according to their identifiers. Also, note that different processors can choose quorums differently.

Definition 12. Let FTS be a fault-tolerant strategy and C be a configuration such that $Fail(C) \neq \phi$. The *cost of failures in C for q* is

$$\mathcal{CF}(q, C, FTS) = \frac{|FTS(q, C)(\{1, \ldots, n\})| - c(\mathcal{Q})}{|fail(C)|}.$$

The cost of failure for a given processor and a given configuration C is equal to the number of extra processors that need to be contacted to choose a quorum due to failures in C divided by the number of failures in C. This definition assumes that if there are no failures, then processors will choose a quorum of minimum size. Note that in general there might be more than one quorum of minimum size. In fact, there are quorum systems in which all quorums have the same size [15].

The following two definitions give the cost of failure of a configuration (for any processor) and the cost of failure of a fault-tolerant strategy (for any configuration).

Definition 13. The *cost of failure of a configuration C* for a given FTS is

$$\mathcal{CF}(C, FTS) = \max_{q \in \mathcal{P}} \mathcal{CF}(q, C, FTS).$$

Definition 14. The *cost of failure of a fault-tolerant strategy FTS* is

$$\mathcal{CF}(FTS) = \max_{C \in \mathcal{C}} \mathcal{CF}(C, FTS).$$

The cost of failures gives an idea of the extra number of processors that need to be contacted if a failure occurs. When a failure occurs, a processor might choose to discard all processors it already contacted to choose a quorum and contact a whole new set of processors. It is clear that this incurs a communication overhead in choosing a quorum. So, it is not enough that the chosen quorum be small, as is commonly suggested, but it important that the total number of contacted processors be small.

5 The Planar Systems

In this section we introduce a new class of quorum systems based on planar graphs. We show that planar quorum systems are coteries and that, for triangulated graphs, planar quorum systems are non-dominated. In what follows, we assume that the basic graph-theoretic terminology is familiar to the reader. We start with some basic definitions.

Definition 15. A set that satisfies property P is minimal if it contains no subset that satisfies P.

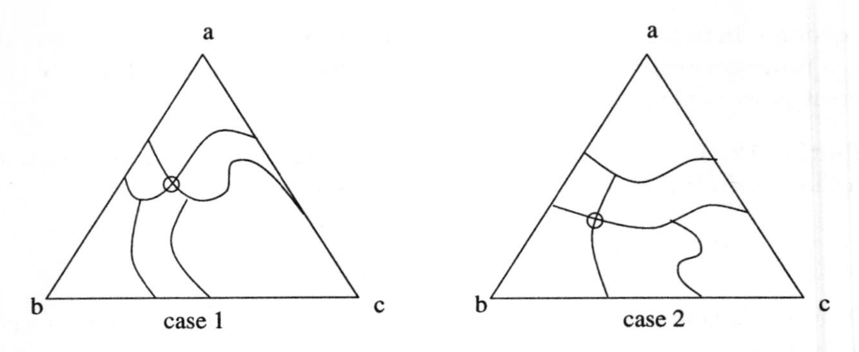

Fig. 1. Intersecting Quorums

Definition 16. Let G be a graph. A set of vertices V of G is connected if the subgraph of G induced by V is connected.

In particular, we say that a path is minimal if it contains no proper subset of vertices connecting its endpoints. A path is identified by a sequence of adjacent vertices and is denoted $(v_1, v_2, \ldots, v_i, \ldots, v_l)$, where v_1 and v_l are the endpoints of the path. Without loss of generality, we assume that all planar embedding are such that edges are embedded into straight lines (this is always possible [2]).

Definition 17. Let G be a connected planar graph that has a planar embedding G^* such that the infinite face of G^* forms a circuit $C = (a, \ldots, b, \ldots, c, \ldots, a)$, where a, b, and c are distinguished vertices of G. A planar quorum on G^* is a minimal connected set of vertices that intersects the three paths $(a, \ldots, b)$, $(a, \ldots, c)$, and $(b, \ldots, c)$ that form the infinite face.

In what follows, we will specify G by specifying an embedding of G and we will say that a planar quorum is on G. Note that there are no vertices on the infinite face other than those in C. We state without proof the following lemma.

Lemma 18. *A planar quorum consists of a minimal path P_{h} that connects $(a, \ldots, b)$ and $(a, \ldots, c)$, and a minimal path P_{v} that connects P_{h} and $(b, \ldots, c)$.*

In the definition of planar quorums, the vertices of the graphs correspond to processors in the system. The edges of the graph are abstract edges and do not necessarily correspond to direct communication links between processors. Nevertheless, it is possible to choose a planar graph that matches some links in the system. This will have the advantage of minimizing the message complexity of protocols using planar quorums.

Theorem 19. *Planar systems are coteries.*

Proof. Let G be a planar graph and Q and Q' be two distinct planar quorums of G. The minimality of Q and Q' implies that neither is a subset of the other. So, all we have to prove is that Q and Q' intersect. Let P_{h} and P_{v} be as defined in Lemma 18 and define ${P'}_{\mathrm{h}}$ and ${P'}_{\mathrm{v}}$ similarly. We have two cases to consider:

1. P_h and P'_h intersect. It follows immediately that Q and Q' intersect.
2. P_h and P'_h do not intersect. We can assume without loss of generality that a is closer to P_h than P'_h. It is not difficult to see that curves of P_v and P'_h intersect at some point x of the plane (Fig. 1). Since the graph is planar, x must be a vertex of the graph because P'_h and P_v are paths of the graph.

In this paper, we say that a graph is *triangulated* if the boundaries of its finite faces are triangles.

Theorem 20. *Planar systems of triangulated graphs are non-dominated.*

Proof. We need to prove that any set of vertices that intersects all quorums must contain a quorum set. Let S be a set that intersects all quorums. It is enough to prove that S contains a path P_s that connects $(a,\ldots,b)$ and $(a,\ldots,c)$ and a path P'_s that connects P_s and $(b,\ldots,c)$. It follows that S contains a connected component that intersects $(a,\ldots,b)$, $(a,\ldots,c)$ and $(b,\ldots,c)$, and that S contains a quorum set. We only prove that S contains P_s. The proof for P'_s is similar. By the definition of S, it must intersect all paths that connect a to $(b,\ldots,c)$ (note that such paths are quorums). The proof is by contradiction. We assume that S does not contain a path from $(a,\ldots,b)$ to $(a,\ldots,c)$, and we show that there must be a path from a to $(b,\ldots,c)$ that does not intersect S. This contradicts the definition of S. Recall that the exterior face is a circuit (Definition 17). If S does not contain a path that connects $(a,\ldots,b)$ to $(a,\ldots,c)$, it follows (using the Jordan Curve Theorem [16] and the fact that S consists of edges of the graph) that there is a continuous curve that does not intersect the edges defined by S or the infinite face and that connects a to some vertex $a' \in (b,\ldots,c)$ (Fig. 2). We will assume for simplicity that the curve does not contain any vertex of the graph other than a and a'. Since the complement of the infinite face is closed and bounded, it follows that it is compact [9]. Then, there must exist such a curve of minimal length [9]. A curve of minimal length must not intersect an edge of a triangle in more than one point (if it does, we can short-circuit it and obtain a shorter curve; see Fig. 2). Every edge that the curve intersects must contain a vertex that does not belong to S. Using this fact, we can construct a path that connects a and a' and that does not intersect S (Fig. 2). This contradicts the definition of S. The path is constructed iteratively as follows. The first vertex of the path is a. The ith vertex of the path, a_i, is a vertex that does not belong to S and that belongs to the ith triangle that the curve intersects. The triangles that the curve intersects are adjacent. Also, the curve does not contain any vertex other than a and a', it follows that a_i is adjacent to a_{i+1}. (It is easy to modify the proof for the case where the curve contains a vertices other than a and a'. In fact, any such vertex v must belong to two triangles t_i and t_{i+1} that the curve intersects, so we take v to be both the ith and $i+1$th points.)

6 The Triangle Lattice

In this section we study the planar quorum on a particular planar graph, the triangle lattice. Sect. 6.1 defines the triangle lattice. Sects. 6.2 , 6.3, and 6.4 study the availability, load, and cost of failure of the triangle lattice.

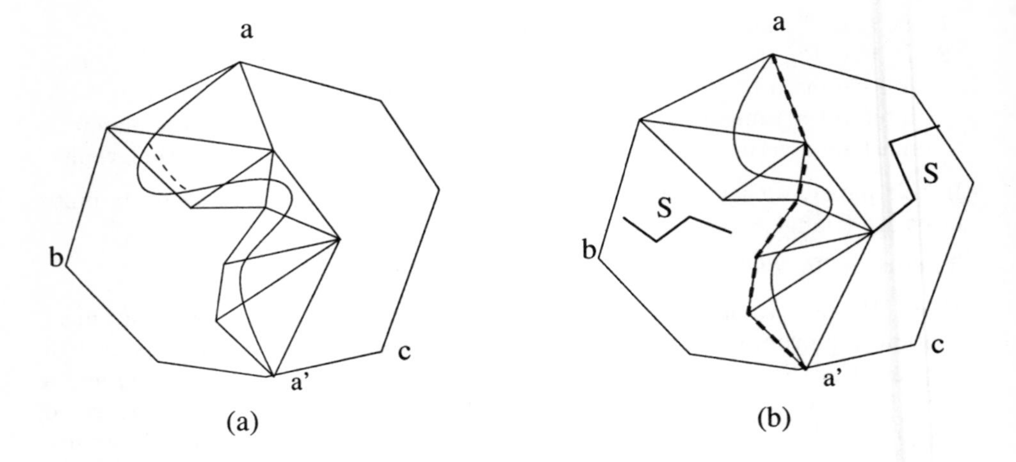

Fig. 2. (a) Short–circuiting a non–optimal curve; (b) Continuous curve and corresponding path.

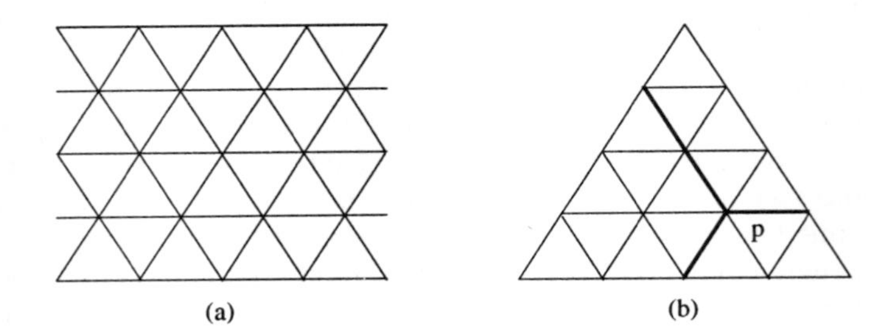

Fig. 3. (a) Infinite triangular lattice; (b) Triangle lattice, $d = 5$.

6.1 Definition

For any integer d, we define a triangle lattice consisting of $(d^2 + d)/2$ vertices and that are connected as shown in Fig. 3. We first define the infinite triangular lattice.

Definition 21. The infinite triangular lattice is the infinite graph whose vertices are points of the form (i, j) or $(i + \frac{1}{2}, j + \frac{1}{2})$ where i and j are integers and such that each vertex (x, y) of the graph has the vertices $(x + 1, y)$,$(x - 1, y)$,$(x + 1/2, y + 1/2)$,$(x - 1/2, y + 1/2)$,$(x + 1/2, y - 1/2)$, and $(x - 1/2, y - 1/2)$ as neighbors.

The triangle lattice is a subgraph of the infinite triangular lattice.

Definition 22. A triangle lattice $\mathcal{T}$ is a finite subgraph of the infinite triangular lattice whose vertices (i,j) satisfy $0 \leq j \leq \frac{d-1}{2}$ and $j \leq i \leq d-1-j$ for some positive integer d.

Fig. 3 shows the triangle lattice for $d = 5$. The three points a, b, and c in the definition of the quorum system are the three end-vertices of the triangle. In this section we will only study the planar quorum system of the triangle lattice and we will use the triangle lattice and its planar quorum system as synonyms. Note that $n = \frac{d^2+d}{2}$ and $d \approx \sqrt{2n}$. It is not difficult to see that d is the size of the smallest quorum of the triangle lattice.

6.2 Availability of the Triangle Lattice

To study the failure probability of the triangle lattice, we need to use results from percolation theory [11]. In percolation theory sites (vertices) or bonds (edges) of a graph are either *occupied* or *vacant*. Percolation problems are divided into site-percolation and bond-percolation problems (hybrid models are also studied). In the site percolation problems, a site is occupied with probability p and vacant with probability $1-p$. In percolation theory, one studies the probability of the existence of an infinite connected component in a graph. A good reference for site problems is [11], and a good reference for bond problems is [7]. In this paper, we use results for site problems to study the availability of the triangle system. In our terminology, the probability of the failure of a processor is p. Therefore, we will say that a failed processor is occupied.

The following definitions are needed for our results.

Definition 23 *i*-crossing. Let $B = \Pi_{j=1,\ldots,m}[a_j, b_j] = \{\mathbf{x} | a_j \leq x[j] \leq b_j\}$ be a m dimensional cube in $\mathbb{R}^m$. An i-crossing of B is a path $(v_0, \ldots, v_k)$ such that

1. $v_l \in B^o = \Pi_{j=1,\ldots,m}(a_j, b_j) = \{\mathbf{x} | a_j < x[j] < b_j\}$, $0 < l < k$
2. $[v_0, v_1]$ intersects $\{\mathbf{x} | x[i] = a_i\} \cap B$ at a point of B^0.
3. $[v_{k-1}, v_k]$ intersects $\{\mathbf{x} | x[i] = b_i\} \cap B$ at a point of B^0.

An i-crossing is a path that is contained in the m-dimensional cube and that intersects the two faces of the cube in the i dimension. In two dimensions, we have two crossings: horizontal and vertical. We define a *faulty i-crossing* to be an i-crossing consisting of faulty processors.

Definition 24. The crossing probability is $\sigma((a_1, \ldots, a_m), i, p, G) = Pr_p\{\exists$ a faulty i-crossing on G of $[0, a_1] \times \ldots \times [0, a_m]\}$.

Definition 25. For a periodic graph G, the critical crossing probability is:

$$P_S = sup\{p \in [0,1] | \lim_{n \to \infty} \sigma((3n, \ldots, 3n, n, 3n, \ldots, 3n), i, p, G) = 0, 1 \leq i \leq m\},$$

where the one component equal to n in $\sigma((3n, \ldots, 3n, n, 3n, \ldots, 3n), i, p, G)$ is the ith component.

Lemma 26 **[11] proposition 3.67.** *For the triangular lattice,* $P_S = \frac{1}{2}$.

This means that, if $p < \frac{1}{2}$, the probability that there exists a faulty horizontal crossing on the triangular lattice of a rectangle whose horizontal width is n and whose vertical height is $3n$ tends to 0 as n increases. A similar statement applies to faulty vertical crossings.

The following Lemma is proved in [22]. We will need it to prove that the failure probability of the triangle lattice is Condorcet.

Lemma 27. *For a non-dominated quorum system, the failure probability is such that* $F_p = 1 - F_{1-p}$.

It follows immediately that $F_{1/2} = 1/2$ for a non-dominated quorum system.

Lemma 28. *The failure probability of the planar quorum of the triangle lattice is Condorcet.*

Proof. By Theorem 20, the triangle lattice is non-dominated. By Lemma 27, it is enough to prove that $\lim_{d\to\infty} F_p(n) = 0$. Consider a rectangle R whose basis is that of the triangle lattice and whose height is $\dfrac{d}{3}$. By Lemma 26 and Definition 25, if $p < \frac{1}{2}$, then, with probability 0, there exists a vertical crossing of R consisting of faulty processors when d goes to ∞. By arguments similar to those used in proving Theorem 20, we can show that there exists no faulty vertical crossing of R if and only if there exists a horizontal crossing of R consisting of correct processors. It follows that, with probability 1, there exists a horizontal crossing of R consisting of correct processors. Similarly, we consider R' as shown in Fig. 4 and prove that, with probability 1, there exists a vertical crossing of R consisting of correct processors. The union of these two paths clearly contains a quorum in the triangle lattice. It follows that, if $p < \frac{1}{2}$, then, with probability 1, there exists a quorum consisting of correct processors when d goes to ∞. This completes the proof.

We can also obtain a direct bound on the availability of the triangle lattice. The proof relies on a counting argument. We recall that a quorum consists of two intersecting minimal paths. Any minimal path cannot contain three points on a triangle (Fig. 5) because one of the three points can always be eliminated by taking a shortcut. It follows that there are at most 3^k minimal paths of length k because there are at most three points that can extend a minimal paths of length $k-1$ into a minimal path of length k. Let H be the set of minimal paths of length k connecting $(a, \ldots, b)$ and $(a, \ldots, c)$. There are at most $d3^k$ (3^k for each starting point) paths in H. There are at most $d3^{k'}$ minimal path of length k' connecting a path in H to $(b, \ldots, c)$. It follows that there are at most $d^2 3^{k+k'}$ quorums of size $k + k'$. The probability that any of these quorums consists of faulty processors is $p^{k+k'}$. Hence, the probability that there exists a quorum consisting of faulty processors is at most $\Sigma_{d \le l \le n} d^2 (3p)^l \le \dfrac{d^2(3p)^d}{1-3p} \le \dfrac{n(3p)^d}{1-3p}$. It follows, with little algebraic manipulation, that, if $p < 1/4$, $F_p(n) = O(e^{-\frac{4\sqrt{n}}{10}})$. For $p < 1/4$, this

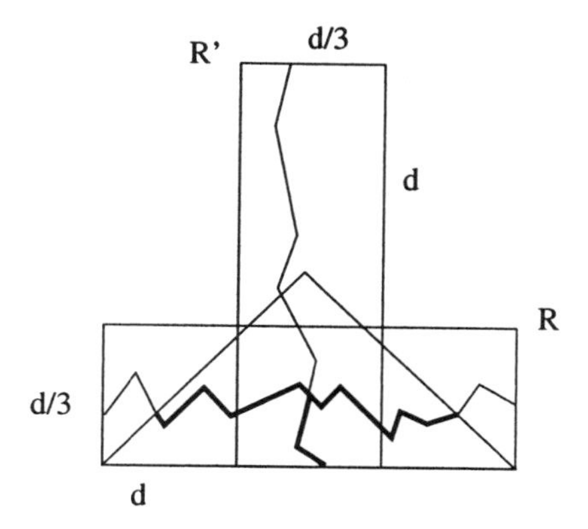

Fig. 4. Horizontal and vertical crossings.

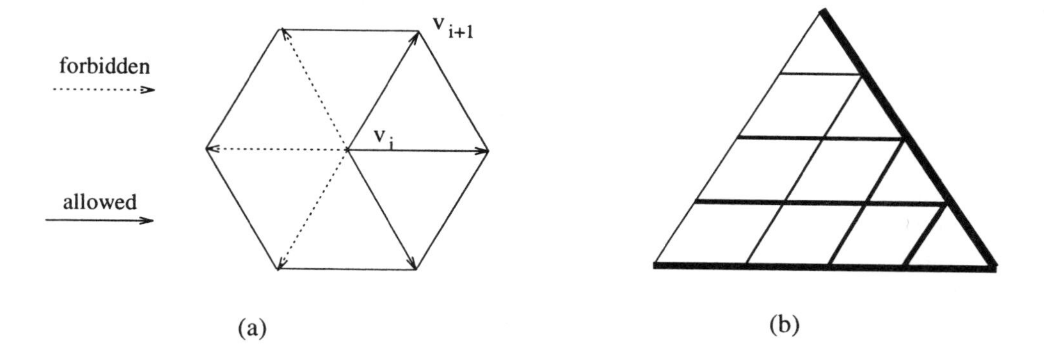

Fig. 5. (a) Forbidden paths; (b) Quorums for a low load.

is better than the availability of the Paths system given in [20]. [4]Also, note that the availability of the triangle lattice is better than the availability of the Paths system for $p = 1/2$.

6.3 Load of the Triangle Lattice

In this section we calculate exactly the load of the triangle lattice.

For the lower bound, we use the following theorem from [20].

Theorem 29. *The load of a quorum system satisfies:* $\mathcal{L}(\mathcal{Q}) \geq \dfrac{c(\mathcal{Q})}{n}$.

[4] A similar calculation yields $F_p(n) = O(e^{-\frac{3\sqrt{n}}{10}})$ for the Paths systems.

The triangle lattice has $c(\mathcal{T}) = d$. It follows that $\mathcal{L}(\mathcal{T}) \geq \frac{d}{(d^2+d)/2} = \frac{2}{d+1}$. For the upper bound, we consider the following $d+1$ quorums: $Q_k = \{(i,j) | (i,j) \in \mathcal{T} \wedge ((j=k) \vee (j = i-(d-1)+2k))\}$ for $0 \leq k \leq d-1$ and $Q_d = \{(i,j)|(i,j) \in \mathcal{T} \wedge j = d-1-i\}$. These quorums are shown in Fig. 5 in which lines of different width represent the different quorums. Note that every vertex in $\mathcal{T}$ belongs to exactly two quorum sets. We consider a strategy that chooses each of $d+1$ quorums with probability $\frac{1}{d+1}$. For this strategy, the load on every processor is $\frac{2}{d+1}$. It follows that $\mathcal{L}(\mathcal{T}) \leq \frac{2}{d+1}$.

So, we conclude that $\mathcal{L}(\mathcal{T}) = \frac{2}{d+1} = \frac{4}{1+\sqrt{8n+1}} \approx \frac{\sqrt{2}}{\sqrt{n}}$. This is the exact value of the load. This is better than the result of [20] which bounds the load of the paths system between $\approx \frac{\sqrt{2}}{\sqrt{n}}$ and $\approx \frac{2\sqrt{2}}{\sqrt{n}}$.

6.4 Cost of Failures of the Triangle Lattice

In this section, we present a fault-tolerant strategy for the triangle lattice. Due to space limitation we do not give a proof of correctness. In the full version of the paper we show that this fault-tolerant strategy yields a constant cost of failures for any triangulated graph with bounded degree. The rest of this section is organized as follows. First, we give an overview of the strategy, then we see how it applies to a particular example.

6.5 A Fault-Tolerant Strategy

We first present a strategy that finds a horizontal path (Figs. 6 and 7). At the end of this section, we explain how to find an available quorum. In the discussion below, we assume that processor p_1 is using the strategy. The strategy consists of two functions: Find_Horiz_Path and Find_Fail_Region. Find_Horiz_Path is the main function and it calls Find_Fail_Region. Find_Horiz_Path finds a horizontal path connecting $(a,\ldots,b)$ and $(a,\ldots,c)$. Find_Fail_Region finds a connected component consisting of faulty vertices (processors) and that contains a given vertex. It also finds the boundary of the failure region.

There are four global variables used by the fault-tolerant strategy. The following is a description of each of the global variables.

1. *Faulty* is a set of vertices. Initially it is empty. If a vertex is detected to be faulty, it is added to this set.
2. *Checked* is the set of vertices that have been contacted. Whenever a vertex is contacted, it is added to *Checked*.
3. *FailureRegion* is a connected region consisting of failed vertices.
4. *FailureBoundary* is the boundary of *FailureRegion*. This means that every vertex in *FailureBoundary* is adjacent to a vertex in *FailureRegion*.

```
Faulty : set of Vertex initially faulty := φ
Checked : set of Vertex initially Checked := φ
FailureBoundary, FailureRegion: Graph

/* The parameter H is of the form H := first = h1, h2, ..., hk = last, */
/* such that first ∈ (a, ..., b), last ∈ (a, ..., c)                  */
Graph Find_Horizontal_Path(H);
    Path: Graph initially Path := φ
    current: Vertex initially current := first
    repeat
        if current not failed do
            Path := Path ∪ {current}
            current := vertex to the right of current in H
        else
            add current to Faulty
            FailureBoundary := φ FailureRegion := φ
            Find_Failure_Region(current)
            current := rightmost vertex of H on FailureBoundary
            Path := Path ∪ FailureBoundary
    until last ∈ Checked
    if Path connects (a, ..., b) to (a, ..., c)
        return A horizontal path in Path
    else
        return φ
```

Fig. 6. Finding a horizontal path

Now we explain how Find_Horiz_Path works. Find_Horiz_Path takes as input a path H that connects $(a, \ldots, b)$ to $(a, \ldots, c)$. The end-vertices of H are $first$ and $last$. If there are no failures, this is the path returned by the strategy. If there are failures, a path that "approximates" H is found. Note that choosing H is part of the strategy, but it is done locally without contacting any processor. There are two local variable used by Find_Horiz_Path.

1. *current* is the vertex currently being examined for inclusion in the path. It is always a vertex of H. Initially, *current* is equal to $first$. Current usually changes by being set to the vertex to the right of it on H. If a failure is detected, current is updated so as to go around the failure.
2. *Path* is the path being constructed. If there is a horizontal path in the system, then, when Find_Horiz_Path returns, *Path* contains a horizontal path. It is continuously updated as *current* progresses on H.

At the start of Find_Horiz_Path, *Path* is initialized to the empty set and *current* is initialized to $first$. After this initialization phase, the processing starts.

The processing is done inside a loop that terminates only when *last* is in *Checked*. In the loop, p_1 does the following. It first checks if *current* failed (note

```
Find_Failure_Region(initial)
    add initial to FailureRegion
    add initial to Faulty
    add initial to Checked
    for q ∈ Neighbors(initial) do
        if q ∉ Checked then
            if q failed or q ∈ Faulty then
                Find_Failure_Region(q)
            else
                add q to FailureBoundary
                add q to Checked
```

Fig. 7. Finding a connected failure region and its boundary

that this requires sending a message). If *current* is not faulty then it is added to *Path* and *current* is updated to the point to the right of it in H. If *current* is faulty, it clearly cannot be added to *Path*. So, p_1 tries to go around the failure. This is achieved by calling Find_Fail_Region which sets *FailureRegion* to be equal to the maximal connected component that contains *current* and that consists solely of faulty vertices. Also, Find_Fail_Region sets *BoundaryRegion* to be equal to the boundary of *FailureRegion*. Find_Fail_Region adds to *Checked* every vertex it sends a message to (to check if it is correct or faulty). Any failed vertex detected by Find_Fail_Region is added to *Faulty*. Note that every vertex is contacted at most once in the strategy. If a vertex is found to be faulty then it need not be contacted again because this information is saved in *Faulty*. After Find_Fail_Region is executed, *current* is set to the rightmost vertex of H in *FailureBoundary*. Also, *FailureBoundary* is added to *Path*. After the loop terminates, p_1 checks if *Path* connects $(a, \ldots, b)$ to $(a, \ldots, c)$. If it does, then a horizontal path, subset of *Path*, is returned, if not, the empty set is returned indicating that there is no path that connects $(a, \ldots, b)$ to $(a, \ldots, c)$.

Now we describe how p_1 can find an available quorum. To find an available quorum, p_1 chooses an optimal quorum and divides it into a horizontal path H and a vertical path V. The strategy is executed with input H and V respectively yielding two paths $Path_H$ and $Path_V$. If $Path_H$ and $Path_V$ are not empty, their union contains a quorum.

6.6 An Example

Figure 8 depicts an execution of the Find_Horiz_Boundary with the input path H consisting of a horizontal line of vertices. In the execution, two *FailureRegion*'s are encountered. Vertices of the failure regions are depicted as black discs, and vertices of failure boundaries are depicted as gray discs. vertices that are contacted by p_1 but that do not belong to a *FailureRegion* or a *FailureBoundary* are depicted as white discs with wide circumferences. Vertices that are not contacted by the strategy are shown as white discs. The first failure region separates

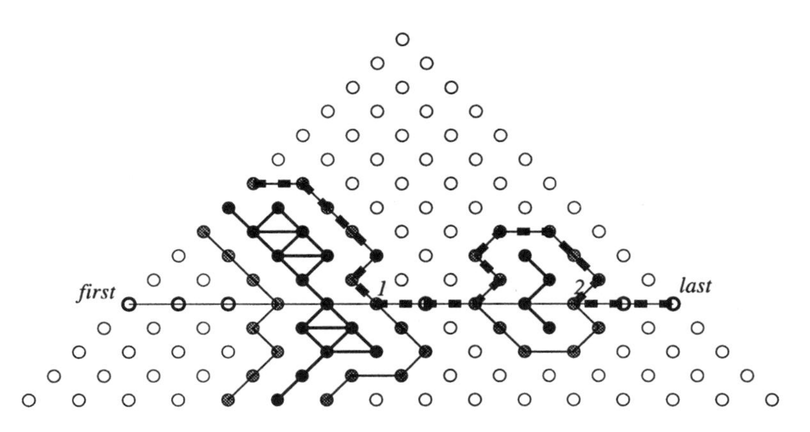

Fig. 8. Finding a horizontal path

first from $(a, \ldots, c)$. After the failure region is encountered *current* is updated to vertex 1 in the figure. After the second failure region is encountered, *current* is updated to vertex 2. Note that at the end of the execution, there is a subset of *Path* that connects $(a, \ldots, b)$ to $(a, \ldots, c)$. This subset is shown as a dotted line in the figure.

One might note that the low cost of failures is achieved by a deterministic strategy, whereas the low load is achieved by a probabilistic strategy. It is reasonable to ask if this is a limitation on the results. As it turns out, one can achieve both a high low and a low cost of failures simultaneously. In fact, the way the fault tolerant strategy works is by starting with a minimum size quorum and finding a live quorum that approximates it. We note that all the quorums needed to achieve a low load are of minimum size. It follows that one can adopt the following strategy to achieve both low load and low cost of failures. First, a processor chooses a quorum at random with probability $\frac{1}{d+1}$ as explained in Sec. 6.3. Next, and using the fault-tolerant strategy, a live quorum is found that approximates the chosen quorum. This combined strategy is not deterministic and therefore one has to change the definition of the cost of failures to accommodate it. If there are no failures, the resulting load is low. If there are failures, the cost of failures is low.

Acknowledgments

I would like to thank the reviewers and Mark Weiss for their useful comments.

References

1. D. Agrawal and A. El–Abbadi. An efficient and fault-tolerant solution for distributed mutual exclusion. *ACM Transactions on Computer Systems*, 9(1):1–20,1991.

2. B. Bollobás. Graph Theory, An Introductory Course *Graduate Texts in Mathematics*, Springer Verlag, 1979.
3. B. Bollobás. *Combinatorics.* Cambridge, 1983.
4. N. Condorcet. Essai sur l'application de l'analyse a la probabilité des decision rendues à la pluralité des voix. Paris, 1785.
5. H. Garcia–Molina and D. Barbara. How to assign votes in a distributed system. *Journal of the ACM*, 32(4):481–860, 1985.
6. D. K. Gifford. Weighted Voting for Replicated Data *Proceeding of 7th ACM Symposium on Operating Systems Principles*, pages 150–162, December 1979.
7. G. R. Grimmett. *Percolation.* Springer Verlag, 1989.
8. M. P. Herlihy. *Replication Methods for Abstract Data Types.* Ph.D. Thesis, Massachusetts Institute of Technology, 1984.
9. J. G. Hocking and G. S. Young *Topology.* Dover, 1988
10. T. Ibaraki and T. Kameda. A theory of Coteries: Mutual Exclusion in Distributed Systems. *IEEE Transactions on Parallel and Distributed Systems*, 4(7):779–749,1993.
11. H. Kesten. *Percolation Theorey for Mathematicians.* Progress in Probability and Statistics, Birkhäuser, 1982.
12. A. Kumar. Hierarchical quorum consensus: A new algorithm for managing replicated data. *IEEE Transactions of Computers*, 40(9):996–1004,1991.
13. A. Kumar., M. Rabinovich, and R. Sinha. A performance study of general grid structures for replicated data. In *Proceedings of International Conference on Distributed Computing Systems*, pages 178–185, May, 1993.
14. L. Lovász. Covering and colorings of hypergraphs. In *Proceedings of 4th Southeastern Conference on Combinatorics, Graph Theory and Computing*, pages 3–12, 1973.
15. M. Maekawa. A $\sqrt{n}$ algorithm for mutual exclusion in decentralized systems. *ACM Transactions on Computer Systems*, 3(2):145–159,1985.
16. C. R. F. Maunder *Algebraic Topology* Van Nostrand, 1970
17. S. J. Mullender and P. M. B. Vitanyi. Distributed Match Making. *Algorithmica*, 3:367–391, 1992.
18. M. L. Neilsen *Quorum Structures in Distributed Systems.* Ph.D. Thesis, Department of Computer and Information Sciences, Kansas State University, 1992.
19. M. L. Neilsen and M. Mizuno. Decentralized Consensus Porotocols In *Proceedings of 10th International Phoenix Conference on Computing and Communications*, pages 257–262, 1991.
20. M. Naor and A. Wool The Load, capacity and availability of quorum systems. In *Proceedings of the 35th IEEE Symposium on Foundations of Computer Science*, pages 214–225. 1994.
21. D. Peleg and A. Wool. The availability of quorum systems. *Information and Computation*, 123(2):210-223, 1995.
22. D. Peleg and A. Wool. Crumbling Walls: A class of high availability quorum systems. In *Processedings 14th ACM Symposium on Principles of Distributed Computing*, pages 120–129, 1995.
23. D. Skeen. A quorum–based commit protocol. In *Proceedings of 6th Berkeley Workshop on Distributed Data Management and Computer Networks*, pages 69–80, 1982.
24. A majority consensus approach to concurrency control for multiple copy database. *ACM Transactions on Database Systems*, 4(2):180–209, 1979.

"Γ−Accurate" Failure Detectors

Rachid Guerraoui and André Schiper

Département d'Informatique
Ecole Polytechnique Fédérale de Lausanne
1015 Lausanne, Switzerland

Abstract. The knowledge about failures needed to solve distributed agreement problems can be expressed in terms of completeness and accuracy properties of failure detectors introduced by Chandra and Toueg. The accuracy properties they have considered restrict the false suspicions that can be made *by all the processes in the system*. In this paper, we define "Γ−accurate" failure detectors, whose accuracy properties *(only)* restrict the false suspicions that can be made *by a subset Γ of the processes*. We discuss the relations between the classes of Γ−accurate failure detectors, and the classes of failure detectors defined by Chandra and Toueg. Then we point out the impact of these relations on the solvability of agreement problems.

1 Introduction

1.1 Restricting accuracy

Chandra and Toueg have expressed the knowledge about failures needed to solve distributed agreement problems in terms of *completeness* and *accuracy* properties of failure detectors [4]. Completeness properties require that every process that crashes is eventually permanently suspected, while accuracy properties restrict the mistakes (false suspicions) that can be made by *all the processes in the system*. We extend these accuracy properties, by considering properties that only restrict the false suspicions that can be made by a *subset Γ of the processes*. The new properties are called "Γ-accuracy" properties. Given a subset Γ of the processes in the system:

- *strong Γ-accuracy* is satisfied if no process p in Γ is suspected by any process in Γ before p crashes, and
- *weak Γ-accuracy* is satisfied if some correct process (not necessarily in Γ) is not suspected by any process in Γ.

1.2 Motivation

Our work is motivated by the observation that, because the accuracy properties defined in [4] span the whole system, the formalism does not apply to systems that can be subject to network partitions. Indeed, as failure suspicions are usually implemented with time-outs, the probability that any accuracy property holds (even weak accuracy) during a partition of the system, can be considered to be zero. For instance, in a system partitioned into Π_1 and Π_2, processes in Π_1 most probably suspect processes in Π_2, and processes in Π_2 most probably suspect processes in Π_1, i.e. there is no correct process not suspected by any process. Even the weak accuracy property does not hold while the system is partitioned. By restricting the accuracy properties to subsets Γ of the system, the Γ-accurate failure detectors address these concerns.

1.3 Reliable vs eventually reliable channels

Failure detectors of [4] have been considered in a system model with reliable channels. A reliable channel ensures that if a message m is sent by a process p to a process q, and q is correct, then m is eventually received by q. A reliable channel does not loose messages. The reliable channel assumption is however incompatible with network partitions, that are considered in the paper. Therefore we consider in the paper channels with a weaker reliability property that is called *eventually reliable channels*: if a message m is sent by a process p to a process q, and *both p and q are correct*, then m is eventually received by q. An eventually reliable channel can loose messages. The definition is close to the one considered in [1].

In the paper, we prove both possibility results (i.e. equivalence of failure detectors), and impossibility results (i.e non-equivalence results). In order for these results to be more general, we consider the eventually reliable channel assumption to prove possibility results, and the stronger reliable channel assumption to prove impossibility results.

1.4 Results

We use $\mathcal{P}(\Gamma)$ to denote the class of failure detectors that satisfy strong completeness and strong Γ-accuracy, $\mathcal{S}(\Gamma)$ to denote the class of failure detectors that satisfy strong completeness and weak Γ-accuracy, and $\mathcal{W}(\Gamma)$ to denote the class of failure detectors that satisfy weak completeness and weak Γ-accuracy. The failure detectors $\Diamond\mathcal{P}(\Gamma)$, $\Diamond\mathcal{S}(\Gamma)$ and $\Diamond\mathcal{W}(\Gamma)$ are similarly defined by requiring, roughly speaking, the corresponding Γ-accuracy property to eventually hold.

Consider a set Ω of processes. Among others, this paper establishes the following interesting relations between the Γ-accurate failure detector classes and the Chandra-Toueg classes:

1. Let f be the maximum number of processes of Ω that can crash. For any $\Gamma \subseteq \Omega$, and with eventually reliable channels, if $|\Gamma| > |\Omega|/2$ and $f < |\Omega|/2$, then any failure detector of $\Diamond\mathcal{S}(\Gamma)$ can be transformed into some failure detector of $\Diamond\mathcal{S}$ (which implies $\Diamond\mathcal{S}(\Gamma) \cong \Diamond\mathcal{S}$ [1]). Hence, given that $|\Gamma| > |\Omega|/2$ and $f < |\Omega|/2$, every problem that can be solved with $\Diamond\mathcal{S}$, (e.g consensus [4], uniform consensus [4], atomic broadcast [4], and non-blocking weak atomic

[1] This result has been informally stated in [3] for reliable channels.

commitment [6]) can also be solved with $\diamond\mathcal{S}(\Gamma)$ [2]. These problems can hence be solved whenever some correct process is, roughly speaking, eventually never suspected by a majority of correct processes.

2. For any $\Gamma \subset \Omega$, and with reliable channels, we cannot transform any failure detector of $\diamond\mathcal{W}(\Gamma)$ into some failure detector of $\diamond\mathcal{W}$ (which implies $\diamond\mathcal{W}(\Gamma) \prec \diamond\mathcal{W}$). Hence, for any $\Gamma \subset \Omega$, problems that need $\diamond\mathcal{W}$ (e.g consensus, uniform consensus, atomic broadcast, and non-blocking weak atomic commitment [3]) cannot be solved with $\diamond\mathcal{W}(\Gamma)$.
3. For any $\Gamma \subset \Omega$, and with reliable channels, we cannot transform any failure detector of $\mathcal{P}(\Gamma)$ into some failure detector of $\mathcal{P}$ (which implies $\mathcal{P}(\Gamma) \prec \mathcal{P}$). Hence, for any $\Gamma \subset \Omega$, problems that need $\mathcal{P}$ (e.g election [10], genuine atomic multicast [7], and non-blocking atomic commitment [6]) cannot be solved with $\mathcal{P}(\Gamma)$.

The rest of the paper is organized as follows. Section 2 defines the system model. Section 3 defines "Γ-accurate" failure detectors. Section 4, where we consider eventually reliable channels, establishes the above result 1. In Sections 5, 6 and 7, we assume reliable channels. Section 5 establishes the result 2, and Section 6 establishes the result 3. Section 7 compares Γ-accurate failure detector classes. Finally, Section 8 uses the results established in the paper to compare the resilience of various atomic commitment protocols.

2 Model

Our model of asynchronous computation with failure detection is similar to the one described in [3].

2.1 Failures

A discrete global clock is assumed, and Φ, the range of the clock's ticks, is the set of natural numbers. Processes do not have access to the global clock. The distributed system consists of a set Ω of processes. Processes fail by *crashing*, and failures are permanent. A correct process is a process that does not crash. A *failure pattern* is a function F from Φ to 2^{Ω}, where $F(t)$ denotes the set of processes that have crashed through time t. We assume, as in [4], that in any failure pattern, there is at least one correct process. A *failure detector history* is a function from $\Omega \times \Phi$ to 2^{Ω}, where $H(p,t)$ denotes the set of processes suspected by process p at time t. A *failure detector* is a function $\mathcal{D}$ that maps each failure pattern F to a set of failure detector histories. The processes are connected through asynchronous, either (1) *reliable*, or (2) *eventually reliable* channels, represented by a *message buffer* (see Sect. 2.2):

[2] It has been shown that these problems are solvable with the specified failure detectors, and reliable channels. It can be shown that these problems are also solvable with eventually reliable channels, see [1].

- a *reliable channel* ensures that every message sent by a process p to a process q is eventually received by q, if q is correct.
- an *eventually reliable channel* ensures that every message sent by a process p to a process q is eventually received by q, if *q and p are both correct.*

The *eventually reliable* channel provides a weaker model than the *reliable channel*: an eventually reliable channel can loose messages. [3]

2.2 Algorithms

An *algorithm* is a collection A of n deterministic automata $A(p)$ (one per process p). Computation proceeds in steps of the algorithm. In each step of an algorithm A, a process p performs atomically the following phases: (1) p receives a message from q, or a "null" message λ; (2) p queries and receives a value d from its failure detector module (d is said to be *seen* by p); (3) p changes its state and sends a message (possibly null) to some process. This third phase is performed according to (a) the automaton $A(p)$, (b) the state of p at the beginning of the step, (c) the message received in phase 1, and (d) the value d seen by p in phase 2. The message received by a process is chosen non-deterministically among the messages in the message buffer destined to p, and the null message λ. A *configuration* is a pair (I, M) where I is a function mapping each process p to its local state, and M is a set of messages currently in the message buffer. A configuration (I, M) is an initial configuration if $M = \emptyset$. A step of an algorithm A is a tuple $e = (p, m, d, A)$, uniquely defined by the algorithm A, the identity of the process p that takes the step, the message m received by p, and the failure detector value d seen by p during the step. A step $e = (p, m, d, A)$ is *applicable to a configuration* (I, M) if and only if $m \in M \cup \{\lambda\}$. The *unique* configuration that results from applying e to $C = (I, M)$, is noted $e(C)$.

2.3 Schedules and runs

A *schedule* of an algorithm A is a (possibly infinite) sequence of steps of A, noted $S = S[1]; S[2]; \ldots S[k]; \ldots$. A schedule is applicable to a configuration C if (1) S is the empty schedule, or (2) $S[1]$ is applicable to C, $S[2]$ is applicable to $S[1](C)$, etc. Given any schedule S, we note $P(S)$ the set of the processes that have at least one step in S.

A *partial run* of A using a failure detector $\mathcal{D}$, is a tuple $R =< F, H, C, S, T >$ where, F is a failure pattern, H is a failure detector history and $H \in \mathcal{D}(F)$, C is an initial configuration of A, T is a finite sequence of increasing time values, and S is a finite schedule of A such that: (1) $|S| = |T|$, (2) S is applicable to

[3] The rational behind the definition is the following. To ensure eventual reception of the message m sent by p to q, the communication library linked to p will have to retransmit the message m, until m is eventually received by q. If p crashes, retransmission will stop, and q might never receive m. Therefore *eventually reliable* channels ensure reception only if the sender and the receiver are both correct.

C, and (3) for all $i \leq |S|$ where $S[i] = (p, m, d, A)$, we have $p \notin F(T[i])$ and $d = H(p, T[i])$.

A *run* of an algorithm A using a failure detector $\mathcal{D}$, is a tuple $R =< F, H, C, S, T >$ where F is a failure pattern, H is a failure detector history and $H \in \mathcal{D}(F)$, C is an initial configuration of A, S is an infinite schedule of A, T is an infinite sequence of increasing time values, and in addition to the conditions above of a partial run ((1), (2) and (3) above), the two following conditions are satisfied: (4) every correct process takes an infinite number of steps, (5) under the *reliable channel* assumption, every message sent by a process to a correct process is eventually received, and under the *eventually reliable channel* assumption, every message sent by a correct process to a correct process is eventually received.

Let $R =< F, H, C, S, T >$ be a partial run of some algorithm A. We say that $R' =< F', H', C', S', T' >$ is *an extension of* R, if R' is either a run or a partial run of A, and $F' = F$, $H' = H$, $C' = C$, $\forall i$ s.t. $T[1] \leq i \leq T[|T|]$: $S'[i] = S[i]$ and $T'[i] = T[i]$.

3 From accurate to "Γ-accurate" failure detectors

3.1 Accurate failure detectors

Failure detectors are abstractly characterized by completeness and accuracy properties. Completeness properties determine the degree to which crashed processes are suspected. Accuracy properties restrict the mistakes (false suspicions) that a process can make. Two completeness properties are defined in [4]: (1) *strong completeness*: eventually every process that crashes is permanently suspected by every correct process, and (2) *weak completeness*: eventually every process that crashes is permanently suspected by some correct process. The following accuracy properties are defined in [4]: (1) *strong accuracy*: no process is suspected before it crashes; (2) *weak accuracy*: some correct process is never suspected; (3) *eventual strong accuracy*: eventually correct processes are not suspected by any correct process, and (4) *eventual weak accuracy*: eventually some correct process is not suspected by any correct process.

A failure detector class is a set of failure detectors defined by some accuracy and some completeness property. Figure 1 shows the notations introduced in [4]. For example, the class $\diamond\mathcal{S}$ contains failure detectors that satisfy strong completeness and eventual weak accuracy.

3.2 Γ-accuracy properties

The accuracy properties defined by Chandra and Toueg restrict the mistakes made by all the processes in the system. We extend these properties, by considering properties that only restrict the mistakes of a subset Γ of the processes. Given $\Gamma \subseteq \Omega$, we define the Γ-accuracy properties as follows:

Completeness	Accuracy			
	Strong	Weak	◇Strong	◇Weak
Strong	P (Perfect)	S (Strong)	◇P	◇S
Weak	Q	W (Weak)	◇Q	◇W

Fig. 1. Accurate failure detector classes

- *Strong Γ-accuracy* is satisfied if no process p in Γ is suspected by any process in Γ, before p crashes.
- *Weak Γ-accuracy* is satisfied if some correct process (not necessarily in Γ) is never suspected by any process in Γ.
- *Eventual strong Γ-accuracy* is satisfied if eventually no correct process in Γ is suspected by any correct process in Γ.
- *Eventual weak Γ-accuracy* is satisfied if eventually some correct process (not necessarily in Γ) is never suspected by any correct process in Γ.

These accuracy properties can be viewed as generalizations of the accuracy properties defined in [4]. The latters correspond to the case where $\Gamma = \Omega$. The asymmetry of the definitions of Γ-accuracy properties ("p in Γ" in the definition of strong Γ-accuracy, "some correct process not necessarily in Γ" in the definition of weak Γ-accuracy) is because we want the Γ-accuracy properties to satisfy the *inclusion* property: given $\Gamma_1 \subset \Gamma_2$, if strong (respt. weak) Γ_2-accuracy holds, then strong (respt. weak) Γ_1-accuracy also holds. We comment on this in Section 3.4.

3.3 Γ-accurate failure detectors

A *Γ-accurate* failure detector is a failure detector defined by some completeness and some Γ-accuracy property. Figure 2 introduces the notations for Γ-accurate failure detector classes. For example, the class $\diamond\mathcal{S}(\Gamma)$ gathers failure detectors that satisfy strong completeness and eventual weak Γ-accuracy.

Completeness	Accuracy			
	Strong Γ-	Weak Γ-	◇ Strong Γ-	◇Weak Γ-
Strong	P(Γ)	S(Γ)	◇P(Γ)	◇S(Γ)
Weak	Q(Γ)	W(Γ)	◇Q(Γ)	◇W(Γ)

Fig. 2. Γ-accurate failure detector classes

3.4 Simple relations between failure detector classes

Given two failure detectors $\mathcal{D}_1$ and $\mathcal{D}_2$, if there is an algorithm that *transforms* $\mathcal{D}_1$ into $\mathcal{D}_2$, then $\mathcal{D}_2$ is said to be *reducible* to $\mathcal{D}_1$, written $\mathcal{D}_2 \preceq \mathcal{D}_1$ [4]. If every failure detector of a class $\mathcal{C}_2$ is reducible to some failure detector of a class $\mathcal{C}_1$, then $\mathcal{C}_2$ is said to be *weaker* than $\mathcal{C}_1$, written $\mathcal{C}_2 \preceq \mathcal{C}_1$. The relation $\preceq$ is an equivalence relation. If $\mathcal{C}_2 \preceq \mathcal{C}_1$ and $\mathcal{C}_1 \preceq \mathcal{C}_2$, then $\mathcal{C}_1$ and $\mathcal{C}_2$ are said to be *equivalent*, written $\mathcal{C}_1 \cong \mathcal{C}_2$. Finally, if $\mathcal{C}_2 \preceq \mathcal{C}_1$ and $\neg(\mathcal{C}_1 \preceq \mathcal{C}_2)$, then $\mathcal{C}_2$ is said to be *strictly weaker* than $\mathcal{C}_1$, written $\mathcal{C}_2 \prec \mathcal{C}_1$.

The inclusion property: The Γ-accuracy properties satisfy the *inclusion* property: given $\Gamma_1 \subset \Gamma_2$, if Γ_2-accuracy holds (strong, weak, eventually strong, or eventually weak), then Γ_1-accuracy also holds (strong, weak, eventually strong, or eventually weak). Roughly speaking, the inclusion property reflects the intuition that reducing the set Γ should not invalidate the accuracy property.

Note that the inclusion property would not hold if weak (respt. eventual weak) Γ-accuracy had been defined as follows: (eventually) some correct process " in Γ " is never suspected by any process (correct process) in Γ. Take this definition, and consider $\Gamma_1 \subset \Gamma_2$. If weak Γ_2-accuracy holds, then there is some process $p \in \Gamma_2$ that is never suspected by any process in Γ_2. However, $p \in \Gamma_2$ does not imply $p \in \Gamma_1$, i.e weak Γ_1-accuracy does not necessarily hold.

The following lemma state simple relations between the classes of accurate failure detectors, and the classes of $\Gamma-$accurate failure detectors.

Lemma 3.1 *Let $\mathcal{C}$ stand for $\mathcal{P}, \mathcal{Q}, \mathcal{S}, \mathcal{W}, \diamond\mathcal{P}, \diamond\mathcal{Q}, \diamond\mathcal{S}$, or $\diamond\mathcal{W}$. For any $\Gamma \subseteq \Omega$, and with both reliable and eventually reliable channels, we have $\mathcal{C}(\Gamma) \preceq \mathcal{C}$.*

PROOF. Follows directly from the definition and the inclusion property. For any $\Gamma \subseteq \Omega$, any failure detector of class $\mathcal{C}$ satisfies the properties of class $\mathcal{C}(\Gamma)$. We thus trivially have $\mathcal{C} \subset \mathcal{C}(\Gamma)$, which implies $\mathcal{C}(\Gamma) \preceq \mathcal{C}$. □

From strong Γ-accuracy to weak Γ-accuracy: For any $\Gamma \subset \Omega$, eventual strong Γ-accuracy implies eventual weak Γ-accuracy. However, strong Γ-accuracy implies weak Γ-accuracy only if $f < |\Gamma|$ (i.e. if there is at least some correct process in Γ) (see Lemma 3.2). Indeed, assume the processes of Γ do not suspect each others, but suspect all the processes outside Γ and then crash. In this case, strong Γ-accuracy is satisfied whereas weak Γ-accuracy is not.

The following lemma states simple relation between the classes of $\Gamma-$accurate failure detectors.

Lemma 3.2 *For any $\Gamma \subseteq \Omega$, and with both reliable and eventually reliable channels, we have (1) $\mathcal{Q}(\Gamma) \preceq \mathcal{P}(\Gamma)$, $\mathcal{W}(\Gamma) \preceq \mathcal{S}(\Gamma)$, $\diamond\mathcal{Q}(\Gamma) \preceq \diamond\mathcal{P}(\Gamma)$, $\diamond\mathcal{W}(\Gamma) \preceq \diamond\mathcal{S}(\Gamma)$, (2) $\diamond\mathcal{S}(\Gamma) \preceq \diamond\mathcal{P}(\Gamma)$, $\diamond\mathcal{W}(\Gamma) \preceq \diamond\mathcal{Q}(\Gamma)$, and (3) if $f < |\Gamma|$, we also have $\mathcal{S}(\Gamma) \preceq \mathcal{P}(\Gamma)$ and $\mathcal{W}(\Gamma) \preceq \mathcal{Q}(\Gamma)$.*

Proof. As strong completeness implies weak completeness then we obviously have results (1). Consider now results (2). Let $\mathcal{D}$ be any failure detector of class $\diamond\mathcal{P}(\Gamma)$ (respt. of class $\diamond\mathcal{Q}(\Gamma)$). $\mathcal{D}$ satisfies strong completeness (respt. weak completeness) and eventual strong $\Gamma-$accuracy. Hence, for every failure pattern, eventually no correct process in Γ suspects any correct process in Γ. If there is some correct process in Γ, then $\mathcal{D}$ trivially satisfies eventual weak $\Gamma-$accuracy. Altogether, $\mathcal{D}$ is of class $\diamond\mathcal{S}(\Gamma)$ (respt. of class $\diamond\mathcal{W}(\Gamma)$), which implies $\diamond\mathcal{S}(\Gamma) \preceq \diamond\mathcal{P}(\Gamma)$ (respt. $\diamond\mathcal{W}(\Gamma) \preceq \diamond\mathcal{Q}(\Gamma)$).

Consider now results (3). Let $\mathcal{D}'$ be a failure detector of class $\mathcal{P}(\Gamma)$ (respt. of class $\mathcal{Q}(\Gamma)$). $\mathcal{D}'$ satisfies strong completeness (respt. weak completeness) and strong $\Gamma-$accuracy. As $f < |\Gamma|$, then for every failure pattern, there is some correct process in Γ, and this process is never suspected by any process in Γ. Hence, $\mathcal{D}$ satisfies weak $\Gamma-$accuracy. Altogether, $\mathcal{D}$ is of class $\mathcal{S}(\Gamma)$ (respt. of class $\mathcal{W}(\Gamma)$), which implies $\mathcal{S}(\Gamma) \preceq \mathcal{P}(\Gamma)$ (respt. $\mathcal{W}(\Gamma) \preceq \mathcal{Q}(\Gamma)$). □

4 About weak accuracy and strong completeness

In this section, we present an algorithm that transforms any failure detector of class $\diamond\mathcal{S}(\Gamma)$ (respt. $\mathcal{S}(\Gamma)$) into some failure detector of class $\diamond\mathcal{S}$ (respt. $\mathcal{S}$). Our transformation algorithm is correct under the assumptions (1) Γ is a majority of Ω ($|\Gamma| > |\Omega|/2$), (2) there is a majority of correct processes in Ω ($f < |\Omega|/2$), and (3) channels are eventually reliable. Under these assumptions, we have $\mathcal{S} \cong \mathcal{S}(\Gamma)$ and $\diamond\mathcal{S} \cong \diamond\mathcal{S}(\Gamma)$.

4.1 The transformation algorithm

Let Γ be a majority of Ω ($|\Gamma| > |\Omega|/2$), and assume a majority of correct processes ($f < |\Omega|/2$) and eventual reliable channels. With these assumptions, the algorithm in Figure 3 transforms any failure detector of $\mathcal{S}(\Gamma)$ (respt. of $\diamond\mathcal{S}(\Gamma)$) into some failure detector of $\mathcal{S}$ (respt. of $\diamond\mathcal{S}$). The algorithm uses any failure detector, say $\mathcal{D}_1$, of $\mathcal{S}(\Gamma)$ (respt. of $\diamond\mathcal{S}(\Gamma)$) to emulate the output of a failure detector $\mathcal{D}_2$ of $\mathcal{S}$ (respt. of $\diamond\mathcal{S}$). The emulation is done in a distributed variable, $output(\mathcal{D}_2)$. Each process p has a local copy of $output(\mathcal{D}_2)$, denoted $output(\mathcal{D}_2)_p$, which provides the information that should be given by the local failure detector module of $\mathcal{D}_2$ at process p (noted $\mathcal{D}_{2_p}$). The value of $output(\mathcal{D}_2)_p$ at time t is denoted $output(\mathcal{D}_2, t)_p$. Informally, the algorithm works as follows.

- Every process p periodically sends the message $(p, suspected_p)$ to all processes (line 3), where $suspected_p$ denotes the set of processes that p suspects according to its local failure detector module $\mathcal{D}_{1_p}$.
- When p receives a message of the form $(q, suspected_q)$ (line 4), it executes the following:
(1) for each r in $suspected_q$, p adds q to $suspecting(r)_p$ (line 7), where $suspecting(r)_p$ denotes the set of processes p thinks are currently suspecting r. If $suspecting(r)_p$ contains a majority of processes, then p adds r to $output(\mathcal{D}_2)_p$ (line 8);

(2) for each r not in $suspected_q$, p removes q from $suspecting(r)_p$ and removes r from $output(\mathcal{D}_2)_p$ (lines 10-11).[4]

```
    /* Every process p executes the following */

    /* Initialisation */
      suspected_p ← ∅;   /* The set of processes suspected by p */
      output(D_2)_p ← ∅;
            /* The local variable emulating the failure detector module D_2_p */
      for each r in Ω: suspecting(r)_p ← ∅;
            /* The set of processes p thinks are currently suspecting r */

   cobegin  /* two concurrent tasks */
   || /* Task 1: */
1       repeat forever
2           suspected_p ← D_1_p   /* p queries its failure detector module D_1_p */
3           send (p, suspected_p) to all ;

   || /* Task 2: */
4      when (q, suspected_q) received from some q
5          for each r in Ω
6              if r in suspected_q then
7                  suspecting(r)_p ← (suspecting(r)_p ∪ {q}) ;
8                  if |suspecting(r)_p| > |Ω|/2
                       then output(D_2)_p ← (output(D_2)_p ∪ {r}) ;
9              else
10                 suspecting(r)_p ← (suspecting(r)_p − {q}) ;
11                 output(D_2)_p ← (output(D_2)_p − {r}) ;
   coend
```

Fig. 3. From $\mathcal{S}(\Gamma)$ (respt. $\diamond\mathcal{S}(\Gamma)$) to $\mathcal{S}$ (respt. $\diamond\mathcal{S}$)

4.2 Correctness of the transformation

By Lemma 4.1 below, if $|\Gamma| > |\Omega|/2$, the algorithm in Figure 3 transforms weak Γ-accuracy into weak accuracy. The proof is by contradiction. Similarly, by Lemma 4.2, if $|\Gamma| > |\Omega|/2$, the algorithm also transforms eventual weak Γ-accuracy into eventual weak accuracy (proof also by contradiction). Finally, by Lemma 4.3, if $f < |\Omega|/2$, the transformation of Figure 3 preserves strong

[4] The "correction phase" (lines 10-11) is needed to transform eventual weak Γ-accuracy into eventual weak accuracy, but is not needed to transform weak Γ-accuracy into weak accuracy.

completeness. Altogether, if $|\Gamma| > |\Omega|/2$ and $f < |\Omega|/2$, we get: $\mathcal{S} \cong \mathcal{S}(\Gamma)$ and $\diamond\mathcal{S} \cong \diamond\mathcal{S}(\Gamma)$ (Proposition 4.4).

If $f < |\Omega|/2$, the consensus [4], uniform consensus [4], atomic broadcast [4], and non-blocking weak atomic commitment [6] problems can be solved with any failure detector of the class $\diamond\mathcal{S}(\Gamma)$ and reliable channels. It can be shown that these problems are also solvable with any failure detector of the class $\diamond\mathcal{S}(\Gamma)$ and eventual reliable channels [1]. Thus from Proposition 4.4, $\forall\Gamma \subset \Omega$ such that $|\Gamma| > |\Omega|/2$, if $f < |\Omega|/2$, the consensus, uniform consensus, atomic broadcast and non-blocking weak atomic commitment problems can be solved with any failure detector of the class $\diamond\mathcal{S}(\Gamma)$.

Lemma 4.1 (from weak Γ-accuracy to weak accuracy) *Let $\mathcal{D}_1$ be a failure detector that satisfies weak Γ-accuracy. If $|\Gamma| > |\Omega|/2$ and with eventual reliable channels, the algorithm of Figure 3 transforms $\mathcal{D}_1$ into a failure detector $\mathcal{D}_2$ that satisfies weak accuracy.*

PROOF. As $\mathcal{D}_1$ satisfies Γ-weak accuracy, there is a correct process r such that no process in Γ suspects r. Assume (by contradiction) that there is a process p such that r is in $output(D_2)_p$. This means that a majority of processes have suspected r (Fig. 2, line 8). As Γ contains a majority of processes, then some process in Γ must have suspected r: a contradiction. □

Lemma 4.2 (from $\diamond$ weak Γ-accuracy to $\diamond$ weak accuracy) *Let $\mathcal{D}_1$ be a failure detector that satisfies eventual weak Γ-accuracy. If $|\Gamma| > |\Omega|/2$, $f < |\Omega|/2$ and with eventual reliable channels, then the algorithm of Figure 3 transforms $\mathcal{D}_1$ into a failure detector $\mathcal{D}_2$ that satisfies eventual weak accuracy.*

PROOF. (By contradiction). As $\mathcal{D}_1$ satisfies eventual weak Γ-accuracy, there is a correct process r, and a time t_1 after which no process in Γ suspects r. Hence, there is a time $t_2 > t_1$, from which no process receives a message $(q, suspected_q)$ from a process q in Γ, such that $suspected_q$ contains r.

As Γ contains at least one correct process (by the assumption $|\Gamma| > |\Omega|/2$ and $f < |\Omega|/2$), there is a process $p \in \Omega$ and a time $t_3 > t_2$, at which p receives a message $(q, suspected_q)$ from a correct process q in Γ, and $r \notin suspected_q$. Thus at t_3, we have $r \notin output(D_2)_p$ (Fig. 2, line 11). Assume (by contradiction) that there is a time $t_4 > t_3$ at which r is (again) in $output(D_2)_p$ (Fig. 2, line 8). This means that p has received, from a majority of processes, messages $(q, suspected_q)$ such that $r \in suspected_q$. As $|\Gamma| > |\Omega|/2$, some process in Γ must have suspected r after time t_1: a contradiction. □

Lemma 4.3 (preserving strong completeness) *Let $\mathcal{D}_1$ be a failure detector that satisfies strong completeness. If $f < |\Omega|/2$ and with eventual reliable channels, the algorithm of Figure 3 transforms $\mathcal{D}_1$ into a failure detector $\mathcal{D}_2$ that also satisfies strong completeness.*

PROOF. Consider time t_1 at which all processes that are not correct have crashed. After t_1, let r be a process that has crashed. As $\mathcal{D}_1$ satisfies strong completeness, there is a time $t_2 > t_1$ after which every correct process p suspects r forever, and sends its suspicion message $(p, suspected_p)$, where $r \in suspected_p$, to all (Fig. 2, line 3). These suspicions are thus sent by correct processes. As there is a majority of correct processes, and channels are eventually reliable, every correct process p eventually receives such a suspicion message from a majority of processes, and puts r into $output(D_2)_p$ forever (Fig. 2, line 8). □

Proposition 4.4 *Let $\Gamma \subset \Omega$ be such that $|\Gamma| > |\Omega|/2$, and consider $\mathcal{S}$, $\mathcal{S}(\Gamma)$, $\diamond\mathcal{S}$ and $\diamond\mathcal{S}(\Gamma)$. If $f < |\Omega|/2$ and with eventual reliable channels, we have $\mathcal{S} \cong \mathcal{S}(\Gamma)$ and $\diamond\mathcal{S} \cong \diamond\mathcal{S}(\Gamma)$.*

PROOF. By Lemma 3.1, we have $\mathcal{S}(\Gamma) \preceq \mathcal{S}$ and $\diamond\mathcal{S}(\Gamma) \preceq \diamond\mathcal{S}$. By Lemma 4.1 and Lemma 4.3, we have $\mathcal{S} \preceq \mathcal{S}(\Gamma)$. Thus $\mathcal{S} \cong \mathcal{S}(\Gamma)$. By Lemma 4.2 and Lemma 4.3, we have $\diamond\mathcal{S} \preceq \diamond\mathcal{S}(\Gamma)$. Thus $\diamond\mathcal{S} \cong \diamond\mathcal{S}(\Gamma)$. □

5 About weak completeness

This section compares the class $\diamond\mathcal{W}(\Gamma)$ (respt. $\mathcal{W}(\Gamma)$) of Γ-accurate failure detectors with the class $\diamond\mathcal{W}$ (respt. $\mathcal{W}$) of accurate failure detectors. We assume $|\Omega| > 2$ and we show that, for any subset $\Gamma \subset \Omega$ and even with reliable channels, we have: $\diamond\mathcal{W}(\Gamma) \prec \diamond\mathcal{W}$ and $\mathcal{W}(\Gamma) \prec \mathcal{W}$ [5].

It has been shown that $\diamond\mathcal{W}$ is the weakest failure detector class that enables to solve consensus, atomic broadcast, uniform consensus [3, 4], and non-blocking weak atomic commitment [6]. A consequence of $\diamond\mathcal{W}(\Gamma) \prec \diamond\mathcal{W}$ is that neither consensus, atomic broadcast, uniform consensus, nor non-blocking weak atomic commitment is solvable with $\diamond\mathcal{W}(\Gamma)$ (for any subset $\Gamma \subset \Omega$).

Failure detector $\mathcal{D}(\Gamma, r)$. The proofs of the above results ($\diamond\mathcal{W}(\Gamma) \prec \diamond\mathcal{W}$ and $\mathcal{W}(\Gamma) \prec \mathcal{W}$) use a specific failure detector, noted $\mathcal{D}(\Gamma, r)$. The specification of $\mathcal{D}(\Gamma, r)$ is based on a failure pattern that we call *1-pattern*: we say that a failure pattern F is a *1-pattern* if at most one process crashes in F. Similarly, we say that a failure pattern F is a *0-pattern* if no process crashes in F. Consider a subset $\Gamma \subset \Omega$ and $r \in \Omega - \Gamma$. We define $\mathcal{D}(\Gamma, r)$ such that (a) in any *1-pattern* F, $\mathcal{D}(\Gamma, r)(F)$ is the set of histories such that: (a.1) as long as r does not crash, r permanently suspects every other process, and (a.2) $\forall r' \neq r$, as long as r' does not crash, r' permanently suspects r, but r' never suspects any other process, and (b) in any pattern F' that is not a 1-pattern, $\mathcal{D}(\Gamma, r)(F')$ is the set of histories that satisfy strong completeness and strong

[5] The assumption $|\Omega| > 2$ is only needed for $\mathcal{W}(\Gamma) \prec \mathcal{W}$ (for $|\Omega| = 2$, we have $\mathcal{W}(\Gamma) \cong \mathcal{W}$). $\diamond\mathcal{W}(\Gamma) \prec \diamond\mathcal{W}$ holds for $|\Omega| > 1$ (for $|\Omega| \leq 1$ all failure detector classes are equivalent). However, for presentation uniformity, we assume in this section that $|\Omega| > 2$.

accuracy (i.e. in any pattern that is not a 1-pattern, $\mathcal{D}(\Gamma, r)$ behaves like a failure detector of the class $\mathcal{P}$).

We show that $\mathcal{D}(\Gamma, r)$ is of class $\mathcal{W}(\Gamma)$ (and also of class $\mathcal{Q}(\Gamma)$, see Lemma 5.1), and no algorithm can transform $\mathcal{D}(\Gamma, r)$ into some failure detector of class $\diamond\mathcal{W}$. More precisely, we show that if there exists an algorithm $A_{\mathcal{D}(\Gamma,r)\rightarrow\Delta}$ that transforms $\mathcal{D}(\Gamma, r)$ into some failure detector Δ that satisfies weak completeness, then Δ cannot satisfy eventual weak accuracy (Lemma 5.2 and Lemma 5.3).

Lemma 5.1 *$\mathcal{D}(\Gamma, r)$ is of the classes $\mathcal{W}(\Gamma)$ and $\mathcal{Q}(\Gamma)$.*

PROOF. In any run of which failure pattern is not a 1-pattern, $\mathcal{D}(\Gamma, r)$ satisfies strong completeness and strong accuracy. Let $R =< F, H_{\mathcal{D}(\Gamma,r)}, C, S, T >$ be any run with F a 1-pattern. If r is correct, then every process that crashes is permanently suspected by r. If r crashes, then all other processes are correct, and they all suspect r. Hence $\mathcal{D}(\Gamma, r)$ satisfies weak completeness in R. Consider now accuracy. As $|\Omega| > 2$, then there is at least some correct process in R that is never suspected by any process of Γ. Thus $\mathcal{D}(\Gamma, r)$ satisfies weak Γ-accuracy. As no process suspects any other process in Γ then $\mathcal{D}(\Gamma, r)$ satisfies strong Γ-accuracy. Hence, $\mathcal{D}(\Gamma, r)$ satisfies weak completeness, weak Γ-accuracy, and strong Γ-accuracy in R. □

Lemma 5.2 *Let $A_{\mathcal{D}(\Gamma,r)\rightarrow\Delta}$ be any algorithm that transforms $\mathcal{D}(\Gamma, r)$ into some failure detector Δ. Let $R =< F, H_{\mathcal{D}(\Gamma,r)}, C, S, T >$ be any partial run of $A_{\mathcal{D}(\Gamma,r)\rightarrow\Delta}$ where F is a 0-pattern. If Δ satisfies weak completeness, then there is an extension $R_\Omega =< F, H_{\mathcal{D}(\Gamma,r)}, C, S_\Omega, T_\Omega >$ of R, where for every correct process p, there is a correct process q, and a time t, $T[|T|] \leq t \leq T_\Omega[|T_\Omega|]$, such that $p \in output(\Delta, t)_q$ in R_Ω.*

PROOF: Consider the partial run $R =< F, H_{\mathcal{D}(\Gamma,r)}, C, S, T >$ where F is a 0−pattern, and let p be any process in Ω. Let $R' =< F', H_{\mathcal{D}(\Gamma,r)}, C, S, T >$ be a partial run such that F' is a 1−pattern, similar to F, except that in F', p crashes at time $T[|T| + 1]$ (immediately after $T[|T|]$). A process such as p does exist as Ω contains at least two processes. As $\mathcal{D}(\Gamma, r)$ provides the same values both for F and F', and S is applicable to C, then R' is a partial run of $A_{\mathcal{D}(\Gamma,r)\rightarrow\Delta}$. By the weak completeness property of Δ, there is an extension $R'_p =< F', H_{\mathcal{D}(\Gamma,r)}, C, S_p, T_p >$ of R', and a correct process $q \in \Omega$, such that $p \in output(\Delta, T_p[|T_p|])_q$. Let $S_{Susp(p)}$ be the schedule of $A_{\mathcal{D}(\Gamma,r)\rightarrow\Delta}$ such that $S_p(C) = S_{Susp(p)}(S(C))$. The schedule $S_{Susp(p)}$ can be viewed as the schedule needed to put p into the $output(\Delta)_q$ of process q.

Consider now the run $R =< F, H_{\mathcal{D}(\Gamma,r)}, C, S, T >$. As $\mathcal{D}(\Gamma, r)$ provides the same values both for F and F', and S_p is applicable to C, then $R_p =< F, H_{\mathcal{D}(\Gamma,r)}, C, S_p, T_p >$ is an extension of R, and there is a correct process q, such that $p \in output(\Delta, T_p[|T_p|])_q$. By iteratively applying the construction of the partial run R_p to every process $p \in \Omega$, the partial run R can be extended to a partial run $R_\Omega =< F, H_{\mathcal{D}(\Gamma,r)}, C, S_\Omega, T_\Omega >$ where *every* process p is put in $output(\Delta)_q$ for some process q. □

Lemma 5.3 *Let $A_{\mathcal{D}(\Gamma,r)\to\Delta}$ be any algorithm that transforms the failure detector $\mathcal{D}(\Gamma, r)$ into some failure detector Δ. If Δ satisfies weak completeness, then there is a run of $A_{\mathcal{D}(\Gamma,r)\to\Delta}$, where Δ does not satisfy eventual weak accuracy.*

PROOF: Consider the partial run $R =< F, H_{\mathcal{D}(\Gamma,r)}, C, S, T >$ with F a 0−pattern. By Lemma 5.2, there is an extension of R, $R_\Omega =< F, H_{\mathcal{D}(\Gamma,r)}, C, S_\Omega, T_\Omega >$, such that for every process $p \in \Omega$, there is a time t, $T[|T|] \leq t \leq T_\Omega[|T_\Omega|]$, and a correct process q, such that $p \in output(\Delta, t)_q$. Let (I, M) be the configuration $S_\Omega(C)$. Consider now a schedule S_{Mess}, of which steps are defined by: the reception by the processes of all messages in M not received in S_Ω, then the reception by every process of the null message λ. The schedule S_{Mess} is by construction applicable to $S(C)$, and we write $S_\Sigma(C) = S_{Mess}(S_\Omega(C))$. There is a sequence of increasing time values T_Σ, such that $R_\Sigma =< F, H_{\mathcal{D}_r}, C, S_\Sigma, T_\Sigma >$ is an extension of R. In R_Σ, all messages sent to p before time $T[|T|]$ are received by p before $T_\Sigma[|T_\Sigma|]$, and p takes at least one step between $T[|T|]$ and $T_\Sigma[|T_\Sigma|]$.

Therefore, given any partial run R of $A_{\mathcal{D}(\Gamma,r)\to\Delta}$, with F a 0−pattern, we can extend R to a partial run R_Σ where every process is suspected at some process by Δ. We note $R^0_\Sigma = R_\Sigma$, R^1_Σ an extension of R obtained by applying the construction above to R^0_Σ, R^i_Σ an extension of R obtained by applying the construction above to R^{i-1}_Σ, etc., and $R^\infty_\Sigma = lim_{i\to\infty} R^i_\Sigma$.

In R^∞_Σ, the properties of a partial run are satisfied, every process takes an infinite number of steps, and every message sent to a process is eventually received. Hence R^∞_Σ is a run of $A_{\mathcal{D}_r\to\Delta}$. Furthermore, for any time t and any process p, there is a time $t' \geq t$ and a process q, such that $p \in output(\Delta, t')_q$. Hence Δ does not satisfy eventual weak accuracy in R^∞_Σ. □

Proposition 5.4 *Let $\Gamma \subset \Omega$, and consider $\diamond\mathcal{W}$ and $\diamond\mathcal{W}(\Gamma)$. We have $\diamond\mathcal{W}(\Gamma) \prec \diamond\mathcal{W}$ and $\mathcal{W}(\Gamma) \prec \mathcal{W}$.*

PROOF. By Lemma 5.1 and Lemma 5.3, no algorithm can transform any failure detector of $\mathcal{W}(\Gamma)$, into some failure detector of $\diamond\mathcal{W}$. In other words, $\neg(\diamond\mathcal{W} \preceq \mathcal{W}(\Gamma))$. As $\diamond\mathcal{W} \preceq \mathcal{W}$ and $\diamond\mathcal{W}(\Gamma) \preceq \mathcal{W}(\Gamma)$, then $\neg(\diamond\mathcal{W} \preceq \diamond\mathcal{W}(\Gamma))$ and $\neg(\mathcal{W} \preceq \mathcal{W}(\Gamma))$. By Lemma 3.1, we have $\diamond\mathcal{W}(\Gamma) \preceq \diamond\mathcal{W}$ and $\mathcal{W}(\Gamma) \preceq \mathcal{W}$. Altogether, we have $\diamond\mathcal{W}(\Gamma) \prec \diamond\mathcal{W}$ and $\mathcal{W}(\Gamma) \prec \mathcal{W}$. □

6 About Strong Accuracy

This section compares the classes of Γ-accurate failure detectors $\mathcal{P}(\Gamma)$, $\mathcal{Q}(\Gamma)$, $\diamond\mathcal{P}(\Gamma)$, and $\diamond\mathcal{Q}(\Gamma)$, with the classes of accurate failure detectors $\mathcal{P}$, $\mathcal{Q}$, $\diamond\mathcal{P}$, and $\diamond\mathcal{Q}$. We assume in this section that $|\Omega| > 1$, and we show in the following (Proposition 6.4) that, for any subset $\Gamma \subset \Omega$ and even with reliable channels, no algorithm can transform any failure detector of $\mathcal{P}(\Gamma)$ into some failure detector of $\Diamond\mathcal{Q}$. Hence, we have: $\mathcal{P}(\Gamma) \prec \mathcal{P}$, $\mathcal{Q}(\Gamma) \prec \mathcal{Q}$, $\Diamond\mathcal{P}(\Gamma) \prec \Diamond\mathcal{P}$, and $\Diamond\mathcal{Q}(\Gamma) \prec \Diamond\mathcal{Q}$.

A consequence of $\mathcal{P}(\Gamma) \prec \mathcal{P}$ is that problems requiring $\mathcal{P}$ (e.g election [10], genuine atomic multicast [7], and non-blocking atomic commitment [6]) cannot be solved with $\mathcal{P}(\Gamma)$.

The proof of $\mathcal{P}(\Gamma) \prec \mathcal{P}$, $\mathcal{Q}(\Gamma) \prec \mathcal{Q}$, $\Diamond\mathcal{P}(\Gamma) \prec \Diamond\mathcal{P}$ and $\Diamond\mathcal{Q}(\Gamma) \prec \Diamond\mathcal{Q}$, is similar to the proof of the previous section. We introduce a specific failure detector $\mathcal{D}'(\Gamma, r)$ of class $\mathcal{P}(\Gamma)$, and we show that $\mathcal{D}'(\Gamma, r)$ cannot be transformed into some failure detector of $\Diamond\mathcal{Q}$. More precisely, we show that if some algorithm $A_{\mathcal{D}'(\Gamma,r)\rightarrow\Delta}$ transforms $\mathcal{D}'(\Gamma, r)$ into some failure detector Δ that satisfies weak completeness, then Δ cannot satisfy eventual strong accuracy.

Failure detector $\mathcal{D}'(\Gamma, r)$. Let r be any process in Ω. We define the failure detector $\mathcal{D}'(\Gamma, r)$ using the notion of $\{r\}-$*pattern*. We say that a failure pattern F is a $\{r\}-$*pattern* if only r can crash in F. Using this notion, we define $\mathcal{D}'(\Gamma, r)$ as follows. Consider $\Gamma \subset \Omega$ and $r \in \Omega - \Gamma$. We define $\mathcal{D}'(\Gamma, r)$ such that (1) in any $\{r\}-$*pattern* F, $\mathcal{D}(\Gamma, r)(F)$ is the set of histories such that (a) r is permanently suspected by every process that has not crashed, and (b) $\forall r' \neq r$, r' is never suspected by any process, and (2) in any pattern F' that is not a $\{r\}-$*pattern*, $\mathcal{D}(\Gamma, r)(F')$ is the set of histories such that $\mathcal{D}'(\Gamma, r)$ satisfies strong completeness and strong accuracy.

Lemma 6.1 *$\mathcal{D}'_k$ is of class $\mathcal{P}(\Gamma)$.*

PROOF: In any $\{r\}-$*pattern*, $\mathcal{D}'(\Gamma, r)$ satisfies strong Γ-accuracy and strong completeness. In any run that is not a $\{r\}-$*pattern*, $\mathcal{D}'(\Gamma, r)$ satisfies strong accuracy and strong completeness. Altogether, $\mathcal{D}'(\Gamma, r)$ is thus of class $\mathcal{P}(\Gamma)$. □

Lemma 6.2 *Let $A_{\mathcal{D}'(\Gamma,r)\rightarrow\Delta}$ be any algorithm that transforms the failure detector $\mathcal{D}'(\Gamma, r)$ into some failure detector Δ. Let $R =< F, H_{\mathcal{D}'(\Gamma,r)}, C, S, T >$ be any partial run of $A_{\mathcal{D}'(\Gamma,r)\rightarrow\Delta}$ where F is a $0-$pattern. If Δ satisfies weak completeness, then there is an extension $R_\Omega =< F, H_{\mathcal{D}'(\Gamma,r)}, C, S_\Omega, T_\Omega >$ of R, a process q, and a time t, $T[|T|] \leq t \leq T_\Omega[|T_\Omega|]$, such that $r \in output_\Delta(q, t)$.*

PROOF: (similar to the proof of Lemma 5.2, see [8])

Lemma 6.3 *Let $A_{\mathcal{D}'(\Gamma,r)\rightarrow\Delta}$ be any algorithm that transforms the failure detector $\mathcal{D}'(\Gamma, r)$ into some failure detector Δ. If Δ satisfies weak completeness, then there is a run of $A_{\mathcal{D}'(\Gamma,r)\rightarrow\Delta}$, where Δ does not satisfy eventual strong accuracy.*

PROOF: (similar to the proof of Lemma 5.3, see [7])

Proposition 6.4 *Let $\Gamma \subset \Omega$, and consider $\mathcal{P}(\Gamma)$, $\mathcal{P}$, $\diamond\mathcal{P}(\Gamma)$, $\diamond\mathcal{P}$, $\mathcal{Q}(\Gamma)$, $\mathcal{Q}$, $\diamond\mathcal{Q}(\Gamma)$, and $\diamond\mathcal{Q}$. We have $\mathcal{P}(\Gamma) \prec \mathcal{P}$, $\diamond\mathcal{P}(\Gamma) \prec \diamond\mathcal{P}$, $\mathcal{Q}(\Gamma) \prec \mathcal{Q}$, and $\diamond\mathcal{Q}(\Gamma) \prec \diamond\mathcal{Q}$.*

PROOF: By Lemma 6.1 and Lemma 6.3, no algorithm can transform any failure detector of $\mathcal{P}(\Gamma)$ into some failure detector of $\diamond\mathcal{Q}$. In other words, $\neg(\diamond\mathcal{Q} \preceq \mathcal{P}(\Gamma))$. As $\mathcal{P} \preceq \diamond\mathcal{Q}$, $\diamond\mathcal{P} \preceq \diamond\mathcal{Q}$, $\diamond\mathcal{Q} \preceq \mathcal{Q}$, $\diamond\mathcal{Q}(\Gamma) \preceq \mathcal{P}(\Gamma)$, $\diamond\mathcal{P}(\Gamma) \preceq \mathcal{P}(\Gamma)$, $\mathcal{Q}(\Gamma) \preceq \mathcal{P}(\Gamma)$, then $\neg(\mathcal{P} \preceq \mathcal{P}(\Gamma))$, $\neg(\diamond\mathcal{P} \preceq \diamond\mathcal{P}(\Gamma))$, $\neg(\mathcal{Q} \preceq \mathcal{Q}(\Gamma))$, $\neg(\diamond\mathcal{Q} \preceq \diamond\mathcal{Q}(\Gamma))$.

By Lemma 3.1, we have $\mathcal{P}(\Gamma) \preceq \mathcal{P}$, $\diamond\mathcal{P}(\Gamma) \preceq \diamond\mathcal{P}$, $\mathcal{Q}(\Gamma) \preceq \mathcal{Q}$, and $\diamond\mathcal{Q}(\Gamma) \preceq \diamond\mathcal{Q}$. Altogether, we have $\mathcal{P}(\Gamma) \prec \mathcal{P}$, $\diamond\mathcal{P}(\Gamma) \prec \diamond\mathcal{P}$, $\mathcal{Q}(\Gamma) \prec \mathcal{Q}$, and $\diamond\mathcal{Q}(\Gamma) \prec \diamond\mathcal{Q}$. □

7 Comparing "Γ-accurate" failure detectors

Chandra and Toueg have shown that, for accurate failure detectors, weak completeness can be transformed into strong completeness while preserving accuracy properties [4]. In other words, $\mathcal{Q} \cong \mathcal{P}$, $\diamond\mathcal{Q} \cong \diamond\mathcal{P}$, $\mathcal{W} \cong \mathcal{S}$, and $\diamond\mathcal{W} \cong \diamond\mathcal{S}$. This section shows that these results do not hold anymore for Γ-accurate failure detectors. More precisely, we show that given $|\Omega| > 2$, for any $\Gamma \subset \Omega$, and even with reliable channels, we have $\mathcal{Q}(\Gamma) \prec \mathcal{P}(\Gamma)$, $\diamond\mathcal{Q}(\Gamma) \prec \diamond\mathcal{P}(\Gamma)$, $\mathcal{W}(\Gamma) \prec \mathcal{S}(\Gamma)$, and $\diamond\mathcal{W}(\Gamma) \prec \diamond\mathcal{S}(\Gamma)$ [6].

Our proof is based on the failure detector $\mathcal{D}(\Gamma, r)$, defined in Section 5, which was shown to be of the classes $\mathcal{Q}(\Gamma)$ and $\mathcal{W}(\Gamma)$. We show that $\mathcal{D}(\Gamma, r)$ cannot be transformed into some failure detector of class $\diamond\mathcal{S}(\Gamma)$. More precisely, we show that if an algorithm $A_{\mathcal{D}(\Gamma,r)\to\Delta}$ transforms the failure detector $\mathcal{D}(\Gamma, r)$ into some failure detector Δ that satisfies strong completeness, then Δ cannot satisfy eventual weak $\Gamma-$accuracy. The proof is similar to those of Sections 5 and 6.

Lemma 7.1 *Let $A_{\mathcal{D}(\Gamma,r)\to\Delta}$ be any algorithm that transforms the failure detector $\mathcal{D}(\Gamma, r)$ into some failure detector Δ. Let $R =< F, H_{\mathcal{D}(\Gamma,r)}, I, S, T >$ be any partial run of $A_{\mathcal{D}(\Gamma,r)\to\Delta}$ where F is a $0-$pattern. If Δ satisfies strong completeness, then there is an extension $R_\Omega =< F, H_{\mathcal{D}(\Gamma,r)}, I, S_\Omega, T_\Omega >$ of R, where for every process p and every process q in Ω, there is a time t, $T[|T|] \leq t \leq T_\Omega[|T_\Omega|]$, such that $p \in output_\Delta(q, t)$.*

PROOF: (similar to the proof of Lemma 5.2, see [8]).

Lemma 7.2 Let $A_{\mathcal{D}(\Gamma,r)\to\Delta}$ be any algorithm that transforms the failure detector $\mathcal{D}(\Gamma, r)$ into some failure detector Δ. If Δ satisfies *strong completeness*, then there is a run of $A_{\mathcal{D}(\Gamma,r)\to\Delta}$, where Δ does not satisfy *eventual weak $\Gamma-$accuracy.*

[6] The assumption $|\Omega| > 2$ is only needed for $\mathcal{W}(\Gamma) \prec \mathcal{S}(\Gamma)$. The other results hold for $|\Omega| > 1$. However, for presentation uniformity we assume that $|\Omega| > 2$.

PROOF: (similar to the proof of Lemma 5.3, see [8])

Proposition 7.3 Let $|\Omega| > 2$, $\Gamma \subset \Omega$, and consider $\mathcal{Q}(\Gamma)$, $\mathcal{P}(\Gamma)$, $\diamond\mathcal{Q}(\Gamma)$, $\diamond\mathcal{P}(\Gamma)$, $\mathcal{W}(\Gamma)$, $\mathcal{S}(\Gamma)$, $\diamond\mathcal{W}(\Gamma)$, and $\diamond\mathcal{S}(\Gamma)$. We have: $\mathcal{Q}(\Gamma) \prec \mathcal{P}(\Gamma)$, $\diamond\mathcal{Q}(\Gamma) \prec \diamond\mathcal{P}(\Gamma)$, $\mathcal{W}(\Gamma) \prec \mathcal{S}(\Gamma)$, and $\diamond\mathcal{W}(\Gamma) \prec \diamond\mathcal{S}(\Gamma)$.

PROOF. By Lemma 7.1, and Lemma 7.2, no algorithm can transform any failure detector of $\mathcal{Q}(\Gamma)$, or $\mathcal{W}(\Gamma)$), into some failure detector of $\diamond\mathcal{S}$. In other words, $\neg(\diamond\mathcal{S} \preceq \mathcal{Q}(\Gamma))$ and $\neg(\diamond\mathcal{S} \preceq \mathcal{W}(\Gamma))$. As $\diamond\mathcal{S}(\Gamma) \preceq \mathcal{P}(\Gamma)$, $\diamond\mathcal{S}(\Gamma) \preceq \diamond\mathcal{P}(\Gamma)$, $\diamond\mathcal{Q}(\Gamma) \preceq \mathcal{Q}(\Gamma)$, and $\diamond\mathcal{W}(\Gamma) \preceq \mathcal{W}(\Gamma)$, then $\neg(\mathcal{P}(\Gamma) \preceq \mathcal{Q}(\Gamma))$, $\neg(\diamond\mathcal{P}(\Gamma) \preceq \diamond\mathcal{Q}(\Gamma))$, $\neg(\mathcal{S}(\Gamma) \preceq \mathcal{W}(\Gamma))$, and $\neg(\diamond\mathcal{S}(\Gamma) \preceq \diamond\mathcal{W}(\Gamma))$.

By Lemma 3.2, $\mathcal{Q}(\Gamma) \preceq \mathcal{P}(\Gamma)$, $\diamond\mathcal{Q}(\Gamma) \preceq \diamond\mathcal{P}(\Gamma)$, $\mathcal{W}(\Gamma) \preceq \mathcal{S}(\Gamma)$, and $\diamond\mathcal{W}(\Gamma) \preceq \diamond\mathcal{S}(\Gamma)$ Altogether, we have, $\mathcal{Q}(\Gamma) \prec \mathcal{P}(\Gamma)$, $\diamond\mathcal{Q}(\Gamma) \prec \diamond\mathcal{P}(\Gamma)$, $\mathcal{W}(\Gamma) \prec \mathcal{S}(\Gamma)$, and $\diamond\mathcal{W}(\Gamma) \prec \diamond\mathcal{S}(\Gamma)$. □

8 Summary and Discussion

We have defined a formalism to express the knowledge about crash failures in a distributed system, in terms of Γ−accurate failure detectors. This formalism can be viewed as a generalization of the formalism of accurate failure detectors introduced in [4]. To reuse the results about the solvability of distributed agreement problems stated in [4], we have stated a set of relations between accurate and Γ−accurate failure detector classes. These relations are summarized in Figure 4. We assume in the figure that $\Gamma \subset \Omega$ and $|\Omega| > 2$. The notation $A \longleftrightarrow B$ means that failure detector classes A and B are equivalent. The notation $A \longmapsto B$ means that A and B are equivalent if $|\Gamma| > |\Omega|/2$ and $f < |\Omega|/2$, and A is strictly weaker than B otherwise. The notation $A \longrightarrow B$ means that A is strictly weaker than B.

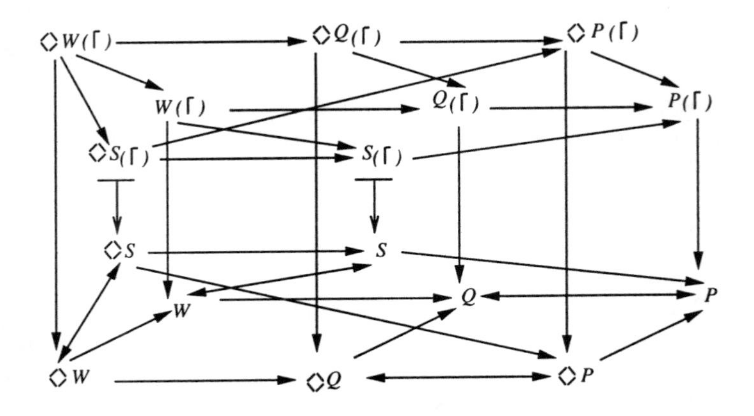

Fig. 4. Relations between failure detector classes

The formalism of Γ-accurate failure detectors enables to characterize and compare some well known distributed protocols designed with network partitions in mind. The initial 3PC protocol (*Three Phase Commit*) proposed by Skeen in 1981 [11] can be seen as requiring the failure detector class $\mathcal{P}$ and reliable channels. In 1982, Skeen proposed a variation of the 3PC protocol, *Quorum Three Phase Commit (Q3PC)* [12], which solves the non-blocking weak atomic commitment problem provided there is a "partition" Γ of correct processes that constitutes a quorum. Stated in our formalism, the Q3PC protocol requires eventual reliable channels, the failure detector class $\mathcal{P}(\Gamma)$, such that Γ is a majority (i.e $|\Gamma| > |\Omega|/2$), and all the processes in Γ correct. As $\mathcal{P}(\Gamma) \prec \mathcal{P}$ (Sect. 6), the Q3PC protocol is an improvement over 3PC.

Later, Keidar and Dolev have defined *Enhanced 3PC (E3PC)* which increases the resilience of Q3PC provided that the quorum exists "eventually" [9]. Stated in our formalism, E3PC assumes a failure detector of the class $\diamond\mathcal{P}(\Gamma)$, such that Γ is a majority (i.e. $|\Gamma| > |\Omega|/2$), and all the processes in Γ correct. As $\diamond\mathcal{P}(\Gamma) \prec \mathcal{P}(\Gamma)$, E3PC is an improvement over Q3PC.

Finally, Guerraoui has shown that the non-blocking weak atomic commitment problem can be solved with $\diamond\mathcal{S}$ [6]. By the result of Section 4, if $f < |\Omega|/2$ and $|\Gamma| > |\Omega|/2$, then $\diamond\mathcal{S} \cong \diamond\mathcal{S}(\Gamma)$. Because $\diamond\mathcal{S}(\Gamma) \prec \diamond\mathcal{P}(\Gamma)$ (Sect. 7), a protocol requiring only $\diamond\mathcal{S}(\Gamma)$ [7] can be seen as an improvement over E3PC. This comparison is somehow unfair as the E3PC protocol is based on a bounded buffer assumption, whereas our eventual reliable channel assumption implicitly requires unbounded buffers (used to store messages to be retansmitted).

The model underlying the E3PC protocol has been described in [5], and the results concerning atomic commitment have been generalized to other consensus-like problems. Babaoğlu et al. have adopted a complementary approach, by discussing the solvability of problems that are weaker than consensus [2], such as *weak-partial group* membership, in a weaker asynchronous system model where channels are not assumed to be eventually reliable. Both models (i.e. [5] and [2]) introduce new failure detector formalisms. Finding out a way to relate our Γ-accurate failure detectors to those introduced in [5] and [2] (and hence the associated results) is still an open issue.

Acknowledgments. We are grateful to Özalp Babaoğlu, Idith Keidar and Aleta Ricciardi, for their valuable comments on earlier drafts of this paper.

[7] Such a protocol can be obtained by combining the protocol of [6] which requires $\diamond\mathcal{S}$, with the protocol of Figure 2 (Sect. 4) which transforms $\diamond\mathcal{S}(\Gamma)$ into $\diamond\mathcal{S}$.

References

1. A. Basu, B. Charron-Bost and S. Toueg. Simulating Reliable Links with Unreliable Links in the Presence of Process Crashes. Proceedings of the *10th International Workshop on Distributed Algorithms*, LNCS, Springer Verlag, October 1996.
2. O. Babaoğlu, R. Davoli and A. Montresor. Failure Detectors, GroupMembership and View-Synchronous Communication in Partitionable Systems. *Technical Report, University of Bologna, Computer Science Department.* November 1995.
3. T. Chandra, V. Hadzilacos and S. Toueg. The weakest failure detector for solving consensus. Journal of the ACM, 43(4), July 1996. A preliminary version appeared in *Proceedings of the 11th ACM Symposium on Principles of Distributed Computing*, pp 147-159. ACM Press. August 1992.
4. T. Chandra and S. Toueg. Unreliable failure detectors for reliable distributed systems. Journal of the ACM, 34(1), pp 225-267, March 1996. A preliminary version appeared in *Proceedings of the 10th ACM Symposium on Principles of Distributed Computing*, pp 325-340. ACM Press. August 1991.
5. R. Friedman, I. Keidar, D. Malki, K. Birman and D. Dolev. Deciding in Partitionable Networks, *Technical Report, Cornell University, Computer Science Department.* November 1995.
6. R. Guerraoui. Revisiting the relationship between Non Blocking Atomic Commitment and Consensus problems. Proceedings of the *9th International Workshop on Distributed Algorithms*, pages 87-100, LNCS 972, Springer Verlag, September 1995.
7. R. Guerraoui and A. Schiper. Atomic Multicast harder than Atomic Broadcast. *Technical Report, Ecole Polytechnique Fédérale de Lausanne, Computer Science Department.* May 1996.
8. R. Guerraoui and A. Schiper. "$\Gamma-$Accurate" Failure Detectors. *Technical Report, Ecole Polytechnique Fédérale de Lausanne, Computer Science Department.* May 1996.
9. I. Keidar and D. Dolev. Increasing the Resilience of Atomic Commit, at No Additional Cost. Proceedings of the *ACM Symposium on Principles of Database Systems*, pages 245-254. ACM Press, May 1994.
10. L. Sabel and K. Marzullo. Election Vs. Consensus in Asynchronous Systems. *Technical Report TR95-1488*, Cornell Univ, 1995.
11. D. Skeen. NonBlocking Commit Protocols. Proceedings of the *ACM SIGMOD International Conference on Management of Data*, pages 133-142. ACM Press, 1981.
12. D. Skeen. A Quorum-Based Commit Protocol. Proceedings of the *Berkeley Workshop on Distributed Data Management and Computer Networks*, pages 69-80, Num 6, 1982.

Fast, Long-Lived Renaming Improved and Simplified

Mark Moir* and Juan A. Garay**

Abstract. In the *long-lived M-renaming problem*, N processes repeatedly acquire and release *names* ranging over $\{0, ..., M-1\}$, where $M < N$. It is assumed that at most k processes concurrently request or hold names. Efficient solutions to the long-lived renaming problem can be used to improve the performance of applications in which processes repeatedly perform computations whose time complexity depends on the size of the name space containing the processes that participate concurrently. In this paper, we consider wait-free solutions to the long-lived M-renaming problem that use only read and write instructions in an asynchronous, shared-memory multiprocessor. A solution to long-lived renaming is *fast* if the time complexity of acquiring and releasing a name once is independent of N. We present a new fast, long-lived $(k(k+1)/2)$-renaming algorithm that significantly improves upon the time and space complexity of similar previous algorithms, while providing a much simpler solution. We also show for the first time that fast, long-lived $(2k-1)$-renaming can be implemented with reads and writes. This result is optimal with respect to the size of the name space.

1 Introduction

In the *one-time M-renaming* problem [2, 3, 4, 10], each of a set of k processes with distinct identifiers ranging over $\{0, ..., N-1\}$ is required to choose a distinct *name* ranging over $\{0, ..., M-1\}$, where $M < N$. The *long-lived M-renaming* problem [2, 5, 6, 10] is a generalisation of one-time renaming, in which N processes repeatedly acquire and release names from $\{0, ..., M-1\}$. It is assumed that at most k processes concurrently request or hold names, and it is required that no two processes hold the same name concurrently. In this paper, we consider wait-free, read/write implementations of long-lived renaming in asynchronous, shared-memory systems. A renaming algorithm is *wait-free* iff each process is guaranteed to acquire a name after a finite number of that process's steps, even if other processes halt undetectably.

* The University of Pittsburgh, Pittsburgh, PA 15260. Email: moir@cs.pitt.edu. This work was carried out while the first author was a graduate student at The University of North Carolina at Chapel Hill, Chapel Hill, North Carolina, and was supported in part by NSF Contract CCR 9216421, by a Young Investigator Award from the U.S. Army Research Office, grant number DAAH04-95-1-0323, and by a UNC Alumni Fellowship.

** Centrum voor Wiskunde en Informatica (CWI), Kruislaan 413, 1098 SJ Amsterdam, The Netherlands, and IBM T.J. Watson Research Center, PO Box 704, Yorktown Heights, New York 10598. E-mail: garay@cwi.nl, garay@watson.ibm.com.

A solution to one-time renaming is useful in performing a computation whose time complexity is dependent on the size of the name space containing the participating processes. By first using an efficient renaming algorithm to reduce the size of the name space, the time complexity of the computation can be made independent of the size of the original name space. Similarly, long-lived renaming is useful if a set of processes repeatedly performs a computation whose time complexity is dependent on the size of the name space containing the processes that participate concurrently. The specific application that first motivated us to study this problem is the implementation of shared objects. The complexity of a shared object implementation is often dependent on the size of the name space containing the processes that access that implementation. For such implementations, performance can be improved by restricting the number of processes that concurrently access the implementation, and by using long-lived renaming to acquire a name from a reduced name space, which can then be used as a process identifier in the shared object implementation. This is the essence of an approach suggested by Anderson and Moir for implementing scalable and resilient shared objects [1]. Because the time complexity of the computations discussed above is often dependent on the size of the name space containing the participating processes, renaming to a smaller name space can result in better overall time complexity. We are therefore motivated to seek renaming protocols whose destination name spaces are as small as possible.

Wait-free, long-lived renaming was first solved for shared-memory systems by Burns and Peterson [6].[3] Like most previous *one-time* renaming algorithms, the time complexity of Burns and Peterson's algorithm is dependent on N, the size of the original name space. Thus, it suffers from the same problem that long-lived renaming is intended to overcome. Recently, Moir and Anderson suggested that a renaming algorithm be called *fast* if its time complexity is independent of N. They presented fast solutions for both one-time and long-lived renaming. Their fast, long-lived renaming algorithms are extremely efficient but depend on strong synchronisation primitives such as `test-and-set`. Later, Buhrman *et al.* showed for the first time that fast, long-lived renaming using only read and write operations is possible [5]. Such algorithms are more portable and more widely applicable than algorithms that rely on special synchronisation primitives. The long-lived renaming solution presented in [5] is quite complicated. In particular, for different values of N and k, different algorithms must be combined, and various parameters chosen to satisfy certain constraints, in order to obtain a fast, long-lived $(k(k+1)/2)$-renaming solution. Furthermore, the space requirements are $\Omega(k^5)$ and, in some cases, are exponential in k.

In this paper, we present a new read/write algorithm for fast, long-lived $(k(k+1)/2)$-renaming and we also observe that this algorithm can be combined with that of Burns and Peterson [6] to achieve a fast, read/write solution to long-lived $(2k-1)$-renaming. Burns and Peterson [6] and Herlihy and Shavit [8]

[3] Actually, Burns and Peterson solved a more general problem, which they called ℓ-assignment. An ℓ-assignment protocol not only assigns names to processes, but also forces some processes to wait if too many request names concurrently. Nonetheless, if at most k processes access an ℓ-assignment protocol that guarantees that, provided at most $k-1$ processes are faulty, every process eventually gets a name, then none have to wait, so ℓ-assignment provides a wait-free solution to the long-lived ℓ-renaming problem. We are grateful to Hagit Attiya for pointing this out to us.

Reference	M	Time Complexity	Space Complexity	Fast?	Long-Lived?
[6]	$2k-1$	$\Theta(Nk^2)$	$\Theta(N^2)$	No	Yes
[10]	$k(k+1)/2$	$\Theta(k)$	$\Theta(k^2)$	Yes	No
[10]	$2k-1$	$\Theta(k^4)$	$\Theta(k^4)$	Yes	No
[10]	$k(k+1)/2$	$\Theta(Nk)$	$\Theta(Nk^2)$	No	Yes
[5]	3^k	$\Theta(k)$	$\Theta(3^k)$	Yes	Yes
[5]	$72k^2$	$\Theta(k \log k)$	$\Theta(k^4)$	Yes	Yes
[5]	$k(k+1)/2$	$\Theta(k^3)$	$\Theta(k^4 \min(3^k, N))$	Yes	Yes
Thm. 1	$k(k+1)/2$	$\Theta(k^2)$	$\Theta(k^3)$	Yes	Yes
Thm. 2	$2k-1$	$\Theta(k^4)$	$\Theta(k^4)$	Yes	Yes

Table 1. A comparison with previous read/write, wait-free M-renaming algorithms that are fast and/or long-lived.

have both shown that long-lived M-renaming cannot be solved in a wait-free manner using atomic reads and writes unless $M \geq 2k-1$. Thus, the latter result is optimal with respect to the size of the name space. These results resolve questions left open in [5] and [10]. Our new long-lived $(k(k+1)/2)$-renaming algorithm improves on the space and time complexity of the algorithm in [5], and is also significantly simpler, as evidenced by the fact that we present full assertional proofs.

Our long-lived $(k(k+1)/2)$-renaming algorithm is similar to the one presented in [10] in that it consists of a grid of building blocks. However, we use a novel technique to make each building block fast. The reason that the previous building block is not fast is that each process p maintains a variable $Y[p]$ that it sets in order to ensure that other processes detect that p is accessing the building block. As a result, it takes $\Theta(N)$ time to determine whether another process is accessing the building block, which means that the building block is not fast. The main idea behind our new building block is to have a total of at most $k+1$ components in the Y-variable of each building block, instead of N.[4] This modification significantly complicates the task of ensuring that some component of Y is set for each process that accesses the building block.

A summary of previous read/write renaming algorithms that are either fast or long-lived appears in Table 1. As shown in this table, we present the most efficient read/write algorithm for fast, long-lived $(k(k+1)/2)$-renaming, and the first read/write algorithm for fast, long-lived $(2k-1)$-renaming. (Two algorithms are presented in [5] that have lower time complexity than ours. However, as shown in Table 1, these algorithms rename to substantially larger name spaces than our algorithm does. As discussed earlier, it is desirable to achieve as small a name space as possible.)

The remainder of the paper is organised as follows. Section 2 contains definitions used in the paper. In Section 3, we present our new long-lived $(k(k+1)/2)$-renaming algorithm, and in Section 4, we prove it correct and prove the two theorems mentioned in Table 1. Concluding remarks appear in Section 5.

[4] Actually, only $k-i-j+1$ Y-variables are used in the building block at grid position (i,j). Thus, if the Y array used in Figure 2 is viewed as a cube, then only half of the cube is actually used.

2 Definitions

Our programming notation should be self-explanatory. Each labeled program fragment in Figure 2 is assumed to be atomic. Some of these fragments are quite long. For example, statement 2 is assumed to atomically read $Y[i, j, h]$, modify j, h, and *moved* accordingly, check the loop condition before line 2, and if that fails, assign zero to h, check the loop condition at line 3, and if that fails, check the loop condition at line 1, and set the program counter to 2, 3, 1, or 8, accordingly. Nonetheless, each labeled program fragment accesses at most one shared variable, and can therefore easily be implemented using only read and write operations. This is because accesses to local variables by some process p do not affect — and are not affected by — steps of other processes. Thus, we can view all of the local steps of statement 2 as being executed atomically with the read of the shared variable $Y[i, j, h]$. A process p is *faulty* if, at some point in time, p is outside its remainder section and p never takes another step; p is *nonfaulty* otherwise.

Notational Conventions: We assume that $1 < k \leq M < N$, and that p and q range over $0, ..., N-1$. Other free variables are assumed to be universally quantified. We use $P^{x_1,x_2,...,x_n}_{y_1,y_2,...,y_n}$ to denote the expression P with each occurrence of x_i replaced by y_i. The predicate $p@s$ holds iff statement s is the next statement to be executed by process p. We use $p@S$ as shorthand for $(\exists s : s \in S :: p@s)$, $p.s$ to denote statement s of process p, and $p.var$ to denote p's local variable var. The following is a list of symbols we use in our proofs, in increasing order of binding power: $\equiv$, $\Rightarrow$, $\vee$, $\wedge$, $(=, \neq, <, >, \leq, \geq)$, $(+, -)$, (multiplication,/), $\neg$, $(., @)$, $(\{, \})$. Symbols in parentheses have the same binding power. We occasionally use parentheses to override these binding rules. We sometimes use Hoare triples [7] to denote the effects of a statement execution. To prove that an assertion I is an invariant, we show that I holds initially, and that no statement falsifies[5] I. When an invariant is in the form of an implication, we often achieve this by showing that the consequent holds after the execution of any statement that establishes the antecedent, and that no statement falsifies the consequent while the antecedent holds. □

In the *long-lived M-renaming* problem, each of N processes repeatedly executes a *remainder section*, acquires a name by executing a *getname section*, uses that name in a *working section*, and then releases the name by executing a *putname section*. It is assumed that each process is initially in its remainder section, and that the remainder section guarantees that at most k processes are outside the remainder section at any time. A solution to the long-lived M-renaming problem consists of wait-free code fragments (and associated shared variables) that implement the getname and putname sections. The getname section for process p is required to assign a value ranging over $\{0, ..., M-1\}$ to $p.name$. Distinct processes in their working sections are required to hold different names. More precisely, for the $(k(k+1)/2)$-renaming algorithm shown in Figure 2, we

[5] A statement execution *falsifies* an expression iff that expression holds before the statement execution and does not hold afterwards. Similarly, a statement execution *establishes* an expression iff that expression does not hold before the statement execution, but holds after.

assume (I1) and are required to prove (I2) and that our algorithm satisfies the wait-freedom property.

invariant $|\{p :: p@\{1..11\}| \leq k$ (I1)
invariant $(p \neq q \ \wedge\ p@9 \ \wedge\ q@9 \Rightarrow p.name \neq q.name) \wedge$
$(p@9 \Rightarrow 0 \leq p.name < k(k+1)/2)$ (I2)

Wait-Freedom: Every nonfaulty process that leaves line 0 eventually reaches line 9, and that every nonfaulty process that leaves line 9 eventually reaches line 0.

Our algorithm requires shared variables of approximately $\log_2 N$ bits. Thus, on a 32-bit shared-memory multiprocessor, these shared variables can be accessed with one shared variable access if $N < 2^{32}$. We measure the time complexity of our algorithms in terms of the worst-case number of steps taken to acquire and release a name once.

3 The Algorithm

We now present our algorithm for fast $(k(k+1)/2)$-renaming. Like the algorithm presented in [10], this algorithm is based on a "grid" of building blocks like the one shown in Figure 1. Each position in the grid has a unique name associated with it. In order to acquire a name, a process p starts at the top-left corner of the grid and accesses the building block there. The result of accessing the building block is "stop", "move right", or "move down". If p stops at a building block, then p acquires the name associated with that building block. It is guaranteed that, if one process stops at a building block, then no other process subsequently stops at that building block until the first process resets it. If p does not stop at a building block, then it moves down or right in the grid, according to the result of accessing the building block. The building blocks are designed to "spread" processes out in the grid so that thay can acquire distinct names. We later show that this spreading is sufficient that, if a process reaches the "edge" of the grid without stopping at any of the building blocks on its path, then it is guaranteed that no other process concurrently reaches the same grid position.

Having described the overall structure of the algorithm, we now concentrate on the implementation of the building block on which it is based. We begin with a description of the building block used in the non-fast algorithm presented in [10].

The long-lived building block used in [10] is based on a "resettable" version of Lamport's fast mutual exclusion technique [9]. To access the building block, a process p first writes its own identifier into a shared variable X, and then reads a shared variable Y. If Y is "set", then p moves right in the grid. Otherwise, p sets the Y-variable and then checks X. If X still contains p's identifier, then p stops at that building block. Otherwise, p "resets" the Y-variable and then moves down in the grid.

Notice that, if p does not stop at a building block (i.e., p moves down or right from that building block), then p detects the presence of another process in that building block, either because the other process has set the Y-variable, or because it writes X after p does. This is the key property that causes the building blocks to "spread" processes out in the grid. As a simple example, suppose that

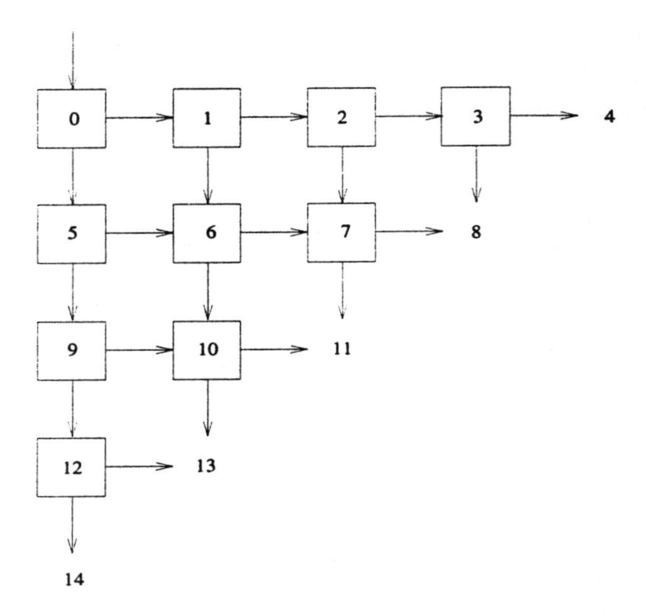

Fig. 1. The grid of building blocks, depicted for $k = 5$. Labels indicate the names associated with each grid position.

each of k processes requests a name once, and consider the building block at position $(0,0)$ in the grid (i.e., the top-left one). It is impossible that all k of these processes go down from building block $(0,0)$. To see this, note that, in order to go down from a building block, a process p must read $X \neq p$ in that building block. However, the last process q to write X reads $X = q$, and therefore does not go down. It is also impossible for all of these k processes to go right from $(0,0)$. To see this, observe that, in order to go right from a building block, a process must detect that some other process has set the Y-variable for that building block. However, as described above, a process that goes right from a building block does not set the Y-variable in that building block. The other building blocks in the grid behave the same way. Therefore, each time a group of processes access a building block in common, at least one of them "breaks away" from the group, either by stopping at that building block, or by going in another direction.

The building block used in our new algorithm follows the same basic structure as the one described above, and the intuition for the grid algorithm is the same. However, we use a new technique for implementing the Y-variable of each building block that results in our algorithm being fast, while the one presented in [10] is not. In the building block used in [10], the Y-variable consists of N bits — one for each process. This ensures that, if a process p sets the Y-variable (by setting $Y[p]$), then the Y-variable stays set until p resets it. This is important to ensure that two processes to not concurrently hold the name at the same building block. Unfortunately, this approach necessitates reading all N Y-bits in order to determine whether the Y-variable is set. This is why the algorithm presented in [10] is not fast. In the algorithm presented in this paper, we use at most $k + 1$ Y-components per building block, allowing the Y-variable to be read in $\Theta(k)$ time. The difficult part of our algorithm lies in ensuring that the following properties are not violated.

(i) If process p sets the Y-variable, then it stays set until p resets it.

(ii) If all processes that have set the Y-variable have since reset it, then the Y-variable is no longer set.

The latter property is important to ensure that a process does not go right from a building block unless some other process is still accessing that building block. If this property is violated, then it is possible for more than one process to reach the same building block at the edge of the grid, thereby acquiring the same name. We now describe our new building block in more detail, and explain why it ensures these properties.

To facilitate formal proofs, we have incorporated all of the building blocks, as well as the code for controlling movement through the grid, into a single algorithm (shown in Figure 2). However, the grid and building block structure should still be apparent. In lines 1 through 7, process p accesses building block (r, c), where $r = p.i$ and $c = p.j$. Building block (r, c) is made up of $X[r, c]$ and $Y[r, c, 0]$ through $Y[r, c, k - r - c]$. The two properties mentioned above are ensured through the use of a new technique for setting and resetting the components of Y. In the loop at lines 3 and 4, process p sets the Y-variable of building block $(p.i, p.j)$ by assigning p to every $Y[p.i, p.j, p.h]$, where $p.h$ ranges over 0 to $k-p.i-p.j$. Before setting each Y-component, p first checks $X[p.i, p.j]$. If $X[p.i, p.j] \neq p$, then p stops writing Y-components for this building block, resets those it has set (lines 5 and 6), and then moves down to the next building block (line 7).

Observe that each Y-component written by p could subsequently be overwritten by another process. Thus, there is a risk that the Y-variable does not stay set while p is accessing the building block. However, if p successfully writes all Y-components of a building block, then *some* component of Y stays set until p resets it. To see why this is so, note that, before p does its last write to Y in building block (r, c) (line 4), p first checks that $X[r, c]$ still contains p (line 3). Thus, $X[r, c]$ holds continuously while p writes all of the Y-components (with the possible exception of the last). If some other process q either resets one of the Y-components (line 6 or 11) or writes its own identifier (line 4) into one of the Y-components, then q previously either reads $X[r, c] \neq p$ or reads q from that Y-component. In either case, it must have done so before p wrote that Y-component. This implies that each process q can "corrupt" at most one of the Y-components p writes. In the next section, we inductively show that at most $k - r - c - 1$ processes other than p concurrently access building block (r, c). Thus, because p writes $k - r - c$ components of Y *before* its last check of $X[r, c]$, it is guaranteed that at least one of the components p sets remains set to p. Also, before process p leaves a building block (either to go to the next one, or because it releases its name), p clears all Y-components that contain p (lines 5 to 6 and lines 10 to 11). Thus, if all processes leave building block (r, c) then the Y-variable for that building block is no longer set. These properties capture the essence of the formal correctness proof, which is presented next.

shared variable X : **array**$[0..k-1, 0..k-1]$ **of** $\{\bot\} \cup \{0..N-1\}$;
Y : **array**$[0..k-1, 0..k-1, 0..k]$ **of** $\{\bot\} \cup \{0..N-1\}$
initially $(\forall r, c, n : 0 \leq r < k-1 \ \wedge\ 0 \leq c < k-1 \ \wedge\ r+c < k-1 \ \wedge$
$0 \leq n \leq k-r-c :: Y[r,c,n] = \bot)$

private variable *name*: $0..k(k+1)/2-1$; *moved*: **boolean**; i, j: $0..k-1$; h: $0..k$
initially $i = 0 \ \wedge\ j = 0 \ \wedge\ h = 0$

```
    while true do
0:    Remainder Section;
      i, j, moved := 0, 0, true;
      while i + j < k − 1 ∧ moved do
1:      X[i, j], h, moved := p, 0, false;
        while h < k − i − j ∧ ¬moved do
2:        if Y[i, j, h] ≠ ⊥ then
            j, h, moved := j + 1, 0, true
          else h := h + 1
          fi
        od;
        h := 0;
        while h ≤ k − i − j ∧ ¬moved do
3:        if X[i, j] = p then
4:          Y[i, j, h], h := p, h + 1
          else
            while h > 0 do
5:            h := h − 1;
              if Y[i, j, h] = p then
6:              Y[i, j, h] := ⊥
              fi
            od;
7:          moved, i := true, i + 1
          fi
        od
      od;
8:    name := ik − i(i − 1)/2 + j;
9:    Working;
      if ¬moved then
        while h > 0 do
10:       h := h − 1;
          if Y[i, j, h] = p then
11:         Y[i, j, h] := ⊥
          fi
        od
      fi
    od
```

Fig. 2. Long-lived renaming with $\Theta(k^2)$ name space, $\Theta(k^2)$ time complexity, and $\Theta(k^3)$ space complexity. Code is given for process p, where $0 \leq p < N$.

4 Correctness Proof

The proofs of the following invariants are straightforward, and are therefore omitted. In particular, (I3) through (I9) follow directly from the program text, (I10) is proved using (I4), (I11) is proved using (I9), and (I8) is proved using (I11). Finally, the proof of (I12) uses (I11), and the proof of (I13) uses (I12).

invariant $p@\{0..11\}$ (I3)
invariant $p@\{5, 10\} \Rightarrow p.h > 0$ (I4)
invariant $p@9 \Rightarrow p.name = (p.i)k - (p.i)(p.i-1)/2 + p.j$ (I5)
invariant $(\forall r, c, n : 0 \leq r < k-1 \ \wedge\ 0 \leq c < k-1 \ \wedge\ r+c < k-1 \ \wedge$

$$0 \leq n \leq k - r - c :: Y[r, c, n] \in \{\bot\} \cup \{0..N-1\}) \qquad \text{(I6)}$$

invariant $p@\{2..7\} \Rightarrow \neg p.moved$ (I7)

invariant $p@2 \Rightarrow p.h < k - p.i - p.j$ (I8)

invariant $p@\{1..7\} \Rightarrow p.i + p.j < k - 1$ (I9)

invariant $p.i \geq 0 \ \wedge \ p.j \geq 0 \ \wedge \ p.h \geq 0$ (I10)

invariant $p.i + p.j \leq k - 1$ (I11)

invariant $p@\{3..7\} \Rightarrow p.h \leq k - p.i - p.j$ (I12)

invariant $p@\{8..11\} \Rightarrow p.h \leq k - p.i - p.j + 1$ (I13)

The following invariant shows that if the nth Y component in building block (r, c) is set, then some process is accessing building block (r, c) at, or beyond, component n.

invariant $(r \geq 0 \ \wedge \ c \geq 0 \ \wedge \ r + c < k - 1 \ \wedge \ n \geq 0 \ \wedge \ n < k - r - c \ \wedge$
$Y[r, c, n] = q) \ \Rightarrow \ (q.i = r \ \wedge \ q.j = c \ \wedge \ \neg q.moved \ \wedge$
$((q@\{3..6, 8..11\} \ \wedge \ q.h > n) \ \vee \ (q@\{6, 11\} \ \wedge \ q.h = n)))$ (I14)

Proof: Assume $r \geq 0 \ \wedge \ c \geq 0 \ \wedge \ r + c < k - 1 \ \wedge \ n \geq 0 \ \wedge \ n < k - r - c$. This implies that $Y[r, c, n] = \bot$ holds initially, so (I14) holds initially. Only statement $q.4$ can establish the antecedent, and it does so only if executed when $q@4 \ \wedge \ q.i = r \ \wedge \ q.j = c \ \wedge \ q.h = n$ holds. By (I7), $\neg q.moved$ also holds in this case. Therefore, because $n < k - r - c$, $q.4$ establishes $q.i = r \ \wedge \ q.j = c \ \wedge \ \neg q.moved \ \wedge \ q@3 \ \wedge \ q.h > n$, thereby establishing the consequent.

No statement modifies $q.i$, $q.j$, or $q.moved$ while the consequent holds. It remains to consider statements that might falsify $((q@\{3..6, 8..11\} \ \wedge \ q.h > n) \ \vee \ (q@\{6, 11\} \ \wedge \ q.h = n))$ while the consequent holds. First, observe that any statement that falsifies the second disjunct while the consequent holds also falsifies the antecedent.

If the first disjunct holds, then $q.h > 0$. Therefore, statement $q.3$ establishes $q@\{4, 5\}$, and does not modify $q.h$. By (I7), statement $q.4$ establishes $q@3$ or $q@8$ and increases $q.h$, so $q.4$ does not falsify first disjunct. If $q.h > n + 1$, then $q@\{5, 6\} \ \wedge \ q.h > n$ holds after $q.5$ is executed. If $q.h = n + 1$, then because $Y[r, c, n] = q$, statement $q.5$ establishes $q@6 \ \wedge \ q.h = n$, thereby establishing the second disjunct listed above. Similarly, $(q@\{10..11\} \wedge q.h > n) \vee (q@11 \wedge q.h = n)$ holds after $q.10$ is executed. Because $q.h > n$ and $n \geq 0$, statement $q.6$ establishes $q@5 \wedge q.h > n$. Similarly, statement $q.11$ establishes $q@10 \wedge q.h > n$. Finally, statements $q.8$ and $q.9$ do not modify $q.h$, $q.8$ establishes $q@9$, and because $\neg q.moved$ and $q.h > n$ holds, statement $q.9$ establishes $q@10$. Thus, no statement falsifies the consequent while the antecedent holds. □

Definitions. For convenience, we define the following predicates. Intuitively, $MOD(q, r, c, n)$ holds if process q is about to modify $Y[r, c, n]$; $SET(q, r, c, n)$ holds if process q has just set $Y[r, c, n]$ to q and has not yet reset it; and $EN(q, r, c)$ holds if process q will access (or has already accessed) a building in the subgrid whose top-left corner is at (r, c).

$$\begin{aligned}
MOD(q, r, c, n) &\equiv q.i = r \ \wedge \ q.j = c \ \wedge \ q.h = n \ \wedge \ q@\{4, 6, 11\} \\
SET(q, r, c, n) &\equiv q.i = r \ \wedge \ q.j = c \ \wedge \\
&\quad ((q@\{3, 5, 8..10\} \ \wedge \ q.h = n + 1) \ \vee \ (q@\{6, 11\} \ \wedge \ q.h = n)) \\
EN(q, r, c) &\equiv q.j \geq c \ \wedge \ ((q.i \geq r \ \wedge \ q@\{1..11\}) \ \vee \ (q.i = r - 1 \ \wedge \\
&\quad ((q@\{2..4\} \ \wedge \ X[r - 1, q.j] \neq q) \ \vee \ q@\{5..7\})))
\end{aligned}$$
□

The following three invariants follow easily from the definitions above. The proof of (I16) uses (I15).

invariant $(SET(p,r,c,n) \vee MOD(p,r,c,n)) \Rightarrow$
$$(\forall m : m \neq n :: \neg SET(p,r,c,m) \wedge \neg MOD(p,r,c,m)) \quad \text{(I15)}$$

invariant $p@2 \wedge EN(p,r,c) \Rightarrow |\{n :: (\exists q :: SET(q,r,c,n) \vee$
$$MOD(q,r,c,n))\}| < |\{q :: EN(q,r,c)\}| \quad \text{(I16)}$$

invariant $EN(p,r,c) \Rightarrow EN(p,r,c-1) \wedge EN(p,r-1,c)$ (I17)

The next invariant implies that at most $k - r - c$ processes access building blocks in the sub-grid whose top left corner is at building block (r, c). In particular, this implies that at most one process at a time occupies each grid position that is $k - 1$ steps from the position origin.

invariant $r \geq 0 \wedge c \geq 0 \wedge r + c \leq k - 1 \Rightarrow$
$$(|\{p :: EN(p,r,c)\}| \leq k - r - c) \quad \text{(I18)}$$

Proof: Initially, $(\forall p :: p@0)$ holds, so (I18) holds. First, observe that (I1) implies that if $r = 0 \wedge c = 0$, then (I18) holds. Henceforth, assume that $r \geq 0 \wedge c \geq 0 \wedge r + c \leq k - 1 \wedge r + c > 0$. (I18) can be falsified only by establishing $EN(q,r,c)$ for some process q. By the definition of EN, this can be achieved only by modifying $q.i$, $q.j$ or $X[r-1, q.j]$, or by establishing $q@\{1..11\}$ or $q@\{2..4\}$ or $q@\{5..7\}$. The statements to check are therefore $q.0$, $q.2$, $q.3$, $q.7$, and $p.1$, where p is any process.

Because $r + c > 0$, statement $q.0$ does not establish $EN(q,r,c)$. Statement $q.3$ potentially establishes $EN(q,r,c)$ only by establishing $q@\{5..7\}$. Thus, $EN(q,r,c)$ holds after $q.3$ is executed only if $q.j \geq c \wedge q.i = r - 1 \wedge q@3 \wedge X[q.i, q.j] \neq q$ holds before, in which case $EN(q,r,c)$ already holds. Statement $q.7$ establishes $q@\{1,8\}$ and could therefore establish $EN(q,r,c)$ only by establishing $q.j \geq c \wedge q.i \geq r$. However, it does so only if executed when $q.j \geq c \wedge q.i = r - 1 \wedge q@7$ holds, in which case $EN(q,r,c)$ already holds. It remains to consider statement $p.1$, where p is any process, and statement $q.2$.

If $p = q \wedge q.i \geq r$, then $p.1$ does not establish $EN(q,r,c)$ because it does not modify $q.i$ or $q.j$ or establish $q@\{1..11\}$. If $p = q \wedge q.i < r$, then by (I10), statement $p.1$ establishes $q@2 \wedge (q.i \neq r - 1 \vee X[r-1, q.j] = q)$ and therefore does not establish $EN(q,r,c)$. If $p \neq q$, then statement $p.1$ can establish $EN(q,r,c)$ only by establishing $X[r-1, q.j] \neq q$ while $q.j \geq c \wedge q.i = r - 1 \wedge q@\{2..4\} \wedge X[r-1, q.j] = q$ holds. However, it does so only if executed when $p@1 \wedge p.i = r - 1 \wedge p.j \geq c$. These assertions imply $EN(q, r-1, c) \wedge \neg EN(q,r,c) \wedge EN(p, r-1, c) \wedge \neg EN(p,r,c)$. Also, $q.i = r - 1 \wedge$ (I10) implies that $r - 1 \geq 0$ and $r + c \leq k - 1$ implies that $r - 1 + c \leq k - 1$. Therefore, because $(\text{I18})^{r,c}_{r-1,c}$ holds before $p.1$ is executed, it follows that $|\{p :: EN(p, r-1, c)\}| \leq k - r - c + 1$ holds before $p.1$ is executed. By (I17), this implies that $|\{p :: EN(p,r,c)\}| \leq k - r - c - 1$ holds before $p.1$ is executed (because $p \neq q$ and $EN(q, r-1, c) \wedge \neg EN(q,r,c) \wedge EN(p, r-1, c) \wedge \neg EN(p,r,c)$ holds), so $p.1$ does not falsify (I18).

Statement $q.2$ can establish $EN(q,r,c)$ only if executed when $q@2 \wedge q.i \geq r \wedge q.j = c - 1 \wedge Y[q.i, c-1, q.h] \neq \bot$ holds. By (I10) and (I8), this implies that $q.i \geq 0 \wedge c - 1 \geq 0 \wedge q.h \geq 0 \wedge q.h < k - q.i - (c - 1)$ holds before $q.2$ is executed. Thus, by (I6), (I9), and $(I14)^{r,c,n,q}_{q.i,c-1,q.h,s}$, it follows that $(\exists s : s \neq q :: EN(s,r,c-1) \wedge \neg EN(s,r,c))$ holds. (Note that $q@2 \wedge s@\{3..6,8..11\}$ implies that $s \neq q$.) Also, $EN(q,r,c-1) \wedge \neg EN(q,r,c)$ holds. Thus, as above, (I10), (I17), and $(I18)^{r,c}_{r,c-1}$ imply that $|\{p :: EN(p,r,c)\}| \leq k-r-c-1$ holds before $q.2$ is executed, so $q.2$ does not falsify (I18). □

The following invariant implies that, while process p is executing the loop at line 2 and $X[p.i,p.j]=p$ still holds, for each component of Y that p has already read, either that component is not set, or some process has just written that component and has not yet cleared it.

invariant $(r \geq 0 \wedge c \geq 0 \wedge r+c < k-1 \wedge p@2 \wedge p.i = r \wedge p.j = c \wedge X[r,c] = p) \Rightarrow (\forall n : 0 \leq n < p.h :: Y[r,c,n] = \perp \vee (\exists q : q \neq p :: Y[r,c,n] = q \wedge \neg q.moved \wedge SET(q,r,c,n)))$ (I19)

Proof: Assume $r \geq 0 \wedge c \geq 0 \wedge r+c < k-1$. Initially, $p@0$ holds, so (I19) holds. Only statement $p.1$ establishes $p@2$. Statement $p.1$ also establishes $p.h = 0$, so the consequent holds vacuously after $p.1$ is executed. No statement modifies $p.i$ while $p@2$ holds, and only statement $p.2$ modifies $p.j$ while $p@2$ holds. However, if $p.2$ modifies $p.j$, then it also establishes $p.moved$ and terminates the loop, thereby falsifying the antecedent. Only statement $p.1$ can establish $X[r,c]=p$, and as shown above, the consequent holds after the execution of $p.1$.

Statements of process p other than $p.2$ are not enabled while the antecedent holds. Statement $p.2$ can affect the consequent only by increasing $p.h$. However, $p.2$ does not falsify the consequent in this case, because it increments $p.h$ only if $Y[r,c,p.h] = \perp$. We now consider steps of process s, where $s \neq p$. Statements of process s can falsify the consequent only by modifying $Y[r,c,n]$ or by falsifying $\neg s.moved \wedge SET(s,r,c,n)$ for some $n < p.h$ while $Y[r,c,n] = s$ holds.

Only statements $s.4$, $s.6$, and $s.11$ modify $Y[r,c,n]$. Statements $s.6$ and $s.11$ establish $Y[s.i,s.j,s.h] = \perp$ and therefore do not falsify the consequent. Statement $s.4$ modifies $Y[r,c,n]$ only if $s@4 \wedge s.i = r \wedge s.j = c \wedge s.h = n$. By (I7), $\neg s.moved$ holds in this case. Therefore, after $s.4$ is executed in this case, $Y[r,c,n] = s \wedge s.i = r \wedge s.j = c \wedge \neg s.moved \wedge s@\{3,8\} \wedge s.h = n+1$ holds, which implies $SET(s,r,c,n)$. Thus, $s.4$ does not falsify the consequent by modifying $Y[r,c,n]$.

We now consider statements that potentially falsify $\neg s.moved \wedge SET(s,r,c,n)$ for some process s and for some $n < p.h$ while $Y[r,c,n] = s$ holds. No statement modifies $s.i$, $s.j$, or $s.moved$ while $SET(s,r,c,n)$ holds. It remains to consider statements that potentially falsify $((s@\{3,5,8..10\} \wedge s.h = n+1) \vee (s@\{6,11\} \wedge s.h = n))$ while $Y[r,c,n] = s \wedge \neg s.moved \wedge SET(s,r,c,n)$ holds. First, observe that if $s.6$ or $s.11$ falsifies the second disjunct while the antecedent and consequent both hold, then it also establishes $Y[r,c,n] = \perp$, thereby preserving the consequent. If the first disjunct holds, then $s.h > 0$. Thus, because the antecedent implies that $X[r,c] \neq s$ (recall that $s \neq p$), statement $s.3$ establishes $s@5$ and does not modify $s.h$. Also, because $Y[r,c,n] = s$, executing $s.5$ establishes $s@6 \wedge s.h = n$, thereby establishing the second disjunct above. Similarly, statement $s.10$ establishes $s@11 \wedge s.h = n$ if executed while the antecedent and consequent both hold. Finally, statements $s.8$ and $s.9$ do not modify $s.h$, $s.8$ establishes $s@9$, and because $\neg s.moved$ and $s.h = n+1$ holds, statement $s.9$ establishes $s@10$. □

The following invariant implies that while process p is executing the loop at lines 3 to 4 and $X[p.i, p.j] = p$ still holds, one of the Y-components that p has set is not overwritten by any other process.

$$\textbf{invariant } r \geq 0 \land c \geq 0 \land r+c < k-1 \land p@\{3..4\} \land p.i = r \land p.j = c \land X[r,c] = p \Rightarrow (\exists n :: (\forall q : q \neq p :: \neg MOD(q,r,c,n)) \land ((Y[r,c,n] = p \land 0 \leq n \land n < p.h) \lor (Y[r,c,n] = \bot \land p.h \leq n \land n < k-r-c)) \quad \text{(I20)}$$

Proof: Assume $r \geq 0 \land c \geq 0 \land r+c < k-1$. Initially, $p@0$ holds, so (I20) holds. Observe that no statement modifies $p.i$ or $p.j$ or establishes $X[r,c] = p$ while $p@\{3..4\}$ holds. Thus, only statements that establish $p@\{3..4\}$ can establish the antecedent. By (I9), statement $p.1$ establishes $p.h < k - p.i - p.j \land \neg p.moved$, so $p.1$ does not establish $p@\{3..4\}$. After statement $p.5$ or statement $p.6$ is executed, $p@\{5..7\}$ holds. Statement $p.7$ establishes $p.moved$, thereby terminating the loop and establishing $p@\{1,8\}$. The following assertions imply that if statement $p.2$ establishes the antecedent, then the consequent holds afterwards.

$$\{p@2 \land (p.i \neq r \lor p.j \neq c \lor Y[p.i,p.j,p.h] \neq \bot \lor X[r,c] \neq p)\}\ p.2\ \{(p.i \neq r \lor p.j \neq c) \lor (p.moved \land p@\{1,8\}) \lor X[r,c] \neq p\}$$

, $p.2$ does not modify $p.i$ or X; if $p.2$ modifies $p.j$ or if $Y[p.i,p.j,p.h] \neq \bot$, then $p.2$ establishes $p.moved \land p@\{1,8\}$.

$$p@2 \land p.i = r \land p.j = c \land Y[p.i,p.j,p.h] = \bot \land X[r,c] = p \land p.h+1 > k-r-c \Rightarrow \mathit{false} \quad \text{, by (I8).}$$

$$\{p@2 \land p.i = r \land p.j = c \land Y[p.i,p.j,p.h] = \bot \land X[r,c] = p \land p.h+1 < k-r-c\}\ p.2\ \{p@2\}$$

, by (I7), the precondition implies $\neg p.moved \land p.h+1 < k - p.i - p.j$, so the loop does not terminate.

$$p@2 \land p.i = r \land p.j = c \land Y[p.i,p.j,p.h] = \bot \land X[r,c] = p \land p.h+1 = k-r-c$$

$$\Rightarrow p@2 \land p.i = r \land p.j = c \land Y[r,c,p.h] = \bot \land X[r,c] = p \land p.h+1 = k-r-c \land |\{n :: (\exists q :: SET(q,r,c,n) \lor MOD(q,r,c,n))\}| < |\{q :: EN(q,r,c)\}| \quad \text{, by (I16) and the definition of } EN(p,r,c).$$

$$\Rightarrow p@2 \land p.i = r \land p.j = c \land Y[r,c,p.h] = \bot \land X[r,c] = p \land p.h+1 = k-r-c \land |\{n :: (\exists q :: SET(q,r,c,n) \lor MOD(q,r,c,n))\}| < k-r-c \quad \text{, by (I18).}$$

$$\Rightarrow p@2 \land p.i = r \land p.j = c \land Y[r,c,p.h] = \bot \land X[r,c] = p \land p.h+1 = k-r-c \land (\exists n : 0 \leq n < k-r-c :: (\forall q :: \neg SET(q,r,c,n) \land \neg MOD(q,r,c,n)))\ \text{, by the pigeonhole principle.}$$

$$\Rightarrow p@2 \land p.i = r \land p.j = c \land Y[r,c,p.h] = \bot \land X[r,c] = p \land$$

$$p.h+1=k-r-c \ \wedge \ (\exists n : 0 \leq n < k-r-c :: Y[r,c,n] = \bot \ \wedge \\ (\forall q :: \neg SET(q,r,c,n) \ \wedge \ \neg MOD(q,r,c,n))) \text{ , by (I19).}$$

$$\{p@2 \ \wedge \ p.i = r \ \wedge \ p.j = c \ \wedge \ Y[r,c,p.h] = \bot \ \wedge \ X[r,c] = p \ \wedge \\ p.h+1 = k-r-c \wedge \text{(I16)} \wedge \text{(I18)} \wedge \text{(I19)}\} \ p.2$$
$$\{(\exists n :: Y[r,c,n] = \bot \ \wedge \ p.h \leq n \ \wedge \ n < k-r-c \ \wedge \\ (\forall q : q \neq p :: \neg MOD(q,r,c,n)))\}$$

, by the preceding derivation and the program text; note that $p.2$ establishes $p.h = 0$ in this case, and does not modify Y or establish $MOD(q,r,c,n)$ for any q. Also observe that the postcondition implies the consequent of (I20). □

The following invariant implies that, if process p is in its working section while occupying an interior building block (r,c), then no other process is in its working section at that building block.

invariant $r \geq 0 \ \wedge \ c \geq 0 \ \wedge \ r+c < k-1 \ \wedge \ p.i = r \ \wedge \ p.j = c \ \wedge$
$$((p@4 \ \wedge \ p.h \geq k-r-c) \ \vee \ p@\{8..9\}) \Rightarrow \\ (\exists n : 0 \leq n < k-r-c :: Y[r,c,n] = p \ \wedge \ (\forall q : q \neq p :: q.i \neq p.i \ \vee \ q.j \neq p.j \\ \vee \ q@\{0..1\} \ \vee \ (q@2 \ \wedge \ q.h \leq n) \ \vee \ ((q@\{2..3\} \ \vee \\ (q@4 \ \wedge \ q.h < k-r-c \ \wedge \ q.h \neq n)) \ \wedge \ X[r,c] \neq q) \ \vee \\ q@\{5,7,10\} \ \vee \ (q@\{6,11\} \ \wedge \ q.h \neq n))) \qquad \text{(I21)}$$

Proof: Assume that $r \geq 0 \ \wedge \ c \geq 0 \ \wedge \ r+c < k-1$. Initially, $p@0$ holds, so (I21) holds. The antecedent of (I21) is only established by modifying $p.i$, $p.j$, or $p.h$, or by establishing $p@4$ or $p@\{8..9\}$. Therefore, statements $p.6$, $p.8$, $p.9$, and $p.11$ do not establish the antecedent. After statements $p.0$, $p.1$, $p.5$, and $p.10$, the antecedent does not hold because $\neg p@\{4,8..9\}$ holds. ((I7) implies that $p.1$ establishes $p@2$.) Statement $p.7$ establishes $p@1 \ \vee \ (p@8 \ \wedge \ p.i+p.j \geq k-1)$, and therefore does not establish the antecedent.

If $p@2 \ \wedge \ Y[p.i,p.j,p.h] = \bot$ holds before $p.2$ is executed, then, by (I7) and (I8), $p@\{2,3\}$ holds afterwards. If $p@2 \ \wedge \ Y[p.i,p.j,p.h] \neq \bot$ holds before $p.2$ is executed, then $p@1 \ \vee \ (p@8 \ \wedge \ p.i+p.j \geq k-1)$ holds afterwards. Thus, $p.2$ does not establish the antecedent. The remaining statements to check are $p.3$ and $p.4$.

By (I7) and (I12), $\neg p.moved \ \wedge \ p.h \leq k-p.i-p.j$ holds before statement $p.4$ is executed. Therefore, the antecedent holds after $p.4$ is executed only if $p.i = r \ \wedge \ p.j = c \ \wedge \ p@4 \ \wedge \ p.h = k-p.i-p.j$ holds before, in which case the antecedent already holds.

Statement $p.3$ establishes the antecedent of (I21) only if executed when $p@3 \ \wedge$ $p.i = r \ \wedge \ p.j = c \ \wedge \ X[p.i,p.j] = p \ \wedge \ p.h \geq k-r-c$ holds. By (I12) and $\text{(I20)}^{r,c}_{p.i,p.j}$, this implies that the following assertion holds before $p.3$ is executed.

$$X[p.i,p.j] = p \ \wedge \ (\exists n : 0 \leq n < k-r-c :: Y[r,c,n] = p \ \wedge \\ (\forall q : q \neq p :: X[p.i,p.j] \neq q \ \wedge \ \neg MOD(q,p.i,p.j,n))) \qquad \text{(A1)}$$

(A1) $\wedge$ (I3) $\wedge$ $\text{(I21)}^{p,q,r,c}_{q,p,q.i,q.j}$ implies the consequent of (I21). To see this, suppose that some $q \neq p$ does not satisfy the universal quantifier in the consequent of (I21). Then (A1) $\wedge$ (I3) implies that $q.i = p.i \ \wedge \ q.j = p.j \ \wedge \ ((q@4 \ \wedge \ q.h \geq$ $k-r-c) \ \vee \ q@\{8..9\})$ holds (because $X[p.i,p.j] \neq q \ \wedge \ \neg MOD(q,p.i,p.j,n)$

holds). In this case, $(\mathrm{I21})^{p,q,r,c}_{q,p,q.i,q.j} \wedge p@3$ implies $X[p.i, p.j] \neq p$, a contradiction. Statement $p.3$ does not falsify the consequent, so the consequent holds after $p.3$ is executed in this case.

We now consider statements that potentially falsify the consequent while the antecedent holds. First, observe that no statement modifies $p.i$, $p.j$, or $Y[r, c, n]$ while the antecedent and consequent both hold. Only statements $q.0$, $q.2$, and $q.7$ potentially falsify the consequent by modifying $q.i$ or $q.j$ for some process q. However, if any of these statements modifies $q.i$ or $q.j$, then $q@1 \vee (q@8 \wedge q.i + q.j \geq k - 1)$ holds afterwards. The latter disjunct implies that $q.i \neq p.i \vee q.j \neq p.j$ holds because the antecedent implies that $p.i + p.j < k - 1$. Thus, no statement that modifies $q.i$ or $q.j$ for any process q falsifies the consequent while the antecedent holds. It therefore remains to consider statements that falsify (A2) below, for some $q \neq p$, while $p.i = r \wedge p.j = c \wedge 0 \leq n < k - r - c \wedge Y[r, c, n] = p \wedge q.i = p.i \wedge q.j = p.j$ holds.

$$q@\{0..1\} \vee (q@2 \wedge q.h \leq n) \vee ((q@\{2..3\} \vee (q@4 \wedge q.h < k - r - c \wedge q.h \neq n)) \wedge X[r, c] \neq q) \vee q@\{5, 7, 10\} \vee (q@\{6, 11\} \wedge q.h \neq n) \qquad \text{(A2)}$$

Only statement $q.1$ falsifies $q@\{0..1\}$, and $q.1$ establishes $q@2 \wedge q.h \leq n$.

Only statements $q.2$ and $q.3$ falsify $q@2 \wedge q.h \leq n$ or $q@\{2..3\}$. As shown above, if $q.2$ modifies $q.j$, then it does not falsify the consequent while the antecedent holds. Also, statement $q.2$ does not falsify $q@2 \wedge q.h \leq n$ by incrementing $q.h$ because $Y[q.i, q.j, n] \neq q$. By (I7) and (I8), statement $q.2$ does not falsify $q@\{2..3\}$ if it increments $q.h$. If statement $q.3$ falsifies $q@\{2..3\}$ while $X[r, c] \neq q$ holds, then $q.3$ establishes $q@\{5, 7\}$. (Recall that $p.i = c \wedge p.j = r \wedge q.i = p.i \wedge q.j = p.j$.)

If statement $q.4$ falsifies $q@4 \wedge q.h < k - r - c \wedge q.h \neq n$, then by (I7), $q.4$ establishes $q@3$. No statement falsifies $X[r, c] \neq q$ while $q@\{2..4\}$ holds. Because $q.i = r \wedge q.j = c \wedge Y[r, c, n] \neq q$ holds, $(q@6 \wedge q.h \neq n) \vee q@\{5, 7\}$ holds after the execution of $q.5$. As shown above, statement $q.7$ does not falsify the consequent while the antecedent holds. Statement $q.10$ establishes $q@\{0, 10\} \vee (q@11 \wedge q.h \neq n)$ (because $q.i = r \wedge q.j = c \wedge Y[r, c, n] = p$). Finally, statement $q.6$ establishes $q@\{5, 7\}$ and statement $q.11$ establishes $q@\{0, 10\}$. Thus, the consequent is not falsified while the antecedent holds. □

The following invariant implies that, if two distinct processes are in their working sections concurrently, then they occupy different grid positions. We ues this fact later to show that distinct processes do not hold the same name concurrently.

invariant $p \neq q \wedge p@\{8..9\} \wedge q@\{8..9\} \Rightarrow p.i \neq q.i \vee q.j \neq q.j$ (I22)

Proof: If $p \neq q \wedge p@\{8..9\} \wedge p.i + p.j < k - 1$ holds, then by (I10) and $(\mathrm{I21})^{r,c}_{p.i,p.j}$, $q.i \neq p.i \vee q.j \neq p.j \vee \neg q@\{8..9\}$ holds, so (I22) holds. If $p \neq q \wedge p@\{8..9\} \wedge p.i + p.j \geq k - 1$ holds, then by (I10), (I11), and $(\mathrm{I18})^{r,c}_{p.i,p.j}$, it follows that $|\{s :: EN(s, p.i, p.j)\}| \leq 1$. By the definition of EN, the antecedent implies $EN(p, p.i, p.j)$. Therefore $\neg EN(q, p.i, p.j)$ holds, which implies that the consequent holds. □

The following claims are proved in [10].

Claim 1: For nonnegative integers c, d, c', and d' satisfying $(c \neq c' \ \vee \ d \neq d') \wedge (c+d \leq k-1) \wedge (c'+d' \leq k-1)$, $ck-c(c-1)/2+d \neq c'k-c'(c'-1)/2+d'$. □

Claim 2: For nonnegative integers c and d satisfying $c+d \leq k-1$, $0 \leq ck - c(c-1)/2+d < k(k+1)/2$. □

The next two invariants show that distinct processes in their working sections hold distinct names from $\{0, ..., k(k+1)/2-1\}$. The first follows easily from (I5), (I10), (I11), (I22), and Claim 1. The second is easily proved using (I5), (I10), (I11), and Claim 2.

invariant $p \neq q \ \wedge \ p@9 \ \wedge \ q@9 \Rightarrow p.name \neq q.name$ (I23)

invariant $p@9 \Rightarrow 0 \leq p.name < k(k+1)/2$ (I24)

(I23) and (I24) imply (I2). To see that the wait-freedom requirement is satisfied, we consider all the loops in the algorithm in Figure 2. By (I10), the loop at line 2 clearly terminates after at most k iterations. Similarly, by (I10), (I12), and (I13), the loop at lines 5 and 6 and the loop at lines 10 to 11 both terminate after at most $k+1$ iterations. Also, note that if the loop at lines 5 to 6 is executed, then statement 7 establishes $p.moved$, so the loop at lines 3 to 7 terminates. Thus, the loop at lines 5 to 6 is executed at most once per execution of the loop at lines 3 to 7. To see that the loop at lines 3 to 7 terminates, consider statement $p.3$. If $X[p.i, p.j] \neq p$ holds before statement $p.3$ is executed, then statement $p.7$ establishes $p.moved$, so the loop terminates. Otherwise, $p.h$ is incremented when statement $p.4$ is executed. Because of the loop condition $p.h \leq k - p.i - p.j$, by (I10), the loop at lines 3 to 7 is executed at most k times. Finally, the loop at lines 1 to 7 executes at most $k-1$ times. To see this, observe that, by (I7), the loop terminates unless some statement establishes $p.moved$. Only statements $p.2$ and $p.7$ establish $p.moved$, and if they do so, they increment either $p.i$ or $p.j$. Thus, because of the loop condition $p.i+p.j < k-1$, (I10) implies that the loop terminates after at most $k-1$ executions. Thus, we have the following result.

Theorem 1: Using *read* and *write*, wait-free, long-lived $(k(k+1)/2)$-renaming can be implemented so that the worst-case time complexity of acquiring and releasing a name once is $\Theta(k^2)$, and the space complexity is $\Theta(k^3)$. □

A fast, long-lived renaming algorithm that yields a name space whose size is independent of N can be combined with any long-lived renaming algorithm — fast or not — to further reduce the size of the name space. This is achieved by each process first accessing the fast, long-lived renaming algorithm to acquire a name, and then using that name as its process identifier in another long-lived renaming algorithm. In particular, by combining our fast, long-lived renaming algorithm with the (non-fast) ℓ-assignment algorithm (with $\ell = 2k-1$) of Burns and Peterson [6], fast, long-lived renaming can be achieved with a name space of size $2k-1$. As explained in Section 1, if at most k processes concurrently access Burns and Peterson's algorithm, then their algorithm is wait-free. The worst-case time complexity of acquiring and releasing a name once is $\Theta(Nk^2)$ [11]. Thus, we have the following result, which is optimal with respect to the size of the name space.

Theorem 2: Using *read* and *write*, wait-free, long-lived $(2k-1)$-renaming can be implemented so that the worst-case time complexity of acquiring and releasing a name once is $\Theta(k^4)$, and the space complexity is $\Theta(k^4)$. □

5 Concluding Remarks

We have presented a new algorithm for long-lived $(k(k+1)/2)$-renaming algorithm that is fast and uses only read and write operations. This algorithm improves on the time and space complexity of similar previous algorithms, while providing a significantly simpler solution. An interesting property of this algorithm is that its time complexity is proportional to contention. That is, if at most $c \leq k$ processes access it concurrently, then the worst-case time complexity is $O(ck)$. We also show for the first time that fast, long-lived $(2k-1)$-renaming can be implemented with reads and writes. This algorithm is optimal with respect to the size of the name space. This resolves questions left open by [5] and [10]. While the latter algorithm is technically "fast" it still has quite high time complexity. It would be interesting to see if this can be improved upon by a more direct solution, as combining renaming algorithms tends to result in high time complexity.

Acknowledgements: This work was partly carried out while Mark Moir was visiting IBM's T.J. Watson Research Center. Mark wishes to thank IBM for its hospitality and Jim Anderson for his support. We are also grateful to Hagit Attiya and Gary Peterson for their help, and to Jaap-Henk Hoepman for pointing out an omission in an earlier version of the paper.

References

1. J. Anderson and M. Moir, "Fast k-Exclusion Algorithms", submitted to Distributed Computing. Preliminary version appeared in *Proceedings of the 13th Annual ACM Symposium on Principles of Distributed Computing*, August 1994, pp. 141-150.
2. H. Attiya, A. Bar-Noy, D. Dolev, D. Koller, D. Peleg, and R. Reischuk, "Achievable Cases in an Asynchronous Environment", *Proceedings of the 28th Annual IEEE Symposium on Foundations of Computer Science*, October 1987, pp. 337-346.
3. A. Bar-Noy and D. Dolev, "Shared Memory versus Message-Passing in an Asynchronous Distributed Environment", *Proceedings of the 8th Annual ACM Symposium on Principles of Distributed Computing*, August 1989, pp. 307-318.
4. E. Borowsky and E. Gafni, "Immediate Atomic Snapshots and Fast Renaming", *Proceedings of the 12th Annual ACM Symposium on Principles of Distributed Computing*, August 1993, pp. 41-50.
5. H. Buhrman, J. Garay, J. Hoepman, and M. Moir, "Long-Lived Renaming Made Fast", *Proceedings of the 14th Annual ACM Symposium on Principles of Distributed Computing*, August 1995, pp. 194-203.
6. J. Burns and G. Peterson, "The Ambiguity of Choosing", *Proceedings of the Eighth Annual ACM Symposium on Principles of Distributed Computing*, ACM, New York, August 1989, pp. 145-157.
7. C. A. R. Hoare, "An Axiomatic Basis for Computer Programming", *Communications of the ACM* 12, October 1969, pp. 576-580,583.

8. M. Herlihy and N. Shavit, "The Asynchronous Computability Theorem for t-Resilient Tasks", *Proceedings of the 25th ACM Symposium on Theory of Computing*, 1993, pp. 111-120.
9. L. Lamport, "A Fast Mutual Exclusion Algorithm", *ACM Transactions on Computer Systems*, Vol. 5, No. 1, February 1987, pp. 1-11.
10. M. Moir and J. Anderson, "Wait-Free Algorithms for Fast, Long-Lived Renaming", *Science of Computer Programming* 25 (1995), pp. 1-39. Preliminary version appeared in *Proceedings of the 8th International Workshop on Distributed Algorithms*, September, 1994, pp. 141-155.
11. G. Peterson, personal communication, November 1995.

A Timestamp Based Transformation of Self-Stabilizing Programs for Distributed Computing Environments

Masaaki Mizuno[1]* and Hirotsugu Kakugawa[2]**

[1] Dept. of Comp. and Info. Sci. Kansas State University, Manhattan, KS 66506
[2] Dept. of Elect. Engineering, Hiroshima University, Higashi-Hiroshima, Japan

Abstract. There are several models for which self-stabilizing (SS) programs have been developed. The distributed model accurately reflects a real distributed computing environment; therefore, programs developed for the model should run directly on a distributed system. However, many SS programs have been developed for the serial model which has the strongest assumptions, because it is much easier to develop and verify a program for the model than one for other models. This paper presents a transformation method that converts a program designed for the serial model to a program for the distributed model. An SS concurrency control protocol is incorporated in a transformed program to guarantee that if the original program is SS, the transformed program is also SS and performs exactly like the original program. We have implemented transformed versions of several serial model SS algorithms and tested them with various initial configurations.

1 Introduction

A distributed system consists of a set of processes and communication links that connect the processes. The processes cooperate with each other by communicating through the links. Dijkstra introduced the notion of *self-stabilization* in the context of distributed systems [4]. A system is defined to be *self-stabilizing* (SS) with respect to a set of legitimate states if regardless of its initial state, the system is guaranteed to arrive at a legitimate state in a finite number of steps and will never leave legitimate states after that. Thus, an SS system need not be initialized and is able to recover from transient failures by itself.

An SS system consists of two components: a program and a run-time support system. A run-time system provides a program with an execution environment and services. Many SS programs have been developed with various assumptions on their execution environments. These assumptions include the semantics of concurrency and communication primitives that the run-time support systems

* This work was supported in part by the National Science Foundation under Grant CCR-9201645

** This work was supported in part by the Telecommunication Advancement Foundation.

provide. Huang *et al.* classified the execution environments into four models based on the assumptions: serial model, synchronous model, synchronous distributed model, and distributed model [6]. Among the four models, the serial and distributed models have the strongest and weakest assumptions on execution environments, respectively.

1. In the serial model, an atomic execution step consists of (1) a read sub-step: which reads the states of its neighbor processes, followed by (2) a write sub-step which modifies its own state (based on the neighbors' current states and its own state). The communication and concurrency semantics are such that each process can always see the current states of its neighbors, and only one process at a time executes an atomic step.
2. In the distributed model, an atomic execution step is either a read sub-step (that reads the states of its neighbors and records them locally) or a write sub-step (that modifies its own state based on the locally recorded neighbors' states and its own state). A process does not always see the current states of its neighbors since local copies of the neighbors' states may hold obsolete values. The degree of concurrency is such that an arbitrary subset of the processes simultaneously execute their atomic steps.

The distributed model accurately reflects a real distributed computing environment. Thus, programs developed for the distributed model should run on a real distributed system, with only minor modifications if any. On the other hand, the other three models have stronger assumptions, and programs developed for these models do not run in a real distributed system without major modifications. However, many proposed SS systems assume the serial model. This is because it is much easier to develop and verify a program for the serial model than one for other models.

Several attempts have been made to develop an automatic transformation that converts a program for one model to a program for another model that has weaker assumptions [5, 6, 7, 10]. Huang *et al.* proposed a transformation method that converts a program for the serial model (denoted by a "serial model program") to a program for the distributed model (denoted by a "distributed model program") [6]. However, in a transformed program, processes run concurrently without any synchronization among them. As a result, not all transformed programs run in the same way as their original programs. Therefore, in their approach, each transformed program needs to be proven correct.

In this paper, we will present a transformation from a serial model program to a distributed model program. A timestamp based concurrency control protocol is incorporated in a transformed program to guarantee that for each execution of a transformed distributed model program, there always exists an *equivalent* execution of the original serial model program. We have defined a correctness criterion for self-stabilizing concurrency control protocols (called "*postfix serializability*") and designed the concurrency control protocol to be self-stabilizing. Therefore, if the original program is correct with respect to self-stabilization, the transformed program is also self-stabilizing. The transformation works with serial model programs that do not terminate or that have fixed-points. In the later case, due to timeout timers, even if a transformed program reaches the fixed-point, it does not terminate.

We have implemented transformed versions of Dijkstra's K-state and three state mutual exclusion algorithms [4] and the minimum spanning tree algorithms [3], on our distributed/parallel algorithm development system [9]. All the implementations were tested with many different initial configurations.

The organization of the paper is as follows: Section 2 gives formal definition of programs for the serial model and for the distributed model. The section also defines the correctness criterion for the transformation. Our transformation and its correctness proofs are presented in Sections 3 and 4, respectively. Finally, Section 5 discusses performance of transformed algorithms.

2 Definitions

2.1 A self-stabilizing program for the serial model

A serial model program is represented by a five-tuple $S_s = (n, \mathcal{P}_s, \mathcal{R}_s, \mathcal{Q}_s, \mathcal{A}_s)$, where

1. n is the number of processes in S_s;
2. $\mathcal{P}_s = \{\rho_1, \rho_2, \cdots, \rho_n\}$ is a set of processes;
3. $\mathcal{R}_s \subseteq \mathcal{P}_s \times \mathcal{P}_s$ is a set of dependency relations such that $(\rho_i, \rho_j) \in \mathcal{R}_s$ if ρ_j refers to the state of ρ_i in its execution;
   If ρ_i and ρ_j refer to each others' states then $(\rho_i, \rho_j), (\rho_j, \rho_i) \in \mathcal{R}_s$.
   If $(\rho_i, \rho_j) \in \mathcal{R}_s$, we say that ρ_i is a *writer* of ρ_j and that ρ_j is a *reader* of ρ_i. We define Writers$(\rho_i) = \{\rho_j | (\rho_j, \rho_i) \in \mathcal{R}_s\}$ and Readers$(\rho_i) = \{\rho_j | (\rho_i, \rho_j) \in \mathcal{R}_s\}$. Processes ρ_i and ρ_j are said to be *neighbors* if $(\rho_i, \rho_j) \in \mathcal{R}_s$ or $(\rho_j, \rho_i) \in \mathcal{R}_s$.
4. $\mathcal{Q}_s = (Qs_1, Qs_2, \cdots, Qs_n)$ is an n-tuple of sets of states such that Qs_i is a set of possible states that ρ_i may be in;
   Process ρ_i maintains variable q_i to store its current state. We use q_i to denote both variable q_i itself and its contents.
   A *global state* of an execution of S_s is defined by an n-tuple of states $\{q_1, q_2, \cdots, q_n\}$.
   The set of all possible global states is denoted by $\mathcal{G}_s$ (*i.e.*, $\mathcal{G}_s = Qs_1 \times Qs_2 \times \cdots \times Qs_n$).
5. $\mathcal{A}_s = (As_1, As_2, \cdots, As_n)$ is an n-tuple of algorithms, where As_i is the algorithm executed by ρ_i.
   Each As_i is described in the following form:

$$*[$$

$$g_{i_1}(q_i, q_{i_1}, q_{i_2}, \cdots, q_{i_k}) \rightarrow q_i := f_{i_1}(q_i, q_{i_1}, q_{i_2}, \cdots, q_{i_k})$$

$$\square$$

$$\vdots$$

$$\square$$

$$g_{i_m}(q_i, q_{i_1}, q_{i_2}, \cdots, q_{i_k}) \rightarrow q_i := f_{i_m}(q_i, q_{i_1}, q_{i_2}, \cdots, q_{i_k})$$
]

where

- Writers(ρ_i) = $\{\rho_{i_1}, \rho_{i_2}, \cdots, \rho_{i_k}\}$ (ρ_{i_j} denotes the jth writer process of ρ_i, and q_{i_j} denotes the ρ_{i_j}'s state).
- Each $g_{i_j}(\cdots) \rightarrow q_i := f_{i_j}(\cdots)$ is called a *guarded command* ($g_{i_j}(\cdots) \rightarrow \cdots$ denotes the jth guarded command in As_i).
- g_{i_j} is a predicate: $Q_i \times Q_{i_1} \times \cdots \times Q_{i_k} \rightarrow$ *boolean*, and $g_{i_j}(\cdots)$ is called a *guard*, and
- f_{i_j} is a function: $Q_i \times Q_{i_1} \times \cdots \times Q_{i_k} \rightarrow Q_i$, $q_i := f_{i_j}(\cdots)$ is called a *command*, and f_{i_j} is called a *command function*.

Each process ρ_i stores Writers(ρ_i), Readers(ρ_i), As_i, and its unique identifier in read only memory. These values are assumed to be incorruptible. In the serial model, each process can obtain the current states of all of its writer processes without delay. Thus, $q_{i_1}, q_{i_2}, \cdots, q_{i_k}$ in a guarded command in any As_i represent the exact current states of ρ_i's writer processes.

If a process ρ_i has a guard g_{i_j} that is true, ρ_i is said to have a *privilege*. At any point in time, there may be more than one process that have privileges.

The system executes the following three steps forever:

1. All the processes evaluate all of their guards.
2. The centralized scheduler, called the *c-daemon*, nondeterministically selects one process, say ρ_i, among all processes that have privileges.
3. Process ρ_i nondeterministically chooses one guarded command out of all the guarded commands whose associated guards are true and executes its command.

A serial model program S_s is self-stabilizing with respect to a predicate $\mathcal{T}$ defined over the set of global states, if it satisfies the following properties [10]:

1. **Closure:** $\mathcal{T}$ is closed under the execution of S_s; that is, once $\mathcal{T}$ is established in the execution, it cannot be falsified; and
2. **Convergence:** Starting from an arbitrary global state, S_s is guaranteed to reach a global state satisfying $\mathcal{T}$ within a finite number of state transitions.

States that satisfy $\mathcal{T}$, $\mathcal{L}_s \subseteq \mathcal{G}_s$, are called *legitimate* states.

For example, Dijkstra's K-state algorithm [4] for n processes is represented in our formalization as follows:

- $(\rho_i, \rho_{((i \bmod n)+1)}) \in \mathcal{R}_s$ for $1 \leq i \leq n$
- $Qs_i = \{1, 2, \cdots, K\}$ for $1 \leq i \leq n$, where $K > n$
- $As_1 : *[q_1 = q_n \rightarrow q_1 := (q_1 \bmod K) + 1]$ $As_i : *[q_i \neq q_{i-1} \rightarrow q_i := q_{i-1}]$ for $2 \leq i \leq n$
- $\mathcal{T}$ is true iff exactly one process in the system has privileges.

2.2 The distributed model

We now consider a transformation that transforms an SS serial model program S_s to an SS distributed model program S_d. The goal of the transformation is that given any S_s, the transformation generates S_d such that for each execution of S_d, say H_d, there exists an execution of S_s that is *equivalent* to H_d. The notion of equivalent executions is formally defined in Section 2.3. Intuitively, if traces of process states in two executions are indistinguishable to outside observers, they are said to be equivalent. In this way, the transformation guarantees that if S_s is self-stabilizing with respect to $\mathcal{T}$, S_d is also self-stabilizing with respect to $\mathcal{T}$.

The following assumptions are made in transformed distributed model programs:

1. The system does not provide a shared memory or a global clock, communication between processes is done only by message passing, and the execution speeds of the processes may vary but are bounded.
2. For each dependency relation $(\rho_j, \rho_i) \in \mathcal{R}_s$ in the serial model, the system provides a bidirectional communication link between the processes of S_d that simulate behaviors of ρ_i and ρ_j. The communication system provides FIFO delivery with unpredictable but bounded communication delay.
3. Any transient failures may occur in the system; however, a process does not fail. For example, a message may be lost; contents of a message may change; the program counter (control point) of a process may be reset to an unexpected address; or a state of a process may change.

Formally, a transformed program is a six-tuple $S_d = (n, \mathcal{P}_d, \mathcal{R}_d, \mathcal{Q}_d, \mathcal{C}_d, \mathcal{A}_d)$, where

1. $\mathcal{P}_d = \{P_1, P_2, \cdots, P_n\}$ is a set of n processes; Process P_i of S_d *simulates* behavior of process ρ_i of S_s.
2. $\mathcal{R}_d \subseteq \mathcal{P}_d \times \mathcal{P}_d$ is a set of dependency relations such that $(P_i, P_j) \in \mathcal{R}_d$ iff $(\rho_i, \rho_j) \in \mathcal{R}_s$; We use terms "writers" and "readers" and notations Writers(P_i) and Readers(P_i) in the same way as those in S_s.
3. $\mathcal{Q}_d = (Qd_1, Qd_2, \cdots, Qd_n)$ is an n-tuple of sets of states such that Qd_i is a set of possible states that process P_i may be in; Qd_i is represented by the Cartesian product of all the variables maintained by P_i.
   Since P_i simulates behavior of ρ_i, one of the variables is q_i. Other variables are used internally to control behavior of P_i. For outside observers, the only interest is a trace of q_i. We define a *simulated global state* of execution of S_d to be the same as that of S_s; that is, $\{q_1, q_2, \cdots, q_n\}$. Furthermore, the set of all possible simulated global states $\mathcal{G}_d$ and the set of legitimate states $\mathcal{L}_d$ are the same as $\mathcal{G}_s$ and $\mathcal{L}_s$, respectively.
4. $\mathcal{C}_d = (\cdots, Cd_{i,j}, \cdots)$ is a tuple of sets of link states such that $Cd_{i,j}$ is a set of possible link states that a link from P_i to P_j may be in; $Cd_{i,j}$ is represented by a sequence of messages in transit from P_i to P_j.
   We define a *global state* of execution of S_d to be a tuple of states, one from each process and link.

5. $\mathcal{A}_d = (Ad_1, Ad_2, \cdots, Ad_n)$ is an n-tuple of algorithms such that Ad_i is executed by P_i $(1 \leq i \leq n)$.
   Each algorithm Ad_i is represented by guarded commands (refer to Fig. 1). Process P_i evaluates all of its guards and the underlying scheduler nondeterministically selects one guarded command out of all the guarded commands whose associated guards are true and executes its command. The underlying scheduler is assumed to guarantee a weakly fair scheduling in selection of guarded commands. That is, when P_i executes statement $*[g_1 \rightarrow C_1 [] \cdots [] g_m \rightarrow C_m]$, if a guard g_k becomes true and thereafter remains true, the associated command C_k will be executed in finite time.

As does the serial model program, each process P_i stores Writers(P_i), Readers (P_i), Ad_i, and its unique identifier in incorruptible read only memory. Each process P_i maintains an array variable of n entries, denoted nq_i. Entry $nq_i[j]$ stores the value of q_j. Whenever P_j updates q_j, each one of P_j's reader processes P_i updates, by message passing, its $nq_i[j]$ to the new value of q_j. Due to message delay, $nq_i[j]$ does not always hold the current value of q_j.

2.3 Transaction view of execution steps

We can view each q_i as a data item. Then, an execution of one iteration of algorithm Ad_i at process P_i, called an "execution step," can be viewed as a transaction that updates q_i in a database of domain $\bigcup_{j=1}^{n} \{q_j\}$; that is,

1. an update of $nq_i[j]$ to a new value of q_j is considered to be a read operation by P_i on q_j, denoted $r_i(q_j)$, and
2. an assignment by P_j, $q_j := f_{j_k}(\cdots)$, is considered to be a write operation by P_j on q_j, denoted $w_j(q_j)$.

Furthermore, we introduce an imaginary commit and abort operations, denoted c_i and a_i, respectively. Each execution step ends with a commit operation. When a commit operation c_j is performed by P_j, its computational effects (*i.e*, updates of q_j and $nq_i[j]$ of $P_i \in$ Readers(P_j)) become permanent. On the other hand, if an abort operation is executed, its computational effects are nullified. In this way, each process can be viewed as a sequence of transactions. In the rest of the paper, we use the terms "(execution) step" and "transaction" interchangeably.

An execution of one iteration of the algorithm As_i of ρ_i that executes a guarded command $g_{i_j}(q_i, q_{i_1}, \cdots, q_{i_k}) \rightarrow q_i := f_{i_j}(q_i, q_{i_1}, \cdots, q_{i_k})$ is simulated by P_i in one execution step consisting of the following 4 substeps:

1. k read operations $r_i(q_{i_l})(1 \leq l \leq k)$ and $r_i(q_i)$,
2. an internal execution (checking guards and an execution of a command f_{i_j}),
3. a write operation $w(q_i)$, and
4. a commit operation c_i.

Note that since ρ_i can only access $nq_i[l]$ and has no direct access to q_l, the following guarded command in As_i

$$g_{i_j}(q_i, q_{i_1}, q_{i_2}, \cdots, q_{i_k}) \rightarrow q_i := f_{i_j}(q_i, q_{i_1}, q_{i_2}, \cdots, q_{i_k})$$

is modified in Ad_i to

$$g_{i_j}(q_i, nq_i[1], nq_i[2], \cdots, nq_i[k]) \rightarrow q_i := f_{i_j}(q_i, nq_i[1], nq_i[2], \cdots, nq_i[k]).$$

Let $\overrightarrow{T_i^t}$ denote a transaction representing the tth step of P_i. We use $w_i^t(q_i)$, $r_i^t(q_j)$, c_i^t, and a_i^t to denote the operations issued by $\overrightarrow{T_i^t}$ when it is important to emphasize that they are issued by the "t^{th}" transaction. Also, the value of q_i written by $\overrightarrow{T_i^t}$ is denoted by q_i^t. Formally, a transaction is defined as:

Definition 1 [1]. An *execution step (transaction)* $\overrightarrow{T_i^t}$ is a partial order $(T_i^t, <_i^t)$, where

1. $T_i^t \subseteq \{r_i^t(q_j) | P_j \in \text{Writers}(P_i)\} \cup \{r_i^t(q_i), w_i^t(q_i)\} \cup \{a_i^t, c_i^t\}$
2. $a_i^t \in T_i^t$ iff $c_i^t \notin T_i^t$
3. if s is c_i^t or a_i^t, for any other operation $p \in T_i^t$, $p <_i^t s$
4. if $r_i^t(x)$ denotes any read operation of T_i^t and $w_i^t(q_i) \in T_i^t$, then $r_i^t(x) <_i^t w_i^t(q_i)$.

Definition 2 [1]. Two operations are said to *conflict* if they are on the same data item and at least one of them is a write operation.

A transaction view of execution (or history) of an SS program is defined as follows:

Definition 3 [1]. Let $T = \{\overrightarrow{T_1^1}, \overrightarrow{T_1^2}, \ldots, \overrightarrow{T_1^{m_1}}, \overrightarrow{T_2^1}, \ldots, \overrightarrow{T_2^{m_2}}, \ldots, \overrightarrow{T_n^1}, \ldots, \overrightarrow{T_n^{m_n}}\}$ be a set of transactions. An *execution* (or *history*) $\overrightarrow{H}$ over T is a partial order $(H, <_H)$ where

1. $H = \bigcup_{i=1}^{n} \bigcup_{j=1}^{m_i} T_i^j$,
2. $<_H \supseteq \bigcup_{i=1}^{n} \bigcup_{j=1}^{m_i} <_i^j$,
3. for operations $p_i^{t_1}$ and $q_i^{t_2}$, $p_i^{t_1} <_H q_i^{t_2}$ if $t_1 < t_2$, and
4. for any pair of conflicting operations p_i and q_j, either $p_i <_H q_j$ or $q_j <_H p_i$.

Definition 4 [1]. $\overrightarrow{T_i^t}$ is *committed* in $\overrightarrow{H}$ if $c_i^t \in H$. The *committed projection*, $C(\overrightarrow{H}) = (C(H), <_{C(H)})$, of an execution $\overrightarrow{H}$ is the restriction of $\overrightarrow{H}$ to the transactions that are committed in $\overrightarrow{H}$.

In the concurrency control theory, the notion of (conflict) *equivalence* is defined as follows:

Definition 5 [1]. Two executions $\overrightarrow{H_1}$ and $\overrightarrow{H_2}$ are *equivalent* if (1) $C(H_1) = C(H_2)$, and (2) for any pair of conflicting operations p and q in $C(\overrightarrow{H_1})$, $p <_{H_1} q$ iff $p <_{H_2} q$.

It has been shown that if two executions are equivalent, from the view of an outside observer, the values in each data item (*i.e.*, q_i) are changed in exactly the same order in both executions, and the two executions are indistinguishable [1].

In the serial model, transactions are executed serially one after another. Such an execution is called a serial execution:

Definition 6 [1]. An execution $\overrightarrow{H}$ is a *serial execution* if for any pair $\overrightarrow{T_i}$ and $\overrightarrow{T_j}$ of $\overrightarrow{H}$, either all operations of $\overrightarrow{T_i}$ appear before any operation of $\overrightarrow{T_j}$ or vice-versa.

In the distributed model, multiple processes may execute their steps concurrently. Thus, without proper control, it is possible for the distributed model to produce executions that do not have equivalent execution in the serial model. For example, assume that P_i and P_j are neighbors of each other. The following execution sequence is possible in the distributed model:

$$r_i(q_j) <_H r_j(q_i) <_H r_i(q_i) <_H r_j(q_j) <_H w_i(q_i) <_H w_j(q_j) <_H c_i <_H c_j$$

We cannot find any serial execution which is equivalent to the above execution.

Some form of concurrency control is required in S_d to make each of its executions equivalent to some execution of S_s in the serial model. The notion of serialization is defined as follows:

Definition 7 [1]. An execution is *serializable* if there exists a serial execution that is equivalent to it.

The goal of the concurrency control protocol in a transformed program S_d is to make executions of S_d serializable. It is important that the concurrency control protocol itself be self-stabilizing to guarantee that S_d is self-stabilizing. Starting with any initial state and given an execution $\overrightarrow{H}$, self-stabilizing concurrency control guarantees that the committed projection of some postfix of $\overrightarrow{H}$ is serializable. Formally, we define such serializability to be *postfix serializability*. First, we introduce the following notations:

1. Given some execution $\overrightarrow{H}$, $|\overrightarrow{H}|$ denotes the number of operations in $\overrightarrow{H}$.
2. Let $\overrightarrow{H_1}$ and $\overrightarrow{H_2}$ be executions. $\overrightarrow{H_1} - \overrightarrow{H_2}$ is defined to be a restriction of $\overrightarrow{H_1}$ to the set of operations $(H_1 - H_2)$.

Let $PRE(\overrightarrow{H})$ denote some prefix of execution $\overrightarrow{H}$.

Definition 8. An execution $\overrightarrow{H}$ is *postfix serializable* if there exists $PRE(\overrightarrow{H})$ such that $|PRE(\overrightarrow{H})| = K$ for some non-negative integer K and $C(\overrightarrow{H} - PRE(\overrightarrow{H}))$ is serializable.

In order to easily determine whether a given execution $\overrightarrow{H}$ is postfix serializable, we use the following serialization graph.

Definition 9 [1]. The *serialization graph* for an execution $\overrightarrow{H}$, denoted $SG(\overrightarrow{H})$, is a directed graph whose nodes are transactions in $C(\overrightarrow{H})$ and whose edges ($\rightarrow$) are defined as follows:

1. wr-edge: if $w_i(x) <_{C(H)} r_j(x)$, then $\overrightarrow{T_i} \rightarrow \overrightarrow{T_j}$.
2. ww-edge: if $w_i(x) <_{C(H)} w_j(x)$, then $\overrightarrow{T_i} \rightarrow \overrightarrow{T_j}$.
3. rw-edge: if $r_i(x) <_{C(H)} w_j(x)$, then $\overrightarrow{T_i} \rightarrow \overrightarrow{T_j}$.

Theorem (Serialization Theorem [1]). *An execution $\overrightarrow{H}$ is serializable iff $SG(\overrightarrow{H})$ is acyclic.* □

It is obvious that the following corollary holds:

Corollary 10. *Given an execution $\overrightarrow{H}$, if $SG(\overrightarrow{H} - PRE(\overrightarrow{H}))$ is acyclic for some $PRE(\overrightarrow{H})$ such that $K = |PRE(\overrightarrow{H})|$ for some non-negative integer K, then $\overrightarrow{H}$ is postfix serializable.* □

The SS concurrency control protocol in S_d guarantees that any execution $\overrightarrow{H}$ produced by S_d is postfix serializable. Consider the simulated global state $\{q_1, \cdots, q_n\}$ of an execution of S_d, right after the execution $PRE(\overrightarrow{H})$ has finished, to be the initial global states of an execution of S_s. Since command functions f_{i_k} are defined in the same way in both S_s and S_d, there exists an execution of S_s that is equivalent to $C(\overrightarrow{H} - PRE(\overrightarrow{H}))$. Therefore, the execution of S_d is guaranteed to stabilize if S_s is self-stabilizing. Such S_d is said to simulate S_s.

Definition 11. *S_d simulates S_s* if for each execution $\overrightarrow{H}$ of S_d, there exists $PRE(\overrightarrow{H})$ such that $|PRE(\overrightarrow{H})| = K$ for some non-negative integer K and there exists an execution of S_s that is equivalent to $C(\overrightarrow{H} - PRE(\overrightarrow{H}))$.

Theorem 12. *S_d* simulates *S_s if every execution produced by S_d is postfix serializable.* □

3 Timestamp-based Transformation

3.1 Overview of timestamp based concurrency control

In this section, we describe transformation from an SS serial model program S_s to an SS distributed model program S_d that simulates S_s. The transformation algorithm transforms each algorithm As_i, which is executed by process ρ_i of S_s, into algorithm Ad_i, which is executed by process P_i of S_d.

In the concurrency control incorporated in a transformed program, each process P_i maintains a Lamport's logical clock *req_ts*$_i$ [8]. The *req_ts*$_i$ value while P_i is executing execution step $\overrightarrow{T_i^x}$ is called the timestamp associated with $\overrightarrow{T_i^x}$.

A process with privileges, say P_i, executes a command as a transaction $\overrightarrow{T_i^x}$ by optimistically assuming that the resulting execution will be serializable. For $\overrightarrow{T_i^x}$ to actually assign a new value to q_i by executing the assignment statement $q_i := f_{i_j}(\cdots)$, it must successfully execute a commit operation c_i^x. In SG, if c_i^x succeeds, a new node corresponding to $\overrightarrow{T_i^x}$ and the associated new edges are created (recall that SG maintains only committed transactions in its node set).

Given an edge E, let $\overrightarrow{src(E)}$ and $\overrightarrow{tgt(E)}$ denote the source node and target node of E, respectively. Given a node $\overrightarrow{T}$ in SG, let $ts(\overrightarrow{T})$ denote the timestamp associated with the node. The idea of a commit operation is to guarantee the following property:

For each edge E in SG, $ts(\overrightarrow{src(E)}) < ts(\overrightarrow{tgt(E)})$.

If this property is maintained, SG is acyclic and the resulting execution is serializable.

3.2 Description of the algorithm

A transformed algorithm Ad_i is given in Fig. 1. Process P_i executes a set of guarded commands forever. A receive statement and a timeout statement may appear in a guard. A receive statement is a boolean function that is evaluated true iff there is a pending message of the specified type in the incoming link. Actual delivery of a message takes place when the guarded command is selected for execution. A timeout statement becomes true when the associated timer (described below) goes off.

A process may be in one of two execution modes: PRIV and NO_PRIV. A process is in PRIV when it has a privilege and is executing toward the commit; in NO_PRIV otherwise.

Five types of messages are exchanged among processes: req_commit, grant, abort, commit, and state messages. When a process P_i has privileges and executes a command, it sends req_commit messages to all of its writer processes. When a process P_j receives a req_commit message from P_i, it replies P_i with either a grant or abort message. If P_i receives grant messages from all of its writer processes, it commits and sends commit messages to all of its reader processes. A state message is used to inform its reader processes of its current state.

Every message carries the current timestamp value req_ts_i. In addition to the timestamp, each grant and abort message returns the timestamp carried by the associated req_commit message. This returned timestamp is used to verify whether the abort or grant message is a reply for the outstanding req_commit message (B02 and C02 in Fig. 1). Each state and commit message by P_i carries its current state q_i.

In addition to q_i, $nq_i[j]$, and req_ts_j, each process P_i maintains the following variables:

- $mode_i$: stores the current mode of P_i, either PRIV or NO_PRIV.
- max_ts_i: maintains the maximum timestamp value that P_i has seen.
- q_tmp_i: temporarily stores a new q_i value that was just computed, until the commit succeeds.
- $granted_i$: maintains a set of writer processes (identifiers) that have already replied with grant messages. This is used to check whether P_i has received grant messages from all of its writer processes.

```
     *[
A01    (mode = NO_PRIV) ∧ (∃j : g_ij(...) = true) →
A02      q_tmp_i := f_ij(...); mode_i := PRIV; granted_i := ∅; req_ts_i := max_ts_i;
A03      multicast req_commit(req_ts_i) to all P_j ∈ Writers(P_i)
A04      commit_timer.start_timer();

B01    [] receive grant(ts_j, ts'_i) from P_j →
B02      if (mode_i = PRIV) ∧ (req_ts_i = ts'_i)
B03        granted_i := granted_i ∪ {j};
B04        if (granted_i = Writers(P_i))
B05          q_i := q_tmp_i; mode_i := NO_PRIV;
B06          multicast commit(req_ts_i, q_i) to all q_j ∈ Readers(P_i);
B07          state_timer.start_timer();
B08          update_timestamps ();   /* max_ts_i := req_ts_i := max_ts_i + delta */
B09        fi
B10      fi

C01    [] receive abort(ts_j, ts'_i) from P_j →
C02      if (ts'_i = req_ts_i) mode_i := NO_PRIV; fi

D01    [] receive commit(ts_j, q_j) from P_j →
D02      nq_i[j] := q_j; mode_i := NO_PRIV;
D03      update_timestamps ();   /* max_ts_i := req_ts_i := max_ts_i + delta */

E01    [] receive state(ts_j, q_j) from P_j →
E02      if (q_j ≠ nq_i[j])
E03        nq_i[j] := q_j; mode_i := NO_PRIV;
E04        update_timestamps ();   /* max_ts_i := req_ts_i := max_ts_i + delta */
E05      fi

F01    [] receive req_commit(ts_j) from P_j →
F02      if (mode_i = NO_PRIV) ∨ (ts_j < req_ts_i)   /* use PIDs to break a tie */
F03        send grant(req_ts_i, ts_j) to P_j;
F04      else
F05        send abort(req_ts_i, ts_j) to P_j;
F06      fi

G01    [] commit_timer goes off →
G02      mode_i := NO_PRIV; commit_timer.start_timer();

H01    [] state_timer goes off →
H02      multicast state(req_ts_i, q_i) to all P_j ∈ Readers(P_i);
H03      state_timer.start_timer();
     ]
```

Fig. 1. Transformation Algorithm

Whenever a process P_i receives a message with timestamp req_ts_j from process P_j, it updates max_ts_i by $max\_ts_i := \max(max\_ts_i, req\_ts_j)$ (this procedure is implicit and not shown in Fig. 1). The value in max_ts_i is used to execute Lamport's logical clock algorithm to update req_ts_i (B08, D03, E04).

Each process P_i maintains two timers:

1. state_timer$_i$: This timer is used to inform P_i's reader processes of the current state of P_i at least once every τ period, where τ is a predefined value and described in more detail in Section 5.
2. commit_timer$_i$: This timer goes off if P_i does not receive all the grant messages or at least one abort message in t_c period after P_i has sent req_commit messages, where t_c is slightly longer than the expected upper bound of the maximum round trip message latency plus execution time for a process to send a grant or abort message.

Each timer has a counter. The value of a counter is decremented by one every unit time (second, millisecond, etc.). When the counter value becomes zero, the timer goes off. The only operation provided with the timers is start_timer(), which assigns a predefined initial counter value (τ or t_c associated with the timer) to the timer counter. Even though a timer counter is not properly initialized at the beginning of an execution, the associated timer will eventually go off.

The following are some remarks on the algorithm:

1. When P_i receives a req_commit message from a reader process P_j (F01), there are two cases to consider:
   (a) if P_i does not have a privilege or if $req\_ts_j < req\_ts_i$ (P_j's execution step has higher priority), P_j's execution step should be allowed to commit; thus, P_i sends a grant message to P_j (F03);
   (b) otherwise (P_i has a privilege and $req\_ts_i < req\_ts_j$), P_i's current execution step should commit before P_j's. Thus, P_i sends an abort message to P_j to force P_j to abort (F05).

   Note that when comparing two timestamp values, process identifiers, which are stored in incorruptible read only memory, are used to break a tie, as suggested in [8]; in this way, all timestamps are totally ordered.
2. When P_i receives a state or commit message, it sets its mode to NO_PRIV. This is because P_i must abandon the current execution step and start a new execution step with the new $nq_i[*]$ values, one of which was just updated. Note that if the message is a state message, it does this only when a value in $nq_i[*]$ is actually changed (E02).
3. Timer state_timer$_i$ is restarted every time it sends commit or state messages (B07, H03). When it goes off, it sends state messages to all of its reader processes (H02). In this way, P_i informs all of its reader processes of P_i's current state at least once every τ time.
4. Timer commit_timer$_i$ should go off only when the process is in PRIV, waiting for grant or abort messages too long. However, since the timer cannot be stopped, it may go off even when the process is in NO_PRIV. It has no effect in such a case (G02).

4 Correctness of the algorithm

4.1 Safety property

To prove that S_d simulates S_s, we will show that every execution of S_d is postfix serializable. First, we define a predicate $\mathcal{T}$ on the global states of an execution $\overrightarrow{H}$ of S_d and show that S_d is self-stabilizing with respect to $\mathcal{T}$. Then, we show that $SG(\overrightarrow{H})$ is acyclic if $\mathcal{T}$ is true; thus, a history produced by S_d is postfix serializable.

We assume that initially, at any process P_i, its communication links may contain finite number of garbage messages and that its variables may hold arbitrary values. A process replies with a grant or abort message when it receives a garbage req_commit message (F03, F05). Such a grant or abort message is also considered to be a garbage message. Note that P_i returns no message when it receives any other types of garbage messages.

We define predicate $\mathcal{T}$ over global states of the transformed program S_d as follows:

Definition (Predicate $\mathcal{T}$). *$\mathcal{T}$ is true in execution $\overrightarrow{H}$ if*

1. *there is no garbage message in communication links; and*
2. *for all processes P_i, either*
   (a) *its mode ($mode_i$) is NO_PRIV and for each $P_j \in Writers(P_i)$,*
      i. *$q_j = nq_i[j]$ and no commit message is in transit from P_j to P_i, or*
      ii. *a commit message that carries q_j is in transit from P_j to P_i; or*
   (b) *$mode_i$ is PRIV and P_i was in NO_PRIV prior to the current mode, and the above i. or ii. was satisfied then*[3].

Lemma 13. *S_d is self-stabilizing with respect to $\mathcal{T}$.*

Proof. We show that the convergence and closure properties hold with respect to $\mathcal{T}$ for S_d. To prove the convergence property, we show that for any execution $\overrightarrow{H}$ of S_d, there exists a finite length of period t_{init} such that after t_{init} in $\overrightarrow{H}$, (1) all the initial garbage messages in the communication links will have been received (and consumed), and then (2) for each process P_i, (2a) $nq_i[j]$ will hold q_j for each $P_j \in Writers(P_j)$ or a non-garbage commit message that carries q_j is in transit from P_j to P_i, and then (2b) P_i will enter NO_PRIV. In the following discussions, since each process executes its steps without being interfered with by any other processes, we may assume that a selected guarded command is executed instantaneously.

First, consider the above (1). The algorithm executed by P_i does not have a loop construct in any of the commands. Thus, whichever guarded command

[3] $\mathcal{T}$ refers to a history of the global states of S_d. It is possible to define $\mathcal{T}$ to refer to only the current global states. However, $\mathcal{T}$ would become very complex since it must capture all the possible legitimate combinations of link and process states. Thus, in this paper, we slightly abuse formality and allow $\mathcal{T}$ to refer to a history of the global states.

that P_i selects, P_i always finishes the execution in finite time and moves to the next execution step. Since we assume a weakly fair scheduling in selection of guarded commands in Fig. 1, all the garbage messages in communication links are received by associated receive statements in finite time, say t_m.

Next, consider the above (2a). Due to the state_timers, each process receives its writers' states periodically (*i.e.*, once every τ period). Thus, in t_v time (τ plus a period required to execute a receive statement) after t_m, it is guaranteed that each $nq_i[j]$ will have held q_j for all processes P_i. During this period, if P_j commits, a commit message is in transit from P_j to P_i, or P_i receives the commit message and $nq_i[j]$ is set to the new q_j.

Finally, consider the above (2b). After $t_m + t_v$, even if $mode_i$ is PRIV for some P_i, (1) its commit_timer will go off (to turn the mode to NO_PRIV), (2) it will receive a commit, state, or abort message, or (3) it will (possibly erroneously) receive enough grant messages and commit. In either case, all processes will enter mode NO_PRIV in finite time, say t_p. As time goes, some process may enter the PRIV mode. Note that the condition for $\mathcal{T}$ holds for such processes.

Let $t_{init} = t_m + t_v + t_p$. Then, $\mathcal{T}$ will hold within t_{init} of the execution. Once $\mathcal{T}$ holds, if P_i is in NO_PRIV, either $nq_i[j] = q_j$ or a commit message that carries q_j is in transit from P_j to P_i since only the time a P_i's writer process P_j changes its state is when it commits. If P_i is in PRIV, it must have entered the PRIV mode by executing the first guarded command (A01); therefore, its previous mode NO_PRIV exists. □

Lemma 14. *When $\mathcal{T}$ is true, whenever an execution step commits, for all edges E created upon the commit in $SG(\overrightarrow{H})$, $ts(\overrightarrow{src(E)}) < ts(\overrightarrow{tgt(E)})$.*

Proof. Consider the execution of $\overrightarrow{H}$ after $\mathcal{T}$ becomes true. Let P_i and P_j be processes such that P_j is a writer process of P_i. Let $\overrightarrow{T_j^y}$ be a committed execution step at P_j (or an imaginary execution step that determines the initial state of P_j). Suppose that $\overrightarrow{T_j^{y+1}}$ is the current executing step at P_j and that $\overrightarrow{T_i^x}$ is the current executing step at P_i such that $nq_i[j]$ of $\overrightarrow{T_i^x}$ holds q_j^y (this is guaranteed by $\mathcal{T}$). Let $req\_ts_i(x)$ denote the timestamp associated with $\overrightarrow{T_i^x}$.

For each edge E in $SG(\overrightarrow{H})$, $ts(\overrightarrow{src(E)})$ and $ts(\overrightarrow{tgt(E)})$ have the following properties[4]:

1. **A wr-edge from $\overrightarrow{T_j^y}$ to $\overrightarrow{T_i^x}$:** This is created when $\overrightarrow{T_i^x}$ commits. When $\overrightarrow{T_j^y}$ committed (before $\overrightarrow{T_i^x}$ commits), it sent a commit message along with $req\_ts_j(y)$ and q_j^y to P_i. When P_i received the commit message, it updated the timestamp to be greater than $req\_ts_j(y)$ (D03). Thus, $req\_ts_j(y) < req\_ts_i(x)$ holds.
   Consider a case in which $\overrightarrow{T_i^x}$ is aborted later. Since req_ts_i never decreases, a new step (also denoted $\overrightarrow{T_i^x}$) is at least as large as the old $\overrightarrow{T_i^x}$. Thus, $req\_ts_j(y) < req\_ts_i(x)$ always holds.

[4] Note that SG defined in Section 2 is a transitive closure of the graph defined here.

2. **A ww-edge from $\overrightarrow{T_j^y}$ to $\overrightarrow{T_j^{y+1}}$:** This is created when $\overrightarrow{T_j^{y+1}}$ commits. When $\overrightarrow{T_j^y}$ committed, it updated the timestamp of $\overrightarrow{T_j^{y+1}}$ to be greater than $req\_ts_j(y)$ (B08). Thus, $req\_ts_j(y) < req\_ts_j(y+1)$ holds.
3. **An rw-edge from $\overrightarrow{T_i^x}$ to $\overrightarrow{T_j^{y+1}}$:** Suppose that $\overrightarrow{T_i^x}$ has committed based on q_j^y. An rw-edge $\overrightarrow{T_i^x} \rightarrow \overrightarrow{T_j^{y+1}}$ is created when $\overrightarrow{T_j^{y+1}}$ commits.
   Before $\overrightarrow{T_i^x}$ committed, it had sent a req_commit message along with $req\_ts_i(x)$ to P_j and obtained a grant message from P_j. There are two cases to consider when P_j received the req_commit message:
   (a) P_j was in NO_PRIV: P_j set max_ts_j to be no smaller than $req\_ts_i(x)$. Later when commit and/or state messages arrive from P_j's writer processes (otherwise P_j's mode could not have become PRIV), P_j sets a timestamp for $\overrightarrow{T_j^{y+1}}$ to be greater than max_ts_j (D03, E04). Thus, $req\_ts_i(x) < req\_ts_j(y+1)$ holds.
   (b) Execution step $\overrightarrow{T_j^{y+1}}$ was under progress at P_j: P_j returned a grant message because its timestamp was greater than $req\_ts_i(x)$ (F02-F03). Thus, $req\_ts_i(x) < req\_ts_j(y+1)$ holds.

In all cases, $ts(\overrightarrow{src(E)}) < ts(\overrightarrow{tgt(E)})$ holds. □

In the scenario given in the proof, $\mathcal{T}$ guarantees that $\overrightarrow{T_i^x}$ reads q_j^y.

Theorem 15. *An execution $\overrightarrow{H}$ of S_d is postfix serializable.*

Proof. Lemma 13 has shown that after t_{init}, $\mathcal{T}$ becomes true and will never become false after that in $\overrightarrow{H}$. Let $PRE(\overrightarrow{H})$ denote the computation of $\overrightarrow{H}$ up to time t_{init}. Let $\overrightarrow{H'}$ be $\overrightarrow{H} - PRE(\overrightarrow{H})$. We show that $SG(\overrightarrow{H'})$ is acyclic. Since $\mathcal{T}$ holds in $C(\overrightarrow{H'})$, from Lemma 14, $ts(\overrightarrow{src(e)}) < ts(\overrightarrow{tgt(e)})$ holds for any edge in $SG(\overrightarrow{H'})$.

Assume, on the contrary, there is a cycle $\overrightarrow{T_1} \rightarrow \cdots \rightarrow \overrightarrow{T_k} \rightarrow \overrightarrow{T_1}$ in $SG(\overrightarrow{H'})$. Then, $ts(\overrightarrow{T_1}) < ts(\overrightarrow{T_2})$ $ts(\overrightarrow{T_2}) < ts(\overrightarrow{T_3})$ $\cdots$ $ts(\overrightarrow{T_k}) < ts(\overrightarrow{T_1})$; that is, $ts(\overrightarrow{T_1}) < ts(\overrightarrow{T_1})$. This is a contradiction. Thus, $SG(\overrightarrow{H'})$ is acyclic. □

From Theorems 12 and 15, the following theorem holds.

Theorem 16. *Every history produced by S_d is postfix serializable; that is, S_d simulates S_s.* □

As we discussed in Section 2, S_d is self-stabilizing if S_s is self-stabilizing and S_d works just like S_s.

4.2 Liveness property

Theorem 17. *Let $\overrightarrow{H}$ be an execution of program S_d. $\overrightarrow{H}$ progresses after t_{init} as long as there exist processes that have privileges (i.e., the program has not reached a fixed point if there is one).*

Proof. Since the program has not reached the fixed point, only the situation in which an execution of S_d would not progress is that privileged processes keep aborting one another forever. A process may be aborted by other processes when it receives a commit or abort message. Note that after t_{init}, each $nq_i[j]$ at P_i holds a valid value when no commit message is in transit to P_i. Therefore, a state message does not abort a process. Let $\mathcal{P}_p$ be a set of processes in $\mathcal{P}_d$ that are in mode PRIV. We show that at least one process in $\mathcal{P}_p$ will commit.

Let P_{p_1} be the process in $\mathcal{P}_p$ that has the smallest timestamp. No process can send an abort message to P_{p_1}. Thus, P_{p_1} is aborted only if it has received a commit message. This implies that either P_{p_1} will commit or there is another process that commits. In either case, at least one process in $\mathcal{P}_p$ will commit.

Therefore, $\overrightarrow{H}$ progresses as long as there exist processes that have privileges. □

Note that Theorem 17 does not guarantee fair scheduling among individual processes. However, many self-stabilizing programs in the serial model do not assume fair scheduling in the c-daemon. These programs include Dijkstra's k-state and three state mutual exclusion algorithms [4] and Burns and Pachl's uniform self-stabilizing mutual exclusion algorithm [2]. They realize fair scheduling in the algorithms themselves, and therefore, their associated transformed programs S_d work without starvation. Furthermore, in most self-stabilizing algorithms, no processes continuously keep holding privileges.

5 Performance evaluation

In this section, we evaluate our algorithm with respect to the number of required messages for each execution step. Once the system is stabilized and no further transient failures occur, state messages do not have to be exchanged. Thus, τ (the timeout value of a state_timer) should be very large for better performance. If τ is longer than the duration between two consecutive execution steps at each process, no state messages are sent. However, as we saw in Section 4, the time taken for the system to stabilize depends on τ.

Let w_i and r_i be the numbers of P_i's reader and writer processes, respectively. For an execution step to commit, P_i needs to exchange $2 * w_i + r_i$ messages; w_i req_commit messages, w_i grant messages, and r_i commit messages. If an execution step is aborted, P_i exchanges maximum of $2 * w_i$ messages; w_i req_commit messages, $w_i - 1$ grant messages, and one abort message.

There is no upper bound on how many times a process is aborted before its execution step is committed. However, in many SS algorithms, conflicts are rare once the system is stabilized. For example, in Dijkstra's K-state and three-state mutual exclusion algorithms, once the system is stabilized, no conflict will occur.

We conducted performance evaluation on our implementation of Dijkstra's K-state algorithm by using a simulation feature of the underlying development system [9]. In the K-state algorithm, $w_i = r_i = 1$ for all processes. After the system is stabilized, aborts will not occur. Thus, if no state messages are exchanged, 3 messages are required for each CS entry.

In the performance evaluation, the following execution parameters are used[5]:

1. the number of nodes (n): 6
2. network transmission latency : (2 time units);
3. execution time inside a CS [denoted by MXTIME]: (10 time units); and
4. execution time of each guard : (0.2 time units).
5. commit_timer counter value t_c : 10.0

We run the program with three τ values: (1) MXTIME(= 10), (2) MXTIME $*$ $(n-1)(=50)$, and (3) MXTIME $*(n-1)*5(=250)$. Table 1 shows the numbers of messages exchanged for each CS entry that are averaged over 5000.0 time units.

Table 1. Average numbers of messages

τ value	10	50	250
messages/entry	12.84	6.01	3.27

Even though the τ value is chosen to be very large (MXTIME$*(n-1)*5$), the number of messages for a CS entry is still slightly greater than 3. This is because (1) extra messages were exchanged at the beginning of computation while the system is stabilizing and (2) commit_timers went off several times.

6 Conclusion

This paper has presented a transformation method that converts an SS serial model program S_s to a distributed model program that simulates S_s. A timestamp based SS postfix serializable concurrency control protocol is incorporated in a transformed program. Therefore, if the original serial model program is correct with respect to self-stabilization, the transformed distributed model program is also self-stabilizing and runs exactly like the original program. We have implemented transformed versions of several SS serial model algorithms and tested the implementations with many different initial configurations.

Acknowledgements

We would like to thank the anonymous referees for their constructive comments.

[5] Parameters (2)–(4) use exponential distribution with the average time units given in parentheses

References

1. Bernstein, P.A, Hadzilacos, V., and Goodman, N. *Concurrency Control and Recovery in Database Systems.* Addison-Wesley Publishing Co., 1987.
2. Burns J.E. and Pachl J. Uniform self-stabilizing rings. *ACM Transactions on Programming Languages and Systems*, 11(2):330–344, April 1989.
3. Chen N.S, Yu F.P, and Huang S.T. A self-stabilizing algorithm for constructing spanning trees. *Inf. Process. Lett.*, 39:147–151, 1991.
4. Dijkstra E.W. Self-stabilizing systems in spite of distributed control. *Communications of the ACM*, 17(11):643–644, November 1974.
5. Dolev S., Israeli A., and Moran S. Self stabilization of dynamic systems assuming only read/write atomicity. *Distributed Computing*, 7:3–16, 1993.
6. Huang S.T., Wuu L.C., and Tsai M.S. Distributed execution model for self-stabilizing systems. In *Proceedings of the 14th International Conference on Distributed Computing Systems*, pages 432–439. IEEE, 1994.
7. Katz S. and Perry K.J. Self-stabilizing extensions for message passing systems. *Distributed Computing*, 7:17–26, 1993.
8. Lamport L. Time, clocks and ordering of events in distributed systems. *Communications of the ACM*, 21(7):558–564, 1978.
9. Mizuno M. *A Distributed/Parallel Algorithm Development System.* Kansas State University, 1994.
10. Schneider M. Self-stabilization. *ACM Computing Surveys*, 25(1):45–67, March 1993.

The Combinatorial Structure of Wait-free Solvable Tasks

(Extended Abstract)

Hagit Attiya[1] and Sergio Rajsbaum[2]

[1] Department of Computer Science, The Technion, Haifa 32000, Israel.***
[2] Instituto de Matemáticas, UNAM, Ciudad Universitaria, D.F. 04510, México.†

Abstract. This paper presents a self-contained study of wait-free solvable tasks. A new necessary and sufficient condition for wait-free solvability is proved, providing a characterization of the wait-free solvable tasks. The necessary condition is used to prove tight bounds on renaming and k-set consensus. The framework is based on topology, but uses only elementary combinatorics, and does not rely on algebraic or geometric arguments.

1 Introduction

This paper studies the tasks that can be solved by a wait-free protocol in shared-memory asynchronous systems. A shared-memory system consists of $n+1$ processes that communicate by reading and writing shared variables; here we assume only atomic read/write registers. We also assume that processes are completely asynchronous, i.e., each process runs at a completely arbitrary speed. Processes start with *inputs* and, after performing some protocol, have to halt with some *outputs*. A *task* specifies the sets of outputs that are allowable for each assignment of inputs to processes. A protocol is *wait-free* if each process halts with an output within a finite number of its own steps, regardless of the behavior of other processes. A task is *wait-free solvable* if there exists a wait-free protocol that solves it.

The study of wait-free solvable tasks has been central to the theory of distributed computing. Early research studied specific tasks and showed them to be solvable (e.g., approximate agreement [9], $2n$-renaming [2], k-set consensus with at most $k-1$ failures [7]) or unsolvable (e.g., consensus [10], $n+1$-renaming [2]). A necessary and sufficient condition for the solvability of a task in the presence of one process failure was presented in [3]. In 1993, a significant advancement

*** Supported by grant No. 92-0233 from the United States-Israel Binational Science Foundation (BSF), Jerusalem, Israel, and the fund for the promotion of research in the Technion. Email: hagit@cs.technion.ac.il.

† Part of this work was done while visiting the MIT Laboratory for Computer Science, and the Cambridge Research Laboratory of DEC. Supported by CONACyT and DGAPA Projects, UNAM. Email: rajsbaum@servidor.unam.mx.

was made in the understanding of this problem with [4, 17, 20]. This advancement yielded new impossibility results for k-set consensus ([4, 17, 20], and later [6, 14, 15]) and renaming ([17, 15]), as well as a necessary and sufficient condition for wait-free solvability ([17, 18]). Of particular interest was the use of topological notions to investigate the problem, suggested in [17, 20]. Yet, much of this development remained inaccessible to many researchers, since it relied on algebraic and geometric tools of topology.

In this paper, we present a self-contained study of wait-free solvable tasks starting from first principles. We introduce a new necessary and sufficient condition for wait-free solvability. This condition is used to prove tight bounds on renaming and k-set consensus. It is also used to derive an extension of the necessary condition of [3]. Our approach borrows critical ideas from previous works in this area (especially, [4, 5, 17, 18, 20]), and integrates them into a unified framework. Below we discuss the relationships between our work and previous work.

To provide a feeling for our results, we present the following rough description of key notions from combinatorial topology. A *colored simplex* is a set, in which each of the elements, called *vertices*, is colored with a process id. A *colored complex* is a collection of colored simplexes which is closed under containment. A mapping from the vertices of one colored complex to the vertices of another is *simplicial* if it maps a simplex to a simplex; it is *color preserving* if a vertex with id p_i is mapped to vertex with id p_i. Finally, a complex whose largest simplex contains m vertices is a *pseudomanifold* if every simplex with $m-1$ vertices is contained in either one or two simplexes with m vertices. Precise definitions appear in Section 3; they do not rely on algebraic or geometric interpretations.

The novel combinatorial concept we use is of a pseudomanifold being a divided image of a simplex. Very roughly, a pseudomanifold is a *divided image* of a simplex if it has the same boundary as the simplex. The divided image preserves some of (but not all) the topological structure of the simplex. We prove a new necessary condition for wait-free solvability (Corollary 13): if a task is wait-free solvable, then there exists a divided image of the complex of possible inputs; it is straightforward to see that the decisions made by the protocol must induce a simplicial map from this divided image to the complex of possible outputs which must agree with the task specification.

We present a necessary and sufficient condition for wait-free solvability, i.e., a characterization of the wait-free solvable tasks. Consider a task, and a wait-free protocol that solves it. We explicitly show that a subset of the protocol's executions, called *immediate snapshot executions* [4, 20], induce a divided image of the complex of possible inputs. We use a solution for the participating set problem ([5]) to show that the above property is also sufficient. Namely, if there exists a simplicial map from a divided image induced by immediate snapshots executions to the output complex which agrees with the task, then the problem is wait-free solvable.

We prove that the divided image induced by immediate snapshot executions is *orientable*. We then prove a combinatorial theorem which extends Sperner's

Lemma (for orientable divided images). This theorem is the key to a completely combinatorial proof that M-renaming is wait-free solvable only if $M \geq 2n$. Using the basic Sperner's Lemma, we also show that k-set consensus is wait-free solvable only if $k > n$. (These bounds are known to be tight, see [2] and [7], respectively.)

Divided images play a role similar to ***spans*** (both the geometric version used in [17, 18, 14], and the algebraic version introduced in [15]). As discussed below (after Definition 1) divided images have weaker mathematical properties than geometric spans, in particular, they may have "holes". We show (in the full version of the paper) that an orientable divided image corresponds in a natural manner to an algebraic span. It was shown that such spans exist (in [17]), but this proof requires a combination of algebraic (homology theory) and geometric (subdivided simplexes) arguments. The existence of algebraic spans with certain properties imply impossibilities of set consensus and renaming [15], without relying on the more involved arguments of [17].

The necessary and sufficient condition we derive is not exactly the same as the one proved by Herlihy and Shavit in [18]. We explicitly construct a specific well-structured divided image (induced by immediate snapshot executions), while Herlihy and Shavit show that an arbitrary span exists ([17]). The notion of immediate snapshot executions was introduced in [4, 20]. The basic ideas needed to show that immediate snapshot executions induce a divided image already appeared in Borowsky and Gafni's paper [4]. However, they were interested in properties of immediate snapshot executions to prove the impossibility result for set consensus. It was not shown that they are orientable (a property used for the renaming impossibility) or that they induce an algebraic span (or our simpler combinatorial notion of a divided image), and no general conditions for wait-free solvability were derived from them.

In the full version of this paper, we derive another necessary condition for wait-free solvability from Corollary 13, of a different nature. This condition is based on connectivity, and is therefore computable. This condition extends the condition for solvability in the presence of one failure [3]. It follows from [11, 16] that there is no computable necessary and sufficient condition for wait-free solvability.

2 Model of Computation

Our model is standard and was used in many papers; we follow [1].

A *system* consists of $n+1$ processes $p_0, \ldots, p_n$. Each process is a deterministic state machine, with a possibly infinite number of states. We associate with each process a set of *local states*. Among the states of each process is a subset called the *initial states* and another subset called the *output states*. Processes communicate by means of a finite number of *single-writer multi-reader atomic registers* (also called *shared variables*). No assumption is made regarding the size of the registers, and therefore we may assume that each process p_i has only one register R_i. Each process p_i has two atomic operations available to it:

- *write*$_i$: p_i writes its entire state to R_i.
- *read*$_i(R)$: p_i reads the shared variable R and returns its value v.

A system configuration consists of the states of the processes and registers. Formally, a *configuration* C is a vector $\langle s_0, \ldots, s_n, v_0, \ldots, v_n \rangle$ where s_i is the local state of process p_i and v_j is the value of the shared variable R_j. Denote $\text{state}_i(C) = s_i$. Each shared variable may attain values from some *domain* which includes a special "undefined" value, $\perp$. An *initial configuration* is a configuration in which every local state is an initial state and all shared variables are set to $\perp$.

We consider an interleaving model of concurrency, where executions are modeled as sequences of steps. Each step is performed by a single process. In each step, a process p_i performs either a *write*$_i$ operation or a *read*$_i(R)$ operation, but not both, performs some local computation, and changes to its next local state. The next configuration is the result of these modifications.

We assume that each process p_i follows a *local protocol* $\mathcal{P}_i$ that deterministically determines p_i's next step: $\mathcal{P}_i$ determines whether p_i is to write or read, and (in case of a read) which variable R to read, as a function of p_i's local state. If p_i reads R, then $\mathcal{P}_i$ determines p_i's next state as a function of p_i's current state and the value v read from R. If p_i writes R, then $\mathcal{P}_i$ determines p_i's next state and as a function of p_i's current state. We assume that all local protocols are identical, i.e., depend only on the state, but not on the process id. A *protocol* is a collection $\mathcal{P}$ of local protocols $\mathcal{P}_0, \ldots, \mathcal{P}_n$.

An *event of* p_i is simply p_i's index i. A *schedule* is a finite or infinite sequence of events. An *execution* is a finite or infinite alternating sequence of configurations and events $C_0, j_1, C_1, \ldots, C_{k-1}, j_k, \ldots$, where C_0 is the initial configuration and C_k is the result of applying the event j_k to C_{k-1}, for all $k \geq 1$. The schedule of this execution is $j_1, \ldots, j_k, \ldots$.

Given an execution $\alpha = C_0, j_1, C_1, \ldots$, and a process p_i, the *view of* p_i *in* α, denoted $\alpha|i$ is the sequence $\text{state}_i(C_0), \text{state}_i(C_1), \ldots$. Intuitively, for example, if p_i decides in α without taking any steps, then the only information contained in $\alpha|i$ is p_i's initial state.

A process p_i is *faulty* in an infinite schedule σ if it takes a finite number of steps (i.e., has a finite number of events) in σ, and *nonfaulty* otherwise. These definitions also apply to executions by means of their schedules.

We assume that each process has two special parts of its state, an *input value* and an *output value*. Initial configurations differ only in the input values of the processes. If we want to have a local protocol which depends on the process id, then the id has to be provided explicitly as part of the input. We assume that the output value is irrevocable, i.e., the protocol cannot over-write the output value. Note that in our definition processes do not halt; they decide by writing the output value, but continue to take steps (which are irrelevant).

A *task* Δ has some domain $\mathcal{I}$ of input values and domain $\mathcal{O}$ of output values; Δ specifies for each assignment of input values to the processes which output values can be written by the processes. A protocol *solves* Δ if for any finite execution, the output values already written by the processes can be completed (in any infinite extension of the execution where all processes are nonfaulty) to output values for all processes that are allowable for the input values in the execution. The protocol is *wait-free* if every nonfaulty process eventually writes an output value.

3 Combinatorial Topology Concepts

In this section, we introduce the basic topological concepts we use in this paper. Previous papers in this area, e.g., [8, 14, 17, 18, 20], used geometric or algebraic interpretations of topological structures; in contrast, our approach is purely combinatorial, abstracting ideas from [12, 19, 21].

Basic Notions: The basis of our definitions is the notion of a complex. A *complex* K is a collection of finite nonempty sets closed under containment; that is, if σ is an element of K, then every nonempty subset of σ is an element of K. A nonempty subset of σ is a *face* of σ. A face of σ is *proper* if it is not equal to σ. Each element of a complex is called a *simplex*. A complex K' is a *subcomplex* of a complex K if $K' \subseteq K$.

The *dimension* of a simplex σ, $dim(\sigma)$, is the number of its elements minus one. A simplex of dimension m (with $m+1$ elements) is called an m-simplex. The *dimension* of a complex K is the maximum dimension of its simplexes; we only consider complexes of finite dimension. A complex of dimension m is called an m-complex. We sometimes use a superscript notation to denote the dimension of simplexes and complexes, e.g., σ^m is an m-simplex and K^m is an m-complex.

The *vertex set* of K is the union of the 0-simplexes of K. We identify the vertex v and the 0-simplex $\{v\}$.

Consider two complexes K and L. Let f be a function from the vertices of K to the vertices of L. f is *simplicial* if for every simplex $\{v_0, \ldots, v_k\}$ of K, $\{f(v_0), \ldots, f(v_k)\}$ is a simplex of L. (Note that $\{f(v_0), \ldots, f(v_k)\}$ is treated as a set, since f need not be one-to-one and there may be repetitions.) This implies that a simplicial map f can be extended to all simplexes of K. Intuitively, a simplicial map f maps every simplex σ of K to a simplex $f(\sigma)$ (perhaps of smaller dimension) of L. We extend f to a set of simplexes of K, S, by defining $f(S)$ to be the set of simplexes $f(\sigma)$ in L, where σ ranges over all simplexes of S. Clearly, if S is a subcomplex of K then $f(S)$ is a subcomplex of L.

Divided Images: An m-complex K^m is *full to dimension* m if every simplex of K^m is contained in some m-simplex of K^m.

Let K^m be a complex full to dimension m. An $(m-1)$-simplex of K^m is *external* if it is contained in exactly one m-simplex; otherwise, it is *internal*. The *boundary complex* of K^m, denoted $bound(K^m)$, is the subcomplex containing all the faces of external simplexes of K^m. Clearly, $bound(K^m)$ is full to dimension $m-1$. Abusing notation, let $bound(\sigma^m)$ be the set of $(m-1)$-faces of simplex σ^m.

A complex K^m is an *m-pseudomanifold*, if it is full to dimension m and every $(m-1)$-simplex is contained in either one or two m-simplexes.[5] An *m-manifold* is an m-pseudomanifold in which every $(m-1)$-simplex is contained in two m-simplexes, i.e., it has no external simplexes.

The following combinatorial definition will play a key role later when we cast the structure of a protocol in the topological framework.

Definition 1. Let L^m be a complex. A complex K^m is a *weak divided image* of L^m if there exists a function ψ that assigns to each simplex of L^m a subcomplex of K^m, such that:

1. for every $\tau \in K^m$ there exists a simplex $\sigma \in L^m$ such that $\tau \in \psi(\sigma)$,
2. for every $\sigma^0 \in L^m$, $\psi(\sigma^0)$ is a single vertex, and
3. for every $\sigma, \sigma' \in L^m$, $\psi(\sigma \cap \sigma') = \psi(\sigma) \cap \psi(\sigma')$. (We assume that $\psi(\emptyset) = \emptyset$.)

K^m is a *divided image* of L^m if it also satisfies the following condition:

4. for every $\sigma \in L^m$, $\psi(\sigma)$ is a $dim(\sigma)$-pseudomanifold with $\psi(bound(\sigma)) = bound(\psi(\sigma))$.[6]

We say that K^m is a divided image of L^m *under* ψ.

Intuitively, a divided image is obtained from L^m by replacing each simplex of L^m with a pseudomanifold, making sure that they "fit together" (in the sense of Condition 3). In addition, Condition 1 guarantees that ψ maps L^m "onto" K^m; Condition 2 guarantees that ψ maps vertices of L^m to vertices of K^m; finally, Condition 4 guarantees that ψ preserves the dimension and the boundary of simplexes in L^m.

Fig. 1 shows an example of the divided image of a complex containing two simplexes. In the figure, solid lines show the boundary of L^2 and their image under ψ, in K^2.

Consider a set σ^m and let $M(\sigma^m)$ be the complex consisting of σ^m and all its proper subsets; $M(\sigma^m)$ is an m-pseudomanifold consisting of a single m-simplex and all its faces. Of particular importance for us is the case where K^m is a divided image of $M(\sigma^m)$. In this case, $\psi(\sigma^m) = K^m$, by Definition 1(4).

Remark. The concept of a divided image is reminiscent of the notion of *acyclic carrier*[7] of [19], in that it associates subcomplexes of one complex to simplexes of another. Munkres uses acyclic carriers to study *subdivisions,* a fundamental concept of algebraic topology (cf. [19, 21]). However, divided images differ from subdivisions, even if the requirement of connectivity is added. For example, a 2-dimensional torus with a triangle removed from its surface is a divided image of a 2-simplex, since its boundary is a 1-dimensional triangle. However it is

[5] In algebraic topology, pseudomanifolds are assumed to have additional properties, which we do not require for our applications.

[6] Notice that $bound(\sigma)$ is a set of simplexes, and $\psi(bound(\sigma))$ is the complex which is the union over these simplexes τ of $\psi(\tau)$.

[7] Not to be confused with the notion of *carrier* defined later.

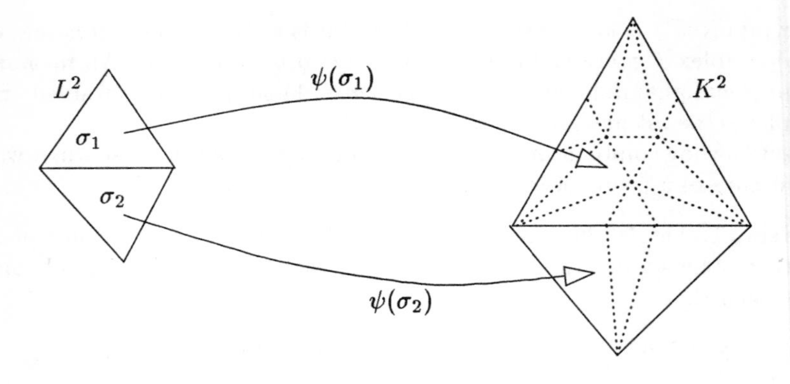

Fig. 1. K^2 is a divided image of L^2 under ψ.

neither an acyclic carrier nor a subdivided simplex since it has "holes" (non-trivial homology groups).

The next proposition states some simple properties of divided images; its proof is left to the full paper.

Proposition 2. *Let K^m be a divided image of L^m under ψ.*

(i) For every σ, $\sigma' \in L^m$, if $\sigma' \subseteq \sigma$, then $\psi(\sigma') \subseteq \psi(\sigma)$.
(ii) For every pair of j-simplexes $\sigma_1^j, \sigma_2^j \in L^m$, if $\sigma_1^j \neq \sigma_2^j$, and $\sigma_1^j \cap \sigma_2^j \neq \emptyset$, then $\psi(\sigma_1^j) \cap \psi(\sigma_2^j)$ is a pseudomanifold of dimension strictly smaller than j.
(iii) For every i-simplex $\sigma^i \in L^m$, $\psi(\sigma^i)$ is a divided image of $M(\sigma^i)$ under $\psi|M(\sigma^i)$.
(iv) A simplex $\tau^{m-1} \in K^m$ is external if and only if for some external simplex $\sigma^{m-1} \in L^m$, $\tau^{m-1} \in \psi(\sigma^{m-1})$.

The *carrier* of a simplex $\tau \in K^m$, denoted $carr(\tau)$, is the simplex $\sigma \in L^m$ of smallest dimension such that $\tau \in \psi(\sigma)$. Intuitively, the carrier of a simplex τ is the "smallest" simplex in L^m which is mapped to τ. By Definition 1(1), every simplex $\tau \in K^m$ is in $\psi(\sigma)$, for some $\sigma \in L^m$. By Proposition 2(ii), the carrier is unique. Therefore, the carrier is well-defined.

Connectivity: For any j, $0 < j \leq m$, the *j-graph of K^m* consists of one vertex for every j-simplex of K^m, and there is an edge between two vertices if and only if their intersection is a $(j-1)$-simplex of K^m. K^m is *j-connected* if its j-graph is connected; K^m is *0-connected* if it consists of a single vertex.

Lemma 3. *Let K^m be a divided image of σ^m under ψ. There exists a complex $\tilde{K}^m$, $\tilde{K}^m \subseteq K^m$, and $\tilde{\psi}$, a restriction of ψ to $\tilde{K}^m$, such that $\tilde{K}^m$ is a divided image of σ^m under $\tilde{\psi}$, $\tilde{\psi}(\sigma^i)$ is i-connected, and $bound(\tilde{\psi}(\sigma^i))$ is $(i-1)$-connected, for every $i > 1$, and every $\sigma^i \in \sigma^m$.*

Colorings: A complex K is *colored* by associating a value from some set of *colors* with each of its vertices. A coloring is *proper* if different vertices in the same simplex have different colors. A simplicial map $f : K \rightarrow L$ is *color preserving* if for every vertex v of K, color(v) = color$(f(v))$. Note that if a coloring is proper and a simplicial map is color preserving, then for any simplex $\{v_0, \ldots, v_k\}$ the vertices $f(v_0), \ldots, f(v_k)$ are different, i.e., $f(\sigma)$ is of the same dimension as σ.

Let K^m be a divided image of L^m. A simplicial map $\chi : K^m \rightarrow L^m$ is a *Sperner coloring* if for every $v \in K^m$, $\chi(v) \in carr(v)$. Intuitively, χ "folds" K^m into L^m with the requirement that each vertex of K^m goes to a vertex of its carrier. The main combinatorial definition we use is:

Definition 4. A complex K^m is a *(weak) chromatic divided image* of L^m, if it is a (weak) divided image of L^m with a proper Sperner coloring χ.

Let K^m be a divided image of $M(\sigma^m)$. The next well-known lemma says that an odd number of m-simplexes of K^m must go to σ^m (and in particular, at least one simplex). This lemma is used in Section 7; it follows from the Index Lemma (Lemma 17), presented later.

Lemma 5 (Sperner Lemma). *Consider a divided image K^m of $M(\sigma^m)$ under ψ, and a Sperner coloring $\chi : K^m \rightarrow M(\sigma^m)$. There exists an odd number of simplexes $\tau \in K^m$ with $\chi(\tau) = \sigma^m$.*

4 Modeling Tasks and Protocols

In this section we model distributed tasks using combinatorial topology; this is an adaptation of [17, 18] to our framework.

Tasks: Denote $ids = \{0, \ldots, n\}$. For some domain of values V, let $P(V)$ be the set of all pairs consisting of an id from ids and a value from V.

For a domain of inputs $\mathcal{I}$, an *input complex*, $\mathcal{I}^n$, is a complex that includes n-simplexes (i.e., subsets of $n+1$ elements) of $P(\mathcal{I})$ and all their faces, such that the vertices in an n-simplex have different id fields. For a domain of outputs $\mathcal{O}$, an *output complex*, $\mathcal{O}^n$, is defined similarly over $\mathcal{O}$. That is, if (i, val) is a vertex of $\mathcal{I}^n$ then val denotes an input value for process p_i, while if (i, val) is a vertex of $\mathcal{O}^n$ then val is an output value for process p_i. Note that $\mathcal{I}^n$ and $\mathcal{O}^n$ are properly colored by the id fields, and are full to dimension n. In addition, each complex is colored (not necessarily properly) by the corresponding domain of values.

Using the combinatorial topology notions, a *task* is identified with a triple $\langle \mathcal{I}^n, \mathcal{O}^n, \Delta \rangle$; $\mathcal{I}^n$ is an input complex, $\mathcal{O}^n$ is an output complex, and Δ maps each n-simplex of $\mathcal{I}^n$ to a non-empty set of n-simplexes in $\mathcal{O}^n$. We sometimes mention only Δ when $\mathcal{I}^n$ and $\mathcal{O}^n$ are clear from the context. The simplexes in $\Delta(\sigma^n)$ are the *admissible* output simplexes for σ^n. Intuitively, if σ^n is an input simplex and $\tau^n \in \Delta(\sigma^n)$ is an admissible simplex, then τ^n is an admissible output configuration when the system starts with input σ^n.

We extend Δ to simplexes of dimension smaller than n, i.e., for executions in which n processes or less take steps, as follows. Recall that it must be possible to complete the outputs of some processes in an execution to outputs for all processes that are allowed for the inputs of the execution. Therefore, Δ maps an input simplex σ of dimension smaller than n to the faces of n-simplexes in $\Delta(\sigma^n)$ with the same dimension and ids, for all input simplexes σ^n that contain σ. Extended in this manner, $\Delta(M(\sigma^n))$ is a subcomplex of $\mathcal{O}^n$. There is another variant of wait-free solvability, which allows to explicitly define Δ for simplexes of dimension smaller than n. This can be captured in our model by adding as part of the input a bit that tells the process whether to participate or not. Non-participating processes are required to output some default value.

Protocol Complexes: We say that a view of a process is *final* if the process has written an output. For an execution α, the set $\{(0, \alpha|0), \ldots, (n, \alpha|n)\}$ is denoted $views(\alpha)$. Given a protocol $\mathcal{P}$, the *protocol complex*, $\mathcal{P}^n$, is defined over the final views reachable in executions of $\mathcal{P}$, as follows. An n-simplex of final views is in $\mathcal{P}^n$ if and only if it is $views(\alpha)$ for some execution α of $\mathcal{P}$. In addition, $\mathcal{P}^n$ contains all the faces of the n-simplexes. The *protocol complex for an input n-simplex* σ, $\mathcal{P}^n(\sigma)$, is the subcomplex of $\mathcal{P}^n$ containing all n-simplexes corresponding to executions of $\mathcal{P}$ where processes start with inputs σ, and all their faces. Intuitively, $\tau \in \mathcal{P}^n(\sigma)$ if and only if there exists an execution α with initial values σ, such that the views of processes in τ are the same as in α. Note however that α is not necessarily unique.

The protocol complex, $\mathcal{P}^n$, is the union of the complexes $\mathcal{P}^n(\sigma)$, over all input n-simplexes σ. If a protocol is wait-free then $\mathcal{P}^n(\sigma)$ is finite, since a process terminates after a finite number of steps. Observe that the protocol complex depends not only on the possible interleavings of steps (schedules), but also on the transitions of processes and their local states. One can regard $\mathcal{P}^n$ as colored with four colors—an id, an input value, a view, and an output value. Note that the ids coloring is proper.

The protocol implies a *decision map* $\delta_{\mathcal{P}} : \mathcal{P}^n \rightarrow \mathcal{O}^n$, which specifies the output value for each final view of a process. When $\mathcal{P}$ solves Δ it holds that if $\tau \in \mathcal{P}^n$ then $\delta_{\mathcal{P}}(\tau)$ corresponds to an output simplex. Therefore, $\delta_{\mathcal{P}}$ is simplicial and preserves the ids coloring. Furthermore, for any input n-simplex σ, $\delta_{\mathcal{P}}(\mathcal{P}^n(\sigma))$ is a complex.

Since the protocol depends only on the input values, if two input n-simplexes σ, σ', have the same input values, i.e., differ only by a permutation of the ids, then $\mathcal{P}^n(\sigma)$ can be obtained from $\mathcal{P}^n(\sigma')$ by applying the same permutation to the ids. Therefore, the decision map must be *anonymous*; i.e., $\delta_{\mathcal{P}}(\mathcal{P}^n(\sigma))$ determines $\delta_{\mathcal{P}}(\mathcal{P}^n(\sigma'))$. If the protocol has to depend on the ids, then they have to be given as part of the inputs.

The above definitions imply:

> $\mathcal{P}$ solves $\langle \mathcal{I}^n, \mathcal{O}^n, \Delta \rangle$ if and only if $\delta_{\mathcal{P}}(\mathcal{P}^n(\sigma)) \subseteq \Delta(M(\sigma))$, for every n-simplex $\sigma \in \mathcal{I}^n$.

We say that $\delta_{\mathcal{P}}$ *agrees* with Δ.

Round #	1	2	3
p_0	w r		
p_1		w r	
p_2	w r		

(a) Execution α_1.

Round #	1	2	3
p_0	w r		
p_1			w r
p_2		w r	

(b) Execution α_2.

Fig. 2. Executions α_1 and α_2 are indistinguishable to p_1 and p_2.

This is the topological interpretation of the operational definition of a protocol solving a task (presented at the end of Section 2).

5 A Condition for Wait-Free Solvability

In this section we define immediate snapshot executions and prove that the subcomplex they induce is a chromatic divided image of the input complex. This implies a necessary condition for tasks which are solvable by a wait-free protocol. This condition is also sufficient since immediate snapshot executions can be emulated in any execution.

An *immediate snapshot* execution (in short, ISE) of a protocol is a sequence of *rounds*, defined as follows. Round k is specified by a *concurrency class* (called a *block* in [20]) of process ids, s_k. The processes in s_k are called *active* in round k. In round k, first each active process performs a write operation (in increasing order of ids), and then each active process reads all the registers, i.e., performs $n+1$ read operations (in increasing order of ids). We assume that the concurrency class is always non-empty. It can be seen that, for a given protocol, an immediate snapshot execution, α, is completely characterized by the sequence of concurrency classes. Therefore, we can write $\alpha = \langle s_1, s_2, \ldots, s_l \rangle$.

Immediate snapshot executions are of interest because they capture the computational power of the model. That is, a task Δ is wait-free solvable if and only if there exists a wait-free protocol which solves Δ in immediate snapshot executions (this is shown as part of the proof of Theorem 14 below). Although they are very well-structured, immediate snapshot executions still contain some degree of uncertainty, since a process does not know exactly which processes are active in the last round round. That is, if p_i is active in round k and observes some other process p_j to be active (i.e., perform a write), p_i does not know whether p_j is active in round $k-1$ or in round k.

Consider for example, Fig. 2. Only p_0 distinguishes between executions α_1 and α_2; p_1 and p_2 have the same views in both executions and cannot distinguish between them. However, as we prove below (in Proposition 8) this is the only uncertainty processes have in immediate snapshot executions.

Denote the subcomplex of the protocol complex which contains all immediate snapshot executions by $\mathcal{E}^n$. For an input simplex $\sigma^n \in \mathcal{I}^n$, $\mathcal{E}^n(\sigma^n)$ is the subcomplex of all immediate snapshot executions starting with σ^n.

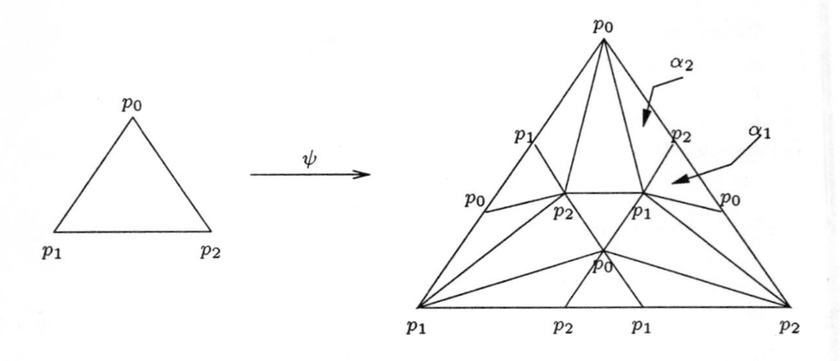

Fig. 3. The ISE complex, when each process takes at most one step.

We now show that if the protocol is wait-free and uses only read/write operations, then the ISE complex is a divided image of the input complex. This is done by defining a function ψ that assigns a subcomplex of $\mathcal{E}^n$ to each simplex of $\mathcal{I}^n$.

Fig. 3 contains an example of an immediate snapshot executions complex for a single input simplex. This is the complex where each process takes at most one step. Note that there are simplexes that correspond to the executions α_1 and α_2 from Fig. 2. Indeed, the vertices that correspond to p_1 and to p_2 are the same in these simplexes, i.e., p_1 and p_2 have the same views.

First, we need a few simple definitions. For a simplex σ of $\mathcal{I}^n$, $\mathcal{O}^n$, or $\mathcal{E}^n$, let $ids(\sigma)$ be the set of ids appearing in vertices of σ. For a simplex σ of $\mathcal{I}^n$ or $\mathcal{E}^n$, let $inputs(\sigma)$ be the set of pairs of inputs with corresponding ids appearing in vertices of σ. Finally, for a simplex σ of $\mathcal{E}^n$, let $views(\sigma)$ be the set of views appearing in vertices of σ and let $observed(\sigma)$ be the set of ids of processes whose operations appear in $views(\sigma)$.[8] Intuitively, if p_i is not in $observed(\sigma)$, then the views in σ are the same as in an execution in which p_i does not take a step. Notice that $ids(\tau) \subseteq observed(\tau)$, since a process always "observes itself."

We can now define ψ. For $\sigma \in \mathcal{I}^n$, $\psi(\sigma)$ is the complex containing all simplexes $\tau \in \mathcal{E}^n$ such that $ids(\tau) = ids(\sigma)$, $inputs(\tau) = inputs(\sigma)$, and $observed(\tau) = ids(\sigma)$, and all their faces. Notice that $\psi(\sigma)$ is full to dimension $dim(\sigma)$. A fact we use later is:

Proposition 6. *For any $\tau \in \mathcal{E}^n$ and $\sigma \in \mathcal{I}^n$, $\tau \in \psi(\sigma)$ if and only if $ids(\tau) \subseteq ids(\sigma)$, $inputs(\tau) \subseteq inputs(\sigma)$, and $observed(\tau) \subseteq ids(\sigma)$.*

Proof. If τ' is a face of τ, then $ids(\tau') \subseteq ids(\tau)$, $inputs(\tau') \subseteq inputs(\tau)$, and $observed(\tau') \subseteq observed(\tau)$. Thus, the definition of ψ implies that if $\tau \in \psi(\sigma)$ then $ids(\tau) \subseteq ids(\sigma)$, $inputs(\tau) \subseteq inputs(\sigma)$, and $observed(\tau) \subseteq ids(\sigma)$.

[8] Recall that these ids are not known by the processes', unless explicitly given in the inputs. To make this definition concrete, a special part of the process' state captures its identity. We defer the exact details to the full version.

Now, assume $ids(\tau) \subseteq ids(\sigma)$, $inputs(\tau) \subseteq inputs(\sigma)$, and $observed(\tau) \subseteq ids(\sigma)$. Since the protocol is wait-free, there exists an execution in which all processes in $ids(\sigma) - ids(\tau)$ (if any) observe only processes in $ids(\sigma)$, and processes in $ids(\tau)$ have the same views as in τ. Let π the simplex in $\mathcal{E}^n$ that corresponds to this execution. Note that $ids(\pi) = ids(\sigma)$, $inputs(\pi) = inputs(\sigma)$, and $observed(\pi) = ids(\sigma)$. Therefore, $\pi' \in \psi(\sigma)$. Since τ is a face of π, the claim follows. □

We first show that the ISE complex is a weak divided image of the input complex. In fact, this property does not depend on the protocol being wait-free or on the type of memory operations used, i.e., that the protocol uses only atomic read/write operations.

Lemma 7. *$\mathcal{E}^n$ is a weak chromatic divided image of $\mathcal{I}^n$ under ψ.*

Proof. Clearly, the process ids are a proper Sperner coloring of $\mathcal{E}^n$. We proceed to prove that the three conditions of weak divided images (Definition 1) hold.
Condition (1): Consider a simplex $\tau \in \mathcal{E}^n$. Let $\tau^n \in \mathcal{E}^n$ be such that $\tau \subseteq \tau^n$. Then there is a simplex $\sigma^n \in \mathcal{I}^n$ with $ids(\tau^n) = ids(\sigma^n)$ and $inputs(\tau^n) = inputs(\sigma^n)$. Since $observed(\tau^n) = ids(\sigma^n)$, $\tau^n \in \psi(\sigma^n)$. Since τ is a face of τ^n, $\tau \in \psi(\sigma^n)$.
Condition (2) follows since the protocol is deterministic.
Condition (3) follows from Proposition 6: $\tau \in \psi(\sigma \cap \sigma')$ if and only if $ids(\tau) \subseteq ids(\sigma \cap \sigma') = ids(\sigma) \cap ids(\sigma')$, $inputs(\tau) \subseteq inputs(\sigma \cap \sigma') = inputs(\sigma) \cap inputs(\sigma')$, and $observed(\tau) \subseteq ids(\sigma \cap \sigma') = ids(\sigma) \cap ids(\sigma')$. This happens if and only if $\tau \in \psi(\sigma) \cap \psi(\sigma')$. □

We say that process p_j is *silent* in an execution α, if there exists a round $k > 0$, such that $p_j \notin s_r$ for every $r < k$ and $s_r = \{p_j\}$ for every round $r \geq k$. Intuitively, this means that no other process ever sees a step by p_j. If p_j is not silent in α, then it is seen in some round. Formally, a process p_j is *seen* in round k if $p_j \in s_r$ for some $r \leq k$, and there exists some process $p'_j \neq p_j$, such that $p'_j \in s_k$. The *last seen round* of p_j is the largest round k in which p_j is seen. These definitions imply:

Proposition 8. *Consider a finite immediate snapshot execution α. If p_j is not silent in α, then k is the last seen round of p_j in α if and only if (a) $s_r = \{p_j\}$ for every round $r > k$, (b) $s_k \neq \{p_j\}$, and (c) either (i) $p_j \in s_k$ or (ii) $s_{k-1} = \{p_j\}$.*

As a consequence, we have the next lemma.

Lemma 9. *Consider an immediate snapshot execution complex $\mathcal{E}^n$. Let τ_1^i be an i-simplex of $\mathcal{E}^n$ corresponding to an execution α, and $p_i \in ids(\tau_1^i)$.*

(i) If p_i is not silent in α, then there exists τ_2^i, another i-simplex of $\mathcal{E}^n$, that differs only in p_i's view, corresponding to α'.
(ii) If there exists τ_2^i, another i-simplex of $\mathcal{E}^n$, that differs only in p_i's view, corresponding to α', then p_j is not silent in α, α'. If k is the last seen round of p_j in α, then, without loss of generality, p_j is in the kth concurrency class of τ_1^i and the kth concurrency class of τ_2^i is $\{p_j\}$.

Lemma 10. *For every simplex* $\sigma^i \in \mathcal{I}^n$, $\psi(bound(\sigma^i)) = bound(\psi(\sigma^i))$.

Proof. Let $\tau \in \psi(bound(\sigma^i))$. Then $\tau \in \psi(\sigma^{i-1})$, for some face σ^{i-1} of σ^i. Let $\tau^{i-1} \in \psi(\sigma^{i-1})$ such that $\tau \subseteq \tau^{i-1}$. It follows from the definition of ψ that $\tau^{i-1} \in \psi(\sigma^i)$. Let $\tau^i \in \psi(\sigma^i)$, such that $\tau^{i-1} \subset \tau^i$. To show that $\tau^{i-1} \in bound(\psi(\sigma^i))$, notice that $observed(\tau^{i-1}) = ids(\sigma^{i-1})$. Let p_j be a process id in $ids(\sigma^i) - ids(\sigma^{i-1})$. Since nobody in τ^{i-1} sees a step by p_j, and in τ^i, p_j does not see a step by any process not in $ids(\sigma^i)$, it follows that p_j's view is determined (because the protocol is deterministic). Namely, τ^{i-1} is contained in a single i-simplex τ^i, and hence $\tau^{i-1} \in bound(\psi(\sigma^i))$. Since τ is a face of τ^{i-1}, by the definition of ψ, $\tau \in bound(\psi(\sigma^i))$.

The other direction of the proof is similar. Since $\tau \in bound(\psi(\sigma^i))$ it follows that τ is a face of some $\tau^{i-1} \in bound(\psi(\sigma^i))$. This implies that τ^{i-1} is a face of a single τ^i. Therefore, if $p_j \in ids(\tau^i) - ids(\tau^{i-1})$, then p_j is not in $observed(\tau^{i-1})$ (by Lemma 9(i)). Hence, $observed(\tau^{i-1}) = ids(\sigma^{i-1})$. It follows that $\tau^{i-1} \in \psi(\sigma^{i-1})$, and thus $\tau^{i-1} \in \psi(bound(\sigma^i))$. This implies that $\tau \in \psi(bound(\sigma^i))$. □

Intuitively, the next lemma implies that once we fix the views of all processes but one, the remaining process may have only one of two views, which correspond to the two options in Proposition 8(c). This shows that the uncertainty about another process is restricted to its last seen round.

Lemma 11. *For every simplex* $\sigma^i \in \mathcal{I}^n$, $\psi(\sigma^i)$ *is an* i*-pseudomanifold.*

Proof. As noted before, $\psi(\sigma^i)$ is full to dimension i. We show that any simplex $\tau^{i-1} \in \psi(\sigma^i)$, is contained in at most two i-simplexes. Let $\tau^i \in \psi(\sigma^i)$ be such that τ^{i-1} is a face of τ^i. Since τ^{i-1} and τ^i are properly colored by the ids, there exists some id p_j, such that p_j appears in τ^i but not in τ^{i-1}. In fact, any i-simplex of $\psi(\sigma^i)$ containing τ^{i-1} includes p_j. Let α be the prefix of an execution with steps by processes in $ids(\sigma^i)$, corresponding to τ^i. We can take such a prefix because $observed(\tau^i) = ids(\sigma^i)$. There are two cases:
Case 1: p_j is silent in α. Then $observed(\tau^{i-1}) = ids(\sigma^{i-1})$. Since p_j does not see an id not in $ids(\sigma^i)$, its view is determined. Hence, τ_i is unique.
Case 2: p_j is not silent in α. Let k be the last seen round of p_j in α. Lemma 9(ii) implies that that there are only two possible views for p_j, compatible with the views in τ^{i-1}: either $p_j \in s_k$ or $s_{k-1} = \{p_j\}$. □

By Lemma 7, $\mathcal{E}^n$ is a weak chromatic divided image of $\mathcal{I}^n$. Lemma 11 and Lemma 10 imply Condition (4) of Definition 1. Therefore, we have:

Theorem 12. $\mathcal{E}^n$ *is a chromatic divided image of* $\mathcal{I}^n$ *under* ψ.

This implies the following necessary condition for wait-free solvability:

Corollary 13. *Let* $\langle \mathcal{I}^n, \mathcal{O}^n, \Delta \rangle$ *be a task. If there exists a wait-free protocol which solves this task then there exists a chromatic divided image* $\mathcal{E}^n$ *of* $\mathcal{I}^n$ *and a color-preserving (on ids), anonymous simplicial map* δ *from* $\mathcal{E}^n$ *to* $\mathcal{O}^n$ *that agrees with* Δ.

We now restrict our attention to *full-information* protocols, in which a process writes its whole state in every write to its shared register. The complex induced by immediate snapshot executions of the full-information protocol for some input complex $\mathcal{I}^n$ is called the ***full divided image*** of $\mathcal{I}^n$. We have the following necessary and sufficient condition for wait-free solvability.

Theorem 14. *Let $\langle \mathcal{I}^n, \mathcal{O}^n, \Delta \rangle$ be a task. There exists a wait-free protocol which solves this task if and only if there exists an full divided image $\mathcal{E}^n$ of $\mathcal{I}^n$ and a color-preserving (on ids), anonymous simplicial map δ from $\mathcal{E}^n$ to $\mathcal{O}^n$ that agrees with Δ.*

Sketch of proof. Assume there exists a protocol $\mathcal{P}$ which solves Δ. Without loss of generality, we may assume that in $\mathcal{P}$ each process operates by writing and then reading the registers $R_0, \ldots, R_n$. Since $\mathcal{P}$ solves Δ, it must solve Δ in immediate snapshot executions. By Theorem 12, the ISE complex, $\mathcal{E}^n$, is a chromatic divided image of $\mathcal{I}^n$. Since the protocol can be simulated by a full-information protocol, the corresponding full divided image is also a chromatic divided image of $\mathcal{I}^n$. Clearly, $\delta_{\mathcal{P}}$ is a color-preserving (on ids), anonymous simplicial map from $\mathcal{E}^n$ to $\mathcal{O}^n$ that agrees with Δ.

Assume there exists an full divided image $\mathcal{E}^n$ of $\mathcal{I}^n$ and a color-preserving (on ids), anonymous simplicial map δ from $\mathcal{E}^n$ to $\mathcal{O}^n$ that agrees with Δ. By using a protocol for the participating set problem ([5]), the immediate snapshot executions can be simulated in a full-information manner. Using δ as the output rule of the protocol, we get the "only if" direction of the theorem. □

Remark. The above theorem ignores the issue of computability. Clearly, the sufficient condition requires that δ is computable; furthermore, if a task is solvable then it implies a way to compute δ. Therefore, we can add the requirement that δ is computable to the necessary and sufficient condition for wait-free solvability.

The previous theorem provides a characterization of wait-free solvable tasks which depends only on the topological properties of $\langle \mathcal{I}^n, \mathcal{O}^n, \Delta \rangle$. To see if a task is solvable, when the input complex is finite, we produce all $\mathcal{E}$-divided images of $\mathcal{I}^n$ and check if a simplicial map δ as required exists. Note that if we are interested only in protocols that are bounded wait-free, i.e., where the protocol has to hold within a predetermined number of steps N, then producing all $\mathcal{E}$-divided images of the input complex (which is finite) is recursive.

Orientability: We now show that the ISE complex, $\mathcal{E}^n$, is an orientable chromatic divided image. This is used to prove that it induces an algebraic span [15]. We leave the proof that an orientable chromatic divided image induces an algebraic span to the full paper, since obviously, it requires the definition of algebraic span, an algebraic concept of a different flavor from the rest of this paper.

Let K^m be an m-pseudomanifold. An *orientation* of a simplex is an equivalence class of orderings of its vertices, consisting of one particular ordering and all even permutations of it. If the vertices are colored with ids, we could consider the *positive* orientation to be the one in which the vertices are ordered

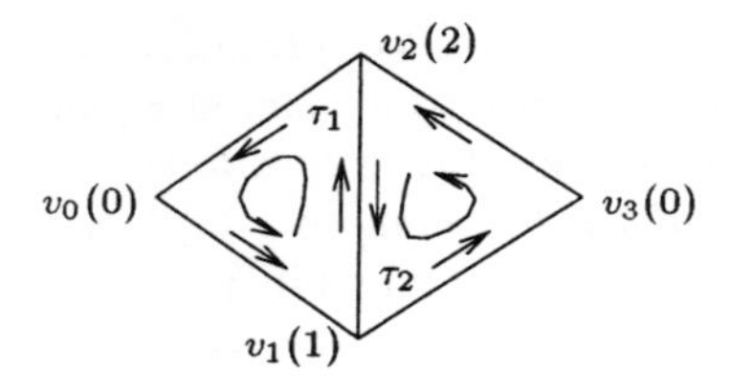

Fig. 4. An oriented 2-pseudomanifold, with a coloring (in brackets).

with the ids from small to large, and the *negative* to be the one where the two vertices with smallest ids are exchanged (each orientation together with all its even permutations). Denote by $\sigma^{(i)}$ the face of σ^m in which the vertex with id i is removed; e.g., $\sigma^{(1)}$ is the face with ids $\{0, 2, \ldots, m\}$. An orientation of an m-simplex induces an orientation on each of its faces, $\sigma^{(i)}$, according to the sign of $(-1)^i$. For example, if σ^2 is oriented $\langle v_0, v_1, v_2 \rangle$, then the induced orientations are $\langle v_0, v_1 \rangle$, $\langle v_1, v_2 \rangle$, and $\langle v_2, v_0 \rangle$.

K^m is *orientable* if there exists an orientation for each of its m-simplexes such that an $m-1$-simplex contained in two m-simplexes gets opposite induced orientations. K^m together with such an orientation is an *oriented* pseudomanifold. (See an example in Fig. 4 of a simple oriented 2-pseudomanifold and the induced orientations.)

In the sequel, we sometimes use a combinatorial notion of orientability. In the full paper, we prove that the previous (usual) definition of orientability is equivalent to the combinatorial definition, for chromatic pseudomanifolds.

Lemma 15. *A chromatic pseudomanifold K^m is orientable if and only if its m-simplexes can be partitioned into two disjoint classes, such that if two m-simplexes share an $(m-1)$-face then they belong to different classes.*

We say that a chromatic divided image of $M(\sigma^m)$ under ψ, K^m, is *orientable* if, for every $\sigma \in M(\sigma^m)$, $\psi(\sigma)$ is orientable.

Theorem 16. *Let $\mathcal{E}^n$ be a chromatic divided image of $M(\sigma^n)$ under ψ, that corresponds to the ISE complex starting with input σ^n, in which any processor takes the same number of steps in every execution. Then $\mathcal{E}^n$ is orientable.*

Proof. Let σ^i be a face of σ^m. We explicitly partition the i-simplexes of $\psi(\sigma^i)$ into two disjoint classes, positive and negative.

Let the *length* of an immediate snapshot execution be the number of concurrency classes in it. An i-simplex $\tau \in \psi(\sigma^i)$ is in positive if the length of the immediate snapshot execution corresponding to τ is even; otherwise, it is in negative. Consider two i-simplexes, τ_1^i and τ_2^i, that share an $(i-1)$-face, and let p_j be the processor whose view is different. By Lemma 9, without loss of generality, p_j is in the kth concurrency class of τ_1^i and the kth concurrency class of τ_2^i is

$\{p_j\}$, where k is the last seen round of p_j in τ_1^i. Furthermore, since the views of all other processors are exactly the same, it follows that the lengths of the corresponding executions differ exactly by one. Hence, the corresponding simplexes are in different classes, i.e., have different orientations. □

6 The Number of Monochromatic Simplexes

In this section we prove a combinatorial lemma about the number of monochromatic simplexes in any binary coloring of an orientable divided image; this lemma is used in the next section to show a lower bound on renaming.

Let K^m be an orientable, chromatic divided image of σ^m under ψ. Fix an orientation of K^m, and an induced orientation on its boundary. K^m is *symmetric* if, for any two i-faces of σ, σ_1^i and σ_2^i, $\psi(\sigma_1^i)$ and $\psi(\sigma_2^i)$ are isomorphic, under a one-to-one simplicial map ζ that is order preserving on the ids: if v and w belong to the same simplex, and $id(v) < id(w)$ then $id(\zeta(v)) < id(\zeta(w))$. A binary coloring, b, of K^m is *symmetric*, if $b(\zeta(v)) = b(v)$, for every vertex $v \in K^m$. This definition is motivated by the notion of comparison-based protocols for renaming, presented in the next section.

Let $\#\text{mono}(K^m)$ be the number of monochromatic m-simplexes of K^m, counted by orientation, i.e., an m-simplex is counted as $+1$ if it is positively oriented, otherwise, it is counted as -1. For example, if K^m consists of just two m-simplexes, both monochromatic, then the count would be 0, since they would have opposite orientations, and hence one would count $+1$ and the other -1.

The main theorem of this section states that, if K^m is a symmetric, oriented chromatic divided image of σ^m under ψ, with a symmetric binary coloring b, then $\#\text{mono}(K^m) \neq 0$. The proof of this theorem relies on the Index Lemma—a classical result of combinatorial topology, generalizing Sperner's Lemma (cf. [12, p. 201]).

To state and prove the Index Lemma, we need the following definitions. Fix a coloring c of K^m with $\{0, \ldots, m\}$. A k-simplex of K^m is *complete* under c, if it is colored with $0, \ldots, k$. The *content*, C, of c is the number of complete m-simplexes, counted by orientation. That is, a complete simplex τ^m is counted $+1$, if the order of the vertices given by the colors agrees with the orientation of τ^m, and -1, otherwise; i.e, it counts $+1$ if the order given by the colors belongs to the equivalence class of orderings of the orientation, and else it counts -1. For example, the 2-simplex τ_1 in Fig. 4 is ordered $\langle v_0, v_1, v_2 \rangle$, and the colors are under this order are $\langle 0, 1, 2 \rangle$, hence, it would count $+1$. On the other hand, the 2-simplex τ_2 in Fig. 4 is ordered $\langle v_1, v_3, v_2 \rangle$, and the colors are under this order are $\langle 1, 0, 2 \rangle$, hence, it would count -1. The *index*, I, of c is the number of complete $(m-1)$-simplexes on the boundary of K^m, also counted by orientation (the orientation induced by the unique m-simplex that contains it).

Lemma 17 (Index Lemma). $I = C$.

Proof. Let S be the number of complete $(m-1)$-simplexes in K^m (colored with $\{0, \ldots, m-1\}$), where the $(m-1)$-simplexes in each m-simplex are considered separately, and counted as +1 or −1, by their induced orientations. We argue that $S = I$ and $S = C$.

To prove that $S = I$, consider the following cases. If an $(m-1)$-face is internal, then it contributes 0 to S, since the contributions of the two m-simplexes containing it cancel each other. Obviously, an internal $(m-1)$-face contributes 0 to I. An external $(m-1)$-face in the boundary of K^m is counted the same, +1 or −1 by orientation, in both S and I. Therefore, $S = I$.

To prove that $C = S$, consider an m-simplex τ^m, and look at the following cases. If τ^m contains two $(m-1)$-faces which are completely colored, then τ^m is not completely colored and contributes 0 to C. Note that τ^m contributes 0 also to S, since the contributions of the two faces cancel each other. If τ^m contains exactly one $(m-1)$-face which is completely colored (with $\{0, \ldots, m-1\}$), then τ^m must be completely colored and contributes +1 or −1, by orientation, to C as well as to S. If τ^m does not contain any $(m-1)$-face which is completely colored, then τ^m is not completely colored and therefore, it contributes 0 to C as well as to S. Finally, note that τ^m cannot contain more than two $(m-1)$-faces which are completely colored. □

Theorem 18 (Binary Coloring Theorem). *Let K^m be a symmetric, oriented chromatic divided image of σ^m under ψ, with a symmetric binary coloring b. Then $\#mono(K^m) \neq 0$.*

Proof. Let ρ be the simplicial map from σ^m to itself that maps the vertex v whose id is i to the vertex whose id is $(i+1) \bmod (m+1)$ (that is, the mapping the rotates the id's). In the rest of the proof, we assume that sub-indices are taken modulo $m+1$.

Define a coloring of K^m, $c(v) = (b(v) + id(v)) \bmod (m+1)$, for every v. Notice that an m-simplex, τ^m, is completely colored by c if and only if τ^m is monochromatic under b. Moreover, for every v, $c(\rho(v)) = (c(v)+1) \bmod (m+1)$. Let C and I be the content and index of K^m under c. Clearly, $C = \#\text{mono}(K^m)$. By the Index Lemma (Lemma 17), $C = I$, and therefore, it suffices to prove that $I \neq 0$. The proof is by induction on m.

The base case is when $m = 0$, K^0 has only one vertex; in this case, trivially, $I \neq 0$.

For the induction step, we consider $bound(K^m)$, and "squeeze" it, by using contractions. A *contraction* of $bound(K^m)$ is obtained by identifying one of its vertices, v', with another vertex, v, with the same color, and deleting any simplex containing both v and v'.

Consider an internal $(m-2)$-simplex, $\tau^{m-2} \in bound(K^m)$, which is contained in two $(m-1)$-simplexes, τ_1 and τ_2. Its *link* vertices are v_1, which is the vertex of τ_1 not in τ^{m-2}, and v_2, which is the vertex of τ_2 not in τ^{m-2}. A binary coloring is *irreducible* if the link vertices of any internal $(m-2)$-simplex simplex of $\psi(\sigma^{(i)})$ have different binary colors.

The first stage of the proof applies a sequence of specific contractions to $bound(K^m)$, to make sure its coloring is irreducible, while preserving all other properties.

The contractions we apply are *symmetric contractions*, in which we choose an internal $(m-2)$-simplex, $\tau^{m-2} \in \psi(\sigma^{(m)})$, to which a contraction can be applied; that is, such that its two link vertices have the same binary coloring. We contract τ^{m-2} and simplexes symmetric to it in $\psi(\sigma^{(i)})$, for all i. (This is a sequence of $m+1$ contractions.) Notice that the simplexes which are symmetric to τ^{m-2} are also internal and their link vertices have the same binary coloring.

A boundary is *proper symmetric* if it is the boundary of a symmetric, oriented chromatic divided image of σ^m under ψ, with a symmetric binary coloring b. In the next claim we show that a symmetric contraction preserves all properties of a proper symmetric boundary.

Claim 19 *Assume we apply a symmetric contraction to $bound(K^m)$, and get a complex $bound'$. Then $bound'$ is a non-empty, proper symmetric boundary under ψ', with $\psi'(bound(\sigma^m)) = bound'$, and $\#mono(\psi'(\sigma^{(i)})) \neq 0$, for every i. Furthermore, $I(bound') = I(bound(K^m))$.*

Proof. Given $\tau^{m-2} \in \psi(\sigma^{(i)})$, with link vertices v_1 and v_2; note that we have that $id(v_1) = id(v_2)$. Therefore, $bound'$ is chromatic. Also, it is easy to see that the orientation on $bound'$ is still well defined: two $(m-1)$-simplexes that did not have an $(m-2)$-face in common before the contraction will have it after the contraction, only if they differ in exactly one vertex, in addition to v_1 and v_2. Thus, two such simplexes have opposite orientations. By the definition of symmetric contraction, $bound'$ remains symmetric.

By induction hypothesis of the theorem, $\#\text{mono}(\psi(\sigma^{(i)})) \neq 0$, for every i. Since a contraction removes simplexes with opposite orientations and the same binary colorings, $\#\text{mono}(\psi(\sigma^{(i)})) = \#\text{mono}(\psi'(\sigma^{(i)})) \neq 0$, for every i, and $I(bound') = I(bound(K^m))$. This implies that $bound'$ is non-empty. □

By Claim 19, for the rest of the proof, we can assume that $\psi(bound(\sigma^m)) = bound(K^m)$ is a non-empty, proper symmetric boundary with an irreducible binary coloring. Recall that $c(v) = (b(v) + id(v)) \bmod (m+1)$.

Claim 20 *All complete $(m-1)$-simplexes on the boundary of K^m are counted with the same sign by I.*

Proof. We first argue that every complete $(m-1)$-simplex in $\psi(\sigma^{(i)})$ is counted with the same sign by I, for any i. To see this, assume, without loss of generality, that $i = m$, and consider an $(m-1)$-simplex with ids $0, \ldots, m-1$, colored with 0, i.e., $\tau_1 = \{(0,0), (1,0), \ldots, (m-1,0)\}$ (the first component of a vertex is the id and the second is its binary color).

Consider a path to any other $(m-1)$-simplex τ_2 colored with the same ids and colors; such a path must exist since, by Lemma 3, we can assume that $\psi(\sigma^{(i)})$ is $(m-1)$-connected. Notice that the colors assigned by c are the same in τ_1 and τ_2, and thus both τ_1 and τ_2 will be counted by I. It remains to show that τ_1 and τ_2 have the same orientation and hence are counted with the same sign by I.

Note that this path consists of a sequence of $(m-1)$-simplexes, each sharing an $(m-2)$-face with the previous simplex, and differing from the previous simplex in exactly one binary color. Thus the path corresponds to a sequence of binary vectors, starting with the all 0's vector and ending with the all 0's vector, and each vector differing from the previous vector in exactly one binary color. That is, the path corresponds to a cycle in a hypercube graph. Since the hypercube graph is bipartite, the length of any cycle in it is even; therefore, the length of the path is even. Clearly, since the complex is oriented, consecutive simplexes on the path have different orientations. Since the length of the path is even, τ_1 and τ_2 have the same orientation. Hence, τ_1 and τ_2 are counted with the same sign by I.

Next, we show that complete $(m-1)$-simplexes in different $\psi(\sigma^{(i)})$'s are also counted with the same sign by I. Again, without loss of generality, assume that $\tau_1 = \{(0,0),(1,0),\ldots,(m-1,0)\} \in \psi(\sigma^{(m)})$, is counted by I. Note that the c coloring of τ_1 is $\{(0,0),(1,1),\ldots,(m-1,m-1)\}$.

We now show that any complete $(m-1)$-simplex $\tau_3 \in \psi(\sigma^{(i)})$ will be counted with the same sign by I. Without loss of generality, assume $i = 0$. Note that τ_3 is complete, with id's $\{1,2,\ldots,m\}$. Thus, the binary color of the vertex with process id m must be 1, in order to get the color 0 under c. This implies that the c coloring of τ_3 is $\{(1,1),(2,2),\ldots,(m,0)\}$ and its binary coloring is $\{(1,0),(2,0),\ldots,(m,1)\}$.

Consider the simplex $\tau_2 \in \psi(\sigma^{(0)})$, which is the image of τ_1 under the symmetry map, ρ. That is, $\tau_2 = \{(1,0),(2,0),\ldots,(m,0)\}$. Consider a path in $\psi(\sigma^{(0)})$ between τ_2 and τ_3. Since the binary coloring vector of τ_3 differs from the binary coloring vector of τ_2 in exactly one position, the length of this path must be odd. Therefore, τ_2 and τ_3 must have different orientations.

The c coloring of τ_3, $\{(1,1),(2,2),\ldots,(m,0)\}$, is rotated w.r.t. its ids, and hence the orderings of τ_2 and τ_3 agree (on the sign of the permutation) if and only if m is odd. E.g., if $m = 2$, they disagree—$\{(1,1),(2,0)\}$, and if $m = 3$, they agree—$\{(1,1),(2,2),(3,0)\}$. Finally, the orientation of τ_1 is $(-1)^m$ times the orientation of τ_2, since they are symmetric simplexes in $\psi(\sigma^{(m)})$ and $\psi(\sigma^{(0)})$. That is, the orientations of τ_1 and τ_2 agree when m is even, and disagree otherwise. Therefore, the orientations of τ_1 and τ_3 agree, and they are counted with the same sign by I. □

Since *bound* is non-empty and contains at least one simplex, Claim 20 implies $I \neq 0$, which proves the theorem. □

7 Applications

In this section, we apply the condition for wait-free solvability presented earlier (Corollary 13) to derive two lower bounds, for renaming and for k-set consensus. The first lower bound also relies on Theorem 18, and therefore, on the fact that the chromatic divided image induced by immediate snapshot executions is orientable. In the full version of the paper we also derive another necessary condition, based on connectivity.

7.1 Renaming

In the *renaming* task ([2]), processes start with an input value (*original name*) from a large domain and are required to decide on distinct output values (*new names*) from a domain which should be as small as possible. Clearly, the task is trivial if processes can access their id; in this case, process p_i decides on i, which yields the smallest possible domain. To avoid trivial solutions, it is required that the processes and the protocol are *anonymous* [2]. That is, process p_i with original name x executes the same protocol as process p_j with original name x.

Captured in our combinatorial topology language, the M-renaming task is the triple $\langle \mathcal{D}^n, M^n, \Delta \rangle$. $\mathcal{D}^n$ contains all subsets of some domain $\mathcal{D}$ (of original names) with different values, M^n contains all subsets of $[0..M]$ (of new names) with different values, and Δ maps each $\sigma^n \in \mathcal{D}^n$ to all n-simplexes of M^n. We use Theorem 12 and Theorem 18 to prove that there is no wait-free anonymous protocol for the M-renaming task, if $M \leq 2n-1$. The bound is tight, since there exists an anonymous wait-free protocol ([2]) for the $2n$-renaming problem.

Theorem 21. *If $M < 2n$, then there is no anonymous wait-free protocol that solves the M-renaming task.*

Proof. Assume, by way of contradiction, that $\mathcal{P}$ is a wait-free protocol for the M-renaming task, $M \leq 2n-1$. Without loss of generality, we assume that every process executes the same number of steps. Also, $\mathcal{P}$ is comparison-based, i.e., the protocol produces the same outputs on inputs which are order-equivalent. (See Herlihy [13], who attributes this observation to Eli Gafni).

Assume that $\mathcal{D} = [0..2n]$; i.e., assume that the original names are only between 0 and $2n$. By Corollary 13, there exists a a chromatic full divided image of the input complex $\mathcal{D}^n$, $\mathcal{S}$; let $\delta_{\mathcal{P}}$ be the decision map implied by $\mathcal{P}$. By Theorem 16, $\mathcal{S}$ is orientable. Since the protocol is comparison-based and anonymous, it follows that for any two i-simplexes, σ_1^i and σ_2^i of $\mathcal{D}^n$, $\delta_{\mathcal{P}}$ maps $\psi(\sigma_1^i)$ and $\psi(\sigma_2^i)$ to simplexes that have the same output values (perhaps with different process ids).

Let δ' be the binary coloring which is the parity of the new names assigned by $\delta_{\mathcal{P}}$. Therefore, the assumption of Theorem 18 is satisfied for $\mathcal{S}(\sigma^n)$, and therefore, at least one simplex of $\mathcal{S}(\sigma^n)$ is monochromatic under δ'.

On the other hand, note that the domain $[0, 2n-1]$ does not include $n+1$ different odd names; similarly, the domain $[0, 2n-1]$ does not include $n+1$ different even names. This implies that δ' cannot color any simplex of $\mathcal{S}$ with all zeroes or with all ones; i.e., no simplex of $\mathcal{S}$ is monochromatic. A contradiction. □

7.2 *k*-Set Consensus

Intuitively, in the *k-set consensus* task ([7]), processes start with input values from some domain and are required to produce at most k different output values. To assure non-triviality, we require all output values to be input values of some processes.

Captured in our combinatorial topology language, the k-set consensus task is the triple $\langle \mathcal{D}^n, \mathcal{D}^n, \Delta \rangle$. $\mathcal{D}^n$ is $P(\mathcal{D})$, for some domain $\mathcal{D}$, and Δ maps each $\sigma^n \in \mathcal{D}^n$ to the subset of n-simplexes in $\mathcal{D}^n$ that contain at most k different values from the values in σ^n. In the full version of the paper, we use Theorem 12 and Sperner's Lemma to prove that any wait-free protocol for this problem must have at least one execution in which $k+1$ different values are output. This implies:

Theorem 22. *If $k \leq n$ then there does not exists a wait-free protocol that solves the k-set consensus task.*

This bound is tight, by the protocol of [7].

8 Discussion

This paper presents a study of wait-free solvability based on combinatorial topology. Informally, we have defined the notion of a chromatic divided image, and proved that a necessary condition for wait-free solvability is the existence of a simplicial chromatic mapping from a divided image of the inputs to the outputs that agrees with the problem specification. We were able to use theorems about combinatorial properties of divided images to derive tight lower bounds for renaming and k-set consensus. Our results do not use homology groups, whose computation may be complicated. We also derive a new necessary and sufficient condition, based on a specific, well structured chromatic divided image.

Many questions remain open. First, it is of interest to find other applications to the necessary and sufficient condition presented here; in particular, can we derive interesting protocols from the sufficient condition? Second, there are several directions to extend our framework, e.g., to allow fewer than n failures (as was done for one failure in [3]), to handle other primitive objects besides read/write registers (cf. [14, 6]), and to incorporate on-going tasks.

Acknowledgments: We would like to thank Javier Bracho, Eli Gafni, Maurice Herlihy, Nir Shavit and Mark Tuttle for comments on the paper and very useful discussions.

References

1. Hagit Attiya, Nancy Lynch and Nir Shavit, "Are Wait-Free Algorithms Fast?" *Journal of the ACM*, 41(4), pages 725-763, July 1994.
2. Hagit Attiya, Amotz Bar-Noy, Danny Dolev, David Peleg, and Rudiger Reischuk, "Renaming in an asynchronous environment," *Journal of the ACM*, July 1990.
3. O. Biran, S. Moran, S. Zaks, "A combinatorial characterization of the distributed 1-solvable tasks," *Journal of Algorithms*, 11, pages 420–440, 1990.
4. E. Borowsky and E. Gafni, "Generalized FLP impossibility result for t-resilient asynchronous computations," in *Proceedings of the 1993 ACM Symposium on Theory of Computing*, pages 91–100, 1993.

5. E. Borowsky and E. Gafni, "Immediate atomic snapshots and fast renaming," in *Proceedings of the 12th Annual ACM Symposium on Principles of Distributed Computing*, pages 41–51, 1993.
6. E. Borowsky and E. Gafni, "The implication of the Borowsky-Gafni simulation on the set consensus hierarchy," Technical Report 930021, UCLA Computer Science Dept., 1993.
7. S. Chaudhuri, "More Choices Allow More Faults: Set Consensus Problems in Totally Asynchronous Systems," *Information and Computation,* 105 (1), pages 132–158, July 1993.
8. S. Chaudhuri, M.P. Herlihy, N. Lynch, and M.R. Tuttle, "A tight lower bound for k-set agreement," in *Proceedings of the 34th IEEE Symposium on Foundations of Computer Science*, October 1993.
9. D. Dolev, N. Lynch, S. Pinter, E. Stark and W. Weihl, "Reaching Approximate Agreement in the Presence of Faults," *Journal of the ACM,* 33 (3), pages 499–516, 1986.
10. M. Fischer, N.A. Lynch, and M.S. Paterson, "Impossibility of distributed commit with one faulty process," *Journal of the ACM*, 32(2), pages 374–382, 1985.
11. E. Gafni, E. Koutsoupias, "3-processor tasks are undecidable," Brief Announcement in *Proceedings of the 14-th Annual ACM Symposium on Principles of Distributed Computing*, page 271, August 1995.
12. M. Henle, *A Combinatorial Introduction to Topology,* Dover Pub, NY, 1994.
13. M.P. Herlihy, *A Tutorial on Algebraic Topology and Distributed Computation,* notes for a tutorial presented in UCLA, August 1995. Manuscript dated December 13, 1994.
14. M.P. Herlihy and S. Rajsbaum, "Set Consensus Using Arbitrary Objects," in *Proceedings of the 13th Annual ACM Symposium on Principles of Distributed Computing*, pages 324–333, August 1994.
15. M.P. Herlihy and S. Rajsbaum, "Algebraic Spans," in *Proceedings of the 14th Annual ACM Symposium on Principles of Distributed Computing*, pages 90-99, August 1995.
16. M.P. Herlihy and S. Rajsbaum, "On the Decidability of Distributed Decision Tasks," *http://www.cs.brown.edu/people/mph/decide.html.* Brief Announcement in *Proceedings of the 15th Annual ACM Symposium on Principles of Distributed Computing*, page 279, May 1996.
17. M.P. Herlihy and N. Shavit, "The asynchronous computability theorem for t-resilient tasks," In *Proceedings of the 1993 ACM Symposium on Theory of Computing*, pages 111-120, May 1993.
18. M.P. Herlihy and N. Shavit, "A simple constructive computability theorem for wait-free computation," In *Proceedings of the 1994 ACM Symposium on Theory of Computing*, May 1994. Full version of [17] and [18] appeared as Brown University Technical Report CS-96-03.
19. J.R. Munkres, *Elements of Algebraic Topology,* Addison-Wesley, 1993.
20. M. Saks and F. Zaharoglou, "Wait-free k-set agreement is impossible: The topology of public knowledge," In *Proceedings of the 1993 ACM Symposium on Theory of Computing*, pages 101–110, May 1993.
21. E.H. Spanier, *Algebraic Topology,* Springer-Verlag, New York, 1966.

On the Robustness of h_m^r *

(Preliminary Version)

Shlomo Moran and Lihu Rappoport

Faculty of Computer Science, Technion, Israel

Abstract. We introduce an N-process deterministic concurrent object for $N \geq 3$ processes, called the *conditional consensus* object. This object, denoted as W, is hard-wired in the sense that each process P can access it using a single fixed *port* (though P can use different ports in different copies of W). We prove that W satisfies the following properties:

- There is no consensus protocol for three processes which uses many shared registers and many copies of W (and does not use any other object); but
- There is a consensus protocol for N processes which uses one copy of W and $\binom{N}{3}$ copies of CO_3, where CO_3 is the standard consensus object for three processes.

This implies that the hierarchy h_m^r is not robust for deterministic hard-wired objects.

1 Introduction

Research on wait-free asynchronous shared memory systems has led to the definition of some fundamental concurrent objects, which are used in the implementation of shared data structures, and for synchronization tasks among concurrent processes. This led to the question of the *power* of different objects—more powerful objects being able to perform harder tasks than less powerful ones.

An important progress in this direction was achieved by Herlihy in [Her91a], where he used *consensus protocols* [FLP85] for the classification of objects, and introduced the notion of the *consensus number* of a concurrent object. The *consensus number* of a concurrent object X is the maximal number k for which a wait-free consensus protocol involving k processes, using a single copy of X and any number of read/write registers, exists. If there is no such maximum, then the consensus number of X is ∞.

In [Her91a], Herlihy computes the consensus number of many fundamental objects (e.g. read/write registers, stacks, queues, etc.), and shows the existence of a *hierarchy of objects* with respect to wait-free implementation (a mapping of objects into $\{1, 2, \ldots\} \cup \{\infty\}$). Herlihy further shows that an object X mapped

* This work was supported by the France–Israel cooperative project: Graph Theoretical Methods in Distributed Computing 4474-2-93.

at level N, together with read/write registers, can be used to construct a wait-free implementation of any concurrent object Y shared by N processes. This last property of Herlihy's hierarchy is called the *universality* property, and it formulates the notion of power conveyed by this hierarchy. A hierarchy of concurrent objects which has the universality property is called a *wait-free hierarchy* [Jay93].

In this paper, as in a previous paper [CM94], we use a slightly more general notation for the term consensus number, which holds for any collection of shared objects. Let the consensus number of a given set $\mathcal{S}$ of shared objects, denoted $\mathcal{CN}(\mathcal{S})$, be the maximum number n such that there is a wait-free consensus protocol for n distinct processes, which communicate only by accessing objects in $\mathcal{S}$. Also, for an object X, let $\mathtt{X}^n$ denote a collection of n copies of X. Note that the consensus number of an object X as defined in [Her91a] is $\mathcal{CN}(\mathtt{X} \cup \mathtt{R}^\infty)$, where $\mathtt{R}^\infty$ denotes an infinite set of read/write registers.

For most "standard" concurrent objects X and Y, it holds that $\mathcal{CN}(\mathtt{X} \cup \mathtt{R}^\infty) = \mathcal{CN}(\mathtt{X}^\infty \cup \mathtt{R}^\infty)$, and $\mathcal{CN}(\mathtt{X} \cup \mathtt{Y} \cup \mathtt{R}^\infty) = \max\{\mathcal{CN}(\mathtt{X} \cup \mathtt{R}^\infty), \mathcal{CN}(\mathtt{Y} \cup \mathtt{R}^\infty)\}$. This led to the belief that an object with a large consensus number cannot be implemented using any combination of objects, each having a smaller consensus number. Informally, such a property means that the consensus number is a *robust* property. The question of robustness was presented and formally defined by Jayanti in [Jay93].

In [Jay93], Jayanti suggests four ways for classifying objects using consensus numbers. An object X is associated with four (not necessarily distinct) numbers: $h_1(\mathtt{X}) = \mathcal{CN}(\mathtt{X})$, $h_m(\mathtt{X}) = \mathcal{CN}(\mathtt{X}^\infty)$, $h_1^r(\mathtt{X}) = \mathcal{CN}(\mathtt{X} \cup \mathtt{R}^\infty)$ and $h_m^r(\mathtt{X}) = \mathcal{CN}(\mathtt{X}^\infty \cup \mathtt{R}^\infty)$. These definitions are naturally extended for sets of objects. Classifying objects by these numbers corresponds to four possible hierarchies, denoted h_1, h_m, h_1^r and h_m^r respectively. Jayanti showed that the first three hierarchies are not robust by showing that in some cases the consensus number of a combination of two objects (according to the appropriate definition) is greater than the consensus number of each of the objects. Consequently, the question whether h_m^r is robust had become of special interest, and was studied in [BGA94, BNP94, CHJT94a, CHJT94b, PBN94].

Jayanti in [Jay95] states that if h_m^r is robust, the difficult problem of computing the combined power of a set of objects reduces to the simpler problem of computing the power of the individual objects in the set. If, on the other hand, h_m^r is not robust, it may be possible to implement a strong object from a set of weak objects. Jayanti also relates to the notion of power captured by h_m^r. If $h_m^r(\mathtt{X}) > h_m^r(\mathtt{Y})$, X is *more powerful* than Y in the following sense: X is universal for a larger number of processes than Y, that is, the maximum number of processes for which X can solve *any* synchronization problem is larger than that of Y. This does not mean that X can solve any problem that Y can. In particular, [Rac94] shows just the opposite.

The robustness of h_m^r can be studied with respect to different classes of objects, e.g., one can either allow non-determinism or not, either allow hard-wired objects or not, and use different ways for processes to be connected to the objects (we discuss the different classes of concurrent objects in detail in the next section). The question of the robustness of h_m^r can be now formulated as follows: given a class $\mathcal{X}$ of of concurrent objects, is h_m^r robust with respect to $\mathcal{X}$? Formally, h_m^r is *robust with respect to* $\mathcal{X}$ if for each subset of objects $\{\mathtt{X}_1, \ldots, \mathtt{X}_n\} \subseteq \mathcal{X}$, $h_m^r(\mathtt{X}_1, \ldots, \mathtt{X}_n) = \max\{h_m^r(\mathtt{X}_1), \ldots, h_m^r(\mathtt{X}_n)\}$. It turns out that different works consider different classes of objects, which lead to different results. We discuss these works in detail in the sequel.

This paper shows for the first time that the h_m^r hierarchy is not robust for deterministic hard-wired objects. We introduce a new deterministic concurrent object for $N \geq 3$ processes, the *conditional consensus* object, denoted as $\mathtt{W}$, such that $\mathcal{CN}(\mathtt{W}^\infty \cup \mathtt{R}^\infty) < 3$, but $\mathcal{CN}(\mathtt{W} \cup \mathtt{CO}_3^\infty) = N$ ($\mathtt{CO}_3$ is the standard consensus object for three processes). This implies that h_m^r is not robust. The conditional consensus object $\mathtt{W}$ is hard-wired, in the sense that each process P can access it using a single fixed port (though P can use different ports for different copies of $\mathtt{W}$). Examples of hard-wired objects found in the literature include single-writer multi-reader registers [Lam86], and *Load Linked* and *Store Conditional* [Her91b]. Our work is closely related to [CM94], where similar ideas are used to present a different aspect of the non-robustness of the consensus number hierarchies.

1.1 An Outline of the Construction

The intuition behind the conditional consensus object, $\mathtt{W}$, is as follows. Each copy of $\mathtt{W}$ is similar to the hard-wired version of the standard consensus object for N processes, $\mathtt{CO_N}$[2], with the following modification: each copy of $\mathtt{W}$ responds with the right consensus value only after it gets a "proof" that the system has the power to reach a consensus for three processes *without* using this specific copy of $\mathtt{W}$. Specifically, $\mathtt{W}$ has four phases of operations, during which each process accesses it six times:

1. $\mathtt{W}$ receives the input values of the processes that access it, and stores the first input as its decision value, D.
2. $\mathtt{W}$ gives each process P_i a *test value*, T_i. In order to make the test values unpredictable by the processes, $\mathtt{W}$ decides on the test values according to the order in which the processes access it (this order is unknown to the processes). In this way, we use the non-determinism of the scheduler to make the object *look* non-deterministic to the processes. As will be explained later on, this phase requires each process to access $\mathtt{W}$ three times.
3. Each process P_i accesses $\mathtt{W}$ with $\binom{N-1}{2}$ decision values, one for each triplet of processes $\{P_i, P_j, P_k\}$ which includes P_i. The decision value, C_{ijk}, for each such triplet, must be a consensus value consistent with the test values, $\{T_i, T_j, T_k\}$, given by $\mathtt{W}$ to the processes in that triplet.
4. Finally, each process accesses $\mathtt{W}$ in order to receive the decision value D that $\mathtt{W}$ stores at phase 1. However, if $\mathtt{W}$ receives an incorrect value, C_{ijk}, in phase 3, $\mathtt{W}$ may return a wrong value, instead of D.

[2] We assume that the "standard" consensus object $\mathtt{CO_N}$ is soft-wired.

The tricky parts in the construction and in the proof are in making the test values unpredictable by the processes, and in making W's responses to the processes reveal as little information as possible on W's internal state. We are then able to show that processes which try to reach a consensus over $\mathtt{W}^\infty \cup \mathtt{R}^\infty$ must provide with one of the copies of W consensus values for all triplets of test values (without using this copy of W). We use this last fact to show that the assumption that there is a consensus protocol for three processes over $\mathtt{W}^m \cup \mathtt{R}^\infty$ for any number m leads to a contradiction, which implies our result.

2 Definitions and Notations

A *concurrent system* consists of a collection of processes and of a shared memory. The shared memory is modeled by a (possibly infinite) set of deterministic or non-deterministic *concurrent objects.* A *concurrent object* is a basic memory unit, which enables a given set of processes to perform predefined operations on it. Processes are modeled by infinite state machine, and they are *asynchronous*— there is no global clock timing them. Each process runs at a different speed, and might be subject to arbitrarily long delays. In particular, a process cannot tell whether another process is halted or is running very slowly. The processes communicate by applying operations on the concurrent objects. We assume that the concurrent objects are *atomic*, hence all operations applied on the objects are totally ordered in time.

2.1 Concurrent Objects

Formally, a (deterministic) *concurrent object* is defined by a 5-tuple (n, S, OP, R, δ), where: n is the number of *ports*, S is a (possibly infinite) set of *states*, OP is a (possibly infinite) set of *operations* (composed of a function name and parameters), R is a (possibly infinite) set of *responses* (output values) which can be returned by the object, and $\delta : \{1..n\} \times S \times OP \rightarrow S \times R$ is a transition function, which specifies for each port in $\{1..n\}$, each state of the object, and each operation applied to it, what is the next state and what is the output value returned by this operation.

Some papers consider also *oblivious* objects, which are special type of a concurrent object, in which the transition function δ does not depend on the port being used: $\delta : S \times OP \rightarrow S \times R$. As we show in the full version of this paper, oblivious objects can be modeled as (non-oblivious) objects with appropriate connection schemes (which we describe next).

In order to completely specify a concurrent object, one should also specify the rules by which a process may connect or disconnect from the object's ports. For this, we augment the definition of a concurrent object to be a pair $((n, S, R, OP, \delta), M)$, where M is a *port-connection scheme* applicable to the object. Following [Jay95], we classify connections schemes by whether or not they satisfy the following rules:

1. Once a process is connected to a port, it cannot disconnect.

2. Each process may be connected to at most one port.
3. At any given moment, at most one process may be connected to a given port.

All connection schemes considered in this paper satisfy rule number 3. Objects which satisfy rule number 1 are called *hard-wired* objects [BGA94]. Notice that in hard-wired objects, a process may be connected to different ports in different copies of the same object. Jayanti in [Jay95] defines three connection schemes, M_1, M_2 and M_3, as follows: In M_1, all three rules apply. In M_2, only rules 1 and 3 apply, that is, a process may be connected to more than one port in a given object. Observe that both M_1 and M_2 characterize hard-wired objects. It is possible to further sub-divide hard-wired objects according to the time in which processes are are allowed to be connected to objects: a *static* connection scheme allows processes to be connected to objects only at the protocol initialization stage, while a *dynamic* connection scheme allows processes to be connected to objects also while the protocol is executed (in run time). A third class of objects is the class of *soft-wired* objects, defined by the M_3 connection scheme. In this class, only rule 3 applies, which means that a process may be disconnected from a port and then be reconnected to (another) port in run time.

Hard-wired objects model objects which are shared by *processors* using hardware that can distinguish among the different processors, while soft-wired objects model objects which are shared either by *processes* or by indistinguishable *processors*. Although it is more accurate to use the term *processors* when relating to hard-wired objects, we will use the term *processes* for both hard-wired and soft-wired objects.

Examples: A (soft-wired) read/write register, over a set of values V, is defined by $(n, S_{reg}, OP_{reg}, R_{reg}, \delta_{reg})$, where $S_{reg} = V$, $R_{reg} = V \cup \{ok\}$, and OP_{reg} consists of a *read* operation and of $|V|$ *write* operations (one *write* operation for each value in V). Let $v, v' \in V$, then $\delta_{reg}(v, read) = (v, v)$, and $\delta_{reg}(v, write_{v'}) = (v', ok)$.

A natural example of an M_1 object, is an object which supports the operations *Load Linked* (*LL*) and *Store Conditional* (*SC*) [Her91b]. Informally, $LL(x)$ reads the value of a concurrent variable x, such that x may be subsequently used in combination with a SC operation. $SC(x, a)$ writes the value a to x and returns $SUCCESS$ if no other successful $SC(x)$ was executed since the last $LL(x)$ operation executed by the process, and returns $FAILURE$ otherwise. Formally, a (hard-wired) *LL-SC* object, over a set of values V, which is shared by n processors, is defined by $(n, S_{LL-SC}, OP_{LL-SC}, R_{LL-SC}, \delta_{LL-SC})$, where: $S_{LL-SC} = V \times 2^{\{1..n\}}$, $R_{LL-SC} = V \cup V \times \{SUCCESS, FAILURE\}$, and OP_{LL-SC} consists of a LL operation and of $|V|$ SC operations (one SC operation for each value in V). Let $v, v' \in V$, $p \in \{1..n\}$, $P \subseteq \{1..n\}$, then $\delta_{LL-SC}(p, (v, P), LL) = ((v, P \cup \{p\}), v)$, and

$$\delta_{LL-SC}(p, (v, P), SC_{v'}) = \begin{cases} ((v', \perp), SUCCESS) & p \in P \\ ((v, P), FAILURE) & p \notin P \end{cases}$$

2.2 Consensus Protocols and Consensus Numbers

A *wait-free consensus protocol* (in short a "consensus protocol") for n processes over a given set of concurrent objects, $\mathcal{S}$, is a protocol for n processes over $\mathcal{S}$, defined as follows: each process receives an input, and in every possible run of the protocol, every process which makes infinitely many steps eventually decides on some input value, such that all the processes which decide, decide on the same value.

A run of a consensus protocol is said to be in a *bivalent configuration*, if the eventual decision value is not yet determined. A run is in a *0-valent* (*1-valent*) configuration, if in all possible extensions of this run, the eventual decision value of the processes is 0 (1). A configuration which is 0-valent or 1-valent is said to be *univalent*. Finally, a univalent configuration is *decisive* if some process had actually decided in this run. See [FLP85, LA87, BMZ90, TKM94, LM95] for further exposition of these terms, and of consensus protocols in general.

The *consensus number* of a set $\mathcal{S}$ of concurrent objects, denoted $\mathcal{CN}(\mathcal{S})$, is defined as follows: if there is an integer n such that there is a consensus protocol for n processes over $\mathcal{S}$, and there is no consensus protocol for $n+1$ processes over $\mathcal{S}$, then $\mathcal{CN}(\mathcal{S}) = n$. If for every n there is such a protocol, then $\mathcal{CN}(\mathcal{S})$ is unbounded. If there is a consensus protocol for an infinite set of processes over $\mathcal{S}$, then $\mathcal{CN}(\mathcal{S}) = \infty$. For an object $\mathtt{X}$ and $n \in \{1, 2, 3, \ldots\} \cup \{\infty\}$, $\mathtt{X}^n$ denotes a set consisting of n objects, each of which is isomorphic to $\mathtt{X}$, and $\mathcal{CN}(\mathtt{X})$ denotes $\mathcal{CN}(\{\mathtt{X}\})$.

3 Related Work

Following [Jay93], the robustness of h_m^r with respect to various classes of concurrent objects was studied in few papers. We survey below the results obtained in [BGA94, CHJT94a, CHJT94b, PBN94, BNP94], and compare them with our result. We observe that in these papers there is no explicit distinction between the M_1 and M_2 models. This distinction was introduced later in [Jay95], following an earlier version of this paper.

3.1 The Results of Chandra, Hadzilacos, Jayanti and Toueg

Two results from [CHJT94b] concern the robustness of h_m^r. The first result hold for M_2 and M_3 objects, and it states that for any set $\mathcal{S}$ of objects (either deterministic or non-deterministic), $h_m^r(\mathcal{S}, \mathtt{CO_N}) \geq N+1 \Longrightarrow h_m^r(\mathcal{S}, \mathtt{CO_{N-1}}) \geq N$. By recursively applying this result, it implies that $h_m^r(\mathcal{S}, \mathtt{CO_{N-1}}) = N \Longrightarrow h_m^r(\mathcal{S}) = N$, which in turn implies that $h_m^r(\mathcal{S}, \mathtt{CO_N}) = \max\{h_m^r(\mathcal{S}), N\}$.

The second result of [CHJT94b] shows that h_m^r is not robust for the class of non-deterministic M_1 objects. This is proved by presenting a non-deterministic M_1 object called `booster`, which has the following property: $h_m^r(\mathtt{booster}) = 1$, but $h_m^r(\mathtt{booster}, \mathtt{CO_2}) = 3$. I.e., the `booster` object shows that levels 2 and 3 of the h_m^r hierarchy collapse, in some precise sense. The `booster` object is described in detail in [CHJT94a]. It is stated in [CHJT94a] that **booster** can be generalized, such that for all $n \geq 3$, one can obtain an n-ported object T_n such that $h_m^r(T_n) = 1$ and $h_m^r(T_n, \mathtt{CO_2}) = n$. Our result shows that this latter result holds also for the class of deterministic M_1 objects.

3.2 The Results of Peterson, Bazzi and Neiger

In [PBN94], Peterson, Bazzi and Neiger study n-ported objects. They show that for that class, level n of the hierarchy is robust. Specifically, they present a result which implies that the following conditions are equivalent for a hard-wired n-ported concurrent object $\mathtt{X}$:

(a) $h_m^r(\mathtt{X}) < n$.
(b) For any n-ported object $\mathtt{Y}$ with $h_m^r(\mathtt{Y}) = n-1$, $\mathtt{X}$ can be implemented by $\mathtt{Y}^\infty \cup \mathtt{R}^\infty$.

((b) $\Rightarrow$ (a) follows from the definition of h_m^r and letting $\mathtt{Y} = \mathtt{CO_{n-1}}$; the main contribution in [PBN94] is in proving (a) $\Rightarrow$ (b)). There is no similar result for n-ported objects with consensus number at most k, for any fixed $k < n-1$. In particular, it is stated in [PBN94] that there exists an n-ported concurrent object $\mathtt{X}$, such that $h_m^r(\mathtt{X}) \leq n-2$, but $\mathtt{X}$ cannot be implemented by some of the n-ported objects with consensus number $n-2$.

Our result implies that the result of [PBN94] does not hold for deterministic (and non-deterministic) M_1 objects. In particular, for $n > 3$, our object $\mathtt{W}$ satisfies $h_m^r(\mathtt{W}) < 3 < n$, but $\mathtt{W}$ cannot be implemented by $\mathtt{CO_{n-1}^\infty} \cup \mathtt{R}^\infty$, since this would imply that $n-1 = h_m^r(\mathtt{CO_{n-1}}) = h_m^r(\mathtt{CO_{n-1}} \cup \mathtt{W}) = n$, a contradiction. Jayanti shows in [Jay95] that, assuming M_2 objects, the result of [PBN94] implies that h_m^r is robust (for all levels).

In [BNP94] it is shown that for deterministic objects, h_m and h_m^r are the same (i.e. for each object $\mathtt{X}$, $h_m^r(\mathtt{X}) = h_m(\mathtt{X})$), which implies that h_m is robust iff h_m^r is robust. It is then shown that Jayanti's use (in [Jay93]) of non-deterministic objects to show that h_m is not robust is a must. Finally, it is claimed that since it was proved in [PBN94] that h_m^r is robust for deterministic objects, this implies that (for deterministic objects) h_m is robust as well.

3.3 The Results of Borowsky, Gafni and Afek

The main result of [BGA94] states that any task over $m \leq 2n$ processes which is solvable using only objects $\mathtt{X}$ with $h_m^r(\mathtt{X}) < n+1$, is also solvable by $\mathtt{CO_n^\infty} \cup \mathtt{R}^\infty$. In particular, any collection of objects whose consensus number is at most n cannot implement $n+1$ consensus, which means that h_m^r is robust. Our result implies that this result does not hold for M_1 objects.

4 The Conditional Consensus Object

In this section, we introduce the conditional consensus object, $\mathtt{W}$, which is an M_1 object for $N \geq 3$ processes, $P_1, \ldots, P_N$. In the following section we use $\mathtt{W}$ to prove our main result.

An operation applied on W can be described as a pair (i, v), where i is a port number (in which the operation is applied), and v is an input value. Each state of W is a sequence of operations applied to it: whenever an operation is applied on W, the state of W is extended by this operation. Thus, if the initial state of (a copy of) W in a given run is ϵ, then the state of W in this run is exactly the sequence of operations performed on W. Note that W "knows" from its state exactly how many times it was accessed from each port. W's response to a given operation depends on W's state, on the port number used and on the input applied to W. W reacts to an access on port i according to the number of times W was accessed on that port so far, as follows:

1. First access:
   Input: I_i. If the state of W is ϵ, W sets its *decision variable*, D, according to I_i—if $I_i = 0$, D is set to 0, otherwise D is set to 1. If the state of W is not ϵ, W ignores I_i.
   Output: *nil*.
2. Second access: Input: ignored. Output: *nil*.
3. Third access: Input: ignored. Output: *nil*.
4. Fourth access: Input: ignored. Output: T_i (defined below).
5. Fifth access:
   Input: $\{C_{ijk} \mid 1 \leq j < k \leq N \ \wedge \ j, k \neq i\}$, where C_{ijk} is a consensus on $\{T_i, T_j, T_k\}$. For an input to be legal, C_{ijk} must be equal to one of $\{T_i, T_j, T_k\}$, and C_{ijk} given in ports i, j, and k must be the same. If W is accessed with an illegal input, W is said to be *stuck*.
   Output: *nil*.
6. Sixth access:
   Input: ignored.
   Output: A consensus on $\{I_1, I_2, ..., I_N\}$, which is the value of W's decision variable, D. If, however, W is stuck, and i is the first port in which W is accessed twice, then W outputs 1.
7. Seventh access and on: Input: ignored. Output: *nil*.

We now describe how W calculates T_i on the fourth access to port i, for $1 \leq i \leq N$. Let i_0 be the first port in W that was accessed three times, and assume that k ports were accessed twice before i_0 was accessed three times $(1 \leq k \leq N)$. Let $s(i)$ be the i'th port which was accessed twice before i_0 was accessed three times, and let $\mathcal{Q} = \{s(1), \ldots, s(k)\}$. Once i_0 is accessed for the third time, all the T_i's, $1 \leq i \leq N$, are set. For any port $j = s(i) \in \mathcal{Q}$, $T_j (= T_{s(i)})$ is defined in two steps, as follows:

Step 1

$$T'_j = \begin{cases} 0 \text{ if } s(i) < s(i+1) \ \wedge \ i < k \\ 1 \text{ otherwise} \end{cases}$$

Step 2

$$T_j = \begin{cases} T'_j & \text{if } j \neq i_0 \\ 1 - T'_j & \text{otherwise} \end{cases}$$

For any port $j \notin \mathcal{Q}$, $T_j = 1$.

The above procedure for computing the values T_i satisfies the following properties:

Proposition 1. *Let (T_1, T_2, T_3) be an arbitrary binary vector of size 3, let process P_i, $i \in \{1,2,3\}$, be connected to port i in a given copy of* W*, and assume that in each run each non-faulty process accesses* W *at least four times. Then there is a scheduling of processes $\{P_1, P_2, P_3\}$ in which process P_i, $i \in \{1,2,3\}$, gets output T_i on its fourth access to* W*. Moreover, the values of the T_i's depend only on the order in which the processes access* W *in the second and third times.*

Proof. Schedule processes P_1, P_2 and P_3, such that all three of them access W twice before any of them accesses W for the third time. Let a, b and c denote the steps in which P_1, P_2 and P_3 access W for the second time, respectively. Then abc, acb, bac, bca, cab and cba denote the possible ways to order the second accesses made by $\{P_1, P_2, P_3\}$ to W. In abc we get $T_1' = 0, T_2' = 0$ and $T_3' = 1$—for short 001. In acb we get 011, in bca we get 101, and in cba we get 111. Now, the vector (T_1, T_2, T_3) is obtained by selecting the process that accesses W first for the third time, causing its entry in (T_1', T_2', T_3') to be flipped. The proof is completed by noting that each binary triplet can be obtained by flipping a bit in one of the vectors {001,011,101,111}. □

5 Proof of Main Result

In this section we prove that the conditional consensus object W satisfies the following properties:

1. $\mathcal{CN}(\mathtt{W}^\infty \cup \mathtt{R}^\infty) < 3$.
2. $\mathcal{CN}(\mathtt{W} \cup \mathtt{CO}_3^\infty) = N$, where $\mathtt{CO}_3$ is the standard consensus object for three processes.

It is easy to show that $\mathcal{CN}(\mathtt{W} \cup \mathtt{CO}_3^\infty) = N$. We describe a wait-free consensus protocol for N processes over $\mathtt{W} \cup \mathtt{CO}_3^\infty$, which uses a single W object and $\binom{N}{3}$ $\mathtt{CO}_3$ objects, tagged $\{\mathtt{CO}_3^{ijk} \mid 1 \leq i < j < k \leq N\}$. Process P_i, $1 \leq i \leq N$, which executes the protocol, accesses W for the first time with P_i's input value I_i, and then accesses W for the second and third times with no input. On its fourth access, P_i receives from W a test value T_i. Then, for each j and k such that $1 \leq j < k \leq N \wedge j, k \neq i$, P_i accesses $\mathtt{CO}_3^{ijk}$ with T_i and receives from $\mathtt{CO}_3^{ijk}$ a value C_{ijk} (which is a consensus value on $\{T_i, T_j, T_k\}$). P_i accesses W for the fifth time with input $\{C_{ijk} \mid 1 \leq j < k \leq N \wedge j, k \neq i\}$, and on its sixth access to W, receives a consensus value on $\{I_1, I_2, ..., I_N\}$. This protocol is wait-free (it has no loops), and it is easy to show that it is correct.

In order to prove that $\mathcal{CN}(\mathtt{W}^\infty \cup \mathtt{R}^\infty) < 3$, we assume by way of contradiction that there exists a wait-free protocol, Pr, that solves the consensus problem for three processes, $\{P_1, P_2, P_3\}$, over $\mathtt{W}^\infty \cup \mathtt{R}^\infty$. Since Pr is wait-free, there exists a *critical run*, ρ_c, in which the system is in a *critical configuration*, C_{cr}: C_{cr}

is a bivalent configuration, but a single step of each of $\{P_1, P_2, P_3\}$ brings the system to a univalent configuration [Her91a]. Since C_{cr} is bivalent, there exists a value $v \in \{0, 1\}$ such that two of the three steps bring the system to a v-valent configuration, and one will bring it to a $(1 - v)$-valent configuration. Also, a standard consideration shows that all the three steps must be applied to the same *critical* concurrent object. Arguments identical to these in [CIL87, LA87, Her91a] show that the critical object cannot be a read/write register, so it must be a copy of $\mathtt{W}$, to be denoted $\mathtt{W_C}$. We check two cases: either $\mathtt{W_C}$'s state in the critical configuration is ϵ, or it is different from ϵ.

5.1 $\mathtt{W_C}$'s state in the critical configuration is ϵ

In this section, we deal with the case where in the critical configuration, C_{cr}, defined above, $\mathtt{W_C}$'s state is ϵ, which implies that no process accesses $\mathtt{W_C}$ before C_{cr}. Since the responses of $\mathtt{W}$ depend only on the relative order of the processes, we may assume that the identities of the three processes which run Pr are P_1, P_2, and P_3, and that they are connected to ports 1, 2, and 3 respectively.

Main Lemma: *Assume that there is a wait-free protocol, Pr, that solves the consensus problem for three processes over $\mathtt{W}^m \cup \mathtt{R}^\infty$, for some $m > 0$. Assume further that Pr has a critical configuration in which the state of the critical object $\mathtt{W_C}$ is ϵ. Then there exists a wait-free protocol, Pr', that solves the consensus problem for three processes over $\mathtt{W}^{m-1} \cup \mathtt{R}^\infty$.*

The Main Lemma is proved using the following lemmas.

Lemma 2. *Let ρ be any finite run of Pr which starts at the critical configuration C_{cr}, such that $\mathtt{W_C}$'s state in C_{cr} is ϵ. Assume that no process takes any step after accessing $\mathtt{W_C}$ six times in ρ, and let $\mathcal{P}$ be the set of processes that have accessed $\mathtt{W_C}$ at least once and at most five times within ρ. Then*

1. *Let $\rho = \rho_1 a \rho_2$, where a is the first step taken by process P_i in ρ. Then, $\rho' = a\rho_1\rho_2$ is also a run of Pr, in which the state of each process in $\mathcal{P}$ is the same as in ρ.*
2. *If process P_i does not take any step in ρ, then $\rho' = a\rho$, where a is the step taken by P_i if it were scheduled immediately after C_{cr}, is also a run of Pr, in which the state of each process in $\mathcal{P}$ is the same as in ρ.*

Proof. Let $|\rho_1|$ and $|\rho_2|$ be the numbers of steps taken in ρ_1 and in ρ_2 respectively. Let ρ_1^j and ρ_2^j be the prefixes of length j of ρ_1 and ρ_2, respectively. Let $\mathcal{P}^j$ be the set of processes that have accessed $\mathtt{W_C}$ less than six times within ρ_1^j, excluding P_i, and let $\mathtt{MEM}$ denote the collection of all the objects in the shared memory excluding $\mathtt{W_C}$. Using an induction on j, we first show that:

- $a\rho_1^j$ is a possible run of Pr.
- For each $j \leq |\rho_1|$, the state of $\mathtt{MEM}$, and the states of the processes in $\mathcal{P}^j$, are the same in ρ_1^j and in $a\rho_1^j$.

Induction base $j = 0$: By the definition of a, a is a possible run of Pr. Since a must be an access to $\mathtt{W_C}$, the state of **MEM** and the states of all processes excluding P_i, are the same in the run $\rho_1^0 = \epsilon$ and in the run $\rho_1^0 a = a$.

Induction step: First we show that $a\rho_1^j$ is a run of Pr. Let Q be the process that takes the step after ρ_1^{j-1}. By our assumptions, $Q \neq P_i$, and Q accesses $\mathtt{W_C}$ at most five times in ρ_1^{j-1} and in $a\rho_1^{j-1}$. Therefore, $Q \in \mathcal{P}^{j-1}$. Since, by the induction assumption, Q is in the same state after ρ_1^{j-1} and after $a\rho_1^{j-1}$, it will take the same step both in ρ_1^{j-1} and in $a\rho_1^{j-1}$. Therefore, $a\rho_1^j$ is also a possible run of Pr. It remains to show that the induction hypothesis holds for the processes in $\mathcal{P}^j$ and for the shared objects in **MEM**. Q can take one of the following step types in both runs:

1. An access to a shared object in **MEM**: both Q and the object accessed by it are in the same state in ρ_1^{j-1} and in $a\rho_1^{j-1}$, and hence Q takes the same step in both runs.
2. One of the first five accesses of Q to $\mathtt{W_C}$: in both cases, Q gets the same response—none in its first, second, third and fifth accesses, and T_i in its fourth access. T_i is the same in both cases, since the orders in which processes access $\mathtt{W_C}$ in the second and third times are the same. Therefore, Q moves to the same state in both runs.
3. Sixth access of Q to $\mathtt{W_C}$: Q may move to different states (in each run).

In cases 1 and 2, the state of Q is the same for both runs, and the induction holds for the processes in $\mathcal{P}^j = \mathcal{P}^{j-1}$. In case 3, the induction holds for $\mathcal{P}^j$ since $\mathcal{P}^j = \mathcal{P}^{j-1} \backslash \{Q\}$, and, by the induction hypothesis for $j-1$, the states of all the processes in $\mathcal{P}^{j-1} \backslash \{Q\}$ are the same in both runs. In cases 2 and 3, the state of **MEM** is not affected, and in case 1 it is affected in the same way in both runs. Therefore, by the induction assumption, **MEM** is in the same state in both runs.

Let $\mathcal{P}'^j$ be the set of processes that have accessed $\mathtt{W_C}$ less than six times within $\rho_1 a \rho_2^j$, then $\mathcal{P}'^j \setminus \{P_i\} \subseteq \mathcal{P}^{|\rho_1|}$. We use this fact to prove that $a\rho_1\rho_2^j$ is a run of Pr, and that for each $j \leq |\rho_2|$ the state of **MEM** and the states of all processes in $\mathcal{P}'^j$ (including P_i) are the same in $a\rho_1\rho_2^j$ and in $\rho_1 a \rho_2^j$. This implies that $a\rho_1\rho_2^j$ is a possible run of Pr, and completes the proof of the first part of the lemma. The proof of the second part of the lemma, where P_i does not take any step in ρ, is similar. □

Lemma 3. Let ρ be defined as in Lemma 2, and let P_i be a process which accesses $\mathtt{W_C}$ at most five times in ρ, then P_i does not reach a decision state within ρ.

Proof. Assume by way of contradiction that there exists a process $P_i \in \{P_1, P_2, P_3\}$ which accesses $\mathtt{W_C}$ at most five times in ρ, and reaches a decision state within ρ. Let $Q_v \in \{P_1, P_2, P_3\}$ be the process that takes the first step in ρ, and assume that this step leads to a v-valent configuration. Since C_{cr} (the initial configuration in ρ) is critical, there exists a process $Q_{1-v} \in \{P_1, P_2, P_3\}$ that, if scheduled after C_{cr}, would lead to a $(1-v)$-valent configuration. Let a be the first step after C_{cr} taken by Q_{1-v}. If Q_{1-v} takes a step in ρ, then let $\rho = \rho_1 a \rho_2$, and let

$\rho' = a\rho_1\rho_2$. If Q_{1-v} does not take a step in ρ, then let $\rho' = a\rho$. By Lemma 2, ρ' is also a possible run of Pr, in which the state of P_i in its decision state is same as in ρ. Therefore, P_i reaches the same decision in ρ' as in ρ, namely, v. This is a contradiction to the fact that the configuration after the first step in ρ' is $(1-v)$-valent. □

Given Pr, we define a set $\mathcal{R}$ of runs, which is the set of all runs of Pr which involve processes P_1, P_2 and P_3, such that for each run $\rho \in \mathcal{R}$, the following conditions hold:

- the initial configuration of ρ is the critical configuration C_{cr}.
- Each of the processes P_1, P_2 and P_3 reaches a decision in ρ.
- Each process which is about to access W_C in the ith time, $i \in \{2,3,4,5\}$, does not take any additional step until the two other processes are about to access W_C for the ith time.
- Each process which accesses W_C in the sixth time does not take any additional step until the two other processes access W_C six times.

Lemma 4. *$\mathcal{R}$ is not empty, and in each run $\rho \in \mathcal{R}$ each one of P_1, P_2 and P_3 accesses W_C (at least) six times.*

Proof. In order to show that $\mathcal{R}$ is not empty, we describe how to schedule the processes P_1, P_2 and P_3, starting from configuration C_{cr}, to obtain a run $\rho \in \mathcal{R}$. Start by scheduling processes arbitrarily. If a process is about to access W_C for the second time, stop scheduling it. Since the protocol is wait-free, each process must reach a decision within a bounded number of its steps, and by Lemma 3 this implies that all processes will eventually access W_C twice. This procedure can now be repeated for the third, fourth and fifth accesses. A similar procedure can be used for the sixth access. Once all three processes access W_C six times, schedule processes arbitrarily, and halt a process which reaches a decision. It is obvious the run obtained by this procedure is indeed in $\mathcal{R}$.

Using a similar consideration, it is easy to show that in each run $\rho \in \mathcal{R}$, each one of the three processes accesses W_C at least six times. □

Lemma 5. *Let ρ be a prefix of a run in $\mathcal{R}$ which ends when all three processes have accessed W_C for the sixth time, then no process may get W_C stuck in ρ.*

Proof. Assume by way of contradiction that there exists a process $P_i \in \{P_1, P_2, P_3\}$ which gets W_C stuck in ρ. Let $Q_v \in \{P_1, P_2, P_3\}$ be the process that takes the first step in ρ, and assume that this step leads to a v-valent configuration. Since C_{cr} (the initial configuration in ρ) is bivalent, there exists a process $Q_{1-v} \in \{P_1, P_2, P_3\}$ that, if scheduled after C_{cr}, would lead to a $(1-v)$-valent configuration. Let a be the first step after C_{cr} taken by Q_{1-v}, let $\rho = \rho_1 a \rho_2$, and let $\rho' = a\rho_1\rho_2$. By Lemma 2, ρ' is also a possible run of Pr.

Process P_i can only get W_C stuck by giving it a wrong value for one of the C_{ijk}'s on its fifth access, therefore, W_C is stuck before any process accesses it for the sixth time. Furthermore, moving a to be right after C_{cr} does not change the T_i's nor the C_{ijk}'s, and thus P_i also gets $\mathtt{W_C}$ stuck in ρ'. Let $P^\diamond$ be the first process that accessed $\mathtt{W_C}$ for the second time in ρ. From the proof of Lemma 2, $P^\diamond$'s state and the states of all the shared objects excluding $\mathtt{W_C}$ is the same before $P^\diamond$'s sixth access to $\mathtt{W_C}$ in ρ and in ρ'. Since both in ρ and in ρ' $\mathtt{W_C}$ is stuck while $P^\diamond$ accesses it for the sixth time, $P^\diamond$ gets the same response on its sixth access to $\mathtt{W_C}$, namely, 1. Therefore, $P^\diamond$'s state is also the same after its sixth access to $\mathtt{W_C}$ in ρ and in ρ'. After $P^\diamond$ accesses $\mathtt{W_C}$ for the sixth time (both in ρ and in ρ'), let $P^\diamond$ run alone until it reaches decision. When $P^\diamond$ reaches decision, it is in the same state in both cases, and thus it reaches the same decision, namely v, which is a contradiction. □

Lemma 6. *Let $\rho \in \mathcal{R}$, and let $P_i \in \{P_1, P_2, P_3\}$. Then the value C_{123} given by process P_i as part of its input to $\mathtt{W_C}$ in P_i's fifth access to $\mathtt{W_C}$, is a consensus value on $\{T_1, T_2, T_3\}$, and this value is equal for P_1, P_2, and P_3.*

Proof. Immediate from Lemma 4 and Lemma 5. □

Lemma 7. *Let $\rho \in \mathcal{R}$, and let ρ' be a prefix of ρ that ends just before all three processes are about to access $\mathtt{W_C}$ in the fourth time. Let $\rho' = \rho_1 a \rho_2 b \rho_3$, where a and b are steps taken by any two processes, in which they access $\mathtt{W_C}$ in the ith time, $i \in \{2, 3\}$. Then $\rho'' = \rho_1 b \rho_2 a \rho_3$ is also a prefix of a run in $\mathcal{R}$, and the states of the shared objects in MEM and the states of the processes are same in ρ' and in ρ''.*

Proof. Immediate, using similar considerations to those used in Lemma 2. □

We can now prove the Main Lemma. We change Pr to obtain a new wait-free protocol, Pr', which solves the consensus problem for P_1, P_2 and P_3 over $\mathtt{W}^{m-1} \cup \mathtt{R}^\infty$.

The collection of shared objects used by Pr' is MEM (notice that MEM contains only $m-1$ objects of type W). The initial configuration of Pr', C_{init}, is defined as follows: let ρ_0 be an arbitrary (but fixed) run in $\mathcal{R}$, then C_{init} is the configuration reached by ρ_0 when P_1, P_2 and P_3 are about to access $\mathtt{W_C}$ for the fourth time. In configuration C_{init}, the next step of process P_i, $1 \leq i \leq 3$, according to Pr, is an access to $\mathtt{W_C}$ in the fourth time (in order to get T_i). In Pr', we replace this step by a step in which P_i reads its input (i.e., T_i is replaced by the private input of P_i). P_i proceeds in Pr' as it should have in Pr, until P_i is instructed by Pr to access $\mathtt{W_C}$ in the fifth time with $\binom{N-1}{2}$ decision values C_{ijk}. Then, instead of accessing $\mathtt{W_C}$, Pr' instructs P_i to decide on C_{123} and halt.

By Lemma 5, in each run of Pr', each non-faulty process P_i must reach a stage where in the corresponding run of Pr it would have access W for the fifth time. Since in Pr' this access is replaced by a step in which P_i decides and halts, protocol Pr' is a wait-free protocol, in which every non-faulty process eventually decides on a binary value.

We prove that Pr' is indeed a consensus protocol for P_1, P_2 and P_3. Assume by way of contradiction that there is a run ρ' of Pr', in which there exists a process which decides on an illegal value, and let T_1, T_2, T_3 be the input values in this run. By Lemma 7, and the fact that $\mathtt{W_C}$ is not in $\mathtt{MEM}$, we have that C_{init} does not depend on the order in which the processes accessed $\mathtt{W_C}$ in the second and third times. This implies, using Proposition 1, that there is a run of Pr, which reaches a configuration in which the state of $\mathtt{MEM}$ and the states of the processes are the same as in C_{init}, and in which T_1, T_2, T_3 are the values returned to P_1, P_2, and P_3 on their fourth accesses to $\mathtt{W_C}$. Hence, there is a run of Pr which starts when the state of the system excluding $\mathtt{W_C}$ is the same as in C_{init}, and is identical to ρ'. The assumption that in ρ' there is a process which returns an illegal decision value implies that in this latter run of Pr the same process accesses $\mathtt{W_C}$ with an incorrect value of C_{123}—contradicting Lemma 6. □

5.2 $\mathtt{W_C}$'s state in the critical configuration is not ϵ

In this section, we show that in the critical configuration the state of the critical object, $\mathtt{W_C}$, must be ϵ.

Lemma 8. *Let Pr be a consensus protocol for three processes over $\mathcal{W}^\infty \cup \mathcal{R}^\infty$; Then there is no critical configuration of Pr in which the state of the critical object, $\mathtt{W_C}$, is not ϵ.*

Proof. For clarity we assume that the three processes are P_1, P_2, and P_3, and that they are connected to ports 1, 2, and 3 of $\mathtt{W_C}$, respectively (the proof for other processes and ports is identical). P_1, P_2 and P_3 are referred to as the *active* processes, while all other processes are referred to as the *inactive* processes. Assume that C_{cr} is a critical configuration for Pr, and let $\mathtt{W_C}$ be the corresponding critical object. Then there is one process, say P, which, if scheduled next, brings the system to a v-valent configuration, while if one of the two other processes is scheduled after C_{cr}, the system reaches a $(1-v)$-valent configuration. This means that in any run which starts from C_{cr}, each non-faulty process will eventually be able to tell whether P was the first to access $\mathtt{W_C}$ after the C_{cr} or not. We reach a contradiction by proving that this last demand cannot hold if the state of $\mathtt{W_C}$ in C_{cr} is not ϵ.

$\mathtt{W_C}$'s state is a sequence of accesses by the N processes (note that $\mathtt{W_C}$ can be initialized to a state in which it "thinks" that some processes already accessed it before the run started). We start by observing that if in C_{cr} there exists a process which accessed $\mathtt{W_C}$ three or more times, then in this configuration the values of the T_i's are already determined; hence, if we let any of the three processes be the first to access $\mathtt{W_C}$ after C_{cr}, and then we let a process P' other then $P^\diamond$ (the first process that accessed $\mathtt{W_C}$ for the second time) to run alone, P' will receive the same responses in all three cases. Therefore, P' will not be able to decide whether P was the first to access $\mathtt{W_C}$, which leads to the desired contradiction.

We partition the possible states of $\mathtt{W_C}$ in C_{cr} to classes, according to the number of accesses made to $\mathtt{W_C}$ by P_1, P_2 and P_3. Each class is described by a tuple (S_1, S_2, S_3), where S_i, $i \in \{1,2,3\}$, denotes the number of times process P_i accessed $\mathtt{W_C}$ before C_{cr}. Each one of $\{S_1, S_2, S_3\}$ can either be a number in the range 0–6, or a sub-range of numbers x–y, within that range. For example,

(0,0–2,0–2) stands for the set of states in which P_1 made no access to $\mathtt{W_C}$ before C_{cr}, while P_2 and P_3 accessed $\mathtt{W_C}$ at most twice before C_{cr}.

By the above, we have to check all possible cases of (S_1, S_2, S_3), where $0 \leq S_i \leq 2$. We must show that in each such configuration, there is no process $P \in \{P_1, P_2, P_3\}$ such that every non-faulty process will eventually be able to tell whether P was the first to take a step after C_{cr}. Let a, b and c denote the first steps taken after the critical configuration by P_1, P_2 and P_3, respectively.

- (0,0–2,0–2): Since the state of $\mathtt{W_C}$ is not ϵ, there is no importance to the order between a first access and all other accesses. Therefore, all the states reached by $\mathtt{W_C}$ after abc, bac and bca are equivalent, and the same for acb, cab and cba. Since abc, bac and bca are equivalent, $P \notin \{P_1, P_2\}$, and since acb, cab and cba are equivalent, $P \notin \{P_1, P_3\}$. The same argument holds for (0–2,0,0–2) and for (0–2,0–2,0).
- (1,1,1): acb, bac and cab are equivalent—in all three of them $(T_1, T_2, T_3) = (0, 1, 1)$. Thus, if we let say P_1, run alone after each of acb, bac and cab, P_1 will not be able to distinguish between them. Therefore, $P \notin \{P_1, P_2, P_3\}$.
- (1,1,2): We show that P_2 cannot distinguish between acb and cab when let to run alone after each one of them, until it reaches decision. This implies that $P \notin \{P_1, P_3\}$. In both acb and cab, $P_2 \notin \mathcal{Q}$, therefore $T_2 = 1$, and thus P_2 gets the same response on its fourth access to $\mathtt{W_C}$. Since $P_2 \neq P^\diamond$, P_2 also gets the same response on its sixth access to $\mathtt{W_C}$. In the same way, it can be shown that P_1 cannot distinguish between bca and cba when let to run alone after each one of them, until it reaches decision. This implies that $P \notin \{P_2, P_3\}$. A similar argument applies to (1,2,1) and (2,1,1).
- (1,2,2): We show that P_1 cannot distinguish between abc, bac and cab when let to run alone after each one of them, until it reaches decision. This implies that $P \notin \{P_1, P_2, P_3\}$. In bac and cab, $P_1 \notin \mathcal{Q}$, therefore, $T_1 = 1$. In abc, $P_1 = P_{s(k)} \neq P^\star$, therefore, $T_1 = 1$. In all three $P_1 \neq P^\diamond$. Therefore, in all three cases P_1 gets the same response on its fourth and sixth accesses to $\mathtt{W_C}$. As before, it can be shown that the same argument applies also to (2,1,2) and (2,2,1).
- (2,2,2): Assume without loss of generality, that $P^\diamond \notin \{P_1, P_2\}$. We show that P_2 cannot distinguish between acb and cab when let to run alone after each one of them, until it reaches decision. This implies that $P \notin \{P_1, P_3\}$. Since in both acb and cab P_2 is not the first process which accesses $\mathtt{W}$ three times, T_2 is the same for both. Thus, P_2 gets the same response on its fourth access to $\mathtt{W_C}$. Since $P_2 \neq P^\diamond$, P_2 also gets the same response on its sixth access to $\mathtt{W_C}$. In the same way, it can be shown that P_1 cannot distinguish between bca and cba when let to run until alone after each one of them,it reaches decision. This implies that $P \notin \{P_2, P_3\}$.

This completes the proof of the lemma. □

5.3 Proof of the main result

In this subsection, we put the pieces together and prove that $\mathcal{CN}(\mathtt{W}^\infty \cup \mathtt{R}^\infty) < 3$. Assume by way of contradiction, that there exists a wait-free consensus protocol, Pr, for three processes over $\mathtt{W}^\infty \cup \mathtt{R}^\infty$. Let the identities of the three processes be P_i, P_j, and P_k. Since Pr is wait-free, there is a number m such that Pr never uses more than m copies of object $\mathtt{W}$. Among all possible protocols, assume that Pr is one for which this number m is minimized. Also, Pr must have a run in which it reaches a critical configuration, C_{cr}.

Let $\mathtt{W_C}$ be the critical object in C_{cr}. By Lemma 8, the state of $\mathtt{W_C}$ in the configuration C_{cr} must be ϵ. But this implies, by the Main Lemma, that there is a consensus protocol for three processes over $\mathtt{W}^\infty \cup \mathtt{R}^\infty$, Pr', which uses only $m-1$ copies of $\mathtt{W}$—a contradiction to the minimality of m. □

Theorem 9. *h^r_m is not robust for M_1 objects.*

Proof. Since we have $\mathcal{CN}(\mathtt{W}^\infty \cup \mathtt{R}^\infty) < 3$ and $\mathcal{CN}(\mathtt{CO}_3^\infty \cup \mathtt{R}^\infty) = 3$, but $\mathcal{CN}(\mathtt{W}^\infty \cup \mathtt{CO}_3^\infty \cup \mathtt{R}^\infty) = N$, h^r_m is not robust. □

6 Conclusion and Further Research

In this paper, we have studied the robustness of the h^r_m hierarchy for the class of hard-wired objects under the M_1 connection scheme. We showed that this hierarchy is not robust, by introducing a deterministic hard-wired conditional consensus object, denoted as $\mathtt{W}$ (whose consensus number is at most 3), which demonstrates that objects at level 3 of h^r_m can implement consensus for any number N of processes. The current knowledge as for the robustness of the various classes of h^r_m is as follows:

1. h^r_m is not robust for the class of hard-wired objects (either deterministic or non-deterministic) using M_1 connection scheme ([CHJT94a] and the current work).
2. The question for non-deterministic soft-wired types and for non-deterministic hard-wired using M_2 or M_3 connection rules is still open.

A natural question is whether a simple modification of our technique can be used to prove that h^r_m is not robust for M_2 or M_3 objects. The following lemma, implied by the result of [CHJT94b], indicates that this is probably not the case.

Lemma 10. *Let X and Y be two M_2 or M_3 objects such that $h^r_m(X) = k_1$, $h^r_m(Y) = k_2$, $h^r_m(X, Y) = k$, and $k > \max\{k_1, k_2\}$. Then, neither X nor Y can be implemented by $\mathtt{CO}_{\mathtt{k-1}}^\infty \cup R^\infty$.*

Proof. Assume by way of contradiction that Y may be implemented by $\mathtt{CO}_{\mathtt{k-1}}^\infty \cup \mathtt{R}^\infty$, then $h^r_m(\mathtt{X}, \mathtt{Y}) \le h^r_m(\mathtt{X}, \mathtt{CO_{k-1}})$. By the result of [CHJT94b], $h^r_m(\mathtt{X}, \mathtt{CO_{k-1}}) = \max\{h^r_m(\mathtt{X}), k-1\}$. Putting the pieces together, we get that $h^r_m(\mathtt{X}, \mathtt{Y}) \le \max\{h^r_m(\mathtt{X}), k-1\} = k-1$, contradicting the assumption that $h^r_m(\mathtt{X}, \mathtt{Y}) = k$. Therefore, Y cannot be implemented by $\mathtt{CO}^{\infty}_{\mathtt{k-1}} \cup \mathtt{R}^{\infty}$. In the same way it can be shown that $\mathtt{X}$ cannot be implemented by $\mathtt{CO}^{\infty}_{\mathtt{k-1}} \cup \mathtt{R}^{\infty}$ as well. □

The above lemma shows that if an M_2 or M_3 object, based on our object $\mathtt{W}$, can be combined with another object $\mathtt{Y}$ with $h^r_m(\mathtt{Y}) < N$ to obtain an N-consensus protocol, then $\mathtt{Y}$ cannot be implemented by $\mathtt{CO}^{\infty}_{\mathtt{N-1}} \cup \mathtt{R}^{\infty}$. (Notice that by the universality result, $\mathtt{Y}$ must be shared by at least N processes.) To the best of our knowledge, no such deterministic object $\mathtt{Y}$ is known. A non-deterministic object with the above property (which is shared by $2N + 1$ processes) is given in [Rac94].

Acknowledgments:

We wish to thank Robert Cori for some very helpful remarks on an earlier version of this paper.

References

[BGA94] E. Borowsky, E. Gafni, and Y. Afek. Consensus power makes (some) sense! In *Proc. 13th ACM Symp. on Principles of Distributed Computing*, August 1994.

[BMZ90] O. Biran, S. Moran, and S. Zaks. A combinatorial characterization of the distributed 1-solvable tasks. *Journal of Algorithm 11*, pages 420–440, 1990. A preliminary versions appeared in Proc. 7th ACM Symp. on Principles of Distributed Computing, August 1988.

[BNP94] R. Bazzi, G. Neiger, and G. Peterson. On the use of registers in achieving wait-free consensus. In *Proc. 13th ACM Symp. on Principles of Distributed Computing*, August 1994.

[CHJT94a] T. Chandra, V. Hadzilacos, P. Jayanti, and S. Toueg. The h^r_m hiererachy is not robust. Manuscript, August 1994.

[CHJT94b] T. Chandra, V. Hadzilacos, P. Jayanti, and S. Toueg. Wait-freedom vs. t-resiliency and the robustness of wait-free hierarchies. In *Proc. 13th ACM Symp. on Principles of Distributed Computing*, August 1994.

[CIL87] B. Chor, A. Israeli, and M. Li. On processor coordination using asynchronous hardware. In *Proc. 6th ACM Symp. on Principles of Distributed Computing*, pages 86–97, 1987.

[CM94] R. Cori and S. Moran. Exotic behaviour of consensus numbers. In *Proceedings of 8-th International Workshop on Distributed Algorithms*, September 1994.

[FLP85] M. J. Fischer, N. A. Lynch, and M. S. Paterson. Impossibility of distributed consensus with one faulty process. *Journal of the ACM*, 32(2):374–382, April 1985.

[Her91a] M. Herlihy. Impossibility results for asynchronous PRAM. In *Proc. 3rd ACM Symp. on Algorithms and Architectures*, pages 327–336, 1991.

[Her91b] M. P. Herlihy. Impossibility results for asynchronous PRAM. In *Proc. 3rd ACM Symp. on Algorithms and Architectures*, pages 327–336, 1991.

[Jay93] P. Jayanti. On the robustness of herlihy hierarchy. In *Proc. 12th ACM Symp. on Principles of Distributed Computing*, August 1993.

[Jay95] P. Jayanti. Wait-free computing. In *Proc. 9th International Workshop on Distributed Algorithms*, September 1995.

[LA87] C. M. Loui and H. Abu-Amara. Memory requirements for agreement among unreliable asynchronous processes. *Advances in Computing Research*, 4:163–183, 1987.

[Lam86] L. Lamport. On interprocess communication, parts I and II. *Distributed Computing*, 1(2):77–101, 1986.

[LM95] R. Lubitch and S. Moran. Closed schedulers: A novel technique for analyzing distributed protocols. *Distributed Computing*, 8(4):203–210, 1995. An extended preliminary version appeared in " Closed Schedulers: Constructions and Applications to Consensus Protocols", Proceedings of 6-th International Workshop on Distributed Algorithms, 1992 and in TR #796, Dept. of Computer Science, Technion, January 1994.

[PBN94] G. Peterson, R. Bazzi, and G. Neiger. A gap theorem for consensus types. In *Proc. 13th ACM Symp. on Principles of Distributed Computing*, August 1994.

[Rac94] Ophir Rachman. Anomalies in the wait-free hierarchy. In *Proc. 8th International Workshop on Distributed Algorithms*, 1994.

[TKM94] G. Taubenfeld, S. Katz, and S. Moran. Impossibility results in the presence of multiple faulty processes. *Information and Computation*, 1994. Preliminary version appeared in 9th FCT-TCS Conference, Bangalore, India, December, 1989, Lecture Notes in Computer Science, vol. 405 (eds.:C.E. Veni Madhavan), Springer Verlag 1989, pages 109-120.

Understanding the Set Consensus Partial Order Using the Borowsky-Gafni Simulation * (Extended Abstract)

Soma Chaudhuri and Paul Reiners

Iowa State University, Ames, IA 50011, USA

Abstract. We present a complete characterization of the Set Consensus Partial Order, a refinement of the Consensus Hierarchy of Herlihy. We define the (n, k)-*set consensus problem* as the k-set consensus problem for n processors. We then answer the question of whether an (n, k)-set consensus object (an object which solves the (n, k)-set consensus problem) can be implemented using a combination of (m, ℓ)-set consensus objects and snapshot objects, for all possible values of n, k, m, ℓ, creating a *partial order* of set consensus objects. The model we consider is the asynchronous shared memory model.
To prove our results, we use the Borowsky-Gafni Simulation technique, a powerful tool which has been used to prove several impossibility results about shared memory algorithms. Lynch and Rajsbaum gave a formal description of the basic technique, along with a proof of its correctness. We extend their results to include simulations of algorithms which access set consensus objects. Our description of the simulation, and its proof of correctness, are also in terms of I/O Automata. We need this stronger version of the simulation algorithm to obtain our results on the Set Consensus Partial Order. We state a general Simulation Theorem which specifies the properties of the simulation, and characterizes all the impossibility results that can be obtained using this technique. Our partial order result can then be derived as a special case of this theorem.

1 Introduction

The CONSENSUS PROBLEM is a fundamental problem in distributed computing, where a set of n processors communicate among each other to decide on a common output value from among their input values. While the problem itself seems simple, in a surprising result, Fischer *et. al.* [6] showed that in a *totally asynchronous* system, the CONSENSUS PROBLEM could not be solved deterministically even in the presence of only *one* fail-stop fault.

The k-SET CONSENSUS PROBLEM, a generalization of CONSENSUS introduced by Chaudhuri [5], requires each processor to decide on some processor's input value as its output (this is called the *validity condition*), and the set of values decided upon must be of size at most k (this is called the *k-agreement condition*).

* Supported in part by NSF grant CCR-93-08103.

Chaudhuri conjectured that the k-SET CONSENSUS problem is unsolvable in the presence of k fail-stop faults, while showing that it was solvable in the presence of $k-1$ faults. Since the 1-SET CONSENSUS problem is equivalent to the CONSENSUS problem, this is a generalization of the FLP impossibility result, and it was later proven correct by three teams of researchers, Borowsky and Gafni [2], Herlihy and Shavit [9], and Saks and Zaharoglu [15].

In his seminal paper on wait-free synchronization [7], Herlihy introduced the consensus hierarchy by defining an n-consensus object, an object which solves the consensus problem for n processors. Herlihy showed a hierarchy among these objects, and his result was later extended by Jayanti and Toueg [13] to show a strict linear hierarchy among the n-consensus objects for all values of n. In particular, they showed that there is a wait-free implementation of an n-consensus object, using any number of read/write objects and m-consensus objects, if and only if $n \leq m$.

Continuing along those lines in trying to refine this hierarchy, we study the (n,k)-set consensus problem (and its corresponding (n,k)-set consensus object), as defined by Borowsky and Gafni [3]. The (n,k)*-set consensus problem* is the k-set consensus problem for n processors. An (n,k)*-set consensus object* (henceforth referred to as an (n,k)*-setcon* object) is a data object which supports a one-time computation of k-set consensus by up to n processors. The object has to be wait-free, meaning that it is $(n-1)$-resilient, where the number of processors that can access it is n.

We answer the question of whether an (n,k)*-setcon* object can be implemented *in a wait-free manner* by any combination of (m,ℓ)*-setcon* objects and snapshot objects. Since snapshot objects can be implemented by read/write registers [1], this restriction imposed on implementations is consistent with the h_m^r hierarchy defined by Jayanti [10], where an object A can be implemented by any combination of objects B and read/write registers. We show the relationship (that is, whether a wait-free implementation is possible or not) between every pair of *setcon* object types. We determine that the relationships define a *Set Consensus Partial Order* rather than a hierarchy because we will show the existence of pairs of objects, neither of which can implement the other. We, thus, completely characterize these relationships in the theorem stated below (the $\preceq$ relation stands for "can be implemented by").

Theorem 1. [Partial Order Theorem] *Let n, k, m, and ℓ be any positive integers. Then*

1. IF $n \leq k$, THEN (n,k)*-setcon* $\preceq$ (m,ℓ)*-setcon*,
2. IF $n > k$ AND $m \leq \ell$, THEN (n,k)*-setcon* $\not\preceq$ (m,ℓ)*-setcon*,
3. IF $n > k$ AND $m > \ell$, THEN
   (a) IF $k \geq \ell\lceil n/m \rceil$, THEN (n,k)*-setcon* $\preceq$ (m,ℓ)*-setcon*,
   (b) IF $k \geq \ell\lfloor n/m \rfloor + (n - m\lfloor n/m \rfloor)$, THEN (n,k)*-setcon* $\preceq$ (m,ℓ)*-setcon*,
   (c) IF $k < \ell\lceil n/m \rceil$ AND $k < \ell\lfloor n/m \rfloor + (n - m\lfloor n/m \rfloor)$, THEN (n,k)*-setcon* $\not\preceq$ (m,ℓ)*-setcon*.

The first two cases are pretty straightforward. The interesting point about the third case is that the intuition behind the impossibility result of the last sub-case is based on the algorithms of the two other sub-cases.

Borowsky and Gafni [3, 4] were the first to consider the question of whether a certain set consensus object can implement another. They obtained some partial results, including both impossibility results and algorithms (cases 3(a) and (b) of the Partial Order Theorem). Their impossibility results use a powerful simulation technique. Our result described above is an extension of their work and uses the same technique. The same result is obtained by Herlihy and Rajsbaum [8] using more complicated topological techniques.

The impossibility results we obtain are derived using the Borowsky-Gafni simulation technique, a powerful tool for studying possibility and impossibility results in the asynchronous shared memory system with failures. It is therefore important to understand exactly what the simulation allows. This was not completely clear in the informal presentation of the technique by Borowsky and Gafni, which left open some questions. To answer these questions, Lynch and Rajsbaum [11] recently studied the simulation technique and came up with a precise, formal description of the basic technique, in terms of I/O automata, along with a formal proof of correctness.

We extend the results of Lynch and Rajsbaum, also using I/O automata, to include simulations of algorithms which access *setcon* objects, since we will be using such simulations in obtaining our results on the Set Consensus Partial Order. We present our results about the simulation as a Simulation Theorem—a general characterization of the properties of the simulation—which characterizes all the results that can be obtained using the technique. Our results can then be derived as a special case of the Simulation Theorem. What makes the Simulation Theorem easy to use is that, while it is proven using I/O automata, it is stated in more general terms, which makes it easy to see in which situations it can be applied.

Lynch and Rajsbaum [11] used their results to come up with a notion of a fault-tolerant reducibility between any two problems. Our Simulation Theorem answers the question of whether a simulation of an algorithm for Problem P_2 *for* a set of n_2 processors with at most f_2 faults *by* a set of n_1 processors with at most f_1 faults will solve Problem P_1 for the set of n_1 processors. We specify what we mean for an n_1-processor system to successfully simulate an algorithm for an n_2-processor system. Our Simulation Theorem differs from the f-reducibility of [11] in several ways. It is more general in that it lets the number of faults allowed in the simulating system be different from the number of faults allowed in the simulated system, due to the additional implementations of set consensus objects. It is therefore not a reducibility in the same sense, since the simulation needs to implement these objects not provided by the simulated system. Also, while f-reducibility focuses on the *decision problems* themselves, our Simulation Theorem looks at the *algorithms* for the decision problems. This makes our result less abstract in this respect, but also gives us some advantages. Specifically, focusing on the algorithm instead of the problem allows our Simulation Theorem to express both the safety and liveness properties of the technique, while f-reducibility is restricted to the safety properties.

2 The Model

We use the I/O Automata model of Lynch and Tuttle [12], which we briefly review here. An I/O automaton is a simple state machine, where the transitions are *actions*, classified as *internal*, *input*, or *output*. There is a *fairness* constraint on the executions, requiring that every enabled non-input action be given a fair turn.

We assume an *asynchronous shared memory system* with *snapshot variables*. The shared memory is accessed by *snap* and *update* operations. We assume that, given a set of n_1 processors, the memory consists of n_1 components. A *snap* operation by any processor will return the value of the entire memory, *i.e.*, all n_1 components. An $update(x)$ operation by the processor i changes the value of the ith component to x. In particular, only processor i can change the value of the ith component of memory while all other processors can read it.

3 The Borowsky-Gafni Simulation

We now give an informal overview of the Borowsky-Gafni simulation technique. The technique allows n_1 real processors to each simulate n_2 programs or sequences of code (the simulated processors), so that the n_1 sets of simulations are consistent with one another, as well as consistent with a real execution of the n_2 programs, even though the programs may invoke nondeterministic objects. We assume that the n_2 programs can solve some decision problem P_2, and a simulation of the n_2 programs, by the n_1 real processors, solves some other decision problem P_1. We allow for fault-tolerance; we assume that the set of n_2 programs is resilient to f_2 fail-stop faults, while the set of n_1 simulating processors may have as many as f_1 such faults.

The n_1 real processors each simulate a shared-memory algorithm P_2 involving n_2 programs. We now define what we mean by a simulation. Specifically, each real processor i has a function g_i which maps its input value x_i to an n_2-length vector of proposed input values for the n_2 programs j. The execution simulated adopts the input value proposed by one of the processors i for each program j. At the termination of the simulated execution, each processor i observes a certain n_2-length vector of the outputs of the programs j (with as many as f_2 null entries). Each of these vectors are consistent with each other in that, for all k, all non-null entries in the kth position of each vector are the same. Also, each output vector consists of a set of valid outputs for the programs j. Now, for each i, a function h_i maps the n_2-length vector observed by processor i to the decision value y_i for processor i. These mapping functions follow the lines of the reducibility defined in [11].

Definition 2. An *f-fault tolerant decision problem $P = \langle \mathcal{I}, \mathcal{R}, \Delta \rangle$ for a distributed system of n processors* is a set, $\mathcal{I}$, of vectors of length n, a set, $\mathcal{R}$, of vectors of length n, each of which may have up to f null entries, and a set $\Delta \subseteq \mathcal{I} \times \mathcal{R}$. The intuitive idea is that $\mathcal{I}$ is the set of possible inputs to the system, $\mathcal{R}$ is the set of possible outputs of the system, with null entries corresponding to processors which have failed, and $(I, R) \in \Delta$, if R is a correct solution of problem P with input I.

Definition 3. An algorithm A *solves* an f-fault tolerant decision problem $P = \langle \mathcal{I}, \mathcal{R}, \Delta \rangle$ for a distributed system of n processors if, given an input $I \in \mathcal{I}$, every execution of A starting at input I in which no more than f processors fail will produce an output $R \in \mathcal{R}$, where $(I, R) \in \Delta$.

In the following, we let $P_1 = \langle \mathcal{I}_1, \mathcal{R}_1, \Delta_1 \rangle$ be an f_1-fault-tolerant decision problem for a system of n_1 processors and let $P_2 = \langle \mathcal{I}_2, \mathcal{R}_2, \Delta_2 \rangle$ be an f_2-fault-tolerant decision problem for a system of n_2 processors. Let A_2 be an algorithm (a set of n_2 programs) that solves P_2. A *simulation of A_2 by a set of n_1 processors* is an algorithm for a system of n_1 processors that simulates an algorithm A_2 of P_2. Both A_2 and the simulation of A_2 are shared-memory algorithms, and we need to specify the systems in which they run. In particular, in the simulations described in [11], both A_2 and its simulation run in systems which contain only snapshot objects. We will also consider simulations where, while the algorithm A_2 may access both snapshot and set consensus objects, its simulation can access only snapshot objects.

Definition 4. We say that the *simulation S of an algorithm for P_2 by n_1 processors with at most f_1 faults solves problem P_1* (Figure 1), if there is a sequence of functions $g_1, \ldots, g_{n_1}$, where, for all i, $g_i : \mathcal{I}_1[i] \rightarrow \mathcal{I}_2$, and a sequence of functions $h_1, \ldots, h_{n_1}$, where, for all i, $h_i : \mathcal{R}_2 \rightarrow \mathcal{R}_1[i]$, and the following conditions hold for each execution of S with no more than f_1 failures. Let the input values be the vector $\langle x_1, \ldots, x_{n_1} \rangle \in \mathcal{I}_1$.

1. The execution terminates with some output vector $\langle y_1, \ldots, y_{n_1} \rangle \in \mathcal{R}_1$. This is the liveness condition.
2. The outputs $y_1, \ldots, y_{n_1}$ are consistent with a valid solution for problem P_1 given inputs $x_1, \ldots, x_{n_1}$, where at most f_1 entries y_i are null. More formally, $(\langle x_1, \ldots, x_{n_1} \rangle, \langle y_1, \ldots, y_{n_1} \rangle) \in \Delta_1$.
3. There exist an n_2-length input vector $\hat{I} \in \mathcal{I}_2$ and n_1 n_2-length output vectors $\hat{R}_1, \ldots, \hat{R}_{n_1} \in \mathcal{R}_2$ such that
   (a) For all i, $\hat{R}_i$ is a valid output of problem P_2 on input $\hat{I}$, *i.e.*, $(\hat{I}, \hat{R}_i) \in \Delta_2$.
   (b) The set of vectors $\hat{R}_i$ are consistent, *i.e.*, for all j, if the jth entries in two different vectors are both non-null, then they must be equal.
   (c) The vector $\hat{I}$ is derived from $\langle x_1, \ldots, x_{n_1} \rangle$ and the functions $g_1, \ldots, g_{n_1}$. Specifically, for all j, the jth entry of $\hat{I}$ is equal to the jth entry of $g_i(x_i)$, for some i.
   (d) The vector $\langle y_1, \ldots, y_{n_1} \rangle$ is derived from the vectors $\hat{R}_1, \ldots, \hat{R}_{n_1}$ and the functions $h_1, \ldots, h_{n_1}$. Specifically, for all i, $h(\hat{R}_i) = y_i$.

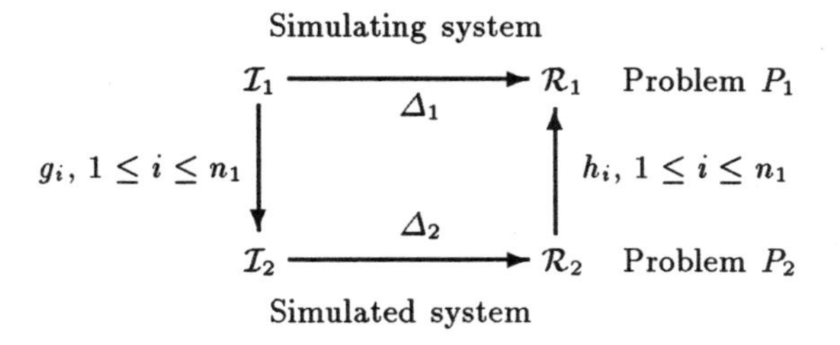

Fig. 1. A simulation of an algorithm which solves problem P_2 being used to solve problem P_1.

We assume that the simulated system $\mathcal{P}$ is an atomic snapshot memory system of n_2 programs, with a single snapshot variable mem' consisting of n_2 components $mem'(j)$. We will refer to the processes whose executions are *being simulated* as *programs* and the *simulating* processes as *processors* to avoid confusion. As in [11], we let the simulating system $\mathcal{Q}$ be an atomic snapshot memory system of n_1 processors, with a single snapshot variable mem consisting of n_1 components $mem(i)$. Each component $mem(i)$ is a copy of the entire memory mem', and reflects processor i's simulation of system $\mathcal{P}$. Along with a copy of $mem'(j)$, for each simulated process j, $mem(i)$ also includes a *sim-steps*(j) counter which records the number of steps that i has simulated for j at the time of its last simulated update for j. This helps the other simulating processors identify the latest updated value, for any particular j, among all simulating processors i. A function *latest* can check, independently for each j, the *sim-steps*(j) value in each $mem(i)$, and choose, for each j, the particular $mem(i).mem'(j)$ so that $mem(i).$*sim-steps*(j) is highest over all i.

To maintain consistency among the real processors, it is important that each real processor agrees on the same snapshot value at each step so that they make the same state transitions. Simulating each snapshot therefore requires an agreement protocol, where different processors submit their individual versions of the snapshot variable mem' and the agreement protocol chooses the version of a specific real processor, a value which is then adopted by all processors. In addition, the fact that each real processor submits snapshot values computed by the *latest* function makes sure that the snapshot values decided upon are consistent with a real run of the simulated processes.

We now consider the simulations of algorithms which access set consensus objects. Here, the simulation is further complicated by the fact that, while the simulated system has set consensus objects, there are no such objects in the simulating system. The simulation therefore also has to simulate the set consensus objects. Suppose the simulated algorithm has an (m, ℓ)-*setcon* object $\mathcal{O}$ which is accessed by m simulated programs $1, \ldots, m$, each with an input value. We need the *setcon* object $\mathcal{O}$ to return at most ℓ distinct values. We simulate the m accesses to object $\mathcal{O}$ as follows. We let the n_1 real processors, each simulating the *setcon* operation of all of the m programs to object $\mathcal{O}$ (this involves $n_1 m$ total such simulated accesses), all participate in the same version of an ℓ-agreement protocol, returning at most ℓ distinct values. We still need to ensure that the n_1 simulations of the *setcon* operation for *each particular program* $j \in 1, \ldots, m$, obtain the same value. We therefore have the n_1 real processors participate in a 1-agreement protocol, one for each program j. Therefore each *setcon* operation is simulated by an ℓ-agreement followed by a 1-agreement. The simulation therefore requires the implementation of both 1-agreement and ℓ-agreement protocols which are discussed in Section 5.

We now consider the fault-tolerance of the simulation. If the simulated algorithm is f_2-resilient, it is important that the simulations of no more than f_2 programs be blocked. As we will show in Section 5, our agreement (respectively, ℓ-agreement) protocol is 0-resilient (respectively, $(\ell - 1)$-resilient). Therefore, a real processor i, simulating a snapshot of a program j, could fail while participating in the agreement algorithm and block the program j from being simulated any further by other processors. Similarly, if ℓ real processors simulating the access to the (m, ℓ)-*setcon* object $\mathcal{O}$ by the programs $1, \ldots, \ell$, respectively, fail in the middle of the ℓ-agreement algorithm, they could potentially block all m programs that can access $\mathcal{O}$. In general, a combination of these can happen, causing the simulation of certain programs to be blocked. To minimize the amount of damage caused by a faulty real processor and still allow progress to be made, we require that each real processor be in a position to block at most one agreement or ℓ-agreement algorithm at any time. We will show how to achieve this in Section 5. We can now argue about the maximum number f_1 of faulty simulating processors that can be tolerated to successfully simulate the f_2-resilient algorithm. The Simulation Theorem, which will be proved formally in Section 6, is stated below.

Theorem 5. [Simulation Theorem] *A set of n_2 programs $j_1, \ldots, j_{n_2}$ solving the f_2-fault tolerant problem P_2, which accesses snapshot objects and (m, ℓ)-setcon objects, can be simulated by a set of n_1 processors $i_1, \ldots, i_{n_1}$ solving the f_1-fault tolerant problem P_1, which only accesses snapshot objects, if*

1. *the sets of functions g and h exist as required by the simulation, and*
2. *$f_1 \geq a\ell + b$ implies $f_2 \geq am + b$, for all positive integers a and b.*

We can now state a corollary to the Simulation Theorem, which applies to simulated systems with no *setcon* objects. It is obtained by setting $m = \ell$ in the Simulation Theorem, since (m, m)-*setcon* objects are trivial.

Corollary 6. *A set of n_2 programs solving the f_2-fault tolerant problem P_2, which accesses snapshot objects, can be simulated by a set of n_1 processors solving the f_1-fault tolerant problem P_1, which also accesses snapshot objects, if the set of functions g and h exist as required by the simulation and if $f_1 \leq f_2$.*

The following theorem, restated in terms of *setcon* objects, was originally proved by three teams of researchers [2, 9, 15]. It will be useful in proving the Partial Order Theorem.

Theorem 7. [Set Consensus Impossibility Theorem] *The $(k+1,k)$-setcon object does not have a wait-free implementation in an asynchronous snapshot shared memory system with $k+1$ processors.*

The impossibility of k-resilient solutions for the set consensus problem in the asynchronous shared memory model with n processors, originally proven by Borowsky and Gafni, now follows from Theorem 7 and Corollary 6.

Theorem 8. *There is no solution to the k-set consensus problem in an asynchronous shared memory model with n processors (where $k < n$), which is resilient to k fail-stop faults.*

4 The Set Consensus Partial Order

By considering wait-free (m,ℓ)-*setcon* objects, the consensus hierarchy can be refined into partial orders. The Partial Order Theorem, proved below, gives the relationships between different (m,ℓ)-*setcon* objects for all values of m and ℓ. It is proven using the Set Consensus Impossibility Theorem and the Simulation Theorem.

Proof. (of Partial Order Theorem) Cases 1 and 2 are straightforward. We prove the other cases below. Cases 3a and 3b are based on the protocols in [3]. Case 3(c) is our main impossibility result.

Case 3(a): We implement a wait-free (n,k)-*setcon* object using only wait-free (m,ℓ)-*setcon* objects and snapshot objects. Divide the set of n processors into $\lceil n/m \rceil$ groups of at most m processors each. Let each group invoke a different (m,ℓ)-*setcon* object and decide on the values returned. The number of different decisions made by the processors in any given group is at most ℓ. This implies that the total number of different values decided by all the processors is at most $\ell\lceil n/m \rceil$. Since, by assumption, $k \geq \ell\lceil n/m \rceil$, at most k values are decided, as required.

Case 3(b): We implement a wait-free (n,k)-*setcon* object using just wait-free (m,ℓ)-*setcon* objects and snapshot objects. Divide the set of n processors into $\lfloor n/m \rfloor$ groups of m processors each. There will be $n - m\lfloor n/m \rfloor$ processors remaining which are not part of any group. Let each group of m processors invoke a different (m,ℓ)-*setcon* object and decide on the values returned. The number of different decisions made by the m processors in each of these groups is at most ℓ. Let each of the remaining $n - m\lfloor n/m \rfloor$ processors decide on its own value. The total number of values decided by all the processors is therefore at most $\ell\lfloor n/m \rfloor + (n - m\lfloor n/m \rfloor)$. Since, by assumption, $k \geq \ell\lfloor n/m \rfloor + (n - m\lfloor n/m \rfloor)$, at most k values are decided, as required.

Case 3(c): Suppose $k < \ell\lceil n/m \rceil$ and $k < \ell\lfloor n/m \rfloor + (n - m\lfloor n/m \rfloor)$. Hence, $k+1 \leq \ell\lceil n/m \rceil$ and $k+1 \leq \ell\lfloor n/m \rfloor + (n - m\lfloor n/m \rfloor)$.

Suppose a wait-free (n,k)-*setcon* object can be implemented using wait-free (m,ℓ)-*setcon* objects and snapshot objects. Then, we will show, using the Simulation Theorem, that $(k+1,k)$-*setcon* can be implemented by snapshot objects, contradicting the Set Consensus Impossibility Theorem.

By the assumption above, there is an n program wait-free algorithm which implements an (n,k)-*setcon* object using wait-free (m,ℓ)-*setcon* objects and snapshot objects. We use the Borowsky-Gafni simulation technique to simulate these $n_2 = n$ programs of code using $n_1 = k+1$ simulating processors $i_1,\ldots,i_{k+1}$, with input values $x_1,\ldots,x_{k+1}$. The simulating processors only use snapshot objects while the programs of code use (m,ℓ)-*setcon* objects and snapshot objects. Since problem A of the simulated system and problem B solved by the simulating processors are both the k-set consensus problem with different numbers of inputs, our set of functions g and h are trivial. Let $g_i(v)$ be the n_2-length vector $(v,\ldots,v)$. The function h_i maps an output vector of the simulated program to *any* non-null value in the vector. Since the set of n programs $j_1,\ldots,j_n$ is wait-free, it is resilient to $n-1$ faults, and, since we would like our simulation to be wait-free, we may have as many as k faulty simulating processors.

Now, by the Simulation Theorem, it follows that the simulation will succeed as long as, for all non-negative integer values of a and b, if $k \geq a\ell + b$ then $n-1 \geq am+b$. Let a and b be non-negative integers such that $k \geq a\ell+b$. We will show $n-1 \geq am+b$, thus proving that every non-faulty simulating processor will terminate.

By assumption, $a\ell+b < k+1$. Since $k+1 \leq \ell\lceil n/m\rceil$, it follows that $a < \lceil n/m\rceil$, implying that $a \leq \lfloor n/m\rfloor$. Also, since $k+1 \leq \ell\lfloor n/m\rfloor + (n - m\lfloor n/m\rfloor)$, it follows that $a\ell+b < \ell\lfloor n/m\rfloor + (n-m\lfloor n/m\rfloor)$.
Now,

$$\begin{aligned}
a\ell+b &< n - m\lfloor n/m\rfloor + \ell\lfloor n/m\rfloor \\
&= n - (m-\ell)\lfloor n/m\rfloor \\
\Longrightarrow \qquad b &< n - (m-\ell)\lfloor n/m\rfloor - a\ell \\
\Longrightarrow am+b &< n - (m-\ell)\lfloor n/m\rfloor - a\ell + am \\
&= n - (m-\ell)\lfloor n/m\rfloor + a(m-\ell) \\
&= n - (m-\ell)(\lfloor n/m\rfloor - a)
\end{aligned}$$

Since $m > \ell$ and $a \leq \lfloor n/m\rfloor$, it follows that $(m-\ell)(\lfloor n/m\rfloor - a) \geq 0$, thus proving that $am+b<n$, as required.

So, the conditions of the Simulation Theorem hold. By Definition 4, each non-faulty real processor $i \in \{i_1,\ldots,i_{k+1}\}$ will terminate with an output value y_i, where y_i is the output of some program $j \in \{j_1,\ldots,j_n\}$ in an execution where the input of each program j is in the set $\{x_1,\ldots,x_{k+1}\}$. Since the set of programs implementing the (n,k)-*setcon* object, satisfies *validity*, the output of each program j is an input of some program j' in the execution, and $y_i \in \{x_1,\ldots,x_{k+1}\}$, where y_i is the output of real processor i. Also, since the set of programs j satisfies k-*agreement*, the number of distinct outputs of the programs is no more than k, and, thus, the number of distinct values y_i is also no more than k, where y is the output of real processor i. Thus, the simulation represents a valid implementation of a $(k+1,k)$-*setcon* object using only snapshot objects. This contradicts the Set Consensus Impossibility Theorem. Therefore, an (n,k)-*setcon* object cannot be implemented by any number of (m,ℓ)-*setcon* objects and snapshot objects. □

The theorem takes care of all possible cases, and completely characterizes the relationship between different *setcon* objects. Note that this does *not* define a total order, since there are pairs of objects, where neither implements the other. $(2, 1)$-*setcon* and $(5, 2)$-*setcon* is an example of such a pair.

5 The Agreement Protocol

We now define the agreement protocol as an I/O automaton, very much the same way as done in [11]. Our contribution here is in extending their definitions to include ℓ-agreement protocols. Each agreement module has N ports, numbered $1, \ldots, N$. The agreement protocol may fail due to failures at the ports (the users of the agreement module); however, this can only happen during an 'unsafe' portion of the execution of a particular port. Each port i supports input actions of the form $propose_i(v)$ and $stop_i$, and output actions of the form $safe_i$ and $agree_i(w)$. A value v is proposed for agreement by user i with the input action $propose_i(v)$, and the output action $safe_i$ tells user i that the unsafe portion of its execution is over. Finally, the agreed upon value w is returned by the output action $agree_i(w)$. If user i fails before receiving $agree_i(w)$, the input action $stop_i$ announces this failure.

Now, clearly a *well-formed* execution would require that any $propose_i$, $safe_i$ and $agree_i$ actions for a given agreement module be in that order. Also, as in [11], we require that the programs preserve well-formedness on every port, that is, there is at most one $propose_i$ for any particular i. In addition, each module must satisfy the VALIDITY condition, which says that any agreement value must be a proposed value.

We will be looking at both 1-agreement and ℓ-agreement modules, where $\ell > 1$. They need to satisfy the AGREEMENT and ℓ-AGREEMENT conditions, respectively. The ℓ-agreement condition requires that there are at most ℓ distinct agreement values. AGREEMENT is defined as 1-agreement.

We have two liveness conditions for each of our agreement modules. The first liveness condition, WAIT-FREEDOM, says that a $propose_i$ action on a non-faulty port will eventually receive a $safe_i$. In other words, every non-faulty port is guaranteed to complete the unsafe portion of its execution, no matter what happens at the other ports. More formally, WAIT-FREEDOM requires that in any execution, for any i, if $propose_i$ occurs and no $stop_i$ occurs, then $safe_i$ occurs.

The second liveness condition, SAFE TERMINATION, deals with the fault-tolerant behavior and depends on the specific agreement module. In the case of 1-agreement, SAFE TERMINATION says that, if no ports remain unsafe, then any *propose* event on a non-faulty port will receive an *agree*. In the case of ℓ-agreement, the condition is stronger; ℓ-SAFE TERMINATION says that, if *no more than* $\ell - 1$ ports remain unsafe, then any *propose* event on a non-faulty port will receive an *agree*. More formally, ℓ-SAFE TERMINATION requires that in any execution, if there are no more than $\ell - 1$ indices j such that $propose_j$ occurs and $safe_j$ does not occur, then, for any i, if $propose_i$ occurs and $stop_i$ does not occur, then $agree_i$ occurs. SAFE TERMINATION is defined as 1-safe termination.

These conditions on the agreement protocol allow the real processors to make sure that they are in a position to block at most one agreement protocol at any time, and still make progress simulating programs. In particular, a real processor can be involved in several 1-agreement (or ℓ-agreement) protocols at the same time, as long as it is not in the unsafe portion of more than one (that is, there is at most one module for which it has made the ***propose*** request but has not received the *safe* announcement). In that case, if a real processor fails, then only one agreement module would be blocked. This is ensured by having a ***status*** variable, for each processor i, which has the value ***unsafe*** when i has a ***propose***$_i$ pending at some agreement module and which is changed to ***safe*** when i receives the ***safe*** announcement. A value ***unsafe*** in the ***status*** variable disallows i from sending another ***propose***$_i$ action to some other agreement module.

The agreement module in [11] works as follows. The snapshot shared memory contains a *val* component and a *level* component, which is initialized t 0, for each port i. When a $propose_i(v)$ is received, the value v is recorded in *val* and *level* is raised to 1. Then a snapshot is taken to determine whether $level = 2$ for any other port i'. If so, $level_i$ is set to 0, and otherwise it is set to 2.

Now, repeated snapshots are taken until there is no i' such that $level_{i'} = 1$, in the case of the 1-agreement module, and until there are no more than $\ell - 1$ ports i' such that $level_{i'} = 1$ in the case of the ℓ-agreement module. Now, the *val*, w corresponding to port i_0, where i_0 is the smallest index such that $level_{i_0} = 2$, is chosen and returned by the output action $agree_i(w)$.

The ℓ-agreement module is a general version of the 1-agreement module of [11]. We do not describe it in detail here due to lack of space. The only difference between the two modules is in the condition required for exiting the repeated snapshot loop described above. In addition, we define the actions ℓ-$propose_i$, ℓ-$safe_i$ and ℓ-$agree_i$ in the ℓ-agreement module, to distinguish them from the equivalent actions $propose_i$, $safe_i$ and $agree_i$, respectively, in the 1-agreement module.

Lemma 9. *ℓ-SafeAgreement satisfies the ℓ-agreement condition.*

Lemma 10. *The ℓ-agreement module defined above satisfies the safety conditions of* VALIDITY *and* ℓ-AGREEMENT *and the liveness conditions of* WAIT-FREEDOM *and* ℓ-SAFE TERMINATION.

6 Proof of the Simulation

We now give the details of our simulation. Given the system $\mathcal{P}$ of n_2 programs with access to the snapshot variable *mem'* and a set O of (m, ℓ)-*setcon* objects, we want to simulate it in the system $\mathcal{Q}$ of n_1 processors with access only to the snapshot variable *mem*. In Figure 2, we give the automaton for $\mathcal{Q}$, an extension of the same in [11].

We define $\mathcal{P}$ and $\mathcal{Q}$ as I/O automata, and show that system $\mathcal{Q}$ simulates system $\mathcal{P}$. We do this in two stages, by first defining a new system $\mathcal{C}$. We then show that $\mathcal{C}$ simulates $\mathcal{P}$, and $\mathcal{Q}$ simulates $\mathcal{C}$, thus obtaining our result. Since

```
Simulation System Q:

Shared variables:

mem, a length n1 snapshot value;
  for each i, mem(i) has components:
    sim-mem, a vector in R^n2, initially everywhere r0
    sim-steps, a vector in N^n2, initially everywhere 0

Actions of i:

Input:
  init(v)_i, v ∈ V
  ℓ-agree(w)_{j,O,i}, O ∈ O and w ∈ W_O
  ℓ-safe_{j,O,i}, O ∈ O
  agree(v)_{j,k,i},
    k = 0 and v ∈ V, or k ∈ N+ and v ∈ R^n2
  agree(w)_{j,O,i}, O ∈ O and w ∈ W_O
  safe_{j,k,i}, k ∈ N
Output:
  decide(v)_i, v ∈ V
  ℓ-propose(w)_{j,O,i}, O ∈ O and w ∈ W_O
  propose(v)_{j,k,i},
    k = 0 and v ∈ V, or k ∈ N+ and v ∈ R^n2
  propose(w)_{j,O,i}, O ∈ O and w ∈ W_O
Internal:
  sim-update_{j,i}
  snap_{j,i}
  sim-local_{j,i}
  sim-decide_{j,i}

States of i:

input ∈ V ∪ {null}, initially null
reported, a Boolean variable, initially false
for each j:
  sim-state(j), a state of j, initially the initial state
  sim-steps(j) ∈ N, initially 0
  sim-snaps(j) ∈ N, initially 0
  status(j) ∈ {idle, propose, unsafe, safe},
    initially idle
  sim-mem-local ∈ R^n2, initially arbitrary
  sim-decision(j) ∈ V ∪ {null}, initially null
  object-val_O(j) ∈ W_O ∪ {null}, initially null, O ∈ O

Transitions of i:

init(v)_i
  Effect:        input(i) := v

propose(v)_{j,0,i}
  Precondition:  status(j) = idle
                 ¬∃k : status(k) = unsafe
                 nextop(sim-state(j)) = "init"
                 input(i) ≠ null
                 v = g_i(input(i))(j)
  Effect:        status(j) := unsafe

safe_{j,k,i}
  Effect:        status(j) := safe

agree(v)_{j,0,i}
  Effect:        sim-state(j)
                   := trans-init(sim-state(j), v)
                 sim-steps(j) := 1
                 status(j) := idle

snap_{j,i}
  Precondition:  nextop(sim-state(j)) = "snap"
                 status(j) = idle
  Effect:        sim-mem-local(j) := latest(mem)
                 status(j) := propose

propose(w)_{j,k,i}, k ∈ N+
  Precondition:  status(j) = propose
                 ¬∃m : status(m) = unsafe
                 sim-snaps(j) = k − 1
                 w = sim-mem-local(j)
  Effect:        status(j) := unsafe

agree(w)_{j,k,i}, k ∈ N+
  Effect:        sim-state(j)
                   := trans-snap(sim-state(j), w)
                 sim-steps(j)
                   := sim-steps(j) + 1
                 sim-snaps(j) := sim-snaps(j) + 1
                 status(j) := idle

ℓ-propose(w)_{j,O,i}, O ∈ O
  Precondition:  nextop(sim-state(j))
                   = ("set-consensus_O", w)
                 status(j) = idle
                 ¬∃k : status(k) = unsafe
  Effect:        status(j) := unsafe

ℓ-safe_{j,O,i}, O ∈ O
  Effect:        status(j) := safe

ℓ-agree(w)_{j,O,i}, O ∈ O
  Effect:        object-val_O(j) := w
                 status(j) := propose

propose(w)_{j,O,i}, O ∈ O
  Precondition:  status(j) = propose
                 ¬∃k : status(k) = unsafe
                 nextop(sim-state(j)) = ("set-consensus_O", v)
                 object-val_O(j) = w
  Effect:        status(j) := unsafe

agree(w)_{j,O,i}, O ∈ O
  Effect:        sim-state(j) := trans-sc_O(sim-state(j), w)
                 sim-steps(j) := sim-steps(j) + 1
                 status(j) := idle

sim-update_{j,i}
  Precondition:  nextop(sim-state(j)) = ("update", r)
  Effect:        sim-state(j) := trans(sim-state(j))
                 sim-steps(j) := sim-steps(j) + 1
                 mem(i).sim-mem(j) := r
                 mem(i).sim-steps(j) := sim-steps(j)

sim-local_{j,i}
  Precondition:  nextop(sim-state(j)) = "local"
  Effect:        sim-state(j) := trans(sim-state(j))
                 sim-steps(j) := sim-steps(j) + 1

sim-decide_{j,i}
  Precondition:  nextop(sim-state(j)) = ("decide", v)
  Effect:        sim-state(j) := trans(sim-state(j))
                 sim-steps(j) := sim-steps(j) + 1
                 sim-decision(j) := v

decide(v)_i
  Precondition:  reported = false
                 |sim-decision| ≥ n2 − f
                 v = h_i(sim-decision)
  Effect:        reported := true

Tasks of i:

{decide(v)_i | v ∈ V}
for each j:
  all non-input actions involving j
```

Fig. 2. Automaton for Q.

our system $\mathcal{P}$ is basically the simulated system described by Lynch and Rajsbaum [11] with the addition of (m, ℓ)-*setcon* objects, we will give an overview of their proof methodology and then focus on our extensions.

The operations of the algorithm described in the system $\mathcal{P}$ include *init*(v), *snap*, *update*(r), *setcons*$(\mathcal{O}, v)$ and *decide*(v), for each program j. Let *trans-init*(v) be the initial state of program j given the input value v (this gives the initial state with which to start the simulation of program j). Now, given any state s of program j, let *nextop*(s) be the next operation in the program. If *nextop*(s) = *snap*, let *trans-snap*(s, w) be the state resulting from a snapshot operation at state s that returns w. Similarly, if *nextop*(s) = *setcons*$(\mathcal{O}, v)$, let *trans-sc*$(s, w, \mathcal{O})$ be the state that results from a call to the *setcon* object $\mathcal{O}$ that returns the value w. Finally, if *nextop*(s) is either *update* or *decide*, *trans*(s) is the state resulting from the corresponding operation.

We now define the system $\mathcal{C}$, using I/O automata, which simulates $\mathcal{P}$ in a centralized manner. This corresponds to *SimpleSpec* in [11], except for the simulation of *setcon* operations. Here, a single processor selects a program j in $\mathcal{P}$ nondeterministically and simulates its next operation. It chooses the input value for each program j by picking *any* simulating processor i and adopting its choice of the input value, *i.e.*, the jth component of the vector $g_i(input(i))$. When at least $n_2 - f_2$ programs have terminated with decision values, the central processor terminates with n_1 output values determined using the functions h_i.

This simulation is relatively straightforward except that each (m, ℓ)-*setcon* object (m and ℓ are fixed) in system $\mathcal{P}$ has to be directly simulated in system $\mathcal{C}$ (since the simulating system does not have access to *setcon* objects). Specifically, we have two new internal actions, *inv-setcons*$(v, \mathcal{O})$ and *ret-setcons*$(w, \mathcal{O})$, which simulate the proposal of a value v to an ℓ-set-consensus object $\mathcal{O}$ and the return of a value w by the same object $\mathcal{O}$, respectively. We have defined these actions in the most general way possible, so that they can correspond to any possible correct implementation of set-consensus objects. That is, we make no assumptions about which of the proposed values can be returned by an object (such as, say, letting the set of returned values be the first ℓ values proposed, or the first value proposed), other than that no more than ℓ different values are returned.

We now describe how the automaton $\mathcal{C}$ works. $\mathcal{P}$ has invocations (by programs j) to a collection of (m, l)-set-consensus objects, indexed by $\mathcal{O}$. The value set of each set-consensus object is W, and the problem it solves is ℓ-set consensus. $\mathcal{C}$ simulates $\mathcal{P}$ in a centralized manner, simulating all snapshots and updates as in [11]. In addition, $\mathcal{C}$ simulates each $setcons_{\mathcal{O}}(v)$ operation by two internal actions, *inv-setcons* and *ret-setcons*, which correspond to the invocation and response of the operation, respectively.
For each object $\mathcal{O}$, the state includes

$$inv\text{-}vals(\mathcal{O}), ret\text{-}vals(\mathcal{O}) \subseteq W, \text{ initially } \emptyset,$$

and, for each object $\mathcal{O}$ and program j,

$$inv\text{-}setcons(j, \mathcal{O}) \in \{yes, no\}, \text{ initially } no.$$

As specified earlier, $i \in \{1, \ldots, n_1\}$ is an index of a simulating processor while $j \in \{1, \ldots, n_2\}$ is an index of a simulated program.

Simulation System $\mathcal{C}$
Transitions:

$init(v)_i$
  Effect: $input(i) := v$

$sim\text{-}init_j$
  Precondition: $nextop(sim\text{-}state(j)) =$ "$init$"
    for some i
      $input(i) \neq null$
      $v = g_i(input(i))(j)$
  Effect: $sim\text{-}state(j) := trans\text{-}init(sim\text{-}state(j), v)$

$sim\text{-}snap_j$
  Precondition: $nextop(sim\text{-}state(j)) =$ "$snap$"
  Effect: $sim\text{-}state(j) := trans\text{-}snap(sim\text{-}state(j), sim\text{-}mem)$

$sim\text{-}update_j$
  Precondition: $nextop(sim\text{-}state(j)) =$ ("$update$", r)
  Effect: $sim\text{-}state(j) := trans(sim\text{-}state(j))$
    $sim\text{-}mem(j) := r$

$sim\text{-}local_j$
  Precondition: $nextop(sim\text{-}state(j)) =$ "$local$"
  Effect: $sim\text{-}state(j) := trans(sim\text{-}state(j))$

$sim\text{-}decide_j$
  Precondition: $nextop(sim\text{-}state(j)) =$ ("$decide$", v)
  Effect: $sim\text{-}state(j) := trans(sim\text{-}state(j))$
    $sim\text{-}decision(j) := v$

$inv\text{-}setcons_j(v, \mathcal{O})$
  Precondition: $nextop(sim\text{-}state(j)) =$ "$setcons_{\mathcal{O}}(v)$"
  Effect: $inv\text{-}vals(\mathcal{O}) := inv\text{-}vals(\mathcal{O}) \cup \{v\}$
    $inv\text{-}setcons(j, \mathcal{O}) := yes$

$ret\text{-}setcons_j(w, \mathcal{O})$
  Precondition: $nextop(sim\text{-}state(j)) =$ "$setcons_{\mathcal{O}}(v)$"
    $inv\text{-}setcons(j, \mathcal{O}) = yes$
    $w \in inv\text{-}vals_{\mathcal{O}}$
    $|ret\text{-}vals(\mathcal{O}) \cup \{w\}| \leq \ell$
  Effect: $ret\text{-}vals(\mathcal{O}) := ret\text{-}vals(\mathcal{O}) \cup \{w\}$
    $sim\text{-}state(j) := trans\text{-}sc_{\mathcal{O}}(sim\text{-}state(j), w)$

$decide(v)_i$
  Precondition: $reported(i) = false$
    w is a 'sub-vector' of $sim\text{-}decision$
    $|w| \geq n_2 - f_2$
    $v = h_i(w)$
  Effect: $reported(i) := true$

It is easy to see now that system $\mathcal{C}$ simulates system $\mathcal{P}$. Clearly, the *inv-setcons* and *ret-setcons* actions do indeed solve ℓ-set consensus, and all other actions are simulated properly. This result is stated, without proof, in the lemma below.

Lemma 11. *There is a relation between states in system $\mathcal{C}$ to states in system $\mathcal{P}$ such that any fair execution of system $\mathcal{C}$ corresponds to a fair execution of system $\mathcal{P}$.*

We now show that $\mathcal{C}$ can be simulated by $\mathcal{Q}$, along with the agreement modules. By the results of Lynch and Rajsbaum, $\mathcal{Q}$ will simulate $\mathcal{C}$, as long as we ignore the *setcon* operations. In particular, a snapshot operation of program j is simulated by the pairs of actions $propose_j$ and $agree_j$, one for each real processor i, where the values proposed are the individual snapshots of each processor i. We now give an informal description of the simulation of the *setcon* operation. We omit the formal proof for lack of space. As mentioned in Section 3, each $inv\text{-}setcons_j(\mathcal{O}, v)$ and $ret\text{-}setcons_j(\mathcal{O}, w)$ pair is simulated by the sequences $\ell\text{-}propose_{j,\mathcal{O},i}(v)$, $\ell\text{-}agree_{j,\mathcal{O},i}(x_i)$, $propose_{j,\mathcal{O},i}(x_i)$, $agree_{j,\mathcal{O},i}(w)$, one for each simulating processor i. Recall that the ℓ-agreement actions are used to limit the total number of different responses to ℓ, while the agreement actions are used to guarantee consistency among the simulations. Specifically, for any particular program j, only the first $\ell\text{-}propose(v)$ action in $\mathcal{Q}$, among all processors i, is mapped to the $inv\text{-}setcons(v)$ action in $\mathcal{C}$. Similarly, for any particular program j, only the first $agree(w)$ action, among all processors i, in $\mathcal{Q}$ is mapped to the $ret\text{-}setcons(w)$ action in $\mathcal{C}$. For readers familiar with the simulation in [11], unlike in the simulation of the snapshot operation, where it is important to identify the winning proposed simulation of the snapshot since each simulation is different, here each simulation of the $inv\text{-}setcons(v)$ action *of a particular program j* will ℓ-*propose* the same value v *and agree* on the same value w. So, the complication of a backward simulation followed by a forward simulation can be avoided here.

We can now prove the safety and liveness conditions of the simulation of $\mathcal{C}$ by $\mathcal{Q}$, leading to the Simulation Theorem. The following safety conditions follow from Lemma 10.

Lemma 12. *For any setcon object $\mathcal{O}$ accessed in an execution of $\mathcal{P}$, let $V_\mathcal{O}$ be the set of values v of all actions of the form $\ell\text{-}propose_{j,\mathcal{O},i}(v)$ in the execution, and let $W_\mathcal{O}$ be the set of values w of all actions of the form $\ell\text{-}agree_{j,\mathcal{O},i}(w)$ in the execution. Then, $W_\mathcal{O} \subseteq V_\mathcal{O}$ and $|W_\mathcal{O}| \leq \ell$. Also, for any particular program j and object $\mathcal{O}$, if $\ell\text{-}agree_{j,\mathcal{O},i_1}(w_1)$ and $\ell\text{-}agree_{j,\mathcal{O},i_2}(w_2)$ are actions by two real processors i_1 and i_2, then $w_1 = w_2$.*

We also have the following liveness condition.

Lemma 13. *Given a set of processors, $i_1, i_2, \ldots, i_{n_1}$, doing a B-G simulation of a set of programs, $j_1, j_2, \ldots, j_{n_2}$ in the system $\mathcal{Q}$, for each processor i, at any time within its simulation, there is at most one program j such that i's simulation of j is within the 'unsafe' portion of a safe agreement module or ℓ-safe agreement module.*

We now sketch the proof of the Simulation Theorem. For the complete proof refer to the full paper [14].

Proof. (of Simulation Theorem) Lemma 12 guarantees that the (m, ℓ)-*setcon* objects are simulated correctly. It remains to be shown that the simulation terminates.

Suppose we have the required functions g and h. In a B-G simulation, a processor halts after successfully completing its simulation of any program. Hence, the only reason the simulation could fail is if a non-faulty processor i is unable to terminate the simulation of any of the programs $j_1, j_2, \ldots, j_{n_2}$, even though the relationship between the number of faulty processors, f_1 and the resiliency of the set of programs, f_2, is as required. It follows that either there is some program j' such that, after some time t_0, i does not simulate any steps of j', or, for all j, i simulates steps of j infinitely often.

Suppose the first case holds. Since the fairness condition requires i to simulate each program j which has a step enabled, it must be true that j' is in the 'unsafe' portion of an agreement algorithm at all times after t_0. This cannot happen, since, by the wait-free condition satisfied by the agreement module, i will eventually execute $safe_i$ (and set $status$ to $safe$), no matter how many other processors fail.

Suppose the second case holds. Then i is blocked within its simulation of j, for each j, after some time t_0. Now, within i's simulation of each j, i is either blocked within a safe agreement or ℓ-safe agreement module simulating a particular snapshot or set-consensus object invocation, or it is able to terminate each simulation of an individual snapshot or set-consensus object invocation statement in which case it is blocked within an infinite loop of the program j, itself. Let L_1 be the set of ℓ_1 programs in which i is blocked within a safe agreement or ℓ-safe agreement module, and let L_2 be the set of the remaining $\ell_2 = n_2 - \ell_1$ programs. Since the set of programs is f_2-resilient, if at least $n_2 - f_2$ programs are allowed to each take sufficiently many steps, they will all terminate. Therefore, it follows that $\ell_2 < n_2 - f_2$, implying that $\ell_1 > f_2$. Now, i must be within the busy-wait section, that is, the $wait_i$ or the ℓ-$wait_i$ section, of a safe agreement or an ℓ-safe agreement module simulating a snapshot or a set-consensus operation for each simulation of programs j, for ℓ_1 programs j. Let b_0 be the number of snapshot operations being blocked, and let a_0 be the number of distinct set-consensus objects, $\mathcal{O}$, whose operations are being blocked. Since each set-consensus object can be accessed by at most m programs, each blocked set-consensus operation can block at most m programs. Therefore, it follows that $\ell_1 \leq a_0 m + b_0$. Now, the safe agreement modules satisfies the safe termination property, as proved in [11]. Hence, there is some processor which failed within the straight line section (the actions between and including $propose_i$ and $safe_i$) of the safe agreement module used in simulating a snapshot in b_0 distinct programs. Also, by Lemma 10, there must be at least ℓ processors which have failed within the straight line section (the actions between and including ℓ-$propose_i$ and ℓ-$safe_i$) of each of the a_0 ℓ-safe agreement modules used in simulating the operations on a_0 distinct set-consensus objects. By Lemma 13, a processor cannot execute the actions between and including ℓ-$propose_i$ and ℓ-$safe_i$ or between and including $propose_i$ and $safe_i$ of more than one program simultaneously. Therefore, each processor can fail while executing the actions between and including ℓ-$propose_i$ and ℓ-$safe_i$ of at most one program, implying that at least $a_0\ell + b_0$ processors must have failed. Therefore, it follows that

$f_1 \geq a_0\ell + b_0$ and, since $\ell_1 > f_2$, we have $a_0m + b_0 > f_2$. Thus, f_1 and f_2 do not satisfy the relationship, as assumed, and we have a contradiction. □

7 Conclusion

We have given a partial order of set consensus objects, refining the consensus hierarchy of Herlihy. To do so, we used the Borowsky-Gafni simulation technique, proving the stronger version of the technique required for our results. We therefore strengthened the reducibility result derived by Lynch and Rajsbaum, including simulations of algorithms which access set consensus objects. However, our notion of reducibility still allows the simulated programs to have access to just *one kind of* set consensus object (the values m and ℓ are fixed for any particular reduction). The theorem could be generalized to allow access to several different kinds of set consensus objects. It would be interesting to see what other extensions of the Borowsky-Gafni simulation technique are possible, possibly including simulations of a wider variety of objects. This could also bring about a stronger, more general notion of reducibility.

Many of the proofs and algorithms are stated informally here due to lack of space. For the complete, more detailed version, refer to [14].

References

1. Yehuda Afek, Hagit Attiya, Danny Dolev, Eli Gafni, Michael Merritt, and Nir Shavit. "Atomic Snapshots of Shared Memory". *Journal of the Association for Computing Machinery,* 40(4):873–890, September 1993.
2. E. Borowsky and E. Gafni, "Generalized FLP Impossibility Result for t-resilient Asynchronous Computations", *ACM STOC*, 1993.
3. E. Borowsky and E. Gafni, "The Implication of the Borowsky-Gafni Simulation on the Set-Consensus Hierarchy", *UCLA Tech Report No. 930021.*
4. E. Borowsky and E. Gafni, pre-conference presentation at PODC 1995.
5. S. Chaudhuri, "More Choices Allow More Faults: Set Consensus Problems in Totally Asynchronous Systems", *Information and Computation 105 (1)*, July 1993. Appeared earlier in *ACM PODC*, 1990.
6. M. Fischer, N. Lynch, and M. Paterson, "Impossibility of Distributed Consensus with One Faulty Process", *JACM 32*, April 1985. Appeared earlier in *ACM PODC*, 1983.
7. "Wait-Free Synchronization", *ACM TOPLAS 11 (1)*, 1991.
8. "Set Consensus Using Arbitrary Objects", In *Thirteenth Annual ACM Symposium on Principles of Distributed Computing,* pages 324–333. Association for Computing Machinery, ACM Press, 1993.
9. M. Herlihy and N. Shavit, "The Asynchronous Computability Theorem for t-Resilient Tasks", *ACM STOC*, 1993.
10. Prasad Jayanti. "On the Robustness of Herlihy's Hierarchy". In *Proceedings of the Thirteenth Annual ACM Symposium on Principles of Distributed Computing,* pages 145–157. Association for Computing Machinery, ACM Press, 1993.
11. N. Lynch and S. Rajsbaum, "On the Borowsky Gafni Simulation Algorithm", *Israel Symposium on Theory of Computing and Systems*, 1996.

12. N. Lynch and M. Tuttle, "An Introduction to Input/Output Automata", TM-373, MIT Laboratory for Computer Science, November 1988.
13. P. Jayanti and S. Toueg, "Some Results on Impossibility, Universality and Decidability of Consensus", *6th WDAG*, Springer Verlag, 1992.
14. P. Reiners, "Understanding the Set Consensus Partial Order Using the Borowsky-Gafni Simulation", *M.S. Thesis*, Iowa State University, 1996.
15. M. Saks and F. Zaharoglou, "Wait-Free k-set Agreement is Impossible: The Topology of Public Knowledge", *ACM STOC*, 1993.

Author Index

Lecture Notes in Computer Science

For information about Vols. 1–1081

please contact your bookseller or Springer-Verlag

Vol. 1082: N.R. Adam, B.K. Bhargava, M. Halem, Y. Yesha (Eds.), Digital Libraries. Proceedings, 1995. Approx. 310 pages. 1996.

Vol. 1083: K. Sparck Jones, J.R. Galliers, Evaluating Natural Language Processing Systems. XV, 228 pages. 1996. (Subseries LNAI).

Vol. 1084: W.H. Cunningham, S.T. McCormick, M. Queyranne (Eds.), Integer Programming and Combinatorial Optimization. Proceedings, 1996. X, 505 pages. 1996.

Vol. 1085: D.M. Gabbay, H.J. Ohlbach (Eds.), Practical Reasoning. Proceedings, 1996. XV, 721 pages. 1996. (Subseries LNAI).

Vol. 1086: C. Frasson, G. Gauthier, A. Lesgold (Eds.), Intelligent Tutoring Systems. Proceedings, 1996. XVII, 688 pages. 1996.

Vol. 1087: C. Zhang, D. Lukose (Eds.), Distributed Artificial Intelliegence. Proceedings, 1995. VIII, 232 pages. 1996. (Subseries LNAI).

Vol. 1088: A. Strohmeier (Ed.), Reliable Software Technologies – Ada-Europe '96. Proceedings, 1996. XI, 513 pages. 1996.

Vol. 1089: G. Ramalingam, Bounded Incremental Computation. XI, 190 pages. 1996.

Vol. 1090: J.-Y. Cai, C.K. Wong (Eds.), Computing and Combinatorics. Proceedings, 1996. X, 421 pages. 1996.

Vol. 1091: J. Billington, W. Reisig (Eds.), Application and Theory of Petri Nets 1996. Proceedings, 1996. VIII, 549 pages. 1996.

Vol. 1092: H. Kleine Büning (Ed.), Computer Science Logic. Proceedings, 1995. VIII, 487 pages. 1996.

Vol. 1093: L. Dorst, M. van Lambalgen, F. Voorbraak (Eds.), Reasoning with Uncertainty in Robotics. Proceedings, 1995. VIII, 387 pages. 1996. (Subseries LNAI).

Vol. 1094: R. Morrison, J. Kennedy (Eds.), Advances in Databases. Proceedings, 1996. XI, 234 pages. 1996.

Vol. 1095: W. McCune, R. Padmanabhan, Automated Deduction in Equational Logic and Cubic Curves. X, 231 pages. 1996. (Subseries LNAI).

Vol. 1096: T. Schäl, Workflow Management Systems for Process Organisations. XII, 200 pages. 1996.

Vol. 1097: R. Karlsson, A. Lingas (Eds.), Algorithm Theory – SWAT '96. Proceedings, 1996. IX, 453 pages. 1996.

Vol. 1098: P. Cointe (Ed.), ECOOP '96 – Object-Oriented Programming. Proceedings, 1996. XI, 502 pages. 1996.

Vol. 1099: F. Meyer auf der Heide, B. Monien (Eds.), Automata, Languages and Programming. Proceedings, 1996. XII, 681 pages. 1996.

Vol. 1100: B. Pfitzmann, Digital Signature Schemes. XVI, 396 pages. 1996.

Vol. 1101: M. Wirsing, M. Nivat (Eds.), Algebraic Methodology and Software Technology. Proceedings, 1996. XII, 641 pages. 1996.

Vol. 1102: R. Alur, T.A. Henzinger (Eds.), Computer Aided Verification. Proceedings, 1996. XII, 472 pages. 1996.

Vol. 1103: H. Ganzinger (Ed.), Rewriting Techniques and Applications. Proceedings, 1996. XI, 437 pages. 1996.

Vol. 1104: M.A. McRobbie, J.K. Slaney (Eds.), Automated Deduction – CADE-13. Proceedings, 1996. XV, 764 pages. 1996. (Subseries LNAI).

Vol. 1105: T.I. Ören, G.J. Klir (Eds.), Computer Aided Systems Theory – CAST '94. Proceedings, 1994. IX, 439 pages. 1996.

Vol. 1106: M. Jampel, E. Freuder, M. Maher (Eds.), Over-Constrained Systems. X, 309 pages. 1996.

Vol. 1107: J.-P. Briot, J.-M. Geib, A. Yonezawa (Eds.), Object-Based Parallel and Distributed Computation. Proceedings, 1995. X, 349 pages. 1996.

Vol. 1108: A. Díaz de Ilarraza Sánchez, I. Fernández de Castro (Eds.), Computer Aided Learning and Instruction in Science and Engineering. Proceedings, 1996. XIV, 480 pages. 1996.

Vol. 1109: N. Koblitz (Ed.), Advances in Cryptology – Crypto '96. Proceedings, 1996. XII, 417 pages. 1996.

Vol. 1110: O. Danvy, R. Glück, P. Thiemann (Eds.), Partial Evaluation. Proceedings, 1996. XII, 514 pages. 1996.

Vol. 1111: J.J. Alferes, L. Moniz Pereira, Reasoning with Logic Programming. XXI, 326 pages. 1996. (Subseries LNAI).

Vol. 1112: C. von der Malsburg, W. von Seelen, J.C. Vorbrüggen, B. Sendhoff (Eds.), Artificial Neural Networks – ICANN 96. Proceedings, 1996. XXV, 922 pages. 1996.

Vol. 1113: W. Penczek, A. Szałas (Eds.), Mathematical Foundations of Computer Science 1996. Proceedings, 1996. X, 592 pages. 1996.

Vol. 1114: N. Foo, R. Goebel (Eds.), PRICAI'96: Topics in Artificial Intelligence. Proceedings, 1996. XXI, 658 pages. 1996. (Subseries LNAI).

Vol. 1115: P.W. Eklund, G. Ellis, G. Mann (Eds.), Conceptual Structures: Knowledge Representation as Interlingua. Proceedings, 1996. XIII, 321 pages. 1996. (Subseries LNAI).

Vol. 1116: J. Hall (Ed.), Management of Telecommunication Systems and Services. XXI, 229 pages. 1996.

Vol. 1117: A. Ferreira, J. Rolim, Y. Saad, T. Yang (Eds.), Parallel Algorithms for Irregularly Structured Problems. Proceedings, 1996. IX, 358 pages. 1996.

Vol. 1118: E.C. Freuder (Ed.), Principles and Practice of Constraint Programming — CP 96. Proceedings, 1996. XIX, 574 pages. 1996.

Vol. 1119: U. Montanari, V. Sassone (Eds.), CONCUR '96: Concurrency Theory. Proceedings, 1996. XII, 751 pages. 1996.

Vol. 1120: M. Deza. R. Euler, I. Manoussakis (Eds.), Combinatorics and Computer Science. Proceedings, 1995. IX, 415 pages. 1996.

Vol. 1121: P. Perner, P. Wang, A. Rosenfeld (Eds.), Advances in Structural and Syntactical Pattern Recognition. Proceedings, 1996. X, 393 pages. 1996.

Vol. 1122: H. Cohen (Ed.), Algorithmic Number Theory. Proceedings, 1996. IX, 405 pages. 1996.

Vol. 1123: L. Bougé, P. Fraigniaud, A. Mignotte, Y. Robert (Eds.), Euro-Par'96. Parallel Processing. Proceedings, 1996, Vol. I. XXXIII, 842 pages. 1996.

Vol. 1124: L. Bougé, P. Fraigniaud, A. Mignotte, Y. Robert (Eds.), Euro-Par'96. Parallel Processing. Proceedings, 1996, Vol. II. XXXIII, 926 pages. 1996.

Vol. 1125: J. von Wright, J. Grundy, J. Harrison (Eds.), Theorem Proving in Higher Order Logics. Proceedings, 1996. VIII, 447 pages. 1996.

Vol. 1126: J.J. Alferes, L. Moniz Pereira, E. Orlowska (Eds.), Logics in Artificial Intelligence. Proceedings, 1996. IX, 417 pages. 1996. (Subseries LNAI).

Vol. 1127: L. Böszörményi (Ed.), Parallel Computation. Proceedings, 1996. XI, 235 pages. 1996.

Vol. 1128: J. Calmet, C. Limongelli (Eds.), Design and Implementation of Symbolic Computation Systems. Proceedings, 1996. IX, 356 pages. 1996.

Vol. 1129: J. Launchbury, E. Meijer, T. Sheard (Eds.), Advanced Functional Programming. Proceedings, 1996. VII, 238 pages. 1996.

Vol. 1130: M. Haveraaen, O. Owe, O.-J. Dahl (Eds.), Recent Trends in Data Type Specification. Proceedings, 1995. VIII, 551 pages. 1996.

Vol. 1131: K.H. Höhne, R. Kikinis (Eds.), Visualization in Biomedical Computing. Proceedings, 1996. XII, 610 pages. 1996.

Vol. 1132: G.-R. Perrin, A. Darte (Eds.), The Data Parallel Programming Model. XV, 284 pages. 1996.

Vol. 1133: J.-Y. Chouinard, P. Fortier, T.A. Gulliver (Eds.), Information Theory and Applications II. Proceedings, 1995. XII, 309 pages. 1996.

Vol. 1134: R. Wagner, H. Thoma (Eds.), Database and Expert Systems Applications. Proceedings, 1996. XV, 921 pages. 1996.

Vol. 1135: B. Jonsson, J. Parrow (Eds.), Formal Techniques in Real-Time and Fault-Tolerant Systems. Proceedings, 1996. X, 479 pages. 1996.

Vol. 1136: J. Diaz, M. Serna (Eds.), Algorithms – ESA '96. Proceedings, 1996. XII, 566 pages. 1996.

Vol. 1137: G. Görz, S. Hölldobler (Eds.), KI-96: Advances in Artificial Intelligence. Proceedings, 1996. XI, 387 pages. 1996. (Subseries LNAI).

Vol. 1138: J. Calmet, J.A. Campbell, J. Pfalzgraf (Eds.), Artificial Intelligence and Symbolic Mathematical Computation. Proceedings, 1996. VIII, 381 pages. 1996.

Vol. 1139: M. Hanus, M. Rogriguez-Artalejo (Eds.), Algebraic and Logic Programming. Proceedings, 1996. VIII, 345 pages. 1996.

Vol. 1140: H. Kuchen, S. Doaitse Swierstra (Eds.), Programming Languages: Implementations, Logics, and Programs. Proceedings, 1996. XI, 479 pages. 1996.

Vol. 1141: H.-M. Voigt, W. Ebeling, I. Rechenberg, H.-P. Schwefel (Eds.), Parallel Problem Solving from Nature – PPSN IV. Proceedings, 1996. XVII, 1.050 pages. 1996.

Vol. 1142: R.W. Hartenstein, M. Glesner (Eds.), Field-Programmable Logic. Proceedings, 1996. X, 432 pages. 1996.

Vol. 1143: T.C. Fogarty (Ed.), Evolutionary Computing. Proceedings, 1996. VIII, 305 pages. 1996.

Vol. 1144: J. Ponce, A. Zisserman, M. Hebert (Eds.), Object Representation in Computer Vision. Proceedings, 1996. VIII, 403 pages. 1996.

Vol. 1145: R. Cousot, D.A. Schmidt (Eds.), Static Analysis. Proceedings, 1996. IX, 389 pages. 1996.

Vol. 1146: E. Bertino, H. Kurth, G. Martella, E. Montolivo (Eds.), Computer Security – ESORICS 96. Proceedings, 1996. X, 365 pages. 1996.

Vol. 1147: L. Miclet, C. de la Higuera (Eds.), Grammatical Inference: Learning Syntax from Sentences. Proceedings, 1996. VIII, 327 pages. 1996. (Subseries LNAI).

Vol. 1148: M.C. Lin, D. Manocha (Eds.), Applied Computational Geometry. Proceedings, 1996. VIII, 223 pages. 1996.

Vol. 1149: C. Montangero (Ed.), Software Process Technology. Proceedings, 1996. IX, 291 pages. 1996.

Vol. 1150: A. Hlawiczka, J.G. Silva, L. Simoncini (Eds.), Dependable Computing – EDCC-2. Proceedings, 1996. XVI, 440 pages. 1996.

Vol. 1151: Ö. Babaoğlu, K. Marzullo (Eds.), Distributed Algorithms. Proceedings, 1996. VIII, 381 pages. 1996.

Vol. 1153: E. Burke, P. Ross (Eds.), Practice and Theory of Automated Timetabling. Proceedings, 1995. XIII, 381 pages. 1996.

Vol. 1154: D. Pedreschi, C. Zaniolo (Eds.), Logic in Databases. Proceedings, 1996. X, 497 pages. 1996.

Vol. 1155: J. Roberts, U. Mocci, J. Virtamo (Eds.), Broadbank Network Teletraffic. XXII, 584 pages. 1996.

Vol. 1156: A. Bode, J. Dongarra, T. Ludwig, V. Sunderam (Eds.), Parallel Virtual Machine – EuroPVM '96. Proceedings, 1996. XIV, 362 pages. 1996.

Vol. 1157: B. Thalheim (Ed.), Entity-Relationship Approach – ER '96. Proceedings, 1996. XII, 489 pages. 1996.

Vol. 1158: S. Berardi, M. Coppo (Eds.), Types for Proofs and Programs. Proceedings, 1995. X, 296 pages. 1996.